AF333345

Also by Edmond A. Murphy, M.D.

Skepsis, Dogma, and Belief: Uses and Abuses in Medicine
Probability in Medicine
The Logic of Medicine
Principles of Genetic Counseling (with G. A. Chase)

EDMOND A. MURPHY, M.D.

# BIOSTATISTICS IN MEDICINE

THE JOHNS HOPKINS UNIVERSITY PRESS • Baltimore and London

The Johns Hopkins University Press, Baltimore, Maryland 21218
The Johns Hopkins Press Ltd., London

Illustrations by the author

Library of Congress Cataloging in Publication Data

Murphy, Edmond A.
    Biostatistics in medicine.

    Bibliography: pp. 519–25
    Includes index.
    1. Medical statistics.  2. Biometry.  I. Title.
[DNLM: 1. Biometry.  2. Statistics.   WA 950 M978b]
RA409.M867     519.5′02461     81-48191
ISBN 0-8018-2727-2          AACR2

To Fraser and Betty,
with affection and gratitude

# CONTENTS

# PREFACE

In this book I have attempted an elementary account of statistics viewed from inside the domain of medicine. Other books have had somewhat the same objective but none exactly the same. Few have been written by physicians, and fewer still by those engaged in clinical practice. This viewpoint has some impact on the material covered and the examples used; but at a deeper level, it has given a different slant to the meaning of statistical inference.

For instance, in many elementary texts Gaussian normality and the means of testing for it are given little emphasis because where sample means are concerned, even rather serious departures from normality may be offset by appeal to the central limit theorem. But the practicing physician is most of the time concerned with individual patients and cannot appeal to the theorem; neither can the clinical pathologists, who must furnish descriptive statistics on the values which may be assumed by individual readings of each test introduced into clinical practice.

A second objective I have had is to portray statistics from an essentially biological point of view, that is, from the standpoint of someone whose primary allegiance is not to analysis as such but to biological mechanisms. There is, for example, a more or less profound difference between what a statistician means by "interaction" (which revolves about the algebraic additivity of measurable effects) and what the biologist means by it (involving a causal relationship among mechanisms that could still exist under quite different systems of measurement or even for nonmetrical ones). The two are not totally unrelated; but neither are they the same thing. One of the important uses of transformation (on which I put unusual emphasis) is to try to reconcile and interrelate the two concepts. This objective suggests that we must take more responsibility for the meaning of our often arbitrary measurements. These epistemological issues are commonly regarded—quite correctly, I dare say—as no business of the statistician as such. But those who invent indices or who take over others' inventions, have to have some concern with them. It is not enough to know that an index has a Gaussian distribution, for the Gaussian character of the distribution may be pure invention, as it is in the intelligence test. In principle there exists one transformation that will convert each continuous variate into a Gaussian one to which one may apply all of the normal theory, including tests for additivity.

But as a biologist I am not sure what to make of additivity or linearity on a transform of which the only warrant is that it makes the variable Gaussian.

Likewise, statisticians are obsessed with certain stereotypes including regression models in which relationships are *either* quantal, *or* everywhere continuous. But it is not difficult to devise other plausible models, which are a great deal more pertinent in many biological cases (1). The presence of discontinuities may be successfully masked in the presence of error by fitting a polynomial regression curve of suitable degree. But it is both bad biology and bad statistics to substitute mindless curve fitting for biological discernment.

Again I would accuse biostatisticians of too much kindness and good will toward investigators doing bad science. It has been my experience that from any sensibly designed experiment I have ever encountered, the analysis is always unambiguously determined by the design: that the notorious "multiple-comparisons" problem only arises in either badly thought out problems or fishing expeditions, neither of which I would call science. A good experimenter automatically balances his design and selects his treatment groups in such a way as to furnish a unique set of orthogonal contrasts; and each contrast automatically allows us to answer one question of intrinsic importance. I do not claim that these statements are universally true, which would be hard to prove. I would be interested to hear of counterexamples where that is not so and where no better design would have been possible. Of course where historical data only are available, it is often a different story; but, in any case, interpretation of such data is quite different from that of data prospectively collected in planned experiments.

A third objective I have set is to capitalize on a small but agile stock of necessary probability algebra to extend the scope of biostatistics. Hitherto, the consumer has been asked to take his choice between applying stereotyped methods without rationale to stock situations; or to "do it properly," learn calculus and what not, and hack his way through grim jungles of multiple integrals and unending general distributions with weird pathologies. It is my intrepid opinion that there is a middle course: that without undue fuss one can explain rationally how one decides on a rejection region and whether we should do one-tailed tests, two-tailed, or neither; why the significance level is not the answer to the question we are really interested in; how we devise an estimator in a peculiar problem and what properties of it we should be interested in; what we really mean by degrees of freedom and partition of variance; how a transformation works, what its justification is, what its properties are, and how we go about choosing one. I have made bold to do so without using calculus, but freely using numerical methods and approximations. I have the warrant of D. J. Finney (2) and my own experience that in practical problems maximum likelihood estimation very seldom leads to estimators in closed form. In any case, commonly the properties ascribed to the method are asymptotic. Where finite samples are concerned the nonmathematician might as well be hanged for a sheep as for a lamb and use an unashamed numerical method. The theorist, too, often has to be content with such a solution. Many of the estimators, notably least squares and maximum likelihood, are widely used by geneticists, often without being cautioned about the pitfalls. To say that all such persons should take a good stiff course in mathematical statistics may be true; but it is a counsel of perfection.

It is possible, and I have thought it profitable, to insert a few pages on matrix algebra which will explain not only how to use it but also what is going on and why. In practice, the tedious parts of the calculations will be done on a computer. But many investigators have extensive need for multivariate analysis and should at least have a clear idea of the logic, leaving the logistics to the computer programmer.

The intent of this theoretical part (which comprises three chapters near the beginning, and an appendix) is not to produce a cut-price statistician. A well-known probabilist has remarked to me, "You always need to know ten times as much as you use." I think this is an excellent principle; the edge of one's knowledge—like the edge of a cliff—is a dangerous place to work, and with closely analogous penalties for making mistakes. Most people I know of who use "cookbooks" lack this *cordon sanitaire* (so to speak) and one is constantly saddened, although not surprised, at the blunders which appear in print in august biological and medical journals.

Finally, details of ideas, as distinct from analysis, call for special treatment. This subject is much more intimately medical and likely to be of less interest to the general statistician. There are now several books of various scopes which the interested reader may wish to consult (3-6).

This text presupposes some elementary knowledge of probability which I have tried to lay out earlier (7) and which is well catered for in more orthodox terms elsewhere, e.g., Hoel (8). But I make no demands on calculus; and the little matrix algebra I appeal to is discussed in Appendix 1.

# ACKNOWLEDGMENTS

The foundation of this book was first laid down some twenty years ago, and I had certainly been collecting material for it earlier than that. At this distance I find it impossible to do justice to all who have helped me in one way or another; my teachers, my colleagues, my critics, and my students. The provenance of many of my examples will be evident from the references cited or from context. Much of the material was prepared for a series of lectures given in happier times under the kindly auspices of Dr. Mohsen Mahloudji, at the Pahlavi University in Shiraz; text, diagrams, and worked examples are reproduced by kind permission of the editor of the *Pahlavi Medical Journal.* Figures 16.1 and 17.1 are reproduced from the *Johns Hopkins Medical Journal*, by kind permission of the editor.

I am indebted to Drs. Helen Abbey and Carl Huether, who have read parts of the text in manuscript and made many perceptive and helpful comments. Mrs. Ella Foster has braved the demanding manuscript, repeatedly rewritten, with unflagging devotion and care.

# NOTATION

Throughout this text we shall employ the standard conventions for algebraic symbols as follows:

*Italic font.* Uppercase letters denote random variables, i.e., quantities that are indefinite because the particular value has not yet been selected. Probabilities, means (or expectations), variances, correlations, etc., can apply to such quantities alone.

Lowercase letters denote particular values that are definite but, for generality, not specified, e.g., actual sample values. They may also be used to denote constants and the size of a sample.

However, force of tradition overrides this distinction for $t$ and $F$ statistics, which are written so regardless of whether they are random variables or particular values of them: which is the case, must be decided from context.

*Greek letters* are used to denote parameters (true values for the distribution constants, which in statistics are usually unknown). A notable exception is the binomial parameter $p$ the Greek counterpart of which ($\pi$) is reserved to another context.

A sample estimate of a parameter is usually denoted by the corresponding roman letter in italic either as an upper- or lowercase letter according to whether or not it is a random variable. Thus, for the Gaussian distribution, $\mu$ denotes the true mean, $M$ the (randomly variable) estimator of it based on a (future) sample of specific size, and $m$ is the actual value from my particular sample.

Maximum likelihood estimates of the parameters are commonly written as the Greek symbol with a circumflex accent ( $\hat{\ }$ ), uppercase if a random variable, otherwise lowercase.

*Boldface* type is reserved for matrix notation, uppercase for matrices, lowercase for vectors. This unfortunate, but deeply entrenched, convention means we cannot distinguish random variables from sample values, which must be distinguished from context. However, the Greek-Roman convention can be preserved.

# A SUGGESTION FOR
# THE PRACTICAL READER

Reading this book is at variance with Lewis Carroll's facetious dictum (which as a mathematician he clearly did not believe): "Begin at the beginning, go to the end and then stop." I have written it, so far as I am able, in systematic expository order. Hence the first part (Chapters 1–4) is somewhat theoretical (although, I hope, not oppressively so). Those who do not already have some experience or knowledge of statistics are advised to begin either at Chapter 5 or Chapter 12, according to whether they are interested in measurements or categories. It is my experience that those who are intending to use statistics conscientiously will sooner or later get beyond the "cookbook" stage and wish to have somewhat better insight into what they are doing and why. In some measure, they are digging the foundations after they have built a preliminary edifice. Although this course is not totally systematic, it is perfectly reasonable: it is the way that I have learnt statistics; and thirty years after doing my first chi-square test, I am still digging foundations. I take comfort from Carroll's further comment that "the oldest rule in the book" is not necessarily at the beginning of the book.

# PART I
# ELEMENTARY STATISTICAL THEORY

# 1
# INTRODUCTION

Biostatistics is the craft of making inferences from finite samples of authentic biological data subject to error of observation representing at least in part the operation of chance factors.

It is a craft in that it involves judgment based on experience in the adaptation of theoretical structures to practical ends, and ingenuity in circumventing the gaps that are frequently encountered. It is one of the most difficult of professions to practice, and even the best biostatisticians rarely give complete satisfaction to all concerned. This vulnerability is in the nature of things, because the biostatistician is attempting to meet at least three requirements.

First he* must understand something of probability algebra and mathematical statistics and conform his methods to them. The theorist has no reciprocal commitment and, it can be cogently argued, it is right that he should not. The theorist's prime concern is with soundness and rigor, which commonly leads him (in the eyes of the biostatistician) to solve the problems that can, rather than those that should, be solved. Nevertheless, if he were to abdicate this function in favor of relentless practicalities, in the long run applied statistics would suffer and the inspiration for new developments dry up. However I do not want to make too much of the contrast between these two specialists; and let me say, once for all, that the biostatistician's prime allegiance is to the quantitative aspects of data. I hope he will have a much wider perspective, and that he will assume serious responsibility for communication. But it is as a judicious manipulator of quantities that the biostatistician is an expert, not as a chemist, physician, or cytologist.

Secondly his analyses must be in accord with the secure truths of science. To this end he must have a sound knowledge of the relevant features of the field of application and more especially must have the skill to discern what are relevant features. It is not difficult to pick up a smattering of facts about a particular field which may create the illusion of adequate under-

---

*Here, and throughout this book, the words *he, him, his,* etc., are to be interpreted as a common gender. A repeated use of *he or she,* etc., imparts an unpleasantly legal tone which could cramp the style of a Shakespeare. I mean no slight. Some of my best statistical friends...

standing; but these facts are commonly insufficient. They are what the scientist believes germane to the issue and may ignore features that are relevant to the statistical problem. The skilled and experienced biostatistician will know where to probe for weaknesses. Again there may be ambiguities in meaning which are common sources of confusion and error. Words like *normal, random, independent, correlated, parameter, skewed* are commonly used by the mathematical statistician and by the scientist in quite different senses, and the biostatistician is obliged to be familiar with both.

A further difficulty is that the scientist commonly has fact imperfectly separated from construct, inference, and conjecture. For instance we have come to expect a good deal of variability in the response of experimental animals to drugs. Of a group of animals (mice, for example) one of them will, as a logical consequence of biological variation, respond least, and one most. So far so good. To account for the difference, however, the scientist often introduces such notions as *susceptibility* and *resistance.* Of course they may be valuable ideas supported from other evidence. But to infer their existence from nothing other than the response to dosage is a much more difficult and tentative matter than many biologists realize; and the alert biostatistician will not permit unusual responses to be treated separately on such tenuous reasoning. It takes some discernment to recognize the difference between genuine scientific knowledge and an extemporaneous construct invented by the scientist to justify excluding results unfavorable to some preconceived idea. In the same way, the experienced are wary about what has been eliminated from the data presented for the analysis. ("We discarded these results: the machine was not working properly that day." "The technician was called away to the telephone in the middle of the experiment and cannot remember whether the specimen was added to the mixture twice." "Obviously you cannot have a serum cholesterol of 1500 mg/dl!" "Somehow a different male must have got into the cage." "I know the nurse who gave the injection, she has a shaky hand"; etc.) To be aware of such sources of possible bias, the biostatistician needs to understand how biologists think. It is doubtless my prejudice that it is easier to make a biostatistician out of a seasoned biologist than out of a mathematical statistician.

Thirdly, the biostatistician, as liaison between theory and data, has the responsibility of setting up correspondences between the axioms of the mathematician and the empirical experience of the biologist. The two are incommensurable and there must be some unease about building the logic of one subject on that of another. By way of illustration consider the difference in the ways in which a mathematical statistician and a practical epidemiologist use the word *distribution.* The statistician thinks of it essentially as a predictive statement about the outcome of one experiment. "The height of some unspecified man to be picked at random at some time in the future from the population of Normal, Illinois, is normally distributed with a mean of 175 cms." From this he may go on to talk about the difference in height of two men with identically distributed heights. But the practical epidemiologist can make no sense out of either of these statements: to him the distribution is a collective attribute of the whole population. He holds that one man alone cannot have a distribution for his height because (unlike the statistician) he is referring to a particular man who has already been selected. Now it might be argued that this difference is merely a matter of definition; however, the incongruity cannot so readily be dismissed by the biostatistician. If the distribution is to be inferred from the data, little useful inference can be made from a single case. But the sample which the epidemiologist has collected consists of individual subjects. How can the biostatistician test the assumption that all of the people yet to be selected in the sample have identically distributed heights? How can he believe that, when he knows that they are of diverse racial groups of which the mean heights differ? To be sure these difficulties can be dealt with by means of the conceptualizations of the mathematical statistician: by appeal to superpopulations and mixed distributions (see Chapter 17); but this appeal furnishes no empirical assurance whatsoever, and leaves the impression that statistical inference has at most theoretical meaning in the world of mathematics, and no scientific meaning. I quote this problem at some length, not to spread disillusionment, but to point out that the annealing of statistical theory and science is no mere formality. It taxes the craft of the matchmaker more than is perhaps explicitly realized. In fact there are many ways of meeting these difficulties, notably appeal to the property of robustness (see Chapter 17).

Other duties of the biostatistician are, to perceive analogies, and to teach. Science is a massive field and almost all branches of it have an inescapable statistical component. In many areas (physics, genetics, anthropology, economics, psychology, demography) competence in the relevant fields of statistics has been an accepted component of professional training, and enough "home-grown" specialists have been produced to make these areas self-sufficient. Even so, the many areas of biology, or even of medicine alone, not provided for are vast, and a formidable challenge to the missionary zeal of the biostatistician. In some instances, the biostatistician has specialized in a particular field, e.g., pharmacology, epidemiology, or genetics; but this solution is something of a luxury. "The harvest is rich but the laborers are few."

Biostatistics, then, must create its own version of the barefoot doctor, the man native to the branch of science in which he works. To him the biostatistician imparts three things: a sound knowledge of simple biostatistics; a certain ingenuity in exploiting the skill he has; and an awareness of when more expert help is required. For this objective to be practicable, the biostatistician needs to realize that a great many problems, superficially quite diverse, are the same small number of basic patterns in many disguises. Hence it is necessary to be able to recognize analogies and to do so with a sufficiently light hand to avoid harmful distortion. The distillate may be compactly and simply presented with considerable attention to practical detail. This is the pattern I have set myself in this book.

The above discussion may be usefully illustrated by reference to the two most commonly used (and, I may add, misused) distributions, the binomial and the Gaussian. At the same time, it will be convenient to review briefly those of their features that are most important to the statistician.

## TYPES OF DISTRIBUTION

The nature and definition of a distribution is discussed simply and in considerable detail elsewhere (7). It suffices here to say that a distribution (in the statistical sense) is a predictive statement about the outcome of *one* measurable random process. It comprises two things: The set or domain of values which the random variable may assume; and the probability or probability density associated with each.

All distributions are of one of three kinds, discrete, continuous, or (rarely) a mixture of the two. As an example we shall take one form of familial polyposis of the colon known as the Gardner syndrome (9–12). This is an autosomal dominant condition with several features other than the polyposis, notably osteomas, desmoid tumors, skin lesions, and mandibular changes. Characteristically, none of these features is present at birth, but polyps first appear at about 20 years of age and if left undisturbed, with some regularity become cancerous (though I should hesitate to say that this outcome is certain).

Consider now the probabilistic aspects of the disease.

1. The number of polyps may range from none to several thousand, and short of looking, I know of no way of determining the number. Then in a person at risk (e.g., the daughter of an affected person) we might state the empirical probabilities of having no polyps, one, two, etc. This prediction of the number found is a discrete distribution because it can take on only a restricted number of values. However much difficulty there may be in defining a polyp, no pathologist would accept that the number can be $2\frac{1}{2}$ polyps for instance.

2. The age of diagnosis, however, is not so constrained. However close two ages are, provided they are not the same (e.g., 20 years, 117 days and 20 years, 118 days), the age of diagnosis could take on intermediate values (e.g., 20 years, 117.5 days). Such a distribution is said to be continuous. An equivalent statement is that a priori the probability of a patient living *exactly* x years is zero. There is merely a probability *density,* associated with living, say, exactly five years. This density multiplied by the width of some small interval is (approximately) the probability that the outcome will fall in that interval. (For a more exact treatment see [7].)

3. The number of years of further survival in a patient with Gardner's

syndrome represents a mixture. Patients fall into two classes, living and dead. The distribution of further survival for the former is clearly continuous, for the latter it can only assume one value, zero years. The combined distribution is mixed. There is a probability associated with zero. But there is no probability (merely a probability density) associated with any precise time greater than zero, say, exactly three years.

## SPHERE OF APPLICATION

In applying a particular distribution to a set of data from the real world the statistician appeals to one of three main arguments: aptness, robustness, or conformity.

### Aptness

The distribution may be apt a priori because it is predicated on a set of assumptions that the process in the real world is known, thought, or hypothesized, to fulfill. Thus, the conditions underlying the binomial distribution (see below) are known to be fulfilled by subjects being randomly assigned to one of two treatment groups. They are thought to be fulfilled by the processes of Mendelian genetics. The formal test of the null hypothesis (see Chapter 2) of no efficacy of a treatment in an experiment with randomly assigned cases, where the outcome is dichotomous, is based on the conjecture that the binomial structure has not been disturbed by treatment.

### Robustness

While the assumptions underlying the model may not be strictly fulfilled, theory may show that the violations have little impact on tests of the mean. For example, the $t$ distribution is based on the assumption that the data we are planning to obtain are normally distributed. But in fact for samples of reasonable size—say thirty or more—even quite serious departures from normality may hardly matter. We shall discuss the subject more fully in Chapter 17.

### Conformity

There may be no kind of theoretical argument at all for using a particular distribution except that it conforms to the empirical behavior of the data. Though one can invent more or less plausible explanations after the fact, it would be a bold scientist who would surmise that height should follow a Gaussian distribution closely, or blood cholesterol levels a lognormal, or plasma thromboplastin time the reciprocal-normal distribution. It is wisest in such cases to regard the distribution as a convenient computational device and not to suppose that it throws any light on the essential structure of the process.

## THE BINOMIAL DISTRIBUTION

This is the archetypal discrete distribution, not because it is so common (as we shall see, it is undoubtedly overused), but because it is most easily grasped and illustrated by artificial models. In essence it is the distribution of the number of outcomes of one type when a random process with two outcomes is repeated. The number of repetitions is prescribed and they must be performed independently and under identical conditions. Thus, consider a man heterozygous for the gene for Gardner syndrome, married to a woman with normal genes only. The man has two types of genes, either of which may be donated to an offspring, each (so Mendelian genetics contends) with a probability of $\frac{1}{2}$. The father's turnover of sperm is so vast and the interval between conceptions relatively so long that the outcome for one impregnation will be unaffected by that for another. If such a man plans to beget ten children, each will have an independent probability, $p$, of 0.5 of being affected, and the total number of affected progeny will follow a binomial distribution with order ($n$) of 10. We write this distribution $B(10, 0.5)$ or more generally, $B(n, p)$. After the event, of course, the number of affected progeny will be some specific number: it will no longer have a distribution.

### Underlying assumptions

Not all experiments for which there are two possible outcomes are binomial. There are in fact six conditions that must be fulfilled, which we shall state and illustrate by counterexamples.

1. For each experiment ("trial") there must be a choice between the same two outcomes. Thus we may classify subjects into "affected" and "unaffected," but not into "unaffected," "affected," and "dead." The latter would be a trinomial distribution.

2. The outcomes of all trials must be mutually independent. Thus identical twins are produced by one fertilization, one trial. To preserve independence it would be necessary to discard one of the twins randomly. What is equivalent, if both twins had the same phenotype we would count it as one trial only. No such problem would arise with nonidentical twins.

3. The scoring system is a simple 0-1 dichotomy. Thus we must not attach different scores to those with the complete phenotype with polyps, desmoids, and osteomas, and another score to those with a few polyps only. Or again we do not count twice a person examined on two occasions and found to be affected on both. (This is a common error in analysis.)

4. The probability of being affected must be the same for each trial. The Gardner syndrome is age-dependent and the probability of having polyposis is greater in a person aged thirty than in one aged five. Thus the number of affected members in a sibship of size ten ranging in ages from ten to thirty-five will not follow a binomial distribution. Instead it follows a more complicated form known as the Lexis distribution (7).

5. The number of trials must be fixed in advance. But family size is

rarely fixed. If we had a collection of fifty sibships of diverse sizes the numbers with 0, 1, . . . affected sibs would not follow a binomial, but rather a compound binomial distribution.

6. The scores must simply be added, not multiplied or interrelated in any way. For example the clinical geneticist (quite rationally) diagnoses disorders "by the company they keep." There are other causes of polyposis of the bowel than Gardner syndrome, many of them, such as ulcerative colitis, not Mendelian. In atypical cases the physician is likely to diagnose the condition in both of affected sibs and much less likely to diagnose it in one isolated case. Thus on the evidence he may give a score of 0 for one case and of $1 + 1 = 2$ for two cases and clearly $0 + 0 \neq 2$. I do not imply that this course is not reasonable; but it modifies the distribution, a fact to be taken into consideration in analysis. There are plenty of examples where additivity is violated by contagion (e.g., in the spread of infectious disease or a social attitude) to which the binomial distribution is commonly misapplied.

### The properties of the binomial distribution

These may be very briefly summarized. (We are concerned here with the probability algebra; in the statistical problem, $n$ is known but not $p$.) We note first the standard notation, more fully expounded elsewhere (7).

$$n! = n(n - 1)(n - 2) \ldots (3)(2)(1) \qquad \text{where } n \text{ is an integer}$$

$$0! = 1 \qquad \text{by convention}$$

$$\binom{n}{x} = \frac{n!}{x!(n - x)!}$$

The order of the binomial, $n$, is the number of trials. The probability of one particular outcome (arbitrarily called a "success") is $p$, that of the other ("failure"), its complement $1 - p = q$.

1. Then the probability of the typical result $x$ is given by

$$P(x) = \binom{n}{x} p^x (1 - p)^{n-x} \qquad x = 0, 1, \ldots, n \tag{1.1}$$

2. The mean, average or expectation of $x$, the number of successes in $n$ trials, denoted by $E(X)$, is

$$E(X) = np$$

3. The variance is

$$\text{Var}(X) = npq$$

Thus if two parents, both carriers for Tay-Sachs disease (an autosomal recessive), have three children, then $n = 3$ and from simple genetics the probability that any one child has Tay-Sachs disease $p = \frac{1}{4}$. Thus the distribution for the number of affected progeny would be

$$P(0) = \left(\frac{3!}{0!3!}\right) (\tfrac{1}{4})^0 \, (\tfrac{3}{4})^3 = 27/64$$

$$P(1) = \left(\frac{3!}{1!2!}\right) (\tfrac{1}{4})^1 \, (\tfrac{3}{4})^2 = 27/64$$

$$P(2) = \left(\frac{3!}{2!1!}\right) (\tfrac{1}{4})^2 \, (\tfrac{3}{4})^1 = \phantom{2}9/64$$

$$P(3) = \left(\frac{3!}{3!0!}\right) (\tfrac{1}{4})^3 \, (\tfrac{3}{4})^0 = \underline{\phantom{2}1/64}$$

Total                                   1

Note that unless the sum of the probabilities is unity there must be some error either in the logic or the calculations.

$$E(X) = 3(\tfrac{1}{4}) = 0.75 \quad \text{children}$$

$$\text{Var}(X) = 3(\tfrac{1}{4})(\tfrac{3}{4}) = 0.5625 \text{ children}^2$$

Hence the standard deviation for this probability distribution, by definition, the square root of the variance, is:

$$\sigma = \sqrt{\text{Var}(X)} = \sqrt{0.5625} = 0.75 \text{ children}$$

4. If both $np$ and $nq$ are large—say 10 or more—the normal (Gaussian) distribution with the same mean and variance may be used as an approximation, in spite of the fact that the latter is a continuous distribution, the former, discrete. Some refinements to allow for this incongruity are commonly applied (see Chapter 12).

### Uses of the binomial distribution

In my opinion this distribution should be used with much circumspection, and principally on the criterion of aptness. Like the chi-square contingency tests (Chapter 13) which are logically derived from it through the Gaussian approximation, it is often carelessly and inappropriately used, e.g., in analyzing incidences of infectious disease where assumptions 2 and 4 above are clearly violated.

## THE GAUSSIAN (NORMAL) DISTRIBUTION

This distribution has the probability density function:

$$f(x) = \frac{1}{\sigma\sqrt{2\pi}} \, e^{-\left[\frac{(x-\mu)}{\sigma}\right]^2 / 2} \qquad -\infty < x < \infty \tag{1.2}$$

It is readily plotted out as a continuous function by inserting selected values of $x$ in equation 1.2. This distribution is in striking contrast to the binomial on several points. First, it is continuous. Second, it has the paradoxical property of describing a great many variates approximately but none exactly. I mean by this that, strictly, two properties of the distribution are such as to exclude almost any measurement in the real world and certainly in biology: that it has infinite limits and that the variate may assume negative values, properties certainly not fulfilled by height, for example. On the other hand there are many variates such as adult height which conform closely to the model over a wide range (say three standard deviations on either side of the mean), so much so that theory based on normality may be invoked with considerable confidence even for individual readings.

A third difference is that the binomial distribution may be generated basically in only one fashion and by a single set of assumptions. The Gaussian distribution can be generated by a great diversity of mechanisms taken to the limit, including the binomial process itself. It is comparatively easy to argue from the model to the Gaussian distribution, but while speculation is easy, to argue cogently from the Gaussian distribution back to the basic model is difficult.

A fourth major difference is that while in theory a Gaussian distribution may be inferred from a basic algebraic model, by far the commonest warrants the biostatistician has for invoking it are: the conformity of the empirical data to the distribution where individual measurements are concerned; or the robustness demonstrated by the central limit theorem (7) where the distribution, not of individuals, but of means of at least moderately sized samples is involved (see Chapter 17). In view of this last fact, it is of some importance to consider how to determine whether the empirical data conform reasonably well to the distribution. "Reasonably well" is a vague term such as a self-respecting mathematician would abhor. But we would do well to remember that what we call reasonable depends on two matters: the size of the sample, and the purposes to which we wish to put the data. Several methods, of increasing sophistication, are available for testing this conformity.

## 1. The histogram

The data points may be sorted out into groups, each with the same width. The number of groups is not critical—somewhere between ten and twenty is convenient to handle, and sufficient for most purposes. The easiest procedure is to begin by identifying the largest and the smallest readings and dividing the difference by the number of categories. I have found it most convenient to "sort" on a lined pad taking a new line for each reading. After all are sorted, the appearance of the page is like an inverted histogram (see Table 1.1).

**Table 1.1 Birthweights (in kg) of Moroccan Jewish babies (Data of Fried and Davies, 13)**

| | | | | | | | | | |
|---|---|---|---|---|---|---|---|---|---|
| *2.50 | *2.60 | *2.80 | *3.00 | 3.20 | *3.40 | 3.65 | 3.80 | *4.05 | *4.25 |
| | *2.77 | *2.80 | 3.06 | 3.25 | 3.42 | *3.70 | *3.85 | 4.05 | |
| | | 2.82 | 3.10 | 3.25 | 3.45 | *3.70 | | 4.10 | |
| | | 2.95 | *3.12 | *3.26 | *3.45 | *3.70 | | | |
| | | | 3.15 | 3.27 | 3.45 | 3.70 | | | |
| | | | | *3.30 | 3.45 | 3.70 | | | |
| | | | | 3.30 | 3.46 | 3.74 | | | |
| | | | | *3.32 | *3.48 | *3.75 | | | |
| | | | | *3.32 | 3.50 | *3.75 | | | |
| | | | | 3.32 | *3.50 | | | | |
| | | | | *3.35 | *3.50 | | | | |
| | | | | 3.35 | *3.55 | | | | |
| | | | | 3.35 | 3.55 | | | | |
| | | | | | *3.58 | | | | |

NOTE: Values marked with an asterisk (*) are the products of uncle-niece marriages.

**Example 1.1.** As an example we may take some data of Fried and Davies on birth weights of Moroccan Jewish babies (13). Half of these babies were from closely inbred marriages, the rest from non-inbred marriages in the same population. This mixing we shall ignore here since the mean birth weights are similar. The pattern is laid out in Table 1.1. and Figure 1.1 *top*. The extreme weights are 2.50 and 4.25 kg and grouping by intervals of 0.2 kg seems appropriate since it gives ten classes. The histogram conforms satisfactorily to the general Gaussian pattern. There is one central peak in the 3.20–3.59 kg range with a fair degree of symmetry about it, there being similar declines in frequency in the tails on either side of it. For many purposes such a pattern would be acceptably close to the Gaussian without any more detailed analysis. We would with little loss of accuracy assume that the means of samples of such a size follow a Gaussian distribution. (see Chapter 17.)

## 2. The empirical distribution function

A somewhat more refined way of dealing with the same problem is to plot the empirical cumulative distribution (Figure 1.1 *middle*). By the (cumulative) distribution function, $F(x)$ of $X$, we mean the probability that the random variable $X$ assumes the value $x$ or less. The empirical distribution function is a plot of the proportion of sample values assuming the value $x$ or less. The main advantage over the first method is that it avoids the need to make arbitrary decisions about grouping. (With many small samples, by playing around with various groupings, the investigator may be able to make the histogram accord more or less with some preconceived idea about the underlying distribution. How far it is legitimate to do so is arguable, though any deliberate exploration of that kind should somehow be taken into ac-

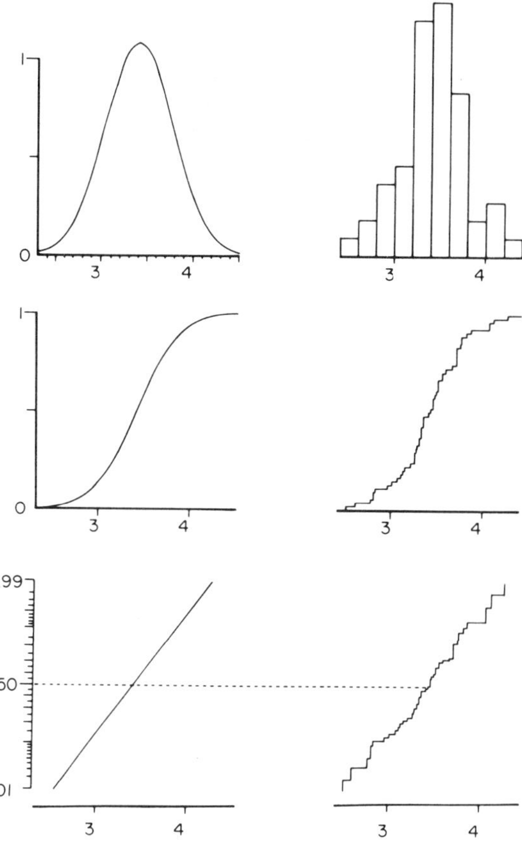

**Figure 1.1.** Weight at birth of 54 Moroccan Jewish babies. In each case the behavior of an ideal Gaussian distribution is shown on the *left* and that of the sample on the *right*. From above down are: *top,* the probability density function and the histogram; *middle,* the cumulative distribution function; *bottom,* the cumulative distribution function plotted on a Gaussian probability scale.

count in analyzing the results.) The details of constructing the curve are given below. Where the distribution is Gaussian, the resulting curve should be more or less sigmoid (flat–increasing slope–decreasing slope–flat). However, not all sigmoid curves are Gaussian, and the curve is not easy to assess even for the practiced eye.

## 3. Gaussian (normal) probability paper

A more time-consuming but also more revealing method is to plot the empirical cumulative probability on paper constructed on a special scale that compresses the middle and stretches the tails of the curve in such a way that an exactly Gaussian curve appears as a straight line (Fig. 1.1 *bottom*). If the distribution is Gaussian and the sample is reasonably large we shall expect the empirical cumulative probability plot to be close to linear, especially in the mid range, say between the 10th and 90th percentiles. Judging how much departure is tolerable for a sample of a given size is largely a matter of experience. One should be most suspicious where there is a *systematic* departure from linearity, e.g., a smooth steady curve, which suggests skewness (i.e., asymmetry), or a sigmoid curve, which suggests that the curve is either too spiked (leptokurtic) or too flat-topped (platykurtic) to be Gaussian (Fig. 1.2). Gaussian probability paper is particularly useful in testing whether or not a particular transformation (see Chapter 4) is normalizing.

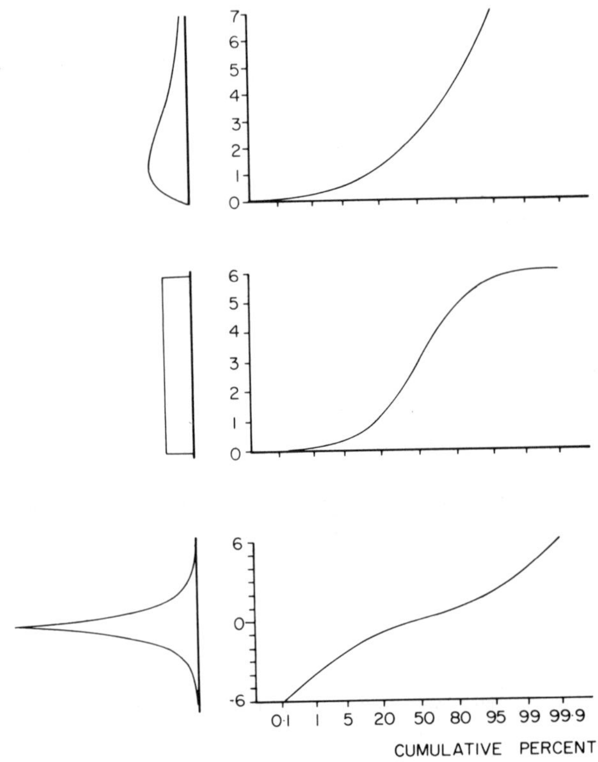


**Figure 1.2.** Plots on Gaussian probability paper of three idealized non-Gaussian distributions. The probability density functions are shown at the left, lying sideways so that they correspond to the conventional way of displaying the probability plots. From above downward are shown: *top*, a gamma function of order 4 with a unitary time parameter, which is positively skewed and which yields a convex curve on the Gaussian probability paper; *middle*, the rectangular distribution in the interval [0, 6] which is platykurtic (flat-topped) but symmetrical, and yields a sigmoid curve; and *bottom*, the Laplace or double-exponential distribution (i.e., an equal mixture of an exponential distribution and its mirror image) which is leptokurtic (excessively spiked) and yields an inverted sigmoid plot.

A straight line drawn by eye through the points furnishes two convenient shortcut estimates: of the mean, which is given by the point at which the line intersects the 50% line, and the standard deviation, which is one-quarter of the distance between the intersections with the 2.24% line and the 97.76% line. (This shortcut applies to Gaussian variates.)

**Practical Details.** The first step is to sort the data into a histogram and within each column arrange the values in ascending order as in Table 1.1. In this example there are 54 readings each of which contributes $\frac{1}{54} = 0.0185$ .... The results are laid out starting with the lowest and then in ascending order, as follows:

| Value of variate ($X$) | Frequency of $x$ | Cumulative frequency | Cumulative proportion | Percentile |
|---|---|---|---|---|
| 2.50 | 1 | 1 | 0.018,52 | 1.852 |
| 2.60 | 1 | 2 | 0.037,04 | 3.704 |
| 2.77 | 1 | 3 | 0.055,56 | 5.556 |
| 2.80 | 2 | 5 | 0.092,59 | 9.259 |
| 2.82 | 1 | 6 | 0.111,11 | 11.111 |

etc.

It is convenient to compute the fourth column by adding the reciprocal of the sample size (here 0.0185 ...) once for each unit in the second column. The values from the last column are then plotted on the appropriate percentile (which is on the probability scale) against the value of the variate on an ordinary linear scale.

Note that whereas we suppose that the theoretical distribution is a smooth continuous function, these sample cumulative proportions constitute a set of discrete jumps and the corresponding plot consists of irregularly spaced steps (Fig. 1.1 *bottom*). The beginning of the first step (0%) and the end of the last step (100%) are at infinity and cannot be plotted on this scale.

## 4. Tests of goodness of fit

The sample mean and standard deviation may be computed by the method of moments (Chapter 5), and the proportions which would be expected to fall in any particular interval may be computed and compared with the observed numbers, and a test of goodness of fit (especially the chi-square test) may then be performed. The computations are straightforward, but discussion of details of the chi-square test will be delayed until Chapter 13.* In the present instance we get the following results, using the grouping shown

---

*I ask the reader to be patient with the difficulty of a coherent exposition of such problems. In order to justify the claim of Gaussian normality for a distribution as a preliminary to expounding certain methods, I am appealing to arguments involving methods that exist but have yet to be explained. The logic is unobjectionable. The system, however, leaves much to be desired. I am describing the process in the order in which I would approach an actual problem. If the reader will only take some detail for granted at this stage, I promise to furnish particulars later.

in Table 1.1. This grouping has not been consciously chosen to make the results appear Gaussian; the reader is free to try other groupings.

| Birth weight | Observed | Expected if distribution were Gaussian |
|---|---|---|
| 2.59 or less | 1 | 0.795 |
| 2.60–2.79 | 2 | 1.961 |
| 2.80–2.99 | 4 | 4.662 |
| 3.00–3.19 | 5 | 8.312 |
| 3.20–3.39 | 13 | 11.119 |
| 3.40–3.59 | 14 | 11.160 |
| 3.60–3.79 | 9 | 8.405 |
| 3.80–3.99 | 2 | 4.749 |
| 4.00–4.19 | 3 | 2.013 |
| 4.20 or greater | 1 | 0.824 |
| Total | 54 | 54.000 |

## 5. Fitting the curve

A similar and rather more vivid method is to overlay the histogram with the Gaussian distribution with mean and variance given by the sample estimates (Fig. 1.3 *top*). Parts lying in the histogram, or in the fitted distribution, but not in both are shown in black. Hence the amount of black is an index of the amount of discrepancy between the two. Since the cells of the histogram are represented as rectangles (we might, I suppose, draw them otherwise, e.g., as trapezoids, but conventionally do not), we never expect a perfect fit to a smooth curve. For comparison (Fig. 1.3 *bottom*) is shown the best possible fit when a Gaussian curve is fitted to a theoretical histogram in which the area of each cell is precisely equal to the area of the Gaussian curve between the same limits. We shall see plenty of examples of bad fits later.

Note that in fitting the Gaussian curve it is necessary that its total area should equal that of the histogram. If the latter is 54 square inches then with a units scale of one inch, the probability density should be multiplied by 54.

## 6. Sample cumulants

Elsewhere (7) I have discussed higher moments. The $k$th central moment of a random variable is the average value when the distance of the variable from its mean ($\mu$) is raised to the power $k$. Related to the central moments are quantities called cumulants. The first four cumulants, and their values for the Gaussian distribution are as follows:

| Cumulants | Description | Gaussian value | Comment |
|---|---|---|---|
| First | $E(X)$ | $\mu$ | The mean |
| Second | $E(X - \mu)^2$ | $\sigma^2$ | The variance |
| Third | $E(X - \mu)^3$ | 0 | Measures skewness (asymmetry) |
| Fourth | $E(X - \mu)^4 - 3\sigma^2$ | 0 | Measures kurtosis |

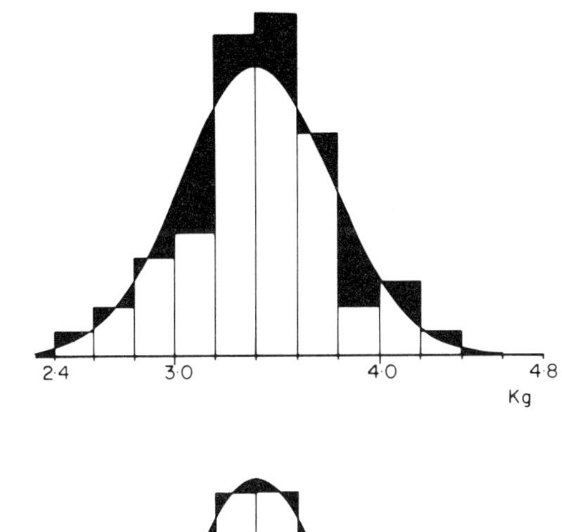


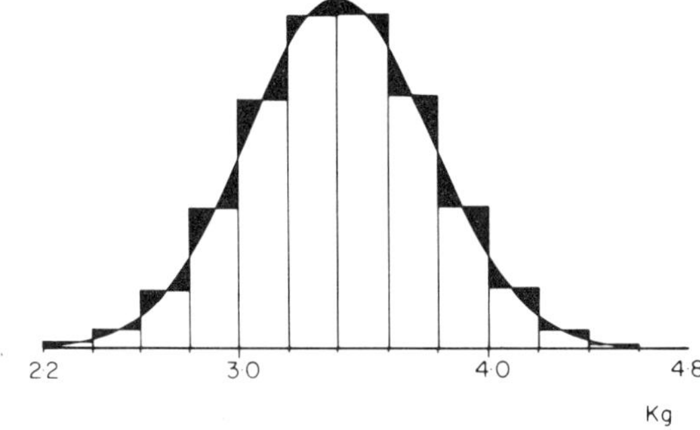


**Figure 1.3.** In the *upper diagram* are shown the histogram from the data of Table 1.1 and the Gaussian probability density function with the mean and variance given by the sample. Discrepancies are shown in black. Because of the rectangular character of the cells of the histogram, we could not expect a perfect agreement under any circumstances. For comparison, in the *lower diagram* is shown the best fitting histogram possible given the same class intervals.

Note that, as above, "E" merely means the expectation or average of the random variable that follows. We may estimate the cumulants from the sample and compare them with the theoretical values as a measure of departure from Gaussian normality. This criterion will be used in Chapters 4 and 14.

## THE PLAN OF THIS BOOK

Following some preliminary theory (chapters 1 to 4) there are four sections.

1. A large section on the use of the Gaussian distribution and distributions derived from it (chapters 5 to 11).

2. A section on categorical data handled mainly by the binomial

distribution and distributions related to it, notably the chi-square and the Poisson (chapters 12 to 14).

3. A section on so-called distribution-free methods (chapters 15 and 16).

4. An apologetic, but I trust helpful, chapter on "makeshift," dealing with data of uncertain provenance, or properties, or both (Chapter 17).

In Appendix 1 is added a discussion of matrix algebra which, although not totally indispensable for multivariate and polynomial methods, is sufficiently helpful (especially where computers are to be used) to merit at least a brief study.

## PROBLEMS

**1.1.** Classify the following distributions according to type (discrete, etc.):
   **a.** The patient's hospital bill.
   **b.** The extent of lymph node involvement in cancer of the breast.
   **c.** Intelligence.
   **d.** Number of previous surgical operations.
   **e.** Genetic linkage (i.e., the degree of independence in the inheritance of genes at difference loci).
   **f.** The extent of lesions in smallpox.
   **g.** The concentration of an aqueous solution at which a person can taste phenylthiocarbamide.

**1.2.** Construct the following distributions:
   **a.** A binomial of order 6 with parameter $\frac{1}{3}$.
   **b.** A Gaussian with mean 10 and variance 25.
   **c.** The winnings of a gambler who tosses a fair coin independently five times. The stakes are 1, 2, 3, 4, and 5 dollars respectively for the five tosses. If he wins the toss he receives the stake; if he loses, he receives nothing.

**1.3.** A colleague, discussing diagnosis in a particular disease, referred to some data "picked up at random in the clinic."
   Comment.

**1.4.** Make a histogram from the data in Table 1.2. Inspect it and comment on the pattern. What do you think the results represent?

**Table 1.2.**

| | | | | | | | | | |
|---|---|---|---|---|---|---|---|---|---|
| 87.7 | 95.8 | 63.9 | 51.3 | 31.2 | 13.6 | 11.3 | 18.1 | 32.8 | 66.8 |
| 7.4 | 36.1 | 58.2 | 71.9 | 83.1 | 47.2 | 46.0 | 98.8 | 89.8 | 33.6 |
| 83.1 | 36.9 | 41.1 | 78.3 | 92.1 | 70.5 | 83.1 | 65.6 | 10.2 | 68.8 |
| 49.8 | 64.0 | 83.1 | 8.3 | 22.4 | 4.3 | 1.8 | 80.3 | 47.7 | 73.6 |
| 12.1 | 79.9 | 73.4 | 97.6 | 37.3 | 60.3 | 46.6 | 1.2 | 63.7 | 59.9 |

# 2

# TESTING OF HYPOTHESES: THEORETICAL ASPECTS

There exist three main fields of applied statistics: testing of hypotheses, estimation, and the making of probabilistic decisions. Scientists try to collect accurate facts and to fit them together into coherent patterns. The accurate facts generate the fields of *estimation* and *classification* which will be dealt with later. The approach to identifying patterns is to formulate propositions to which assent is either given or withheld in the light of the data. It is with such propositions that we shall be concerned in the present chapter. First however it will be worth while to explore the familiar classical method of proof and a modern probabilistic (or "stochastic") generalization of it.

## PROOF

The earliest approaches to science, and the most rigorous standards that are still imposed, call for complete certainty in making statements about propositions. This was the spirit of Aristotelian (deductive) logic, which had clear but limited success, and which later inductive logic attempted in vain to emulate. It is only with the development of scientific biology beyond the descriptive phase and, somewhat more recently, with the study of particle physics, that the scientist has been prepared to settle for something less than complete certainty. It is at this stage that the necessity for stochastic proof arises.

**Example 2.1.** As a point of departure consider the classical syllogism. The form *modus tollens* is illustrated by the following:

1. All living men need oxygen.
2. George does not need oxygen.
3. Therefore George is not a living man.

In the structure of this argument the first is a general statement about living men and is termed the *major premise;* the second is a particular statement about an individual and is called the *minor premise;* and the third is a *conclusion.* The above argument is (*formally*) *valid.* We can at once see the correspondence of the above to Bayes's theorem. (See [7] for details.) The major premise can be seen as a background statement of general convictions before we confront a particular case. In this respect it is like a prior probability. But the discussion is about George, and George might be any one of a number of things—a man, a horse, a sailing boat, a computer. We start then with some prior probability that an entity selected at random is a human being. In the present case it does not matter precisely what this probability is, so that it may be denoted by some quantity $a$ that is greater than 0 and less than 1; then the prior probability that George is not a living man is the complement of this quantity $(1 - a)$. Now look at the conditional probability of the fact that George needs oxygen. If George is a living man this probability (according to the first premise) is 1; in consequence the probability of his *not* needing oxygen is 0. If all living men need oxygen, then no living men do not need oxygen. If George is not a man, the probability is less clear. Some of the things called George, such as animals, would cease to exist as such if they did not have oxygen. On the other hand this result would not be true of computers or sailing boats. So, for the class of things that are not men, the conditional probability that they do not need oxygen will again be some probability lying between 0 and 1, represented by some quantity $b$, the precise value of which is again of no importance in the present case. We may now set up the usual Bayesian table (Table 2.1). The posterior probability that George is a man is 0; and hence it is certain that he is not a living man.

Such certainty is all too rarely achieved. It is attained in the present case because provided both $(1 - a)$ and $b$ are strictly greater than 0, the joint probability that George is not a living man and does not need oxygen is greater than 0. Hence it is an infinitely more plausible explanation for the fact that George needs no oxygen than the impossible event that George could be a living man and not need oxygen. It is only because the conditional probability that George will not need oxygen if he is a living man is 0, that

**Table 2.1. A classical *modus tollens* argument in Bayesian terms**

| | | George is a living man | George is not a living man |
|---|---|---|---|
| i. | Hypothesis | | |
| ii. | Prior probability | $a$ | $(1 - a)$ |
| iii. | Conditional probability of not needing oxygen | $0$ | $b$ |
| iv. | Joint probability (product of [ii] and [iii]) | $0$ | $b(1 - a)$ |
| v. | Posterior probability* | $\dfrac{0}{0 + b(1 - a)} = 0$ | $\dfrac{b(1 - a)}{0 + b(1 - a)} = 1$ |

*Note that the posterior probabilities must—
1. add to unity;
2. bear the same ratios to each other as the corresponding joint probabilities.

(provided that both are greater than zero) the precise values of $(1 - a)$ and $b$ do not matter.

In the scientific world one can rarely make flat statements about impossibility. We can assert with complete assurance that a man cannot live without oxygen because scientists understand much about the mechanisms from which the dependency arises. The only way in which this idea could be attacked would be from the philosophical standpoint as to what the nature of man is. (A theologian, for example, might well argue as to whether in the afterlife man needs oxygen or not. In the present text I shall be content to deal with the problem on a purely natural plane.)

Even in the best medical examples we usually abandon the idea of certainties for probabilities. Sometimes they may be stated very precisely.

**Example 2.2.** Suppose, for instance, that Betty has two colorblind brothers. She has two sons only; both have normal color vision and so do both her parents. What is the probability she is a carrier for colorblindness?

Note first that colorblindness is borne on the $X$ chromosome so that the paternity of an affected male may be ignored. Also colorblindness is common, so we may ignore new mutations. Her mother is clearly a carrier (since Betty has two affected brothers). Betty is thus at a 50% risk of being a carrier. We set up the Bayesian table as before (Table 2.2). The conditional probabilities follow from the genetics. If she is a carrier each son has a probability of $\frac{1}{2}$ that he will be normal, and thus the probability that two independent sons will be normal is $(\frac{1}{2})^2 = \frac{1}{4}$. If she is not a carrier the probability that her sons will be normal is almost exactly one. Comparison of the joint probabilities makes the latter explanation four times more plausible than the former.

This kind of reasoning is commonly encountered in genetic counseling and is discussed in detail elsewhere (14).

But such precise probability statements are rarely available. Consider a common type of diagnostic problem in medicine.

**Example 2.3.** A patient has symptoms which point toward cirrhosis of the liver, though it is not clear of what type. The choice lies, perhaps, between portal cirrhosis and biliary cirrhosis. If there were nothing more to it

**Table 2.2 A problem in genetic counseling**

| | Hypothesis | Betty is a carrier of colorblindness | Betty is not a carrier of colorblindness |
|---|---|---|---|
| ii. | Prior probability | 0.5 | 0.5 |
| iii. | Conditional probability of two normal sons | $(0.5)^2 = 0.25$ | $1^2 = 1$ |
| iv. | Joint probability | 0.125 | 0.5 |
| v. | Posterior probability | $\dfrac{0.125}{0.125 + 0.5} = 0.2$ | $\dfrac{0.5}{0.125 + 0.5} = 0.8$ |

than that, the clinician might be prepared to settle for the former diagnosis, which is a much more common condition. Seen in these terms the prior probability that the condition is portal cirrhosis is overwhelmingly the greater. But the particular patient has such features as a very large liver, very large spleen and xanthomatosis, common manifestations of biliary cirrhosis but rare in portal cirrhosis. Looking exclusively at the conditional probabilities (a common preoccupation of the inexperienced clinician), the evidence would be strongly in favor of biliary cirrhosis. The final formal evaluation of the problem will be a deliberative choice between a common condition in which the manifestations are atypical, and a rare condition of which they are typical. It is not at all clear how the joint probabilities would compare in this particular case and hence what the judgment of the experienced clinician would be.

If the problem could be resolved to this extent and in these terms, the scientist might be content with a (subjective) probability statement about the truths of the hypotheses, in place of the certainty required in classical deductive logic. But it will not have escaped the reader that though the ideas are clear enough, my formulation of the problem of diagnosis of cirrhosis of the liver is vague in detail. The difficulty is that sufficiently precise information about the relative frequencies of the two disorders and the probabilities of the particular physical findings in each have not been available to allow an exact solution to the problem. But at least it is clear how the necessary information could be collected. Given that the two conditions can be defined precisely enough to be investigated, with sufficient time and work there is no problem in principle in determining any of the above facts.

But the difficulties are often much more serious.

**Example 2.4.** For instance, the so-called Weber-Fechner law says that the intensity of a sensation varies as the logarithm of the intensity of the stimulus. In recent years it has been suggested that the relationship would be better described by a power function (15). If there were no errors of measurement and no variability in the behavior of subjects, the matter presumably could be resolved beyond dispute by devising a sufficiently accurate system of measurement and comparing the observations with the predictions under the two laws. But the data are imperfect, because of variability of both the observer and subject; and in the nature of things certainty is not to be attained. Because of error of measurement, both hypotheses could presumably account more or less plausibly for what is observed. Much weight, then, will be attached to the relative plausibilities of the two hypotheses as ways of explaining the observations. But what is one to use for prior probabilities? I personally know of no adequate theory suggesting that the relationship should be either a logarithmic or a power function. In this respect it differs from Example 2.2, where we have a well-developed theory from which we can compare the a priori plausibilities of the two competing hypotheses.

There are many such situations in which the prior probabilities for the various hypotheses are not available: either because there is no rational basis for arriving at them or there is an empirical basis which has not been pursued. Lacking this information the scientist cannot solve the problem. If the formal aspect of inference were taken to a fanatical extreme, the progress of science would be permanently arrested. A fundamental problem of statistical inference is how this defect in our knowledge is to be remedied. The various methods will be discussed later in this chapter.

A fundamental change of outlook from that of classical physics in our approach to scientific truths has resulted from the foregoing probabilistic formulation. It is here that the distinctions between probability and probability density are of fundamental importance (7). As we have noted in Chapter 1, a continuous distribution is one in which the probability of the random variable taking on *exactly* any particular value is a priori zero.

**Example 2.5.** Suppose that in the light of this formulation Newton were exploring his famous law for gravitation that the attraction between two masses varies inversely as the distance between their centers of gravity raised to the power 2. Now a priori why does the law have to take this form? Why could it not vary inversely as the distance raised to the power 1.99? Or 2.01? Or any one of an uncountably infinite number of values in the neighborhood of 2? Doubtless the idea of the inverse-square law was suggested to Newton by the inverse square law for the intensity of light, which in turn emerged from the fact that in Euclidean space (which in eighteenth-century physics was mistakenly supposed a necessity of thought) the surface of a sphere varies as the square of its radius. Thus the same quantity of energy (or gravitation) could be imagined as being spread uniformly over the surface of a sphere, and hence the further the distance from the origin the less its intensity, in accordance with the inverse-square law. But if the prior distribution for the power to which the distance has to be raised is continuous, then it is obvious that the *prior* probability of any *particular* value is 0, and hence what we have called in Table 2.2 the joint probability is 0. The scientist is no longer making the somewhat naive choice between supposing gravitation actually follows the inverse-square exactly or not at all. Rather he is trying to decide among an infinite number of competing hypotheses as to what the true value of the power is.

These speculations may not be here very appealing, because our beliefs in the inverse square laws are deeply ingrained. But consider the relationship between surface area of the body and height or weight. The formula usually used is a power function. However the appropriate power comes not from theory but from empirical observation alone. It is also true—a matter we shall explore more fully in discussing estimation—that however many data there are, the problem can never be solved precisely and with certainty. This view implies that we have abandoned the idea of finding scientific truth exactly and must be content to pursue indefinitely the most exact representation of what happens. We no longer talk of "true" and "false" statements but rather of "truer" and "falser" statements.

But this conclusion does not imply that in scientific inference the scientist keeps forever a totally uncommitted mind. If one abandons the claim that any particular theory is correct and is content to say merely that it is the best theory so far advanced, then the idea emerges of a theory that is "in possession," which will not be discarded simply because it is not exact, but only when a theory is advanced that performs consistently better over a wide experience. As Stevens (15) writes, "It retires from the field of scientific contest only when pushed aside by a stronger theory."

All this bears directly on hypothesis testing in statistics, as I shall now attempt to show. To see whether a drug has an effect on blood pressure, a scientist begins with the hypothesis, conventionally considered "in possession," that the drug has no effect whatsoever. It is commonly referred to as the *null hypothesis*. To this is imparted a momentum of belief which he is prepared to set aside only in the face of substantial evidence to the contrary. Always he has in mind not "Is this hypothesis false?" but the logical conjunction "Is this hypothesis untenable *and* is some *alternative* hypothesis (specified a priori) plausible?" An analogy may help.

The classical symmetrical method of comparing two hypotheses (as in Examples 2.2 and 2.3) is to treat neither hypothesis as being "in possession." We compute the priors and the conditionals and compare the joint probabilities as measures of plausibility. This symmetry is the relationship between two litigants disputing ownership of, let us say, a wallet that has been found by a third party. Neither of the litigants is "in possession," so the judge (or jury) must treat both parties symmetrically. There may be prior evidence of a general character: one litigant might be a rich and well-dressed company director, the other a beggar in rags. These appearances would make the prior probabilities heavier in favor of the rich man as the one who lost the wallet. On the other hand evidence taken may show the beggar knows the contents of the wallet in great detail whereas the rich man has no idea what is in it. This (as it were) "conditional" evidence would point in the opposite direction. The judge would do the equivalent of multiplying the probabilities of these two pieces of evidence together to get a "joint" probability, and the decision would go in favor of the larger joint probability.

Now an action at criminal law is a different matter. The hypotheses of "guilty" and "not guilty" are not treated symmetrically. The hypothesis "in possession" (at least under Anglo-Saxon law) is that the man is innocent. Even if the accused man offers no evidence whatsoever, the prosecution may not succeed in a conviction. To revert to the pharmacological problem, even if there is some evidence that the drug indeed has an effect on blood pressure, the experimental results may be insufficient to overthrow the null hypothesis of no effect.

But it is of the highest importance to understand the exact logical nature of the conclusion. When a judge delivers a verdict of "not guilty," he does *not* mean that the accused did not commit the crime. He merely means that the prosecution (on whom the onus of proof rests) has not succeeded in showing that he *did* commit the crime. In exactly the same way, to say that the effect of the drug on the blood pressure is "not statistically significant" does *not* mean that the drug has no effect on blood pressure. It merely means that in this limited study the pharmacologist (on whom the onus of proof conventionally lies) has not succeeded in showing that there *is* an effect.

Now in Medicine, as in Law, absolute certainty is not called for. In general the prior presumption is that a patient picked at random does not have a perforated gastric ulcer (and therefore should not be operated on for it). This view we shall treat as the null hypothesis. A surgeon does not require absolutely unshakable evidence that a patient has a perforated gastric ulcer before he will consider operating for it. He is prepared, in the light of other matters which will be discussed in Chapter 11, to take some risk that the patient has not a perforated ulcer but, say, acute pancreatitis. If he required absolute certainty he would miss a great many people who did require operation. In exactly the same way, conviction of an accused man does not usually demand absolutely unshakable proof because, if this were so, the vast majority of criminals would not be convicted and the law would be rendered impotent. The judge is prepared to run the risk of occasionally convicting an innocent man.

Whatever decision the judge may make he runs a risk: the risk of convicting an innocent man, and the risk of acquitting a guilty one. Both mistakes impose penalties. In exactly the same way, a surgeon, whatever his decision may be, runs a risk: of operating on a patient with acute pancreatitis which is not indicated, or not operating on a patient with a perforated peptic ulcer which is. The scientist has exactly analogous risks. One of them is to reject the null hypotheses when it is true: this is referred to as *an error of the first kind.* The other risk is that he will "accept" (that is fail to reject) the null hypothesis when it is false: *an error of the second kind.* To convict an innocent man, or to diagnose perforation in a patient with acute pancreatitis, is an error of the first kind; to acquit a guilty man, or to diagnose acute pancreatitis in somebody with a perforated peptic ulcer* is an error of the second kind.

It is obvious that the person making the decision cannot have it both ways. The more stringent the evidence he requires to reject the null hypoth-

---

*Of course, which is the null and which the alternative hypothesis might be reversed. A pediatrician would ordinarily argue that an infant does not have immunity to measles (and therefore should be artificially immunized against it). But this null hypothesis would be overthrown and immunization rendered unnecessary if the child could be shown to have had measles. If there were reason to fear that the child has an allergy to the vaccine, the null hypothesis might be that the child has had measles and the onus of proof lies on those who wish to vaccinate.

esis, the smaller the risk of an error of the first kind but the greater the risk of an error of the second kind. Conversely, the less stringent the evidence required to reject the null hypothesis, the smaller the risk of an error of the second kind but the greater the risk of an error of the first kind. It is rarely possible to eliminate both errors simultaneously (*never* for Gaussian data), and extensive statistical theory about minimizing the errors has been developed.

A great variety of methods could be devised for choosing the relative sizes of the two errors. The one that is usually adopted involves a number of steps and considerations.

### 1. Precision and certainty

If a scientist is preoccupied, to the exclusion of all other considerations, with not rejecting the null hypotheses when it is true, he is unlikely to be able to make any useful statement. A forensic pathologist attempting to reconstruct the height of a man by examining fragments of his bones can be quite certain he is making no mistake if he says the man was somewhere between 18 inches and 10 feet tall. The statement is absolutely unassailable, but clearly not of the slightest use in identifying the man.

### 2. The size of the test

The scientist, then, decides what risk he is prepared to take of rejecting the null hypothesis when it is true. This risk is referred to as *the size of the test*. Note that *the size of the test* must not be confused with the *size of the sample* (to which, other things being equal, it is inversely related). It is determined, in principle at least, before the data have been collected. Depending on how stringent he wishes the proof to be, he may set the size at 0.05, 0.01, 0.001, etc. It is important to note that there is nothing magical about the choice of the particular figure: the chosen size might as well be 0.0499 as 0.05. But usually certain conventional figures are chosen *because these are the values tabled in reference books*. Now that calculators enable one to compute many statistics for tests of any particular size, the scientist is not nearly so restricted in his choices.

Just how small a size should a scientist use for his test? Broadly speaking, the answer depends on three factors:

a. How sure does he want to be about his result? If from his result he merely wishes to decide what small line of inquiry should be pursued next he may not require very severe standards of proof. On the other hand, if his scientific conclusion were to be the basis for some vast public health policy (as for example a program of immunization involving risks) or some decision about safety which might have disastrous consequences to an industry, then high standards of certainty would be demanded.

b. Hypothesis testing of this type is used when adequate prior probabilities and distributions are not available and no explicit attempt is made to provide them. Nevertheless, there is a kind of "ghost" of a prior which colors the standards of stringency demanded. To establish a thesis that does not seem very plausible, a scientist is wise to use a test of very small size which, other things being equal, will call for a large sample. In this way he cannot be accused of laxity in his proof. Conversely if the point being explored is obvious to the point of being trivial, very little proof is called for. Proof, let us say, that experimental animals grow faster if given supplements of amino acids in their diet does not have to be stringent because the idea seems a priori very plausible. On the other hand, the notion that a high dietary intake of RNA improves the intelligence does not seem very plausible and would call for a cogent standard of proof, that is, a test of small size.

c. The scientist commonly chooses his size of test according to what he thinks he will succeed in proving. If there is reason to believe beforehand that a big difference will be observed from the treatment and that the null hypothesis would be rejected at about the one-in-one-million level, then he may choose this size of test. If on the other hand he believes that the hypothesis would barely be rejected at the 1 in 20 level, then he is content with this much larger size of test.

## 3. The power of the test

Having decided on the size of test and how big an experiment it is practicable to perform (for example how many test animals, or subjects, or what size of sample he can handle), he then finds that part of the sample space for the outcome of his experiment that will minimize the error of the *second* kind (by a criterion that will be discussed in detail later). The complement of the error of the second kind is referred to as the *power* of the test: that is, the probability of rejecting the null hypothesis when it is false or of accepting the alternate hypothesis when it is true. This concept will be discussed and illustrated in detail later. Then the scientist's objective is to maximize the power. That part of the sample space that (in accordance with these criteria) is selected to correspond to errors of the first kind is referred to as *the rejection region*. The rest of the sample space is referred to as the *acceptance region*.

## 4. The execution of the experiment

The experiment is now performed, the data collected and analyzed, and the appropriate test statistic is computed. If the resulting value is within the rejection region, then the null hypothesis is "rejected at the 0.01 level" (or whatever size of test has been chosen). If the test statistic falls inside the acceptance region, then the null hypothesis is not rejected (is "accepted") at this particular size of test.

Note, as stated above, that to say that the null hypothesis is "accepted" does not necessarily imply a strong belief that it is true: like a verdict of "not guilty," it merely means that the evidence has not been sufficient to reject the null hypothesis. Whether or not further evidence would lead to rejection of the hypothesis is not a matter about which it is useful or even proper to speculate in the absence of this further evidence. This point is of some importance. Some investigators tend to do a small experiment, fail to reject the null hypothesis and end with some such sentence as "Although the drug lowers the systolic blood pressure on average by 20 millimeters of mercury, the experiment did not lead to rejection of the null hypothesis of no effect at the 0.05 level. Nevertheless a larger sample would probably have led to a rejection of the null hypothesis." Statements of this kind cannot be justified on the basis of the sample data, and can rarely be justified on other grounds.

## 5. Modifications in logic

Many investigators do not choose the size of the test before they do the experiment. They wait until the data have been analyzed, find the size of test at which the null hypothesis would just have been rejected, and quote that figure, referring to it as "the *level of significance.*"* There are theoretical objections to the logic underlying this procedure. It obviously makes the size of the test a function of the data, and hence a random variable. A repeat of the experiment would lead to a different level of significance. According to strict theory, the size of the test is not supposed to be a random variable. On the other hand as an experimenter myself, I have a great deal of sympathy with this procedure. When there is no prior information as to what the outcome of the test is likely to be, it seems obvious that one wants a measure not merely of whether or not a particular hypothesis has been rejected, but also of how forcefully. If the scientist starts with some conventional level such as 0.001 and the data would have been rejected at a level of 1 in $10^8$, it is not doing justice to the data to report it at the size of test fixed beforehand. On the other hand, when one uses this approach it is difficult to deal with the notion of the power of the test.

Note that after completing a particular experiment the investigator is not concerned with *both* size and power. If he rejects the null hypothesis, then he is not interested in power. If he "accepts" the null hypothesis he is not particularly interested in the significance level. In the same way, after he has delivered a verdict in a particular case, a judge cannot be worried *both* that he has acquitted a guilty man and convicted an innocent one.

The point is made because it illustrates the important distinction between testing a hypothesis and testing a model. The pharmacologist who is trying to decide whether or not a drug has an influence on blood pressure concentrates most of his attention on the size of the test, because if he is pursuing a fruitful line of inquiry he is much more likely to be rejecting the null hypothesis than to be accepting it. Of course when the scientist is screening a

---

*Some writers use the terms *size of the test* and *level of significance* interchangeably.

vast number of compounds for impact on cancer in experimental animals, he is anxious to dispose of false claims for many of them. However, he does not want to miss a drug that is highly effective, a risk which may be considerable if for reasons of economy the drug can be tested on a small number of experimental animals only. Nevertheless such massive screening of large numbers of compounds, indiscriminately collected, does not conform to classical patterns of scientific investigation; and while there have been some successes from this technique—for example in the discovery of antibiotics—by and large this approach has not been very fruitful.

On the other hand, if the scientist thinks that he is on a useful tack in exploring models for the description of natural phenomena, a fruitful line will lead him to accept the null hypothesis even at a test of large size. But this objective leads to a somewhat anomalous result: if he wishes to accept the null hypothesis, the fewer data he collects the better! This statement is true only insofar as the *size* of the test is concerned. But of course the standing in which any scientific theory is held depends much more on the assaults which it has withstood from penetrating tests than on how well it fits previously collected data. If the scientist can say that a particular hypothesis has withstood a test with power of 0.99, it is in much better standing than one that has withstood a test with power 0.4. The more probability the hypothesis will be rejected if false, the more convincing it is if accepted. Just in the same way, a person whose reputation has been damaged by an accusation will insist on a searching trial to make it clear that the accusation has been shown unsound and not merely dropped for technical reasons or, say, political expediency. This point is of major importance although, for example, it is embarrassingly neglected in such fields as genetic analysis. The geneticist attempting to sustain the hypothesis that a particular condition is inherited as an autosomal recessive, should certainly test to see whether or not his hypothesis is rejected at an appropriate size of test. But he should be at least as much concerned with describing how much *opportunity* the data have furnished to reject his hypothesis if it is false; in other words, he should be concerned just as much with the power of his test. But to make this idea usable we must lay out the logic more precisely.

Before analyzing the statistical details, it is wise to look a little more closely at the epistemological aspects of this form of hypothesis testing (with which the names of Neyman and Pearson are associated).

First, the terms of the argument are not direct statements about the hypothesis, but about the data. However, the conclusions are about the hypothesis. Consider a sample from which is computed some statistic, $S$, (e.g., the mean). We have some hypothesis, $H$ (e.g., that the sample is from a particular population, or that a treatment is ineffective, or that a Mendelian trait is involved). Thus the scientist wishes to arrive at probabilistic (if not certain) conclusions about $H$, and $S$ is to be the basis for doing so. Thus he wants to make a statement of the form "in view of the data, the probability that the hypothesis is true is 0.9," i.e.,

$$P(H|S) = 0.9$$

Now what the usual analysis tells is something like

$$P(S|H) < 0.05$$

(That rather loose statement will be tidied up later.) We are, in effect, starting on line (iii) of the Bayesian table (e.g., Tables 2.1 and 2.2). Now Bayes' theorem (see [7]) describes how to convert statements of the form $P(S|H)$ into the form $P(H|S)$. We require information on $P(H)$ and the probability of the statistic given the hypothesis, for each competing hypothesis. Too often such information is not available and one cannot answer the question that is of real interest to the scientist. The only logically rigorous conclusion which emerges is about the *sample*, and the probability of obtaining such a result given the hypothesis. The scientific conclusion about the *hypothesis* is much more suspect.

Secondly, in a rational world our evaluation of $P(H)$, by which we mean the prior probability that the hypothesis is true before we examine the data from the current experiment, would depend on several factors.

a. What does the accumulated empirical evidence show so far? There may be forty papers published on this topic making the hypothesis more or less plausible.

b. What does general scientific insight tell us about the hypothesis? Sound conceptional relationships call for less evidence than those based on empirical evidence alone. We understand why the tetralogy of Fallot produces cyanosis much bettter than why it produces clubbing of the fingers, and for some newly discovered congenital heart lesion we would make firm statements about cyanosis more readily than about clubbing. For some hypotheses, e.g., about the influence of the constellation Cancer on its namesake, we would demand very stringent standards of proof before rejecting the null hypothesis of no influence.

c. There is an incurably personal element in convictions about hypotheses which masquerade as "hunches," "insights," and even "philosophy." Einstein, for instance, rejected physical indeterminism on the grounds that "God does not roll dice." Clearly we can do very little about this statement scientifically. But one important issue is whether these transrational convictions come before or after the data are available. I suspect that many probabilists would argue that timing is quite irrelevant to a probabilistic statement. I would not agree; but perhaps they would not admit these "prior" convictions as germane.

All such information would be incorporated, as appropriate, into the Bayesian table under "prior probability." But there is no way in which it can be directly accommodated in the Neyman-Pearson argument at all. It would be catered for indirectly by the size and power demanded of the test.

Thirdly, there is an inherent logical weakness in the argument that because $S$ is an improbable outcome if $H$ is true, therefore $H$ must be false. It may be that under any alternative hypothesis, $S$ is even more improbable. Furthermore in a great many experiments $S$ may *always* prove an unlikely outcome given $H$; and yet the scientist may go on believing $H$ is true.

**Example 2.6.** Visible new mutations at any particular genetic locus, for instance, are very rare events: something of the order of 1 in 100,000 births. When we see a child with Apert's acrocephalosyndactyly, the probability of this child occurring on the hypothesis of new mutation is 1 in 100,000. Clearly $P(S|H)$ is 0.000,01 which is a far lower significance level than is usually required for rejecting a hypothesis. Yet what is the alternative? This genetic disorder is for all purposes lethal, so that inheritance is almost out of the question—and indeed it may be ruled out where the parents are normal, since this is an autosomal dominant. Thus we persist in our belief despite the impressive level of significance. What is worse, we will act in the same way every time we see such a patient, although for every patient the combined significance level decreases by a factor of $10^{-5}$.

Now the reader may feel at this stage that this argument must be false: that no system of inference could have wide acceptance in the face of such defects.* Indeed the statistician has protected his system against some of these criticisms (3); yet no adjustment, however subtle, can compensate for lack of the necessary facts to answer the question we really wish to answer—whether or not the interpretation that the condition is due to a new mutation is true. Whatever the reader gets out of this discussion let him never lose sight of the fact that the Neyman-Pearson system is *parochial* (it does not incorporate empirical evidence or convictions from other sources) and it is *narrow* (adjudicating between two hypotheses only). With these caveats, let us reexamine the problem of mutation.

Recall, first (p 1), my view that the express field of operation of the biostatistician is quantitative. Statistical hypotheses are about *quantities* not *states*. We talk loosely about testing the hypothesis that Gardner syndrome is an autosomal dominant. But to make that a statistical hypothesis we must convert it into a statement about segregation ratios. "Apert syndrome" is not a quantity and we cannot test an immediate *statistical* hypothesis about it. We may set the hypothesis up that the probability of an affected child being born to normal parents is 1 in 100,000 (or some such figure) and compare it with what is found in the population. This comparison might yield a very good fit and lend support for the hypothesis of mutation.

Then, second, the Neyman-Pearson method requires that *a specific*

---

*The reader is advised to try to resolve this problem before reading further.

*alternative hypothesis must always be stated.* A great part of the problem of testing hypotheses in genetics is that it is difficult to specify a satisfactory alternative hypothesis. The surmise "Gardner syndrome is not a Mendelian autosomal dominant" is too vague to make any useful statement about the segregation ratios. It is from the alternative hypothesis, or possibly the class of alternative hypotheses, that one picks the rejection region, as we shall now proceed to discuss.

## CHOICE OF THE REJECTION REGION

Let us suppose the scientist decides to do a test of size 0.05. This decision implies that he will accept a risk of 1 in 20 of mistakenly rejecting the null hypothesis if it is true. If he is not prepared to run any risk of making this mistake at all, he will never reject the null hypothesis even if it is false, and is at a stalemate. Granted that he takes this risk, he wants to get the best possible value for his money: he will want to reject the null hypothesis in favor of the alternative when the alternative hypothesis is most likely to be true. In a word he wishes to maximize the power of the test.

**Example 2.7.** For example, suppose that the sample estimate of the mean change in blood pressure in a group of patients when put on treatment has a Gaussian distribution with mean $\mu$ and a standard deviation* of 10 mm Hg. The null hypothesis ($H_0$) "The treatment has no effect" is to be compared with the alternative hypothesis ($H_A$) "The treatment raises the blood pressure on average 20 mm Hg."
Formally we write

$$H_0: \mu = 0$$

$$H_A: \mu = 20.$$

We may thus draw the distributions of outcomes under the two hypotheses (Fig. 2.1 *upper part*). Now we wish to select one or more segments of the base line such that the corresponding areas under the null hypothesis add to 0.05 of the whole. If there were no more to be considered we might pick any one of a number of sections, e.g., areas I, II, or III (in Fig. 2.1 *upper part*) or, of course, many others. The proportions of the distribution under the null hypothesis in each rejection region are indicated by vertical shading. The areas are (by definition) the same in all three. The area of the distribution under the alternative hypothesis is shown by horizontal shading. It varies. But the aim is to reject the null hypothesis when the probability of the alternative hypothesis is maximum; i.e., to maximize power, the probability of rejecting the null hypothesis when the alternative is true.

---

*Note that here as elsewhere (7) I shall use the term *standard deviation of the (sample) mean* (which is exactly what it is) rather than *standard error of the mean.* The latter term in my experience creates unnecessary confusion.

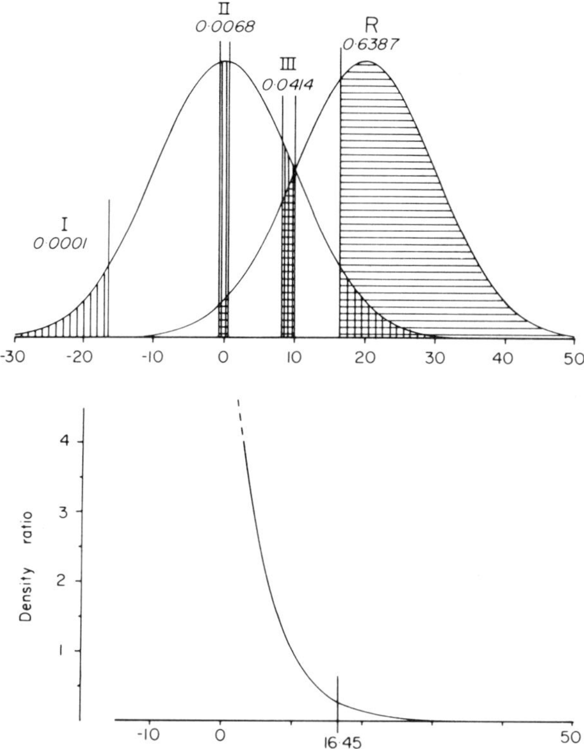


**Figure 2.1.** Choice among rejection regions. The curve to the *left* of the *upper* diagram is the probability density function under the null hypothesis, the curve to the *right* that under the alternative. In the *bottom* diagram is the ratio of the densities (null to alternative) a monotonic decreasing function. Errors of the first kind (sizes) are represented by areas with *vertical* shading, power by areas with *horizontal* shading. The sizes for the four rejection regions (I, II, III, and *R*) are all 5%; but the powers (indicated in italic script and by the extent of the horizontal shading in each) vary widely. If *R* is chosen as the rejection region, the power is maximum. This choice corresponds to those parts of the sample space with the lowest density ratio in the *bottom* diagram.

Neyman and Pearson showed in general that the best choice of a rejection region is based on the ratio of the densities (Fig. 2.1 *lower part*), i.e., in this case

$$\frac{\dfrac{1}{10\sqrt{2\pi}}\, e^{-\left(\frac{x-0}{10}\right)^2\!/2}}{\dfrac{1}{10\sqrt{2\pi}}\, e^{-\left(\frac{x-20}{10}\right)^2\!/2}} \;=\; e^{2-(x/5)} \tag{2.1}$$

where $x$ represents the value that the sample mean assumes. The "best" rejection region is then that part of the domain of $x$ which 1. contains 5% of the distribution of $X$ under the null hypothesis; and 2. comprises the smallest values of the ratio of the densities (likelihood ratio). It is evident from the exponential shape of the likelihood ratio that these two goals are achieved by putting the entire rejection region in the right hand tail of the curve. To achieve the size of 5% it is necessary that from infinity it should extend down to 1.645 standard deviations above the mean under the null hypothesis, i.e., values of 16.45 mm Hg and over. This lies $20 - 16.45 = 3.55$ mm Hg below the mean for the distribution under the alternative hypothesis, i.e., 0.355 standard deviations. Thus clearly the power of the test is that proportion of the distribution under the alternative hypothesis that lies above 0.355 standard deviations below its mean; and consulting tables of the standard Gaussian distribution* we find this to be 0.639. This may be compared with the powers given for regions, I, II, and III in Figure 2.1 *upper part*.

Some nonstatisticians find the implications of the Neyman-Pearson lemma somewhat easier to grasp with the aid of a mechanical model.

**Example 2.8.** In Figure 2.2 is shown a complicated distribution function for the test statistic under the null and alternative hypotheses. That for the null consists of a mixture of three Gaussian distributions: one-quarter of $N(0, 0.64)$, half of $N(3, 0.25)$, and one-quarter of $N(6, 1.44)$. The distribution under the alternative hypothesis has the same structure except that it is bodily moved 0.8 units to the right; that is, the means of the three subpopulations are 0.8, 3.8, and 6.8 respectively, but their forms and variances are the same. This example might represent, for instance, the distribution of a sample value from a Mendelian trait as measured by enzyme activity, with the null hypothesis that a coenzyme is not present against the alternative that it is. Or again (as we shall see in Chapter 17), something like this distribution is produced in the means of small samples from a $U$-shaped distribution.

The likelihood ratio is the rather complicated curve in the lower part of the diagram. Now imagine this likelihood curve as the profile of a building which is gradually being flooded by a rising level of water. The first part to be immersed will be the right-hand tail. However, eventually part of the deeper of the two depressions between peaks is immersed and then part of the shallower. Constructing the most powerful rejection region consists of letting the water rise until the areas in the upper diagram under the null hypothesis (corresponding to the immersed parts) constitute the appropriate size of the test. For instance, the parts of the lower diagram that lie beneath the lower of the two lines drawn in it correspond to the 5% level of test. The corresponding parts of the sample space in the upper diagram are represented in solid black and the correspondences are shown by continuous lines. They involve part of the tail of the curve and part of the body. If we increase the size of test to 10%, we "immerse" three segments of the bottom diagram, and in the up-

*A standard Gaussian variate is one with mean zero and a variance of unity.

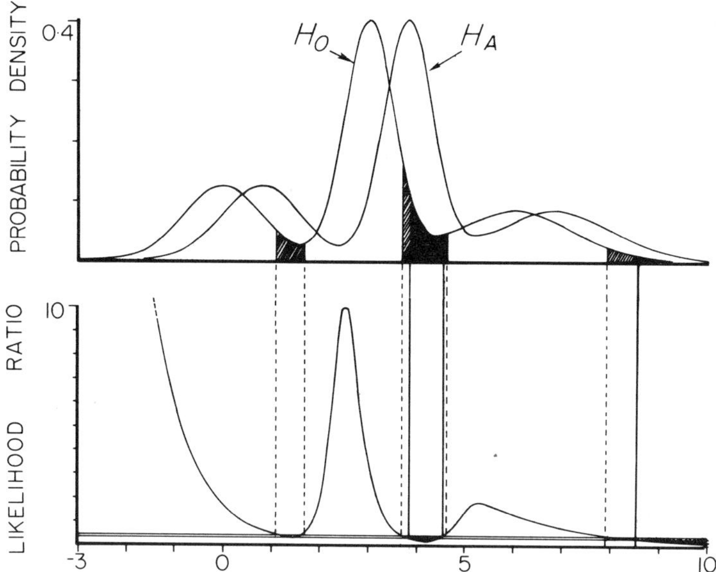


**Figure 2.2.** In the *upper* diagram are the probability densities under the null ($H_O$) and alternative ($H_A$) hypotheses, and in the *lower* diagram the ratio of the probability densities ("Likelihood ratio"). The areas in solid black comprise a region containing 5% of the distribution under the null hypothesis. This choice of region gives the most powerful test at this size. It also corresponds to the "most dependent" parts of the lower diagram. The *hatched* and *black* areas combined indicate the most powerful rejection region when the size of test is increased to 10%. Because the density ratio is not a monotonic function of the random variable, the rejection region is not necessarily connected.

per diagram there are now three unconnected parts of the distribution which comprise the rejection region under the null hypothesis. The parts of the rejection region added when we increase the size from 5% to 10% are shown hatched, and the lines of correspondence as interrupted lines. A test of size 1% (not illustrated) would lie entirely in the right-hand tail.*

## INDEFINITE ALTERNATIVE HYPOTHESES

### One-tailed tests

Most commonly, the investigator expects that the means in the groups he is comparing will show a difference, and even that it will lie in a particular

---

*This horrendous example (the like of which I have never encountered) is given to illustrate both the universality of the Neyman-Pearson criterion and its usefulness in solving the problem of an optimal rejection region where "common sense" has nothing to offer. The reader may be assured that choosing the most powerful rejection reading is usually vastly simpler.

direction, but he does not know how big the difference is. Thus he has to compare the hypotheses about a normally distributed sample of the type

$$H_0: \mu = 0 \qquad \sigma_M = 10$$

$$H_A: \mu > 0 \qquad \sigma_M = 10$$

where $\sigma_M$ is the standard deviation of the sample estimate of the mean.

As in (2.1) above, for any particular alternative value of $\mu > 0$ the ratio of the densities will be an exponential decreasing function of $x$, so that the rejection region will always be in the upper tail of the distribution given $H_0$, and for a test of size 0.05, it will always comprise those sample values greater than 16.45. The power will vary with $\mu$, but in this rejection region, it will always be maximal. Thus we can solve this class of problems in complete generality.

**Example 2.9.** Consider the hypotheses

$$H_0: \mu = 5 \qquad \sigma_M = 1$$

$$H_A: \mu > 5 \qquad \sigma_M = 1$$

In Figure 2.3 is shown the null distribution and a particular alternative, $\mu = 7$ (*top left*). Below is shown the likelihood (density) ratio as before and the appropriate rejection region. At *top right* are shown the likelihood ratios for the two alternative hypotheses $\mu = 7$ and $\mu = 8$. An infinite number of such curves could be drawn and the power for each computed. At *bottom right* is shown power as a continuous function of the mean under the alternative hypothesis. The larger this mean, the greater the power; the closer it comes to 5, the closer the power approaches the size of the test. (Why?)

### Two-tailed tests

Sometimes, however, the investigator has no idea whether the mean difference would be greater or less than zero. Then there is no unique solution to the problem of maximizing power. The common practice is to suppose that the difference is as likely to be positive as negative, and therefore to put part of the rejection region in each tail, a so-called two-tailed test. However, there is no particular reason why these separate components of rejection regions should contain equal proportions under the null hypothesis; what matters is that their sum should equal whatever size of test is chosen.

In Table 2.3 are given for various sizes of test, complementary rejection regions for standard Gaussian variates.


## ADMISSIBILITY OF TESTS

In general, the more that is known, or may be assumed, about a distribution, the more powerful the tests are. If a distribution is known to be Gaussian, the appropriate tests on sample statistics are more powerful than

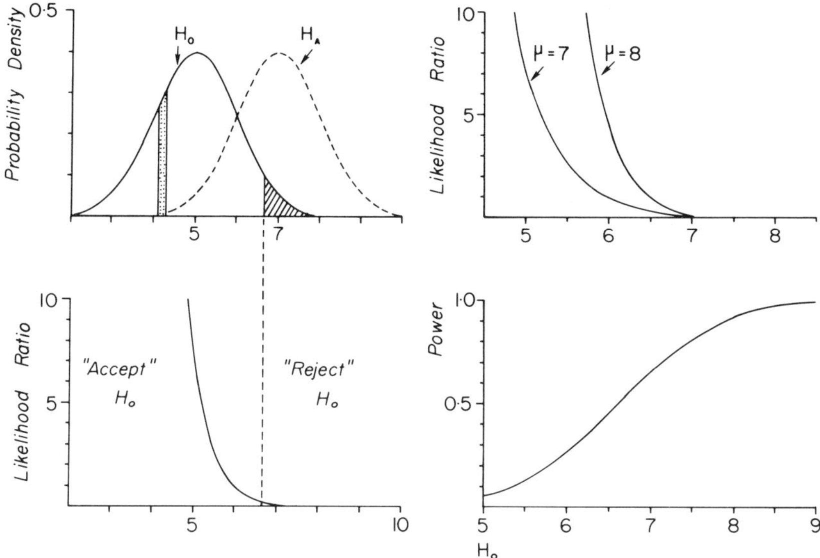

**Figure 2.3.** The probability density curves at the *top left* correspond to the null ($H_O$) hypothesis that the mean is 5, and to the alternative ($H_A$) that it is 7. The rejection region at the 5% level (*hatched*) is seen to conform to the lowest likelihood ratios in the diagram at *bottom left*. At the *top right* are shown the likelihood ratios for two alternative hypotheses: that the mean is 7, and 8, respectively. Obviously the rejection region will be the same for the two alternatives, but the power will be different. At the *bottom right* the power of the test is shown as a continuous function for values of the mean (under the alternative hypothesis) between 5 and 9.

if nothing is assumed about the exact distribution and a nonparametric method is used instead (see Chapter 15). If the variance does not have to be estimated from the data, a still more powerful test can be obtained. Likewise one-tailed tests are more powerful than two-tailed, because they assume more about the alternate hypothesis. I am not stating this relationship between definition and power as a universal: there are well-known exceptions. But it is a broad statement with a plausible ring; and the exceptions appear paradoxical.

Where there are several possible tests of the hypothesis, and over the range of interest one test is more powerful than any other for all possible values of the alternative hypothesis ("uniformly most powerful"), broad policy dictates that we should use that test. The other tests are said to be "inadmissible."

Now we must make a careful distinction here. The interests of inference require that the uniformly most powerful test should be applied. But the test

**Table 2.3. Complementary rejection regions for standard Gaussian variates**

| Lower limit | Probability in lower tail | Complementary upper limit for test of size | | |
| --- | --- | --- | --- | --- |
| | | 0.05 | 0.01 | 0.001 |
| $-1.6449$ | 0.0500 | $\infty$ | | |
| $-1.7$ | 0.0446 | 2.5469 | | |
| $-1.8$ | 0.0359 | 2.1953 | | |
| $-1.9$ | 0.0287 | 2.0279 | | |
| $-1.96$ | 0.0250 | 1.9600 | | |
| $-2.0$ | 0.0228 | 1.9228 | | |
| $-2.1$ | 0.0179 | 1.8503 | | |
| $-2.2$ | 0.0139 | 1.7979 | | |
| $-2.3$ | 0.0107 | 1.7592 | | |
| $-2.3263$ | 0.0100 | 1.7507 | $\infty$ | |
| $-2.4$ | 0.0082 | 1.7301 | 2.9108 | |
| $-2.5$ | 0.0062 | 1.7083 | 2.6702 | |
| $-2.5758$ | 0.0050 | 1.6954 | 2.5758 | |
| $-2.6$ | 0.0047 | 1.6918 | 2.5531 | |
| $-2.7$ | 0.0035 | 1.6794 | 2.4820 | |
| $-2.8$ | 0.0026 | 1.6701 | 2.4350 | |
| $-2.9$ | 0.0019 | 1.6632 | 2.4028 | |
| $-3.0$ | 0.0013 | 1.6581 | 2.3803 | |
| $-3.0903$ | 0.0010 | 1.6546 | 2.3656 | $\infty$ |
| $-3.1$ | 0.0010 | 1.6543 | 2.3643 | 3.9951 |
| $-3.2$ | 0.0007 | 1.6516 | 2.3529 | 3.4203 |
| $-3.2905$ | 0.0005 | 1.6497 | 2.3455 | 3.2905 |
| $-3.3$ | 0.0005 | 1.6496 | 2.3449 | 3.2814 |
| $-3.4$ | 0.0003 | 1.6481 | 2.3392 | 3.2103 |
| $-3.5$ | 0.0002 | 1.6471 | 2.3352 | 3.1681 |
| $-3.6$ | 0.0002 | 1.6464 | 2.3324 | 3.1414 |
| $-3.7$ | 0.0001 | 1.6459 | 2.3304 | 3.1240 |
| $-3.8$ | 0.0001 | 1.6456 | 2.3291 | 3.1125 |
| $-3.9$ | 0.0000 | 1.6453 | 2.3282 | 3.1049 |
| $-4.0$ | 0.0000 | 1.6452 | 2.3275 | 3.0998 |

may be very complicated to perform and we can indulge to some degree the investigator who uses a test somewhat less powerful and much simpler to perform. But if the choice of the inadmissible test is made *after the data have been collected,* and *because* it leads to rejection of the null hypothesis whereas the admissible test would not, then it is not difficult to show that a bias is being introduced and that the true size of the arbitrary choice of test (or, where appropriate, the "level of significance") is exaggerated. Thus any use of inadmissible tests should always arouse our suspicions (at least of the investigator's industry, if not of his honesty) and make us demand a specific disclaimer that the decision to use it was not made after the experiment was carried out.

## PROBLEMS

**2.1.** Consider the following hypotheses about $\mu$, the true mean of the distribution of $M$, the sample mean of a sample of size $n$ from a Gaussian distribution:

$$H_0: \mu = 0 \qquad \sigma_M = 10$$

$$H_A: \mu = 15 \qquad \sigma_M = 20$$

**a.** Find an algebraic expression for the density ratio. Plot it as a function of $x$ from $-60$ to $50$. Where, according to the Neyman-Pearson lemma, should the rejection region be?

**b.** Find numerically by trial and error the most powerful rejection region for a test of size 0.05 and calculate its power.

**2.2.** Clarke et al. (16) in studies on the protective effect of serum on formation of maternal antibodies against Rh-incompatible fetal cells, attempted to conserve supplies of antibody by finding the minimum effective dose. With half the known effective dose they expected an intermediate degree of protection. Instead it *enhanced* antibody formation, the significance level being 1 in 200. Discuss the implications of this finding from a statistical and scientific standpoint.

**2.3.** An investigator has recorded only the sample mean of a set of Gaussian measurements in which the true mean is known to be zero. The standard deviation of the mean $\sigma_M$ is thought to be 10 cms. However, there is some confusion about scale and he wishes to test this hypothesis against the alternative that $\sigma_M = 5$ cms. Construct the appropriate test on the sample mean with size 0.05. Comment on the result.

**2.4.** An investigator may analyze a result by either of two tests, only one of which is admissible. In fact, he publishes the result of whichever test gives the lower significance level. Find under what conditions this policy gives an unbiased estimate of the significance level. Devise examples.

# 3 ESTIMATION: THEORETICAL ASPECTS

Hypothesis testing, which has been discussed in the previous chapter, is concerned with answering questions of the type "Is the true mean of a distribution equal to $\mu$?" Before using this methodology the scientist must be prepared to commit himself to testing a very specific statement. Estimation, with which we shall now deal, answers such questions as "How much … ?" and "How often … ?" His state of mind when the scientist asks questions of this kind is much less committed than when discussing hypotheses.

Nevertheless, the two activities are not so widely separated as might at first be thought. Both of them appeal to data; both use a criterion of plausibility; both compare various explanations of what has been observed. Furthermore, in both cases the scientist's interpretation of his data after the statistical analysis has been performed is at least colored by his prior conceptions. In designing a test for a very implausible hypothesis he would probably insist both on small size and great power. For some hypothesis that seemed intrinsically plausible, he would be content with much less stringent evidence. Much the same kind of attitude is adopted in estimation. If a strong prior conviction exists as to what value some parameter should assume, estimates very different from this value would be accepted only in the face of strong evidence.

For instance, among population geneticists there is a firm conviction as to how large the mutation rates encountered in nature are. The typical figure usually quoted is something like 1 in 100,000 per locus per generation. When data on fibrocystic disease yielded estimates for the mutation rate something like 10 to 20 times greater, geneticists went through all manner of contortions to account for it, invoking multiplicity of loci, heterozygote advantage, the absence of a genetic equilibrium, etc.

Both in hypothesis testing and estimation there are methods devised for incorporating these prior convictions formally into the statistical procedure.

Proponents of these methods constitute the so-called Bayesian group of statisticians. While they are widely criticized for the often arbitrary nature of the prior probabilities used in their formal analyses, there is no serious dispute that, in principle at least, they have a point, and that neither the Neyman-Pearson type of hypothesis testing nor the estimation procedures that we will now discuss are altogether satisfactory. We shall not devote much space to Bayesian methods, which are controversial and often difficult to handle. However, in due course we shall illustrate them with one well-defined problem, "regression toward the mean" (see Chapter 5).

**Example 3.1.** The distinction between estimation and testing hypotheses was nicely brought home to us (14) in genetic counseling for X-linked recessive traits where the mother is phenotypically normal and her only child is an affected son. The common choice is between the hypothesis that the mother has only normal genes and the son is a new mutation; and the hypothesis that the mother is a carrier. But a third surmise is that the mother may be a gonadal mosaic, that is, that her ova are derived from a mixture of normal and carrier cells. To explore this possibility we must ask the question "What *proportion* of ova are normal?" That is, we convert the problem into one of estimation; and the risks that her son will be affected that are surmised by the two hypotheses (0 and 0.5 respectively) are special cases of the answer that estimation might furnish.

**Example 3.2.** An excellent and subtle example of the relationship between testing a hypothesis and estimating a parameter is provided by the genetic ideas of linkage and synteny. *Synteny* is an anatomical term: loci carried on the same chromosome pair are syntenic. This is an unambiguous issue of fact, two states only being possible. *Linkage* describes how syntenic loci behave, whether independently, completely in unison, or exhibiting some intermediate degree of dependence. *Nonsynteny* implies independent assortment, but not conversely. The geneticist thus has two types of questions. Is there synteny? The answer involves testing a hypothesis. How closely are the loci linked? The answer is some quantity between 0 and $1/2$ and is to be determined by estimation.

One important respect in which estimation differs from probability algebra is that it does not lead to a unique solution of a particular problem. There are several methods of estimation (to be discussed later), which do not necessarily give the same answer. At first sight, this is both surprising and disturbing. It is surprising because it may not at first be evident why the apparent reciprocity between the two fields breaks down. It is disturbing because those who seek, if not a certainty, at least a sure degree of uncertainty, may feel that they have been betrayed by a discipline in which they had such faith. Reflection will provide some reassurance.

We shall return to this problem in discussing likelihood (see below). Meanwhile it will be appropriate to start with some very elementary and familiar ideas, which will set the scene for what follows.

## SOME DEFINITIONS

In probability algebra the pattern of inference is prediction of the outcome from a knowledge of the distribution and the parameters.

**Definition 3.1.** *A sample space is the set of outcomes for a probabilistic process.*

The word *space* has much the same meaning as *set* or *universe*.

In estimation the inference goes in the opposite direction to probability algebra—from the outcome to the parameters. By analogy the set of possible answers constitutes the parameter space. So also, one may speak of a hypothesis space, though it is mostly used by theorists only.

**Definition 3.2.** *The parameter space is the set of all values that the parameter(s) may conceivably assume.*

**Example 3.3.** A parameter space may, like a sample space, be finite or infinite, discrete or continuous or mixed; just as a random variable has a mean and a variance, so does the estimate of a parameter. But the degree of closeness between the two must not be pushed too far. The same algebraic expressions may apply equally well to a sample space or a parameter space but the one may be discrete and the other continuous. For example in Figure 3.1 is represented the binomial process of order 3 which has the general form

$$f(x) = \binom{3}{x} p^x (1 - p)^{3-x} \qquad 0 \le p \le 1 \qquad x = 0, 1, 2, 3$$

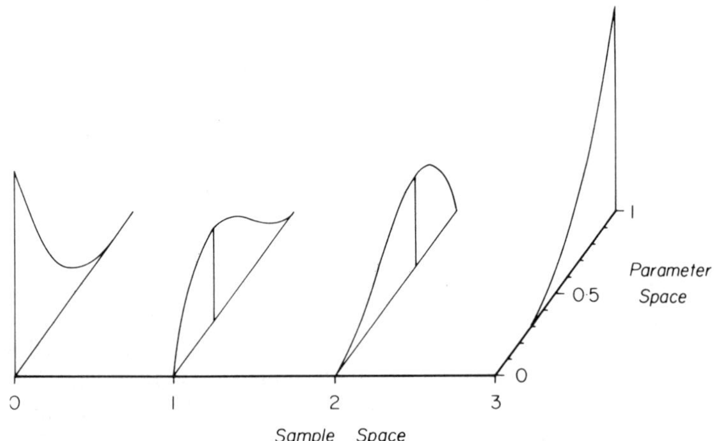


**Figure 3.1.** The binomial process of order 3. The sample space (the horizontal axis) of the number of heads is *discrete,* comprising four points. (For the sake of generality, their densities have not been shown, because, of course, they depend on the true value of the parameter.) The parameter space (fore and aft) is the continuous interval 0 to 1. The likelihood (vertical scale) may range from 0 to 1. For each of the four possible outcomes, the likelihood curve is drawn and the maximum value is indicated by a vertical line.

This expression—which applies equally well to probability algebra and to statistics—may be viewed as a function of $x$, the number of "successes" in a sample space which is discrete (there being only four possible outcomes, all integers). It may also be viewed as a function of $p$, the probability of success, in a parameter space which is continuous since $p$ may conceivably have the values zero, unity, or any quantity in between. To be sure the sample space of our estimate, $\hat{P}$, is still discrete since in this instance there are only four values it can take: 0, $\frac{1}{3}$, $\frac{2}{3}$, or 1. Furthermore, however large the sample, the estimate will always be rational, i.e., the ratio of two integers, and therefore discrete; but the value that $p$ may assume is continuous. The relationships between the probabilities and the statistical aspects are summarized in Table 3.1.

**Definition 3.3.** *An estimate is a conjecture, arrived at from analyzing data, as to what the true value of a parameter is.* It is a "guess" at the value of the parameter yielded by the method of estimation. (It must not be confused with a *statistic* such as the sample mean, which may be computed without any relationship to parameters.)

**Definition 3.4.** *An estimator is a systematic procedure for analyzing the data to obtain an estimate.* For instance in estimating the mean, the procedure "Add the sample values and divide by the number of values" is an estimator.

**Definition 3.5.** *A point estimate is a single value which represents a particular parameter.* Thus the sample mean (which is a single quantity) may be used as a point estimate of the true mean.

**Definition 3.6.** *An interval estimate is a segment of the parameter space in which (in some specified sense) the statistician asserts that the parameter lies.*

**Table 3.1. Comparison between the probability distribution and likelihood function for a binomial variable**

|  | Probability distribution | Likelihood |
|---|---|---|
| Relationship to realization (outcome) | Antecedent | Subsequent |
| Parameters | Known | Unknown |
| Outcome from the random experiment ("data") | Unknown | Known |
| Variable quantity | Number of successes $(x)$ | Parameter $(p)$ |
| Domain of conjecture | Sample space | Parameter space |
| Nature of domain | Discrete: $(n + 1)$ points | Continuous (uncountably many points) |
| Constraints | Sum of probabilities must be unity | Sum of likelihoods has no special meaning |
| Common measure of uncertainty | Variance of variate (see 7) | Curvature of the log likelihood function (see p. 66) |

In general, a statistician is not content merely to make a "best" guess at a parameter: he is also concerned with how accurate the guess is, that is, whether the difference between the estimate and the true value is likely to be small or large. A common measure used is the standard deviation of the estimate, which is generally not known but is itself an estimate based on the data. From this estimated measure of accuracy is derived an interval estimate.

A true *probability* statement about whether or not an interval contains a parameter would require at the least a prior probability. (Even so, there would still be philosophical problems over what such a probability statement could mean.) But just as in testing hypotheses we dealt with the lack of a prior probability by changing the nature of our conclusion, so in estimation we replace such an interval statement by a confidence statement, which we shall now discuss.

### Confidence interval

In the typical situation, then, estimation yields estimates both of the parameter and of the variance of the estimate of the parameter. If the form of the distribution of the estimate is also known—for example if it is known to be Gaussian—it is commonly possible to set up an interval (details of the calculation will be given subsequently) within which the statistician asserts, with some specified degree of assurance, the true value lies. Such an interval is referred to as a confidence interval.

Some considerable precision of thought is required in interpreting what the implications of this term are. First, it is clear that this interval, being based on data from a finite sample is, a priori, a random variable; hence multiple experiments will not in general give the same confidence intervals. Before the experiment is conducted, if the distribution is known, it is possible to make predictive statements about the probability that the confidence interval computed in a particular way will embrace the parameter. After the data have been collected, however, the true mean either does or does not lie in this interval: the probability that it lies in this interval is either 0 or 1, and no objective indeterminacy (7) remains. However, since the true mean and variance are not known, it is impossible to determine which of these two states is correct. In other words, subjective indeterminacy (7) remains. The confidence interval may be set up in such a way that the subjective judgment of the investigator is that he is 95% confident that the true value lies in this interval. What does this statement mean?

The most noncontroversial way of interpreting it is as follows. If the statistician plans beforehand to collect a sample of data, to estimate the appropriate parameters, and to set up 95% confidence limits in an appropriate way, and having done this he proposes to say that the true mean lies within this interval, then a priori he stands a 95% chance of being correct. In somewhat more practical terms, if in a long life of statistical analysis he pursues the policy of determining confidence intervals by this method and making the corresponding statement that it contains the true parameter, he would be correct about 95% of the time, although he will not know in which 95%.

Some statistical theorists attempt another interpretation of a confidence interval which allows them to claim that after the event a true probability statement is being made; but this interpretation is highly controversial and raises fundamental philosophical questions about the nature of indeterminacy and the relationships of philosphical positivism to metaphysical truth. Such a discussion would take us far beyond the domain of this book. The meaning of the phrase used above, "the subjective judgment of the statistician," calls for comment. It must not be understood to imply any whim or caprice. I do not use it to mean a *personal* judgment (although some statisticians employ it in this sense). All statisticians operating by the same rules will reach the same answer. This would be equally true of analysis by a computer. The judgment is subjective because (after the event, at least) it reflects the unresolved ignorance of the statistician. There is no *objective* indeterminacy, and what I have called an objective probability statement has here no meaning at all. The correct value of the parameter is what it is, and it has no distribution. But the reader is warned that both terminology and the philosphical analysis of these terms are disputed.

Differences among statisticians arise over which rules are appropriate and their theories may lead them to different answers. Thus one might say that a *personal* judgment implies individual idiosyncrasy, whereas a *subjective* judgment is a collective opinion of a school of thought.


## MOMENTS AND PARAMETERS

Before we discuss estimation in detail, it is appropriate here to anticipate the confusion that occurs if we do not clearly distinguish among the quantities that are being estimated. Just as, in discussing distributions (7), we may distinguish general properties from particular, so we also must do so when considering estimation. We may take it that the distribution of every natural measurement has moments—a mean, a variance, and so forth. Where the random variable may assume an infinite value, problems may arise; but in the real world this does not occur with measurements, only with complicated operations on measurements (e.g., taking the reciprocal of a measurement that may be zero). I know of no distribution of *measurements* in practice that do not have moments of all positive orders.

On the assumption that moments exist, they can be estimated from data. The estimator, then, does not necessarily require that we know the form of the distribution, although if we do, the estimator may have highly convenient properties. As we showed earlier (7) these simple estimators have properties that may be stated in generality. The sample raw moment of any variate is always an unbiased estimate of the true raw moment.

Estimating central moments presents more difficulty. For a random sample from any distribution the quantity

$$S^2 = \frac{\sum_i (X_i - M)^2}{(n - 1)} \tag{3.1}$$

(where $M$ is the sample mean and $n$ the number of independent observations) is always an unbiased estimator (see below) of the true variance. Similar but more complicated statements about other central moments may be made. A listing of common estimators of this kind is given in Appendix 3.

Now it so happens that for the three distributions most commonly used in practice—the Gaussian, the Poisson, and the binomial—the parameters, i.e., the characteristic constants of the particular distribution, are also simple moments. The Poisson distribution has one parameter only, $\lambda$, which is also the mean and the variance. For the binomial, there is one parameter $p$ which is also the mean number of successes per trial. For the Gaussian there are two parameters $\mu$ and $\sigma^2$, which happen also to be the mean and variance respectively. There is thus some risk that these two entities, moments and parameters, may be confused and that the reader may suppose because the formula (3.1) above gives an unbiased estimate of the variance (and hence also the dispersion parameter of the normal distribution) that it is a good measure of some fundamental parameter for other distributions as well. As pointed out (7), this is most emphatically not true of the lognormal distribution (see below). In general the moments of a distribution are easy to estimate; the parameters may or may not be. Also, distributions of variates with a finite range have infinitely many moments, each of which may be estimated; but they may have only a few parameters. The moments may be expressible as functions of the parameters and conversely; and, as we shall see in the method of moments, sample estimates of moments may be used to find estimates of the parameters. Theoretically in such cases, there are always infinitely many ways of estimating parameters from sample moments, although for convenience we may restrict ourselves to a few of the simpler moments only.

To estimate parameters we will need to know much more about the form of the distribution than to estimate the moments and, in any practical situation I can think of, this means that the *form* (other than the parameters we wish to estimate) must be explicitly known.

However, approaching this issue as a biologist I would like to make a distinction, which is ordinarily ignored, between two types of parameters that I shall call *ad hoc* and *real* respectively. The issue is not one that involves statistics itself, but it has fundamental importance from the standpoint of interpretation.

Consider two problems.

**Example 3.4.** The decay of a radionuclide follows an exponential distribution that is characterized by a single parameter known as the "hazard intensity." $^{32}P$, for instance has a true mean survival of 20.9 days, the reciprocal of which, 0.0478, is the parameter of interest. Here the parameter is a real one; that is, it describes a quality which has universal significance for atoms of this unstable isotope. A philosopher would be unhappy about this use of the word *real*; and certainly I do not intend it to mean that it is a glimpse of a metaphysical truth. But in the present state of physical theory we believe that it is a logically stable descriptor in the same way that height of a man is, or the mass of the moon.

**Example 3.5.** In contrast, consider the relationship between blood pressure and age. From abundant evidence, the investigator may represent the relationship as a polynomial equation. If $y$ is the blood pressure and $x$ the age, then, perhaps, they may be aptly related by the equation

$$y = a + bx + cx^2 + dx^3$$

We shall consider in Chapter 9 how the quantities $a$, $b$, $c$, and $d$ may be estimated from the data. The observed blood pressure in a man aged $x$ may be very close to $y$, so that the equation is an excellent predictor. But how seriously would a *biologist* take the parameters $a$, $b$, $c$, and $d$? Is it useful to try to think of them as having biological counterparts? In studies on blood pressure and age (17) we found not only that the estimates of these regression coefficients varied from one group to another, but even the number of them, i.e., the number of terms in the equation. This behavior throws grave doubt on their biological interpretability. Perhaps the biological facts are that blood pressure depends on a mixture of additive processes each of which can be expressed as a polynomial equation of age, and each of the above coefficients is the sum of several different quantities. Perhaps the relationship represents logically intricate interactions, some additive, some multiplicative, some parastatic, which (it so happens) can all be artificially expanded in a power series. Of course the statistician is perfectly right to disclaim all responsibility for the interpretation of the parameters: after all, it may not have been his decision that a polynomial of this type should be used in the first place. Moreover, the use of the polynomial is a general all-purpose method that does not have to make any sense of this type in the particular case. The main broad claims to be made for it are that it has considerable flexibility, that it provides in general a good means of interpolation, and that analytically it is comparatively easy to handle. The true relationships may be of a totally different form and involve both more complicated and less elegant means of analysis; and even the best statistical treatment may not produce any more precise predictor than the polynomial method. But the biologist is, naturally, much more comfortable with the true relationships. I distinguish parameters with descriptive value but no literal biological meaning as ad hoc rather than real. But of course any serious contention that they had interpretability would convert them, at least provisionally, into "real" parameters.

The distinction is in some degree artificial, since eventually many of our biological constructs will turn out to be of this ad hoc quality. What experts in infectious disease call "infectivity," for example, may be the algebraic sum of a dozen quite different processes. The most helpful distinction is, I suspect, utilitarian. A parameter that (in its own right, and not as an index of some vague tendency) has heuristic value—i.e., that prompts further lines of inquiry—may usefully be called a real parameter. A parameter that has permanent and stable descriptive value may also be seen as real. But a parameter which every reasonable opinion regards as a pure descriptor, with neither universal value nor intellectual provocativeness, is ad hoc. The scientist is warned not to take the formal conclusions (real or supposed) of the statistician literally.


## THE FUNDAMENTAL PROBLEM OF ESTIMATION

Students of elementary algebra are familiar with the problem of having two equations involving two unknowns, and with the techniques used for solving them. In general, to solve for $n$ unknowns it is necessary to have $n$ simultaneous equations.* But this method of finding $n$ unknowns by solving $n$ simultaneous equations of appropriate form leads to unambiguous results precisely because the statements embodied in the equations are exact.

To make statements about parameters on the basis of results that are *not* exact presents a problem which cannot be uniquely solved by the methods of simple algebra. The statistician is caught between two difficulties. If the problem is to find the mean of the distribution, one exact equation containing the mean and no other unknown quantity would lead to an exact and unique solution. There would be no point in further information, which would be either redundant, or inconsistent with the results from the first equation. It is rather like the argument in favor of destroying books: either they agree with authority, in which case they are superfluous, or they disagree with authority, in which case they are heretical.

In the attempt to estimate an unknown parameter from a single datum that is inexact, the degree of exactitude being not known precisely, there are two unknowns which are to be determined from one equation.

**Example 3.6.** For instance, suppose it were required to determine the height of a man where, for simplicity, we shall suppose that height is unambiguously defined. If there were no error of measurement there would be no problem in answering this question. But observers measuring such quantities without consulting each other do not in general agree exactly: there is some

---

*Even then, solution is possible only if the equations exhibit "linear independence," which implies that the matrix of coefficients is of full rank. The issue is dealt with in Appendix I on matrix algebra.

error of measurement. Let the true height of the individual be denoted by $\mu$ and suppose that the measurement of the first observer denoted by $y_1$ involves an error $h_1$. Then we have the equation

$$y_1 = \mu + h_1$$

Of the three quantities, only one, $y_1$, is known and we cannot solve one equation for two unknowns uniquely. One obvious device is to have a second observer furnish a measurement $y_2$ which yields a second equation of like form. But this too contains an error term $h_2$ so that there are now in total two equations to be solved for three unknowns ($\mu$, $h_1$ and $h_2$); and in general, however many observers make measurements there will always be one more unknown than there are observers. A system such as this, in which there are more unknowns than there are independent equations, is said to be algebraically indeterminate. It is capable of an infinite number of solutions. Only if at least one of the errors is known exactly can a unique solution for $\mu$ be found.

The purpose of estimation is to determine which of the infinite number of possible solutions for this set of $n$ simultaneous equations is, in some sense, "best." Since *best* is not a uniquely defined word, it will come as no great surprise to the reader that there is an infinite number of estimators for solving such problems. Most practicing statisticians are accustomed to using seven or eight different statistical procedures for estimating unknown parameters. In general the estimate of the parameter will be associated with some degree of uncertainty which could be expressed as the standard deviation of the estimate. But which estimator actually may be the most useful will depend on what considerations are uppermost in the statistician's mind. Some or all of the following factors are taken into consideration in making the choice.

**1. Simplicity of operation.** Some estimators can be performed very simply; others may call for extremely elaborate and time-consuming computation.

**2. Consistency.** An estimator is said to be consistent if the estimate comes closer to the true value the larger the sample available. If for example an estimator systematically underestimates the true parameter by a factor of one-half, it could not be consistent.* (Consistency is important in practice because it is required in a definition of "efficiency"; see below.)

---

*A proper mathematical definition of consistency involves us in stochastic limits. For practical purposes we may regard an estimator as consistent if, as $n$ becomes indefinitely large,

    a. The estimator is unbiased, at least asymptotically.

    b. Its variance goes to zero.

But these practical criteria (especially the first) are excessively strict.

Those interested in details may consult texts on mathematical statistics. The exceptions I know of involve bizarre ("pathological") assumptions.

The controversy about the relative merits of Bayesian and non-Bayesian estimation (see Chapter 5) is not an issue of consistency. Sound estimates of both types will, in the long run, give the same answer; rather, the question is *how quickly* the estimates approach the true value with increasing sample size.

**3. Efficiency.** Other things being equal, the smaller the standard deviation of a consistent estimate the more efficient the estimator is said to be. Theory of the likelihood curve (see below) shows that, in most instances of interest, there is an asymptotic lower limit (which is a function of sample size) below which the standard deviation of a consistent estimate cannot fall. If the standard deviation of a consistent estimate attains this lower limit, the estimator is said to be "absolutely efficient."

**4. Elasticity.** Some estimators may be readily adapted to almost any problem of estimation; others can be applied to stereotyped situations only. But that is not to say that the more elastic estimator is always the more convenient to use.

**5. Unbiasedness.** For all estimates there is some error, at least so long as the sample size is finite and less than that of the reference population. Should some of this error be random and some be systematic nonrandom error, so that, on average, the answer either overestimates or underestimates the true parameter, then the estimator is positively or negatively *biased,* respectively. If there is no such *systematic* error so that even for samples of very small size the expectation of the estimate is equal to the true parameter, then the estimator is unbiased.

**6. Asymptotic normality.** For some methods, as the sample size becomes indefinitely large the estimate approaches a Gaussian distribution more and more closely. As we shall see, this property has considerable advantages in interval estimation.


## ESTIMATORS

It is possible to devise an unlimited number of estimators. A considerable theory has been developed about their nature and the techniques of optimizing them. Broadly speaking the scientist may take two approaches.

1. He may follow well-trodden paths using any one of a number of well-known systematic procedures: method-of-moments, least squares, weighted least squares, minimum chi-square, modified minimum chi-square, maximum likelihood, minimum variance, diverse Bayesian estimators, etc. The properties of these estimators are explored in detail and in great generality in textbooks on statistical theory. Under certain clearly specified conditions, for example, maximum likelihood estimators are fully efficient, consistent, and asymptotically normal. Thus the statistician in using these methods has considerable information beforehand as to how well he may expect to fare with them.

2. In certain cases it may not be necessary or desirable to pursue this course, because, for example, the estimation procedures are extremely cumbersome, or do not lead to exact solutions, or cannot be adapted to a particular problem. The investigator may then devise, by guess work perhaps, what looks like a reasonable estimator in this particular case, and

then explore systematically its properties—whether it is unbiased, consistent, efficient, etc.

## THE METHOD OF MOMENTS

We have already met this method casually (for instance in equation 3.1, above). Here we shall state the procedure formally. The method of moments prescribes the following estimator:

1. Express some true moment(s) for the population as function(s) of the unknown parameter(s).

2. Compute the sample moment(s) as function(s) of the data.

3. Set each true moment equal to the corresponding sample moment. As we do so, we put "hats" (circumflex accents) on the parameters to indicate they are not true values but estimates.

4. Solve the equation(s) for the estimates of the unknown parameter(s) in terms of the data.

This procedure sounds complicated but, in some cases at least, it is precisely what common sense would dictate.

**Example 3.7.** Consider the data on body temperature in acute appendicitis (Problem 5.2). We may compute the sample mean and variance in the usual fashion.

$$m = \frac{\Sigma x_i}{n} = \frac{449.8}{12} = 37.48$$

$$s^2 = \frac{\Sigma (x_i - m)^2}{n - 1} = \frac{\Sigma x_i^2 - m \Sigma x_i}{n - 1} = \frac{\Sigma x_i^2 - (\Sigma x_i)^2/n}{n - 1}$$

$$= \frac{16{,}864.8 - 16{,}860.0}{11} = 0.436 \ldots$$

Now granted that they are from a Gaussian distribution we set

$$\hat{\mu} = m = 37.48$$

$$\hat{\sigma}^2 = s^2 = 0.436$$

and solve. But here there is no solving to be done, since the functions of the parameters are simply the parameters themselves.

Let us take a less straightforward problem.

**Example 3.8** Suppose that the survival of an experimental animal follows a gamma distribution (7). We have the following data points (in months): 15, 22, 24, 25, and 29. The gamma distribution, which has the probability density

$$f(x) = \frac{a^n}{(n - 1)!} e^{-ax} x^{n - 1} \qquad x \geq 0 \qquad a, n > 0$$

has two parameters $n$, its order, and $a$, the risk intensity. The mean is $n/a$ and the variance $n/a^2$.

So first we compute the sample moments:

$$m = \frac{15 + 22 + 24 + 25 + 29}{5} = \frac{115}{5} = 23$$

$$s^2 = \frac{15^2 + 22^2 + 24^2 + 25^2 + 29^2 - 23 \times 115}{5 - 1} = 26.5$$

Then equating true and sample moments we get

$$\frac{\hat{n}}{\hat{a}} = 23$$

$$\frac{\hat{n}}{\hat{a}^2} = 26.5$$

Now

$$\frac{\hat{n}}{\hat{a}} \div \frac{\hat{n}}{\hat{a}^2} = \hat{a}$$

whence

$$\hat{a} = \frac{23}{26.5} = 0.867,9\ldots$$

Likewise

$$\left(\frac{\hat{n}}{\hat{a}}\right)^2 \div \frac{\hat{n}}{\hat{a}^2} = \hat{n}$$

whence

$$\hat{n} = \frac{(23)^2}{26.5} = 19.962\ldots$$

**Example 3.9.** In rabbits, as in man, the distribution of the serum cholesterol level is close to lognormal (7), i.e., the logarithm (to any base) is approximately Gaussian. This distribution has the density function

$$f(x) = \frac{1}{x\sigma\sqrt{2\pi}}\, e^{-\left(\frac{\log_e x - \mu}{\sigma}\right)^2 \big/ 2} \qquad x > 0$$

The mean can be shown to be

$$e^{\mu + \sigma^2/2}$$

and the variance

$$e^{2\mu + \sigma^2}(e^{\sigma^2} - 1)$$

Note two points. First, like the Gaussian distribution it contains only two parameters, a "centrality" parameter, $\mu$, and a "dispersion" parameter $\sigma^2$. Second, it differs from the Gaussian distribution in that its mean is not equal to $\mu$ nor its variance equal to $\sigma^2$.

Now consider the following readings quoted elsewhere (7), in 12 untreated rabbits: 40, 61, 50, 60, 38, 86, 60, 33, 58, 62, 112, and 66. Then by the usual calculations

$$m = \frac{726}{12} = 60.5$$

$$s^2 = \frac{49{,}058 - 60.5(726)}{12 - 1} = \frac{5135}{11} = 466.818\ldots$$

We set these two qualities equal to the respective true moments

$$e^{\hat{\mu}+\hat{\sigma}^2/2} = 60.5$$

$$e^{2\hat{\mu}+\hat{\sigma}^2}(e^{\sigma^2} - 1) = 466.818\ldots$$

Dividing the square of the first term into the second, we have

$$e^{\hat{\sigma}^2} - 1 = 0.127{,}537{,}239{,}8$$

$$e^{\hat{\sigma}^2} = 1.127{,}537{,}239$$

Taking logarithms to the base $e$

$$\hat{\sigma}^2 = 0.120{,}035{,}820{,}4$$
$$\hat{\sigma} = 0.346{,}461{,}860{,}1$$

Taking logarithms for the first equation

$$\hat{\mu} + \hat{\sigma}^2/2 = 4.102{,}643{,}365$$

$$\hat{\mu} = 4.102{,}643{,}365 - \frac{0.120{,}035{,}820{,}4}{2} = 4.042{,}625{,}454$$

Clearly these estimates of $\mu$ and $\sigma^2$ are very different from the sample moments, $m$ and $s^2$. Here, then, we have a clear instance where the moments and the parameters are quite different, and where method-of-moments estimation is not a totally straightforward procedure.*

*This somewhat cumbersome example is included for three reasons. First, I have repeatedly seen papers in which the authors see fit to publish only the (method-of-moments) estimates of the mean and standard deviation, and do not (as in Example 3.9) publish the individual results in detail. There may be good reason for supposing that the variate follows a lognormal distribution closely. One such instance is the serum level of creatine phosphokinase (see p. 127). Thus if we wish to estimate and compare the *parameters* we have no option but to use the method-of-moments estimators of them also. Second, the example illustrates that while the estimation of moments by method of moments is straightforward, the estimation of parameters may not be. Third, it shows that two estimators may give different estimates. Maximum likelihood estimates (found by taking logarithms of the data in Example 3.9 and treating the results as Gaussian) are $\hat{\mu} = 4.0487$ and $\hat{\sigma} = 0.33971$.

### The advantages and disadvantages of method-of-moments estimation

This estimator is the most elementary of all formal methods, with such an appeal to commonsense that the uninitiated may have difficulty in imagining any other. The method involves no assumptions when used for the estimation of moments except, naturally, that the moments exist. But, as with all estimators, something more of the form of the distribution must be known if parameters are to be estimated. The procedure is usually comparatively easy, though occasionally, as in estimating the parameters of the lognormal, it is somewhat more messy than other methods. Its principal shortcomings are that it is not in general fully efficient and it is somewhat lacking in elasticity.

## LEAST SQUARES ESTIMATION

Least squares estimation can be viewed as a generalization of the method-of-moments, and it has the advantage of being more elastic. The expected value is expressed as a function of the parameters and the difference between this function and the observed value is the error term which we shall denote by $h$. The method consists of finding those values of the parameters that minimize $\Sigma h^2$.

**Example 3.10.** Consider a random sample of values from a population with mean to be estimated: 3, 4, 5, 5, 6, 7, 8, and 10. Then the individual unknown errors are $(3 - \mu)$, $(4 - \mu)$, $\ldots$, $(10 - \mu)$. The least squares estimator is to find that value of $\mu$ that minimizes the sum of the squares of these quantities. Let us first tackle the problem by trial and error. Setting $\mu$ equal to various trial values we get

| $\hat{\mu}$ | $\Sigma h^2$ |
|---|---|
| 4 | 68 |
| 5 | 44 |
| 6 | 36 |
| 7 | 44 |
| 8 | 68 |

This suggests that the minimum value is approximately at $\mu = 6$ and as a check we will evaluate two values close by. At $\mu = 5.9$ the value is 36.08 and at $\mu = 6.1$ it is also 36.08, so that it looks as if our conjecture is correct that $\mu = 6.0$. We may pursue this argument by taking values closer yet. At $\mu = 5.99$ and at $\mu = 6.01$ we get 36.0008 and at 5.999 and 6.001 we get 36.000,008. This is rather a tedious method of finding the minimum value, however, and a more direct approach is desirable. In this particular case, curiously enough, the least squares estimator is $m$, the mean of the sample, which is 6.0; we can readily show that this must be so. Instead of $m$, take some other value $(m + d)$ where $d$ is either positive or negative.

Then

$$\Sigma[x_i - (m + d)]^2 = \Sigma[(x_i - m) - d]^2$$

$$= \Sigma(x_i - m)^2 - 2d\Sigma(x_i - m) + nd^2$$

Of these terms $2d\Sigma(x_i - m) = 0$, since for any sample $\Sigma(x - m) = 0$ because $\Sigma x = nm$; and whether $d$ is positive or negative, $nd^2$ is greater than zero. Thus

$$\Sigma[x_i - (m + d)]^2 = \Sigma(x_i - m)^2 + nd^2 > \Sigma(x_i - m)^2$$

which is true for all $|d| > 0$. Hence $\Sigma(x_i - m)^2$ is the minimum such sum of squares and $m$ is the least squares estimator of $\mu$.

This example is somewhat disappointing in that the answer tells us nothing new. With more complicated cases, the method is more rewarding. In particular the method is extensively used in regression analysis (see Chapters 8 and 9) and its general application will be the objective of Appendix 1 on matrix algebra. We shall postpone detailed discussion, but as a foretaste we shall consider a simple illustration.

**Example 3.11.** Suppose one measurement $f(x)$ is in part fixed and in part proportional to $x$. Thus

$$f(x) = a + bx$$

For example,* the temperature in degrees Fahrenheit, $f$, may be expressed as a function of that in degrees centigrade, $c$, in the form

$$f = 32 + 1.8c$$

Commonly we have specified values of $x$; but $a$ and $b$ are unknown. We do not know $f(x)$ but we have measurements of it (denoted by $y$) more or less in error, so that for each $i$

$$y_i = a + bx_i + h_i$$

where the $h_i$ are (unknown) errors either positive or negative. Then

$$h_i = y_i - a - bx_i$$

and the least squares estimator tells us to pick as estimates of $a$ and $b$ those quantities that minimize $\Sigma h_i^2$, that is, minimize

$$\Sigma(y_i - a - bx_i)^2$$

How to do this systematically is best deferred to Chapter 8.

---

*A note to the cognoscenti. This illustration is somewhat unorthodox in that $x$ and $y$ are structurally related, i.e., the true relationship between them is deterministic and all the variance is due to error of measurement. The main attraction of the illustration, its compelling reasonableness, is my excuse for using it.

The practically minded reader should not be dismayed at the theory involved. When all else fails he may find the least squares estimates of the parameters by searching techniques, i.e., by trial and error, just as we have done in Example 3.10. And despite a false emphasis given in the textbooks, in the real world "all else" fails quite commonly. When a systematic procedure exists it usually saves a good deal of time to use it. But the scope of systematic methods is limited and it takes a little experience to know where they are appropriate. However, least squares estimators may still be used in such forms (said to be "nonlinear in the parameters") as

$$y_i = e^{-bx_i} + h_i$$

where $b$ is unknown, a common problem in physiology, for which there is no systematic solution.

**Assumptions underlying the procedure**

1. We suppose as a minimum that the average value of the errors is zero; that the errors are the only source of random variation; and that the only unknown quantities are the parameters and the errors.

2. If in addition we assume that the errors all have the same variance and are uncorrelated one with another, we are assured that the least squares estimates have a smaller variance than any other linear combination of the $x_i$ and $y_i$ (see below).

3. If we make the further assumption that the errors are normally distributed, the least squares estimators are fully efficient, i.e. (for large samples at least), they give the minimum variance among all consistent estimators. (In fact it is exactly the same as the maximum likelihood estimator in this particular case. See below.)

The properties for the estimator have not been well worked out for nonlinear estimation.

The least squares estimator has the advantage of conceptual simplicity and is much more elastic than method-of-moments, which would not (for instance) be readily applicable to linear regression. Moreover the foregoing assumptions may, with a little elaboration, be circumvented. One such instance is homogeneity of variance (*homoscedasticity*) of errors. As a preliminary to attacking this problem, the question of combining estimates will be explored.

**Linear combination of unbiased estimates**

**Example 3.12.** Suppose that two measurements are available on the serum cholesterol level in a specimen of blood. One, $x$, has been done by hand, in the physician's office, by a method with an error of 5 mg/100 ml; and the other, $y$, is done on an aliquot sent to the hospital laboratory where an autoanalyzer is used which has an error of 2.5 mg/100 ml. What value should the physician use? We shall suppose the technical errors of measurement are uncorrelated and that both methods are unbiased.

Obviously $y$ is a priori a more precise measure than $x$ since its error is only half as much. It has a variance

$$(2.5)^2 = 6.25$$

Many would be content to use $y$ and leave it at that. However they could do better by not ignoring $x$ altogether. Suppose we compute a new quantity

$$z = ax + by$$

From unbiasedness of the estimates

$$E(Z) = a\mu + b\mu = (a + b)\mu$$

where $\mu$ is the true quantity. To ensure $Z$ is unbiased, then we must set $(a + b) = 1$; or, what is equivalent, we use

$$z = \frac{ax + by}{a + b}$$

If $b = 0$, this reduces simply to $x$; and if $a = 0$, it reduces to $y$. Now

$$\text{var}(Z) = \frac{1}{(a + b)^2}[a^2\text{var}(X) + b^2\text{var}(Y)]$$

It can be shown in general that the variance of the fraction becomes minimum if $a = 1/\text{var}(X)$ and $b = 1/\text{var}(Y)$. If $a = 1/25$ and $b = 1/6.25$ then the variance is

$$\frac{1}{\left[\dfrac{1}{25} + \dfrac{1}{6.25}\right]^2} \cdot \left[\frac{25}{25^2} + \frac{6.25}{6.25^2}\right] = 5$$

which is smaller than the variance of either estimate individually.

In general, if we have a set of unbiased estimates $T_1$, $T_2$, ..., $T_n$ of a parameter $\theta$ with variances $\sigma_1^2$, $\sigma_2^2$, ..., $\sigma_n^2$, then the unbiased linear combination which gives the minimum variance is

$$T_p = \frac{\Sigma\left[\dfrac{T_i}{\sigma_i^2}\right]}{\Sigma\left[\dfrac{1}{\sigma_i^2}\right]}$$

Its variance is

$$\text{var}(T_p) = \left(\Sigma\frac{1}{\sigma_i^2}\right)^{-1}$$

These facts lay the groundwork for least squares estimation when the errors have unequal variances and it is necessary to weight the different observations.

## WEIGHTED LEAST SQUARES ESTIMATION

**Example 3.13.**  Suppose that the objective is to estimate the effect of age on the blood cholesterol level from pooled data provided by several hospitals. Suppose further that all are using the same method and that *on average* they get the same results, but that they have different experimental errors. If the relative sizes of the errors are known we may proceed exactly as before except that the observations are weighted. Consider the relationship

$$y_i = f(x_i) + h_i$$

where $x_i$ is the age of the patient and $y_i$ is his height; $f(x)$ is a function of $x$ involving several unknown parameters. If the $i$th observation has an experimental error of $\sigma_i^2$ then we minimize the quantity

$$\Sigma \frac{[y_i - f(x_i)]^2}{\sigma_i^2}$$

with respect to all the parameters. For example, in the simple straight-line regression discussed previously the weighted least squares estimator would consist of finding those values of $a$ and $b$ that minimize

$$\Sigma \frac{(y_i - a - bx_i)^2}{\sigma_i^2}$$

This estimator has all the properties that apply to the ordinary least squares estimator where errors are *homoscedastic* (i.e., all the $\sigma_i^2$ are the same).

We may have estimates of the relative sizes of the errors (for example in counting errors of the Poisson type; see Chapter 14). More often there is no such information. If there are not multiple readings for at least some points, then there are difficulties and it may be necessary to be content with unweighted least squares estimation. If there are multiple readings, and errors seem to follow some well-defined trend, then a suitable system of weights may be worked out or at least assumed. If there is no such trend, the observed errors may be used as weights, though no exact treatment of this problem has been worked out in generality. The problem may be handled by a transformation that stabilizes the variance but which also changes the nature of the function $f(x)$. We shall discuss this in Chapter 4.

In weighted least squares procedures where the weights are less well characterized, there are no tidy answers. However, we may take courage from the fact that the final answers are not acutely sensitive to the weights, and unless the variances of the errors show considerable heterogeneity, weighting makes little difference to the estimates, though we may have some uneasiness about finding significance levels.


## MINIMUM CHI-SQUARE ESTIMATION

In the special case of weighted least squares estimation where the error terms are normally distributed, then for any particular value $y$ with variance $\sigma_i^2$,

$$\frac{[Y_i - f(x_i)]^2}{\sigma_i^2} = \frac{S_i^2}{\sigma_i^2}$$

is the square of a standardized Gaussian deviate, which by definition follows a chi-square distribution with one degree of freedom. The sum of $n$ such independent terms follows a chi-square with $n$ degrees of freedom. Minimizing the weighted sum of squares is equivalent to minimizing a chi-square statistic. It will be evident from what has been said about least squares estimation when the error terms have a Gaussian distribution, that this procedure is efficient.

This is an unorthodox presentation of the minimum chi-square method which I include as a conceptual bridge between the weighted least squares method and the rather more abstract and forbidding method of maximum likelihood. Any estimator that minimizes a chi-square statistic might be called a minimum chi-square estimator. What the statistician usually means by the term, however, is an estimator directed to minimizing a chi-square approximation to multinomial categorical data. The statistical aspects of the latter are sufficiently different to merit a separate treatment, and we shall delay illustrations until Chapter 13. We shall also discuss at that stage the modified minimum chi-square procedure.

The last method of estimation we shall discuss here is the method of maximum likelihood. As a preliminary we shall give some further attention to a fundamental concept.


## THE LIKELIHOOD PRINCIPLE

Earlier in this chapter (Fig. 3.1), likelihood was introduced *formally* as the same mathematical formula as a probability density function, but viewed as a function of the parameter(s) rather than of the random variable. The distinction seems rather pedantic, even though it helps to distinguish a sample space from a parameter space. The idea of a likelihood may now be reintroduced *conceptually* as a reconstruction, after the event, of the probability or probability density we would have predicted for the outcome (which we now know). The use to which it will now be put should make the motivation clear.

What makes the likelihood of such importance is that it is a paradigm of the data: enthusiasts maintain that all the information contained in the sample that is relevant to the parameters is contained in the likelihood ("the likelihood principle"). From time to time the critics of this view come up with counterexamples; but they are usually pathological exceptions, which suggests that the likelihood principle is sound and may be expected to stand up under almost all reasonable assaults.

## INVARIANCE PROPERTIES OF THE LIKELIHOOD FUNCTION

As we shall see presently, there is a special interest in the maximum (the highest value) of the likelihood function, and this is a convenient point at which to mention two invaluable properties of the curve. Let the parameter be denoted by $\theta$ and the estimate of it by $\hat{\theta}$. We shall denote the likelihood function by $L(\theta)$, which will, of course, contain the data.

(1) Let $\phi(\theta)$ be a positive monotonic function of $\theta$, i.e., any increase in $\theta$ leads to an increase in $\phi(\theta)$ and conversely. Then in two graphs we plot the likelihood $L(\theta)$ against $\theta$ and also against $\phi(\theta)$ for all possible $\theta$. If $\hat{\theta}$ maximizes $L(\theta)$ in the first graph $\phi(\hat{\theta})$ will maximize it in the second.

**Example 3.14.**  Consider the function

$$\phi(x) = x^3$$

If $d$ is any positive quantity however small or large

$$(x + d)^3 > x^3$$

and

$$(x - d)^3 < x^3$$

Hence $\phi(x)$ is clearly a monotonic function of $x$. Suppose that we plot $L(\theta)$ against $\theta$ and obtain a maximum at $\theta = 4$. Then when we plot $L(\theta)$ against $\theta^3$ the maximum will be at $(4)^3 = 64$. The reader may allay any doubts by calculation.

(2) Let $g[L(\theta)]$ be a monotonic function of $L(\theta)$. Whatever value of $\theta$ maximizes $L(\theta)$ also maximizes $g[L(\theta)]$.

**Example 3.15.**  Thus a particular and (as we shall see) common and useful function

$$g[L(\theta)] = \log_e[L(\theta)]$$

is a monotonic function which we shall use extensively. If $\theta = 4$ maximizes $L(\theta)$ it also maximizes $\log_e[L(\theta)]$.

These statements sound very involved; in fact they are almost trivially simple, as is evident from Figure 3.2. The second invariance property is merely saying that if $y_2$ is greater than both of two points in its immediate neighborhood, $y_1$ and $y_3$, then

$$\log_e y_2 > \log_e y_1 \text{ and } \log_e y_3$$

which the reader can verify to his heart's content in any number of particular cases.

For those who like familiar analogies, it may help to think of a relief map of the world made according to any of the various projections (all of which produce distortions of scale or shape) in which the vertical scale is also distorting in various ways but preserves the *order* of heights. Then on all of these maps the highest point will always be the same—the top of Mount Everest—although its perceived location on the map, or its height on the map above sea level may vary.

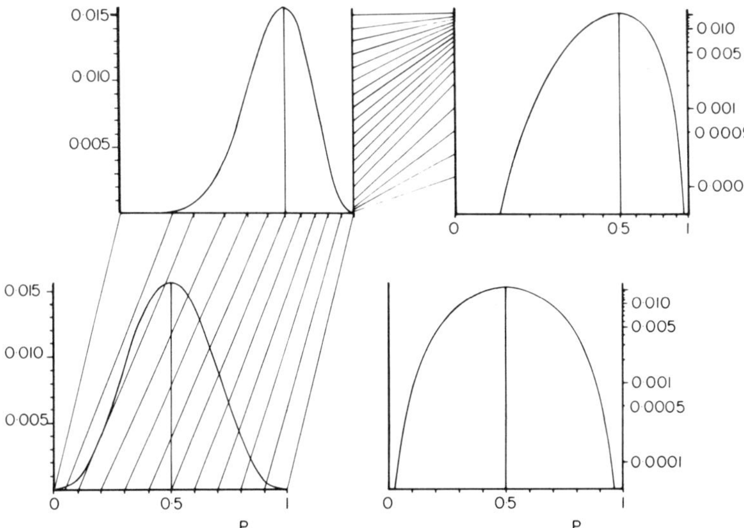

**Figure 3.2.** Invariance properties of maximum likelihood estimation. At *bottom left* is shown the simple likelihood function of a sample of data plotted against the parameter, *p*. The MLE lies at 0.5. At *top left* is shown the same function with the distances on the horizontal axis proportional to the square roots of those on the original scale. (Corresponding points are connected by lines.) The *shape* of the likelihood curve has changed; but the maximum lies opposite $p = 0.5$. At *top right*, the horizontal scale is distorted in the same way. As well, the vertical scale is distorted by converting it into logarithms. (Corresponding points on the vertical scale are again connected by lines.) Again the maximum is at 0.5. At *bottom right*, the vertical scale only has been transformed. The four curves differ considerably in shape; but the maximum occurs at the same corresponding points for all.

## THE METHOD OF MAXIMUM LIKELIHOOD

Without more ado, we may now formulate the maximum likelihood estimator (*MLE*). To construct the likelihood curve one does not need to know the entire set of data in order to have the information about the parameters.

Given the distribution of the variate and the parameters we may, in certain cases, be content to use the mean and the variance as epitomizing characteristics. Analogously for a function which is the natural logarithm of the likelihood we may use two features:

1. The value(s) of the parameter(s) at which occurs the maximum or (if there are multiple maxima) the greatest of them within the acceptable parameter space.

2. The radius of curvature of the function in the immediate neighborhood of this maximum (provided that the values of the function immediately

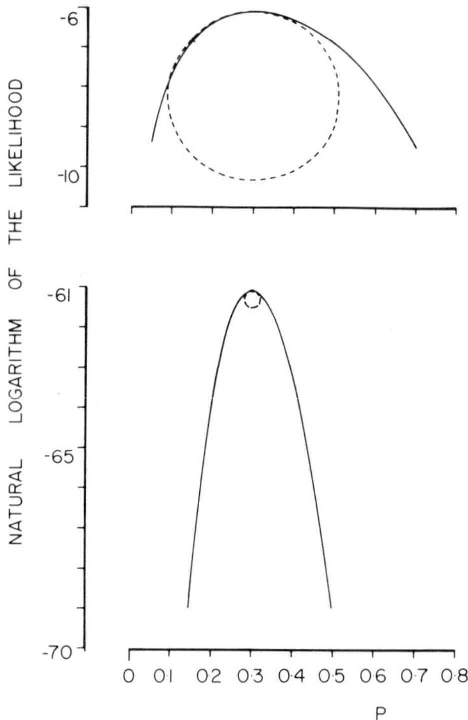

**Figure 3.3.** Approximate variance for the MLE. The *upper* curve is the natural-log-likelihood curve for 3 successes in 10 Bernoulli trials, the *lower,* that for 30 successes in 100 trials. The same scales are used on the respective axes for both. In each, the circle that just touches the curve in the neighborhood of the MLE is also drawn (the interrupted curves). The MLE are the same (0.3) but the circle in the upper case has a radius ten times as large as that in the lower, indicating a variance ten times as large.

on both [or all] sides of the maximum are somewhat less than that at the maximum).

The value of the parameter at the maximum is the maximum likelihood estimate of the parameter. The radius of curvature—that is the radius of the circle which just touches the log-likelihood curve at the maximum—is the estimated variance of the estimate (Fig. 3.3).

Finding this estimator may at times be very complicated, but the procedure is in principle quite simple.

1. Write down the joint probability (or in this context, the likelihood) of the sample values as an algebraic function of the (known) data and the (unknown) parameters.

2. Find those values of the parameters that make this algebraic expression as large as possible.

3. Then these maximizing values are the maximum likelihood estimates of the parameters.

Maximum likelihood estimation is usually more clumsy than method of moments; but under very general conditions the estimator has the cardinal virtues of elasticity, efficiency, and consistency, and the estimate is approximately Gaussian* for large samples. Consider a common example.

**Example 3.16.** Suppose that we have a set of $n$ independent values $(x_1, \ldots, x_n)$ which are derived from the same exponential distribution (i.e., a gamma distribution of order 1) with parameter $a$, which has the probability density function

$$f(x) = ae^{-ax}$$

Then the joint likelihood is given by the product of the individual densities

$$L = f(x_1)f(x_2) \cdots f(x_n)$$
$$= \Pi ae^{-ax_i}$$
$$= a^n e^{-a\Sigma x_i}$$

To find the maximum likelihood estimate of $a$, we use that value of $a$ that makes $L$ as large as possible.

For instance let $x$ be the waiting time until the first radioactive emission from a small piece of radioactive tissue occurs. This time follows the exponential distribution. The observed times in seconds for the first 10 emissions are 0.088, 0.160, 0.059, 0.506, 0.208, 0.142, 0.252, 0.018, 0.128 and 0.108. The sum is 1.669. The aim is then to find $a$ which maximizes

$$L(a) = a^{10}e^{-1.669a}$$

---

*There has been some confusion about this property. It does not imply that *the shape of the likelihood curve* becomes Gaussian (although suitably scaled it may). In statistical terms, if repeated large samples of equal size were taken under identical conditions and the single value that is the MLE found for each, the histogram of answers would conform approximately to a Gaussian distribution.

Some trial values are given in Table 3.2. To two decimal places the answer is 5.99. Theory shows (7) that the mean waiting time is the reciprocal of $a$ and because of the first invariance property (see above) the MLE is the reciprocal of the MLE of $a$ which is 0.1669. Note that to four decimal places this equals the method-of-moments estimate, i.e.,

$$\frac{\Sigma x_i}{n} = \frac{1.669}{10} = 0.1669$$

Theory shows that the two estimators are in fact identical.

### An alternative approach

As we have seen above, we might just as well have found the MLE of $a$ in Example 3.16 by maximizing the logarithm of the likelihood. Accordingly in the last column of Table 3.2 is given the natural logarithm of the likelihood. Natural, rather than common, logarithms have been used for reasons that will be explained later. As is evident, the maximum of the logarithm of the likelihood is attained at the same value of $a$, i.e., 5.99. This illustrates the second invariance property (see p. 60).

In many cases the likelihood function must be solved by just such numerical methods as we have employed: exact solution (by the calculus) is intractable. It will therefore be useful at this stage to see how an approximate

**Table 3.2. The maximum value of $a^{10}e^{-1.669a}$**

| Trial value of $a$ | $L = a^{10}e^{-1.669a}$ | Log $L$ |
|---|---|---|
| First digit | | |
| 0 | 0 | $-\infty$ |
| 2 | 36 | 3.593 |
| 4 | 1,322 | 7.187 |
| 6 | 2,707 | 7.904 |
| 8 | 1,707 | 7.442 |
| Second digit | | |
| 5.6 | 2,647.22 | 7.881,27 |
| 5.8 | 2,692.91 | 7,898,38 |
| 6.0 | 2,707.00 | 7.903,59 |
| 6.2 | 2,691.07 | 7,897,69 |
| Third digit | | |
| 5.96 | 2,706.644 | 7.903,465 |
| 5.98 | 2,706.971 | 7.903,586 |
| 6.00 | 2,706.996 | 7.903,595 |
| 6.02 | 2,706.719 | 7.903,493 |
| 5.99 | 2,707.021 | 7.903,604 |

solution can be obtained somewhat more systematically and with it an estimate of the variance of the maximum likelihood estimator.

The basic idea underlying approximate calculations of the MLE is that over a small interval we may fit a second-degree curve, treat it as the true curve, and then from its properties obtain the necessary approximate estimates. This process may be elaborated as desired. I shall give a simple procedure involving a set of three equally spaced trial values. A rationale for the method involves the calculus and will be omitted.

Suppose we take three trial values of the parameter in the neighborhood of the maximum, $t_1$, $t_2$, and $t_3$ with $(t_2 - t_1) = (t_3 - t_2) = d$; and let the natural logarithms of the corresponding likelihoods be $y_1$, $y_2$, and $y_3$. The second degree curve passing through the points has the form

$$f(y) = y_2 + \frac{y_3 - y_1}{2d} t + \frac{y_1 + y_3 - 2y_2}{2d^2} t^2$$

Then the maximum value of this function is at $\hat{t}$ where

$$\hat{t} = t_2 + \frac{(y_1 - y_3)d}{2(y_1 + y_3 - 2y_2)} \tag{3.2}$$

The accuracy of this solution is improved by making $d$ smaller. However, note two points.

1. The trial points should be in the neighborhood of the maximum which for preference should be between $t_1$ and $t_3$. Thus it is wise to pursue the maximum in several stages, as we shall illustrate in an example.

2. If $d$ is allowed to become very small the differences among the $y$'s become very small also and the coefficient of $d$ in equation (3.2) is the ratio of small quantities; unless great accuracy in the calculations is maintained, the second term may do more harm than good.

**Example 3.17.** The method will be illustrated with a familiar problem, the binomial. In eight trials, there are five successes. First note that the method of moments estimate is

$$\hat{p} = \frac{5}{8} = 0.625$$

And for the associated estimate of the variance we might reasonably use the usual binomial variance evaluated at the estimate of $p$:

$$\frac{\hat{p}(1 - \hat{p})}{n} = \frac{(0.625)(0.375)}{8} = 0.029,296,875$$

In Table 3.3 the approximation to the MLE (3.2) is used, when several sets of three equally spaced values are taken. As $d$ gets smaller, the estimate comes closer to the true method-of-moments estimate. (Theory shows that in fact the estimators are identical in the binomial.)

**Table 3.3. The use of the approximate method of finding the MLE from equally spaced trial values. The function here is a binomial with 5 successes in 8 trials.**

| $d$ | Trial values of $p$ | $\log L$ | Approximate MLE | Approximate variance |
|-----|------|------|------|------|
| .2 | .4 | $-6.113{,}930{,}531$ | | |
|  | .6 | $-5.303{,}000{,}314$ | 0.611,7 | 0.027,5 |
|  | .8 | $-5.944{,}031{,}494$ | | |
| .02 | .6 | $-5.303{,}000{,}314$ | | |
|  | .62 | $-5.292{,}931{,}083$ | 0.624,89 | 0.029,57 |
|  | .64 | $-5.296{,}389{,}256$ | | |
| .001 | .624 | $-5.292{,}522{,}960$ | | |
|  | .625 | $-5.292{,}505{,}905$ | 0.625,000 | 0.029,296,783 |
|  | .626 | $-5.292{,}522{,}984$ | | |
| Exact | | | 0.625 | 0.029,296,875 |

NOTE: The likelihood is $p^5(1 - p)^3$. The combinatorial factor, 56, is omitted.

### Estimate of the variance

The reader may have been struck both that we are giving undue stress to maximum likelihood methods and that we have complicated matters unnecessarily by using natural logarithms rather than the more familiar logarithms to the base ten. We shall now proceed to justify both.

First the likelihood curve has wider implications than maximum likelihood estimation. In fact it tells us very precisely how much information about the parameter the ideal estimator could extract from the data. It can be shown under very general conditions (see below) that, for samples at least, there is a lower limit to the variance of any estimate of the parameter based on a consistent estimator. The estimators that attain this lower limit are said to be fully efficient. The estimator used in Example 3.17 is of this type; and it is pointless trying to find a more precise consistent estimator. This lower limit can be found from the likelihood, provided the curve used is the *natural* logarithm of the likelihood. Quite simply it is the average value of the radius of curvature of this function at the maximum likelihood estimate. What does this mean? Consider the two log-likelihood curves shown in Figure 3.3. The upper one is very flat. This means that quite a wide range of values give almost equally high likelihoods, hence we must be rather indifferent as to which we pick. Intuitively we feel that such a curve would give a vague estimate. The contrary is the case with the lower curve: small departures from the MLE involve a precipitate fall in the likelihood and hence in the plausibility of the estimate. These differences can be represented by fitting circles that touch the likelihood in the neighborhood of the maximum. The upper curve needs a large circle, the lower, a small one. If we take the radius of this circle, theory shows for large samples at least that its average value gives the lower limit on the variance for a consistent estimator.

Of course, this is a very loose statement and does not equip the non-mathematical reader to use this criterion. Practical guidelines may be easily set, however.

1. Using the same notation as before, an adequate approximation to the lower bound on the variance of an efficient estimator, $T$, is given by the formula

$$\operatorname{var}(T) = \frac{d^2}{2y_2 - y_1 - y_3} \tag{3.3}$$

Appropriate calculations are given in the last column of Table 3.3 together with the exact (binomial) variance.

The purist may object that the formula (3.3) is only an approximation; but in most cases the lower limit is only a large-sample result, and hence only an approximation.

2. The lower limit is the *average* value of the reciprocal of the radius of curvature. So in this, as for any sampling procedure, we do not know the variance of the estimate, merely an estimate of the variance of the estimate. The bigger the sample, the better the estimate of the variance.

3. This method only works where the maximum has downward slopes on both sides. For instance, if there are no successes in five trials, it is easily verified that the MLE is 0; but the variance of the estimate at zero is not given by this method since there is no downward slope on the lower side of the MLE. (See Figure 3.4.)

4. The method, and MLE in general, should not be used *where the data set limits on the parameter space*. This is a rare pathology, but important.

**Example 3.18.** Suppose that a variate has the density function

$$f(x) = \frac{1}{(x + a)\sigma\sqrt{2\pi}}e^{-\left|\frac{\log(a+x)-\mu}{\sigma}\right|^2/2}$$

This so-called three-parameter lognormal distribution has been found appropriate to describe systolic blood pressure (18, 19); and the level of carcinoembryonic antigen in serum (20). There are the usual parameters of centrality ($\mu$) and dispersion ($\sigma^2$); but there is a third parameter, $a$, added to each datum. It may be positive or negative. But the use of MLE to estimate it is inappropriate because the range of values it may assume depends on the data. Suppose, for instance, the lowest systolic pressure in the sample is 75 mm Hg. Then $a$ cannot assume the value $-75$ or less, because

$$\log[75 + (-75)] = \log(0) = -\infty$$

and

$$\log[75 + (-76)] = \log(-1)$$

which is an imaginary number.

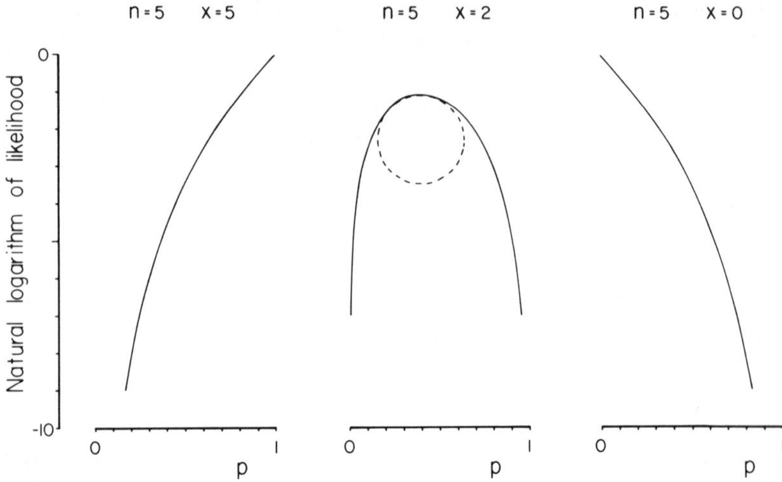



**Figure 3.4.** Likelihood curves for three binomial variates of order 5. The approximate formula for the variance is appropriate where the maximum is clearly inside the parameter space (*central diagram*). Where all the trials are successes (*left*) or all failures (*right*) the maximum is at the edge of the parameter space and at that point the curve is still rising. Then the approximate estimate of the variance does not apply.

## PROBLEMS

**3.1.** The rectangular distribution, $R(0, b)$ is a symmetrical continuous distribution for which all points between 0 and $b$ have the same probability density

$$f(x) = \frac{1}{b}$$

and points elsewhere have zero density. We are given the following four sample values: 4, 17, 3, 0.8

    **a.** Find the method-of-moments-estimate of the mean.

    **b.** Find the maximum likelihood estimate of $b$ and hence of the mean.

    **c.** Find the corresponding estimates when the lower limit is not 0 but also an unknown quantity, $a$. (The density is then $1/(b - a)$.)

Comment on the results.

**3.2.** In Example 3.9 and Table 3.3 the combinatorial coefficient $\binom{8}{5} = 56$ is ignored, that is, we represented the likelihood as $p^5(1 - p)^3$ rather than $56p^5(1 - p)^3$. Prove that this makes no difference to either the MLE of $p$ or the estimate of its variance.

**3.3.** In genetics, it is well known that under certain general conditions, if at one locus there are two alleles, $A_1$ with frequency $p$, and $A_2$ with frequency $q$, so that $p + q = 1$, then assuming that at this locus each patient has two genes randomly and independently assembled we may expect three combinations (known as genotypes):

$A_1A_1$ with probability $p^2$

$A_1A_2$ with probability $2pq$

$A_2A_2$ with probability $q^2$

All three genotypes are distinguishable. In a sample we have 7 of the first group, 9 of the second, and 9 of the third.

**a.** Write down the likelihood function and reduce it to its simplest terms. Write down the MLE of $p$ by inspection.

**b.** Could you have devised a more direct way of arriving at this conclusion?

**3.4.** Suppose that in the last problem the first two genotypes have the same phenotypes; i.e., they cannot be distinguished by appearance. Thus we have 16 of type $(A_1A_1$ or $A_1A_2)$ and 9 of type $A_2A_2$.

**a.** Write down the likelihood function and from it write down by inspection the MLE of $q^2$. What is its estimated variance?

**b.** From this find the MLE of $q$ by inspection. Justify your answer.

**c.** Verify numerically that this is the MLE and find its approximate variance.

**d.** Compare the MLE with the method-of-moments estimator.

**3.5.** Prove from the results given in the section on linear combination of unbiased estimates, that if we have several sets of data obtained by the same method with the same properties and from the same population, the combined estimate of their common means should be a linear combination of their individual sample means weighted according to the sizes of the samples.

# 4
# TRANSFORMATIONS

Statistical analysis always involves a certain number of assumptions about how the data from the study behave. It will have been evident from Chapter 3 that the more efficient an estimator, in general, the more specifications about the form of the variate are involved.

The biologist is right to be critical of Procrustean statistics—the attempt to make the data fit some particular method of analysis by hook or by crook. Truth must never be subordinated to convenience. At the same time, one should realize that certain distributions, for reasons which have been discussed (7), have astonishingly wide application. This is notably true of the Gaussian; and in the next seven chapters we shall develop the appropriate methodology. Its widespread use and the unfortunate term *normal* have created the false impression that "normal" measurements follow this distribution. Such is not the case. A great many variates can be shown both a priori and empirically to follow quite different distributions. Thus an investigator whose entire statistical stock in trade is an understanding of the Gaussian distribution is at something of a loss whenever this distribution cannot be readily invoked. He has several courses open to him.

First, from theoretical or other considerations he may determine the form of the distribution suitable to his observations and, if necessary, develop a whole new theory for dealing with it. This solution may be time-consuming and not very attractive unless the results are to have a fairly general application. It also requires considerable statistical resourcefulness.

Secondly, he may abandon most assumptions (except for a few which are not very exacting) and use the so-called distribution-free or nonparametric methods that are discussed in Chapter 15. For the most part these nonparametric methods are new and unfamiliar to the general reader, and in special cases they do not make fully efficient use of the data. However, these defects may not be serious; and since for the most part the tests are easy to do, they certainly deserve rather more attention than they usually receive.

The third approach is for the investigator to find some way of converting the data into such a form that they follow a familiar distribution with known properties; then statistical procedures based on the familiar distribution may be applied to the converted data. Such a procedure, called transformation, will be the subject of the present chapter. Many people have an irrational suspicion that transformations must in some way be dishonest. Such an attitude is unwarranted. Consider a familiar example, which illustrates one of several purposes of transformation. Suppose that the volume of the erythrocyte has a Gaussian distribution. Because of simplicity the measurement made is not the volume but the diameter of the cell. There is no reason to expect the diameter to follow a Gaussian distribution: since volume is a cubic function of corresponding dimensions it might be much more reasonable to suppose that the cube of the diameter would have a Gaussian distribution. Other illustrations will be given in which the transformation has such a plausible justification. The problem would be just as real if it were the diameter that had the Gaussian distribution and we wished to analyze the volume of the cell (a topic of interest in automatic cell counters).

## OBJECTIVES OF TRANSFORMATION

**1. Normalization,** i.e., converting the variate into Gaussian form or, more generally, making it conform to some familiar distribution, is the main reason for transformation. It will be amply illustrated later.

**2. Stabilization of the variance.** Many tests that we discuss later (for example analysis of variance, comparison of correlation coefficients) are based on the assumption that the variance of the random variable is stable. The commonest and most important violation is that the variance is related to the mean. Thus, rejecting the null hypothesis (that the means of two distributions are equal) may imply that the assumption that the variances (on which the test was based) are equal is unsound. There are well-known instances where such a relationship holds—for example the Poisson distribution (see Chapter 14). But where a few sample values only are known, the statistician may have a very imperfect idea of what the mean is, and therefore what the variance is. It may be possible to transform the data in such a way that the variance is known—or a good approximation is known—and independent of the mean. This is worth while in comparing groups. These uses will also be discussed.

**3. To make true effects linear and additive.** Often experimental error, or the change produced by some manipulation, may be a function of the mean, and the effect of one treatment may be influenced by the effect of another. Thus the effect of hydralazine and reserpine given together on the blood pressure is greater than the sum of the effects of the same dosages given separately. (A pharmacologist would say they are "synergistic.") This may be due to the fact that the drugs act at different sites. There may be some system of measurement, which differs from that used, which makes results additive (see Chapter 17). Again the technical error of measurement in blood cholesterol levels is not the same for all values, but seems to be proportional to the mean. A suitable transformation may make the effects additive on the new scale. In measuring blood cholesterol levels, it has been shown that after a logarithmic transformation, the error of measurement is much the same for high values and low.

**4. To make the mean a good measurement of "the typical value."** This can be well illustrated from the data of Kincaid-Smith and others (21) on survival in malignant hypertension, discussed in some detail later. The estimate of the ordinary (or arithmetic) mean is dominated by a small group of exceptionally long-lived subjects, and only a small fraction of the total live the mean number of years or more. It is thus a poor measure of the general behavior of the group.

**5. To linearize a relationship between two or more variables.** Thus decay of radioactivity over time is represented by an exponential function. Fitting a curve through a set of observations is often a complicated process. However, such a function plotted with activity on a logarithmic scale is linear, and curve fitting less complicated.

**6. Remodeling the distribution.** A common objective of transformation is to convert a less familiar distribution into a more familiar one. For instance Morton (22) showed in genetic linkage that the inverse Kosambi transform

$$y = \frac{e^{2x} - e^{-2x}}{e^{2x} + e^{-2x}} \qquad x \geq 0$$

when applied to the triangular distribution

$$f(x) = \frac{2}{a}\left(1 - \frac{x}{a}\right) \qquad 0 \leq x \leq a$$

conforms well to the rectangular distribution, provided $a$ is not too large. (This relationship is of considerable importance in genetic linkage analysis and the theory of chromosome mapping. One theory proposes the inverse Kosambi function as the relationship between the recombination fraction [which empirically is found to be approximately rectangular] and the map distance, which is approximately triangular. It is somewhat easier to handle the rectangular distribution [7].)

These aims of transformation obviously may be in conflict. It is surprising, however, how often two or more objectives can be realized in the same transformation, and even when such is not the case, the benefits to the one feature may be marked, while the losses to the other are trivial.


## LINEAR TRANSFORMATIONS

The simplest possible examples of transformations are those in which a constant term is added to each measurement or each measurement is multiplied by a constant factor: for example, if we measure the heights of barefooted men and by adding a constant (say ¾ inch) convert this to "heights wearing shoes"; or if we measure a variable in inches and convert it into centimeters by multiplying by 2.54. Neither of these types of transformation changes the shape of the distribution—they merely change the location or the scale. If Americans are on average no taller than Canadians when measured in inches, we would not expect them to be any taller when measured in centimeters. So much is common sense. Transformations of these types (or a combination of the two), i.e.,

$$Y = a + bX$$

are called "linear" transformations because if we were to graph $Y$, the transformed value (or transform), against the original value $X$, a straight line would result.

Linear transformations are of little value in helping to rectify departures from the assumptions underlying a test: they will not convert an asymmetrical distribution into a symmetrical one. But they do contribute to simplicity in tabulation. Thus areas under the Gaussian curve have been extensively tabled for $Z$, the "standard" Gaussian variate, which has a mean of zero and a variance of unity. Most actual distributions have a mean ($\mu$) not equal to zero and a variance ($\sigma^2$) not equal to unity. It would be prohibitive to compute tables for all possible values of $\mu$ and $\sigma^2$. However, if both are known the random variable $X$ may be converted into $Z$ by the linear transformation

$$Z = \frac{X - \mu}{\sigma}$$

The reader can readily satisfy himself that this is a linear transformation by plotting $X$ against $Z$ for any particular values of $\mu$ and $\sigma$. With a little rearrangement we have

$$X = \mu + Z\sigma$$

Thus individual values on the $Z$ scale can be converted to values on the $X$ scale in exactly the same way that a temperature on the Celsius scale can be converted into a temperature Fahrenheit, the counterpart of $\mu$ being 32 and of $\sigma$ being 1.8.


## NONLINEAR TRANSFORMATIONS

These are more helpful in general: countless numbers of them exist. Those commonly employed involve taking logarithms, square roots, raising to various powers, taking trigonometric functions, etc. With such a wealth of choice it is of some importance to consider on what qualities the decision is to be made.

## CHARACTERISTICS DESIRABLE IN A TRANSFORMATION

**1. Simplicity.** For convenience, it is best that the transformation be simple. If there are good theoretical grounds for an elaborate transformation, then its use is justified, though tedious calculations may be involved. If there is no rationale, and it is used only because it is found empirically to work, then the more complicated it is to be, the more detailed and extensive evidence of its utility should be demanded.

**2. Monotonicity.** A monotonic function is one such that either every increase, however small, in the original variable will be associated with an increase (also perhaps small) in the transform, or conversely, that every increase in the original produces a decrease in the transform. Thus the functions

$$Y = \log X \qquad \text{for } x \geq 0$$
$$Y = X^3$$
$$Y = e^X$$

are all monotonic increasing functions. Conversely

$$Y = 1/X$$

is a monotonic decreasing function.

In contrast

$$Y = X^2$$

is not monotonic, as will be evident from the following sets of values. Here $X$ increases throughout but $Y$ does not:

| X | $-3$ | $-2$ | $-1$ | 0 | 1 | 2 | 3 |
|---|---|---|---|---|---|---|---|
| Y | 9 | 4 | 1 | 0 | 1 | 4 | 9 |

$\leftarrow$     Y   decreases $\rightarrow$ $\leftarrow$ Y increases$\rightarrow$

Of course if this transformation were applied only to values of $X \geq 0$ it would be monotonic. In general the transformation consisting of raising $X$ to an even power is not monotonic. However, in many instances (including almost all measurements), $X$ may assume positive values only. For example height, weight, blood sugar, intelligence, etc., must be greater than zero, while others such as number of times married, years of formal education, income, the dose of a drug, cannot be negative. Even-power transformations of such quantities will be monotonic over the domain of interest. Nevertheless there may still be problems (see Example 4.1).

Monotonicity is an advantage because even if the orders of magnitude are changed by it, information about the relative sizes of the values is not entirely discarded and it is possible, by inverting the transform, to determine what the original value must have been. But for example, given that $X^2$ is 9, we cannot tell whether the original value of $X$ was $-3$ or $+3$ except where negative values are excluded.

We have seen in Chapter 3 that monotonicity is an important property of maximum likelihood estimation. Many problems in estimation may be much simplified by exploiting monotonic transformations.

**3. Achievement of multiple purposes.** A transformation may achieve more or less adequately two purposes at once—for example it may normalize (that is, convert into Gaussian form a distribution of quite a different shape) and stabilize the variance (see below). Which of these effects is achieved may help to decide between two transformations, e.g., between taking the logarithms or taking the square roots, both of which may convert the data to something close to a Gaussian distribution but have different degrees of success in stabilizing the variance.

The use of transformations should not be slavish. There may be only a trivial departure from normality, which makes little difference to a procedure. Or a transformation may nearly, but not quite, normalize a very nonnormal distribution. Any big improvement is better than none, even if the transform is still imperfect. Conversely, a reasonable circumspection is required in the use of transformations. Almost any positively skewed distributions will be rendered more nearly normal—or at least symmetrical—by a suitable logarithmic transformation. But if the skewness is due to the presence of a second population in the right-hand tail (3), the transformation to normal is meaningless; if the transformed results are then to be manipulated as if they came from a single homogeneous population, the transformation may do more harm than good.

**4. Eliminating interaction.** We shall defer this topic until Chapter 7.


## NORMALIZATION

The assumption that a distribution is Gaussian is involved to some extent in a great many statistical procedures. The means of samples of reasonable size (say 50 or more) from most distributions have approximately normal distributions as the central limit theorem states (see Chapter 17). In such cases transformation may be scarcely worthwhile. The need for normality of the parent population is most pressing in dealing with single observations individually ("samples of size 1") since, in them, the central limit theorem cannot be invoked. For instance, in medicine the practice has grown up of defining the "normal range" for some measurement as those values which lie within two standard deviations of the mean, the parameters being estimated from a suitable sample of subjects. I will not make an attempt here to criticize this practice, which is discussed later (Chapter 11). It is a common enough argument, and the impression is left with the investigator that this range encompasses ninety-five percent of the population sampled. With small samples this may not be so, because of sampling error in the estimates of the mean and variance; but for even large samples it is in general true* only if the parent distribution from which the sample was drawn is normally distributed.

**Example 4.1.** Consider an example where the distribution is clearly not Gaussian. The data are those of Kincaid-Smith and others (21) on survival in malignant hypertension. The actual values from the time of diagnosis (in many cases the time of admission to hospital) are given in Table 4.1. The distribution shows a high peak to the left and a long tail to the right, i.e., it is positively skewed (Figure 4.1). This skewness is so marked as to make it almost impossible to construct an accurate histogram to scale.

In Figure 4.2 the sample cumulative distribution (see Chapter 1) is shown as a step function, and the theoretical cumulative Gaussian curve (having the mean and variance estimated from the data) as a smooth curve. The discrepancies between them are, as before, in black.

A few calculations will point out the inadequacy of applying Gaussian theory to these values. The sample mean is 6.62 months. (Incidentally, it will be noted that the sample mean is a poor figure to use as representative of the behavior of the group, since less than one-fifth of the population survived as long as this or longer.) The estimated standard deviation is 25.06 months. With a Gaussian model, 0 months gives a $z$ value of

$$\frac{0 - 6.62}{25.06} = -0.264$$

Zero thus lies 0.264 standard deviations below the mean, so that about 40% of the subjects should have a negative survival from the time of diagnosis, obviously an absurd conclusion. A more detailed comparison of the observed with the expected can be made by referring to the table of the Gaussian integral, reading the 10, 20, 30, ..., 90 percentile points and finding the allocation of observed values to these classes. The expected number in each class is, of course, one-tenth of the total, i.e., 10.5. The results are given in Table 4.2. A formal test follows.†

It is clear that the main way in which the distribution differs from the normal is in being grossly skewed positively. Any normalizing transformation, if it is to meet with success, must have the effect of reducing this skewness. Two kinds at once suggest themselves: the square root transformation and the logarithmic.

---

*I say "in general": it is not inconceivable that the statement may also be true for some other natural distribution which is not Gaussian.

†Again I must apologize that I cannot devise a completely orderly arrangement of chapters. These formal tests of goodness of fit require some knowledge of the analysis of categorical data (see Chapter 13). Those who at this stage have no idea at all of the chi-square procedure may come back to these details later, and be content here to make judgments by eye on the basis of the diagrams.

**Table 4.1. Survival in months from diagnosis in malignant hypertension**

| Survival (x) | Number of subjects | Cumulative % | $\sqrt{x}$ | $\log_{10}(x \times 1000)$ |
|---|---|---|---|---|
| 0.0028 | 1 | 0.95 | 0.05 | 0.444 |
| 0.0083 | 2 | 2.86 | 0.09 | 0.921 |
| 0.0333 | 4 | 6.67 | 0.18 | 1.523 |
| 0.100 | 4 | 10.48 | 0.32 | 2.000 |
| 0.133 | 3 | 13.33 | 0.37 | 2.125 |
| 0.167 | 5 | 18.10 | 0.41 | 2.222 |
| 0.200 | 2 | 20.00 | 0.45 | 2.301 |
| 0.233 | 3 | 22.86 | 0.48 | 2.368 |
| 0.267 | 1 | 23.81 | 0.52 | 2.426 |
| 0.300 | 1 | 24.76 | 0.55 | 2.477 |
| 0.333 | 1 | 25.71 | 0.58 | 2.523 |
| 0.367 | 1 | 26.67 | 0.61 | 2.564 |
| 0.400 | 1 | 27.62 | 0.63 | 2.602 |
| 0.467 | 2 | 29.52 | 0.68 | 2.669 |
| 0.500 | 3 | 32.38 | 0.71 | 2.699 |
| 0.533 | 2 | 34.29 | 0.73 | 2.727 |
| 0.567 | 3 | 37.14 | 0.75 | 2.753 |
| 0.600 | 2 | 39.05 | 0.77 | 2.778 |
| 0.750 | 2 | 40.95 | 0.87 | 2.875 |
| 0.867 | 1 | 41.90 | 0.93 | 2.938 |
| 1.00 | 4 | 45.71 | 1.00 | 3.00 |
| 1.25 | 3 | 48.57 | 1.12 | 3.097 |
| 1.40 | 1 | 49.52 | 1.18 | 3.146 |
| 1.50 | 4 | 53.33 | 1.22 | 3.176 |
| 2.00 | 3 | 56.19 | 1.41 | 3.301 |
| 2.25 | 3 | 59.05 | 1.50 | 3.352 |
| 2.50 | 1 | 60.00 | 1.58 | 3.398 |
| 2.75 | 1 | 60.95 | 1.66 | 3.439 |
| 3.00 | 5 | 65.71 | 1.73 | 3.477 |
| 3.50 | 3 | 68.57 | 1.87 | 3.544 |
| 3.75 | 1 | 69.52 | 1.94 | 3.574 |
| 4.0 | 4 | 73.33 | 2.00 | 3.602 |
| 5.0 | 5 | 78.10 | 2.24 | 3.699 |
| 5.5 | 2 | 80.00 | 2.35 | 3.740 |
| 6.0 | 1 | 80.95 | 2.45 | 3.778 |
| 7.0 | 1 | 81.90 | 2.65 | 3.845 |
| 8.0 | 1 | 82.86 | 2.83 | 3.903 |
| 9.0 | 4 | 86.67 | 3.00 | 3.954 |
| 9.5 | 2 | 88.57 | 3.08 | 3.978 |
| 10.5 | 1 | 89.52 | 3.24 | 4.021 |
| 12.0 | 1 | 90.48 | 3.46 | 4.079 |
| 12.6 | 1 | 91.43 | 3.55 | 4.100 |
| 13.0 | 2 | 93.33 | 3.61 | 4.114 |
| 14.0 | 1 | 94.29 | 3.74 | 4.146 |
| 19.0 | 1 | 95.24 | 4.36 | 4.279 |
| 27.5 | 1 | 96.19 | 5.24 | 4.439 |
| 33 | 1 | 97.14 | 5.74 | 4.519 |
| 34 | 1 | 98.10 | 5.83 | 4.531 |
| 55 | 1 | 99.05 | 7.42 | 4.740 |
| 248 | 1 | 100.00 | 15.75 | 5.394 |
| Total | 105 | | | |

The histogram of square roots of the survival times (given in the penultimate column of Table 4.1) is analyzed in Table 4.3. The mean of the transformed values is 1.706. The standard deviation is 1.937 and the goodness-of-fit test gives a much smaller chi-square statistic. Nevertheless, the chi-square statistic still leads to rejection (at $p < 0.001$,) of the hypothesis that the transformed results are normally distributed. Furthermore, an uncomfortably large component of the chi-square value is contributed by the second class, which appears to be systematically overrepresented. Note that the mean of the transformed values, 1.706, may be transformed back to the original scale (in months) by squaring, giving 2.91 months. This latter figure is a reasonably representative one: 66 subjects survive less than this figure, 39 more. The square root transformation appears to be in the right direction and it might be of interest to try taking higher roots, e.g., the fourth or the sixth. The fit of the normal distribution to the transform is shown in Figure 4.3a. There is proportionately somewhat less black than in Figure 4.2. However, although the *data* are all positive and the transform monotonic, part of the estimated distribution of the transformed value would be negative, which is both biologically meaningless and mathematically cumbersome.

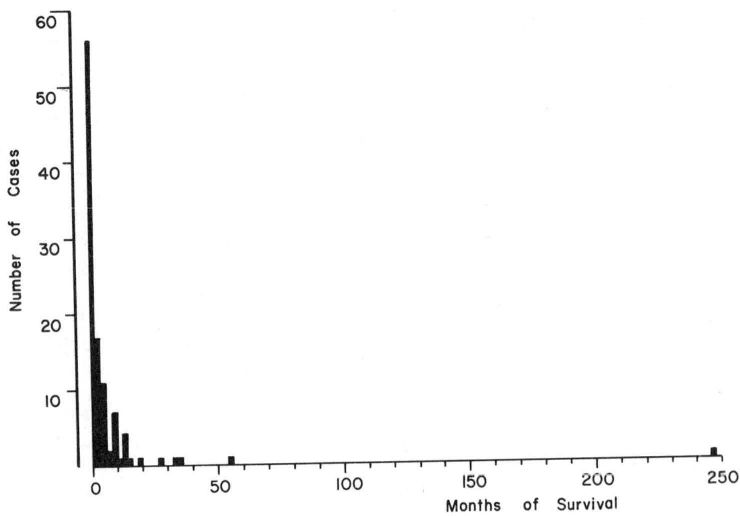


**Figure 4.1.** Histogram of length of survival from diagnosis in 105 cases of malignant hypertension. This variate has a distribution with marked positive skewness. (The value on the extreme right, at 246 months, is always singled out by casual critics for preferential scrutiny, for no better reason than that it is so large. I have logical objections to this kind of uneven editing. However, in this case the diagnosis was unusually well authenticated, passing the test of professorial opinions in both cardiology and ophthalmology.)

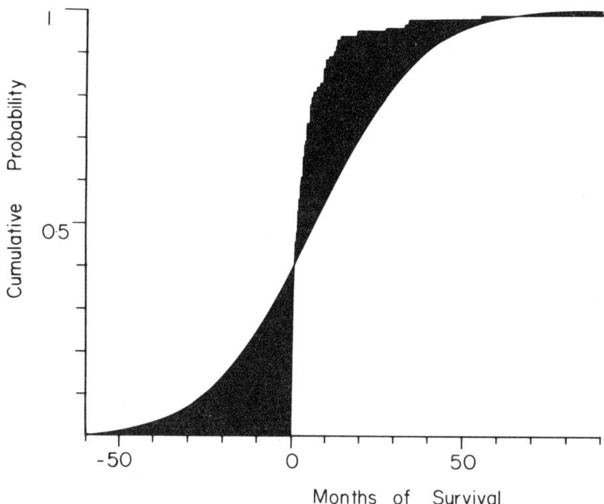



**Figure 4.2.** Cumulative proportion of the cases in Figure 4.1 with a given survival or shorter (step function) and the theoretical cumulative Gaussian distribution function with the mean and variance from the sample estimates (smooth curve). Discrepancies (shown in black) are massive. (Survivals of less than 0 from diagnosis are, of course, biologically meaningless.)

The behavior of the data under logarithmic transformation (given in the last column of Table 4.1) is shown in Table 4.4. (In the intermediate calculations it is convenient to keep all the logarithms positive by multiplying the original data by some number which will ensure that the lowest value is not less than unity. In this case a factor of one thousand was used and this converted the untransformed scale into thousandths of a month.) The sample mean of the logarithms to the base 10 proves to be 0.0780 with a standard deviation of 0.864. The mean converted back into the original scale is 1.1968 months and 48 of the values lie below it, 57 above, the best fit so far. The goodness of fit test yields the smallest chi-square of the three and is the only one of the three which could be considered adequate ($p > 0.1$). The fit is a considerable improvement (Fig. 4.3b). Taking logarithms has in fact slightly overcorrected the skewness. We could try fitting a third parameter $a$ and use the transformation

$$y = \log(x + a)$$

There is no simple method by which $a$ can be found.* The statistician would have to have recourse to trial and error until the distribution is most nearly Gaussian. However, such a procedure is probably unnecessary for all ordinary purposes, even to deal with single values (see Chapter 17).

---

*It will be recalled from p. 67 that, because one of the conditions of MLE is violated, it cannot be employed to find $a$.

**Table 4.2. Histogram of the observed duration of survival by deciles under the hypothesis of Gaussian normality in the data from Table 4.1**

| Group (months) | Observed | Expected | $\chi^2$ |
|---|---|---|---|
| $< -25.50$ | 0 | 10.5 | 10.500 |
| $-25.50$ to $-14.47$ | 0 | 10.5 | 10.500 |
| $-14.47$ to $-6.54$ | 0 | 10.5 | 10.500 |
| $-6.54$ to $+0.27$ | 25 | 10.5 | 20.024 |
| 0.27 to 6.62 | 60 | 10.5 | 233.357 |
| 6.62 to 12.97 | 11 | 10.5 | 0.024 |
| 12.97 to 19.76 | 4 | 10.5 | 4.024 |
| 19.76 to 27.71 | 1 | 10.5 | 8.595 |
| 27.71 to 38.74 | 2 | 10.5 | 6.881 |
| $> 38.74$ | 2 | 10.5 | 6.881 |
| Total | 105 | 105 | 311.286 |
| | | | d.f. $= 7$ |
| | | | $p \ll 10^{-6}$* |

We have dealt at some length with one particular problem in which there is no theoretical basis for the transformation, and justification is empirical. It will perhaps be illuminating to consider another example where there is some kind of background which will make the transformation seem reasonable. In this particular case, it may be noted that the transformation was originally arrived at empirically; only afterwards did the possible significance of the effectiveness of the transformation become evident.

**Table 4.3. Histogram of the square roots of the observed duration of survival from diagnosis in malignant hypertension by deciles under the hypothesis that this transform is Gaussian**

| Survival (months) | Observed | Expected | $\chi^2$ |
|---|---|---|---|
| $<0.13$ | 11 | 10.5 | 0.024 |
| 0.13–0.52 | 23 | 10.5 | 14.881 |
| 0.52–1.18 | 14 | 10.5 | 1.167 |
| 1.18–2.15 | 11 | 10.5 | 0.024 |
| 2.15–3.45 | 10 | 10.5 | 0.024 |
| 3.45–5.26 | 13 | 10.5 | 0.595 |
| 5.26–7.72 | 4 | 10.5 | 4.024 |
| 7.72–11.33 | 8 | 10.5 | 0.595 |
| 11.33–17.64 | 5 | 10.5 | 2.881 |
| $>17.64$ | 6 | 10.5 | 1.929 |
| Total | 105 | 105 | 26.143 |
| | | | d.f. $= 7$ |
| | | | $p < 0.001$ |

SOURCE: Data of Kincaid-Smith et al. (21).

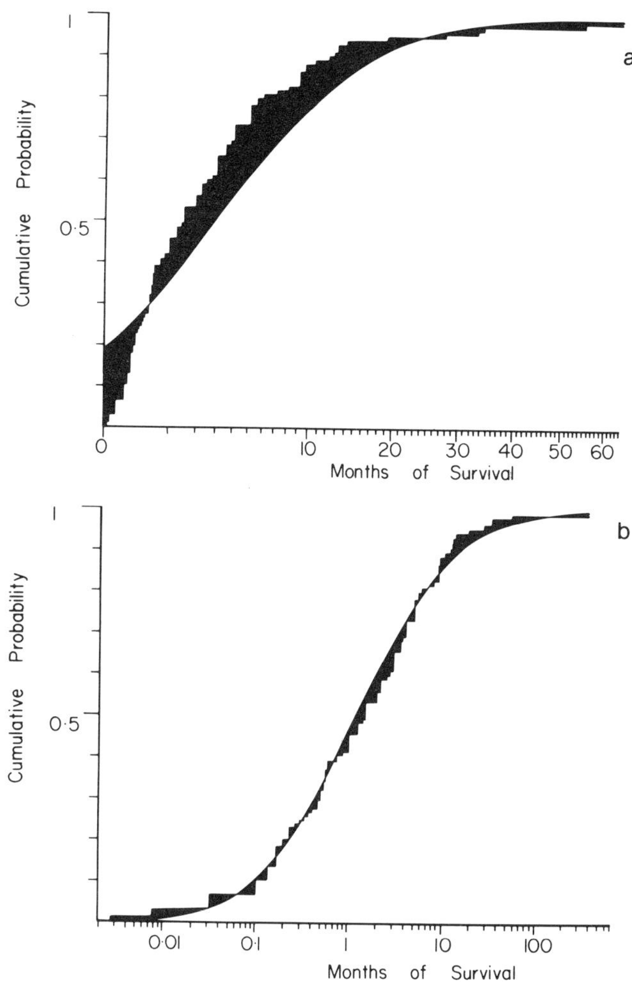

**Figure 4.3.** The same data as in Figure 4.2 plotted with transformed time scales. In each case the empirical (step function) and Gaussian (smooth curve) cumulative functions are shown. (*a*) The square root transform. The fit is closer than in Figure 4.2. However note that the theoretical curve must be truncated at zero because negative numbers have no real square roots. (*b*) Logarithmic time scale, which gives an excellent fit.

**Example 4.2.** In the course of a series of studies on thrombus formation in flowing blood (23) a device was developed for quantitating the rate at which thrombus forms under standardized conditions. The apparatus is connected to the common carotid artery of a pig at one end, and its contralateral internal jugular vein at the other. Flow is allowed to continue for 20 minutes and then stopped, the flow chamber washed, and the amount of thrombus

**Table 4.4. Histogram of the logarithms to the base 10 of the observed duration of survival from diagnosis in malignant hypertension under the null hypothesis that this transform is Gaussian**

| Survival (months) | Observed | Expected | $\chi^2$ |
|---|---|---|---|
| $< -1.029$ | 7 | 10.5 | 1.1667 |
| $-1.029 - -0.649$ | 14 | 10.5 | 1.1667 |
| $-0.649 - -0.375$ | 8 | 10.5 | 0.5952 |
| $-0.375 - -0.141$ | 12 | 10.5 | 0.2143 |
| $-0.141 - +0.078$ | 7 | 10.5 | 1.1667 |
| $+0.078 - +0.297$ | 8 | 10.5 | 0.5952 |
| $+0.297 - +0.531$ | 13 | 10.5 | 0.5952 |
| $+0.531 - +0.805$ | 16 | 10.5 | 2.8810 |
| $+0.805 - +1.185$ | 14 | 10.5 | 1.1667 |
| $> 1.185$ | 6 | 10.5 | 1.9286 |
| Total | 105 | 105 | 11.476 |
| | | | d.f. $= 7$ |

SOURCE: Data of Kincaid-Smith et al. (21).

formed is collected, dried, and weighed. A histogram of the results of 150 such experiments is given in Figure 4.4. It is obviously grossly skewed (Fig. 4.4a) and again there has been difficulty in accurately drawing this huge range of scale. Taking logarithms renders the distribution fairly symmetrical (Fig. 4.4b). The normalizing effect of taking logarithms in this context has subsequently been confirmed by Begent and Born (24).

To interpret these findings biologically, we recall that the formation of a thrombus involves many steps: the clumping and fusion of platelets with the release of adenosine diphosphate, further platelet change, and then liberation of thromboplastin, which leads to thrombin formation and eventually fibrin. Fibrin itself has no effect on platelets, but thrombin leads to further platelet clumping, and the disturbance of blood flow due to the presence of the thrombus may also lead to further platelet clumping. Thus thrombosis tends to be self-perpetuating and the augmenting effect tends to increase, the larger the thrombus. Rate of growth of a thrombus, like that of a crystal, will tend to be proportionate, and therefore linearized by taking logarithms. It can be shown that in general, granted a random factor in growth, the distribution should tend to follow a lognormal distribution, that is, a distribution made normal by taking logarithms (7). We have encountered it in Chapter 3.

One final point about normalizing transformations is of some importance. Under a wide range of circumstances, a transformation will have little effect on the shape of a distribution if the standard deviation is small relative to the mean (see 25 for proof). This may be illustrated by a common problem.

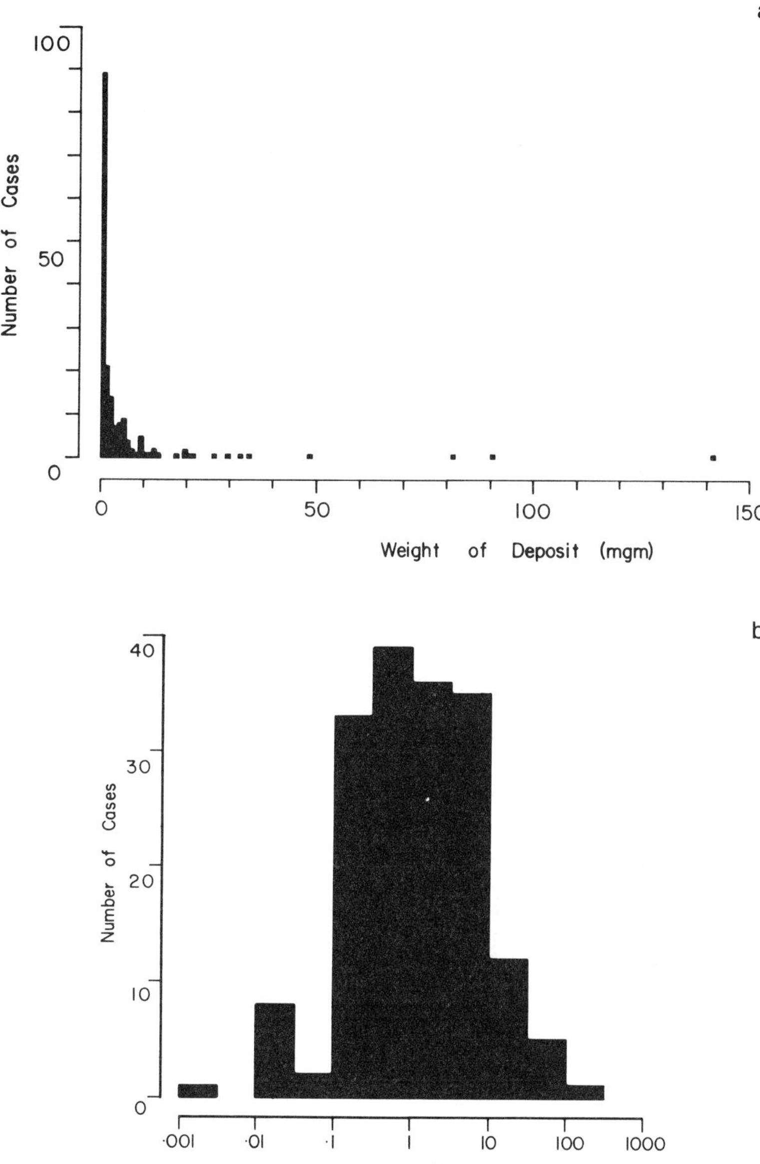


**Figure 4.4.** (*a*) Histogram of the mass of solid deposit formed in a standard flow chamber in twenty minutes in 150 experiments in pigs. (*b*) The same data plotted on a logarithmic scale. The more compact distribution makes representation much tidier and the scale different. This fact must not distract from the main issue which is the *shape* of the histogram as compared with that in (*a*).

**Example 4.3.** In assessing whether or not a patient is overweight, some observers have had recourse to the "ponderal index," $P$, where

$$P = \frac{\sqrt[3]{\text{weight}}}{\text{height}}$$

arguing, reasonably, that weight is a function of volume, which (other things being equal) would be a cubic function of any fixed body dimension such as height; on this conjecture there should be a fixed proportion between the height and the cube root of the weight.* How much difference does it make to the discriminating value of the test whether the height or the cube of the height is used? We take for our illustration the data of the National Center for Health Statistics (28) on the heights of men aged twenty-five to thirty-four. The distribution of the values is given in the upper part of the Figure 4.5; in the bottom half is the distribution of the cubes of the heights.

**Note.** An important refinement is in place here. The number of values in a particular class is represented by the *area* of the histogram plotted opposite the corresponding class interval on the baseline. As the heights get larger in the present example, the class intervals on the transformed scale become larger. Hence, if the area of the cell is to represent the number of cases, it is necessary to scale down the height of the corresponding cell by dividing the number of subjects by the size of the segment. Thus the interval $60''$-$61''$, on the transformed scale (the cubes of the heights) becomes 216,000 to 226,981, an interval of 10,981. On the other hand the interval (on the original scale, the equal interval) $74''$ to $75''$ becomes 405,224 to 421, 875, an interval of 16,651. Thus the height of the cell in the latter class would be scaled down by the factor 16,651/10,981 relative to the former.†

With all these adjustments, however, it will be evident that the general configuration of the distribution curve is little affected by the transformation, reflecting the fact that the standard deviation of the original distribution (2.67 inches) divided by the mean (68.56 inches) is small.


## STABILIZATION OF THE VARIANCE

In certain procedures—least squares estimation, or comparison of group means such as in analysis of variance—it is desirable that the variance should be known, or at least stable, and often neither is the case. For exam-

---

*The fact that there is no such proportionality (26, 27) and that the use of such indices is not justified is not our immediate concern here.

†Note that this adjustment was not required in Figure 4.4b because in that case the individual readings were sorted into classes *after* the classes had been set up with equal intervals on the transformed scale. Two values in the same cell of Figure 4.4b may have been in difficult cells in Figure 4.4a. But here the class intervals are not equal and (for lack of more detailed information) each class of heights is being transformed as a whole.

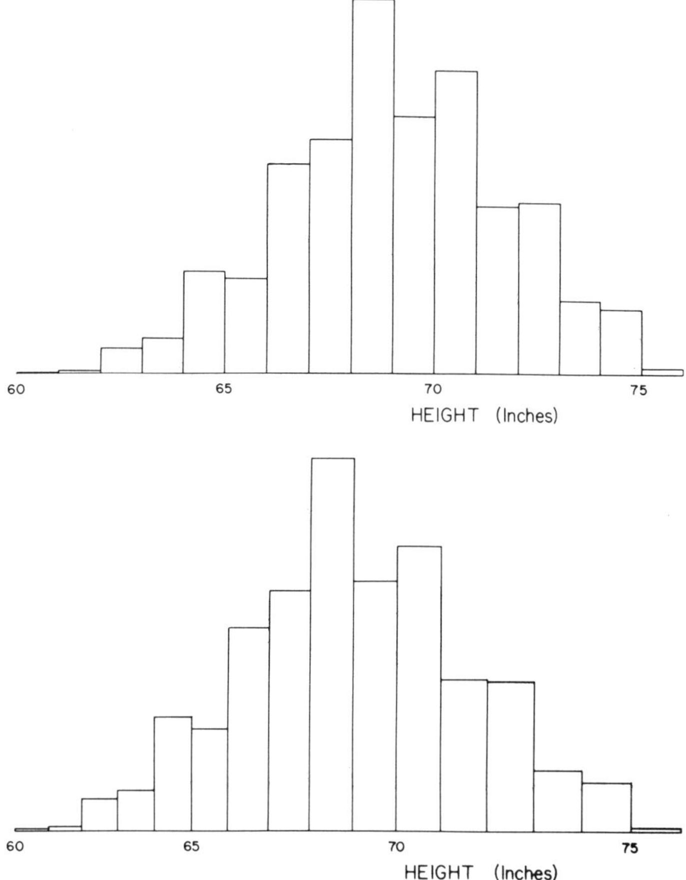


**Figure 4.5.** Histogram of heights in men (28): *top*, on a linear scale; *bottom,* on a cubic scale.

ple, in the Poisson distribution (see Chapter 14) the variance is equal to the mean: if the latter is known there is no problem—but of course such a state would be exceptional. In many cases precisely the point at issue is whether or not the means in two or more samples are different. It is then useful to be able to manipulate the values in such a way that the variance is known or, at least, that it would no longer depend on the mean.

Transformations that convert the variance into a constant value, known, or at least independent of the mean, are termed "variance-stabilizing." There are two possible approaches to the problem of finding how the variance is related to the mean.

It may be known from theory precisely what the relationship should be.

For example, for the Poisson the variance of the random variable equals its expectation or mean:

$$\mathrm{Var}(X) = E(X)$$

Or for the binomial, if $X$ is the number of successes:

$$\mathrm{Var}(X) = np(1 - p) = nE(X)[1 - E(X)]$$

Alternatively when sample values are plotted on graph paper it may turn out, for example, that the standard deviation is more or less proportional to the mean. In either case there are standard ways of stabilizing the variance and obtaining an estimate of the variance on the transformed scale. The rationale of these operations cannot be discussed without using the calculus, and for particular cases the reader is referred to more technical books (e.g., 29). Fortunately, in practice there are standard methods of coping with particular relationships (Table 4.5).

**The relationship between the variance and the mean is known from theory**

A variance proportional to the mean is encountered in two common situations: the Poisson, where it is equal to the mean; and the chi-square distribution (for which the variance is twice the mean) and, more generally, the gamma distribution. (See Chapters 4 and 5, especially Tables 4.4 and 5.4, in [7].)

**Example 4.4. The Poisson distribution.** A convenient example of this process is the rate of emission by a radionuclide with a long half-life. The

**Table 4.5. Some standard transformations used to stabilize the variance where it is a function of some unknown parameter**

| Variance (as a function of some unknown parameters $\mu$, $p$ or $\rho$) | Example of distribution | Transformation | The approximate variance of the transform |
|---|---|---|---|
| $a\mu$ | Poisson | $Y = \sqrt{X} + \sqrt{X + 1}$ | $a^2$ |
| | Chi-Square | $Y = \sqrt{X}$ | $\dfrac{a^2}{4}$ |
| $a\mu^2$ | Lognormal | $Y = \log_e X$ | $a$ |
| $\dfrac{ap(1 - p)}{n}$ | Binomial | $Y = \sin^{-1}\sqrt{\dfrac{X}{n}}$ | $\dfrac{a}{4n}$ |
| $\dfrac{(1 - \rho^2)^2{}^*}{n - 1}$ | Gaussian correlation coefficient | $Y = (\tfrac{1}{2})\log_e\left(\dfrac{1 + r}{1 - r}\right)$ | $\dfrac{1}{n - 3}$ |

NOTE: $a$ is any constant; $\mu$ is the mean; $p$ is the binomial parameter; $\rho$ is the correlation coefficient.

*This is an asymptotic variance.

procedure may be illustrated by reference to some data specially collected for the purpose from a radioactive source consisting of a mass of cesium-137 in a well counter. By suitable variation of the baseline, window, and duration of counting, or background, it is possible to get a wide range of mean counts, each following its own Poisson distribution. At each setting, twenty replicate counts were taken, and the results are shown in Table 4.6. For simplicity we first use the square root transformation, and the appropriate analysis is shown in Table 4.6. For the untransformed data, in each case the sample variance is approximately equal to the sample mean: discrepancies as large as those shown here, or larger, can readily be explained by chance (see Chapter 14). For the transformed values, on the other hand, the variances are all in the neighborhood of $\frac{1}{4}$. It can be shown that for large values of the Poisson parameter (i.e., the true mean) the variance of the transform is close to $\frac{1}{4}$ and for an infinite mean exactly equal to it. There can be no doubt that the transformed data have vastly more stable variances.

### A comment on detail in calculation

Note that in doing the calculations there is a tendency to think that the number of decimal places carried in the calculations should be proportional to the size of the value; for example, that if the value is 2 and the transform used 1.414, for a value of 2,000,000 we should not attempt to carry the transform any further than 1,414. It becomes evident when one actually does the calculations, however, that this viewpoint is incorrect. The variance of the transform is, naturally, estimated by

$$E(\sqrt{X})^2 - \{E(\sqrt{X})\}^2 = E(X) - \{E(\sqrt{X})\}^2$$

The precision of this estimate depends on the precision with which computing the *difference* between the two terms is carried out. When it is recalled that whatever $E(\sqrt{X})$, the sample variance is estimating much the same quantity, then it is clear that in the transformation, the same number of decimal places, not the same number of significant figures, should be retained.

### A formal approach

The reader may feel that this example is only one instance, and wish to know something about performance in general. It is straightforward (though tedious) to express the variance of the transformed values as a function of the Poisson parameter. Calculation shows (Fig. 4.6a) that for a mean of about 8 or greater the variance of $\sqrt{X}$ is within ten per cent or less of 0.25, the asymptotic value, which might be thought acceptable. The maximum value for the variance of the transform is about 0.4125 (when the mean is a little over 1.3).

Various refinements of the transform have been proposed that will speed up the approach of the variance to $\frac{1}{4}$, and are thus more appropriate for smaller values of the Poisson parameter. The transformation $\sqrt{(X + \frac{1}{2})}$

**Table 4.6. Variance-stabilizing transforms of radioactive data**

| Data | | 2 | 6 | 15 | 20 | 59 | 434 |
|---|---|---|---|---|---|---|---|
| | | 0 | 6 | 11 | 24 | 47 | 441 |
| | | 1 | 8 | 12 | 23 | 52 | 384 |
| | | 0 | 6 | 10 | 28 | 58 | 412 |
| | | 2 | 8 | 11 | 14 | 67 | 415 |
| | | 0 | 4 | 12 | 27 | 68 | 427 |
| | | 1 | 8 | 11 | 21 | 75 | 395 |
| | | 0 | 6 | 12 | 17 | 66 | 425 |
| | | 0 | 4 | 11 | 29 | 67 | 414 |
| | | 0 | 4 | 16 | 21 | 52 | 381 |
| | | 1 | 9 | 15 | 21 | 54 | 455 |
| | | 1 | 7 | 6 | 26 | 61 | 435 |
| | | 1 | 9 | 10 | 20 | 49 | 425 |
| | | 0 | 4 | 19 | 27 | 66 | 391 |
| | | 1 | 5 | 13 | 17 | 54 | 429 |
| | | 1 | 9 | 12 | 21 | 66 | 419 |
| | | 1 | 11 | 16 | 18 | 61 | 417 |
| | | 0 | 6 | 13 | 22 | 65 | 462 |
| | | 1 | 9 | 19 | 24 | 72 | 426 |
| | | 0 | 8 | 11 | 23 | 57 | 407 |
| Analysis | $\hat{\mu}$ | .65 | 6.85 | 12.75 | 22.15 | 60.80 | 419.70 |
| | $s^2$ | .45 | 4.24 | 9.88 | 15.92 | 60.91 | 455.38 |
| Transform | | | | *Sample variance* | | | |
| $\sqrt{X}$ | | .3160 | .1604 | .1969 | .1851 | .2541 | .2720 |
| $\sqrt{(X + \frac{3}{8})}$ | | .1142 | .1514 | .1908 | .1819 | .2525 | .2718 |
| $\dfrac{\sqrt{X} + \sqrt{(x + 1)}}{2}$ | | .1675 | .1492 | .1891 | .1809 | .2519 | .2717 |

suggested by Bartlett on the idea that the data are discrete and call for a continuity correction* gives a considerably more rapid approach (Figure 4.6b, lower curve) and might be trusted for a value of the parameter greater than about 2.5. Anscombe suggested $\sqrt{(X + 0.375)}$ which approaches still more rapidly, though, like Bartlett's transform, the variance is always less than 0.25. If one is prepared to accept a little "overshoot," $\sqrt{(X + 0.2)}$ attains a plateau even faster (for values of the parameter greater than about 1.5), though the subsequent approach to the limit is somewhat slower (Fig. 4.6b). It might be argued that one is much more concerned with minimizing absolute deviations from the asymptotic value than with speed of convergence. Note that for any value $c$, the difference between $\sqrt{x}$ and $\sqrt{(x + c)}$ becomes smaller as $x$ increases. For Poisson distributions with large means, these refinements may still be retained, but they would then make little difference to the answer.

*Continuity corrections will be discussed in Chapter 12.

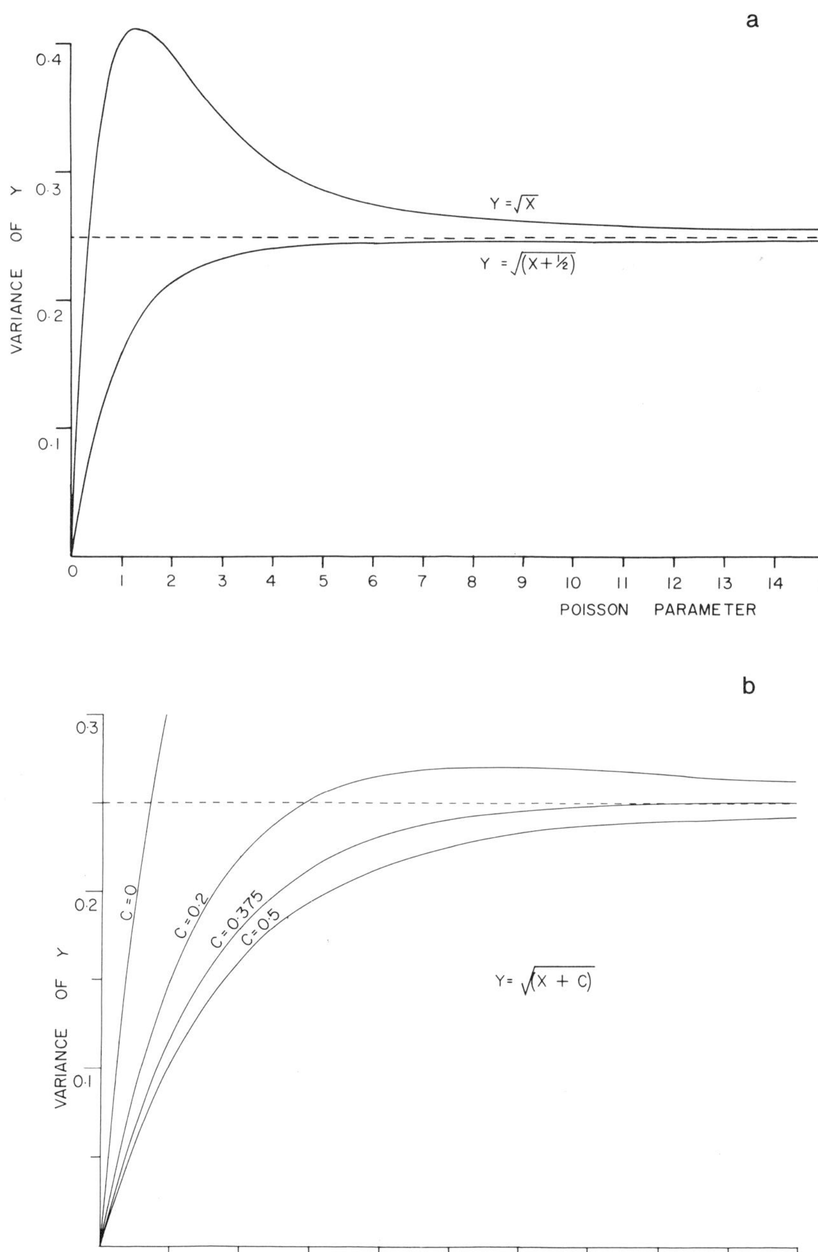



**Figure 4.6.** (*a*) Comparison of the variances of the simple square root transform, and Bartlett's transform, of Poisson variates. (*b*) Comparison of the variances of several simple transforms of the Bartlett type for Poisson variates.

89

A more satisfactory transformation still is that due to Freeman and Tukey (30) who suggest

$$Y = \sqrt{X} + \sqrt{X + 1}$$

The transform, $Y$, has three convenient properties:

1. The variance lies within 10 percent of the asymptotic value for all values of the parameter greater than about 0.9.

2. The asymptotic variance is unity.

3. The transform can be conveniently found by adding adjacent values obtained from an ordinary table of square roots. Since the asymptotic variance of the Freeman-Tukey transform is 1, four times as great as for the square root transform (0.25), for purposes of comparison the square root transform will be doubled, increasing its variance fourfold (Fig. 4.7). The asymptotic variance, unity, is indicated by the arrow. The main disadvantage of the Freeman-Tukey transform is that inverting it is somewhat messy.

There are, of course, aspects to these transforms other than stability of the variance. The Poisson is discrete, so that it never can be made Gaussian. Nevertheless, as the parameter, $\lambda$, becomes larger, the distribution may be more and more closely approximated by the Gaussian. However, the distribution is always positively skewed and it exhibits an excess of kurtosis. As we noted earlier in the chapter, taking the square root tends to reduce positive skewness, and we might wonder whether these various transforms

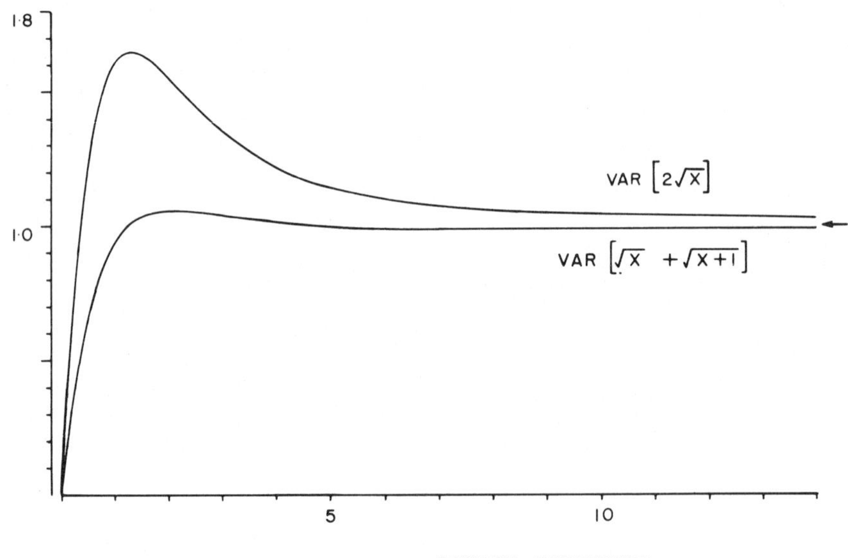


**Figure 4.7.** Comparison of the variances of the simple square root transform (scaled by a factor of 2) and the Freeman-Tukey transform.

make the Poisson variable closer to the Gaussian. In Table 4.7 are shown results for the simple transformation, Anscombe's transformation, and the Freeman-Tukey transformation. The quantities, shown as a function of $\lambda$, are the second, and the standardized third and fourth cumulants that were discussed in Chapter 1.

For the untransformed values the variance is $\lambda$, the third standardized cumulant ($\gamma_1$) is $1/\sqrt{\lambda}$ and the fourth ($\gamma_2$) is $1/\lambda$.

If we take the absolute values of the results, it appears that, by and large, both skewness and kurtosis approach values for the Gaussian curve (zero) more rapidly than for the untransformed variable. The Anscombe transformation is designed to minimize skewness, so that it is not surprising that the $\gamma_1$ for this transform should go to zero the most rapidly of the three transforms listed. A fair generalization would be that the pursuit of a stable variance has had no very undesirable effects and has even produced some benefit to skewness and kurtosis.

**Comment.** Those who have massive data are apt to regard the foregoing details of only theoretical importance: that any blemishes in the higher cumulants are rapidly abolished by averaging in accordance with the central limit theorem (see Chapter 17). This, though true, is of little comfort to those who have a single observation only.

When these transforms are applied to the radioactive decay experiments (Table 4.6) the following sample variances result for the group with the lowest mean:

| Transformation | Sample Variance |
|---|---|
| $\sqrt{X}$ | 0.3160 |
| $\sqrt{X + \frac{1}{2}}$ | 0.0994 |
| $\sqrt{X + \frac{3}{8}}$ | 0.1142 |
| $[\sqrt{X} + \sqrt{X + 1}\,]/2$ | 0.1675 |

The results come out very much as predicated from the curves (figures 4.6 and 4.7), given that the parameter is equal to the sample mean.

**Example 4.5. The binomial.** The proportion of successes in a binomial distribution of order $n$ with a probability of success at each trial of $p$ has the variance

$$\frac{p(1 - p)}{n}$$

Theory gives the following transformation, which stabilizes the variance approximately (how well, we shall see below):

$$Y = \sin^{-1}\sqrt{X/n}$$

where $X$ is the number of successes and $n$ the number of trials. The symbol "$\sin^{-1}$," also written as "arscin," means "find the angle of which this number is the sine." Obviously $X$ must lie between 0 and 1. For example,

**Table 4.7. The variance and standardized third and fourth cumulants of three transforms of Poisson variables**

| Parameter | Variance ($\sigma_x^2$) | | | $\gamma_1 = k_3/\sigma^3$ | | | | $\gamma_2 = k_4/\sigma^4$ | | | |
| | C | A | F-T | Untrans-formed | C | A | F-T | Untrans-formed | C | A | F-T |
|---|---|---|---|---|---|---|---|---|---|---|---|
| 0.1 | 0.382 | 0.188 | 0.184 | 3.162 | 2.817 | 2.917 | 2.853 | 10.000 | 6.138 | 7.150 | 6.493 |
| 0.3 | 0.896 | 0.314 | 0.469 | 1.826 | 1.225 | 1.410 | 1.296 | 3.333 | −0.231 | 0.726 | 0.125 |
| 0.5 | 1.240 | 0.467 | 0.670 | 1.414 | 0.636 | 0.891 | 0.740 | 2.000 | −1.240 | −0.368 | −0.897 |
| 1.0 | 1.609 | 0.717 | 0.940 | 1.000 | −0.088 | 0.312 | 0.097 | 1.000 | −1.335 | −0.801 | −1.101 |
| 1.5 | 1.641 | 0.852 | 1.036 | 0.816 | −0.461 | 0.044 | −0.201 | 0.667 | −0.704 | −0.638 | −0.685 |
| 2 | 1.560 | 0.924 | 1.058 | 0.707 | −0.664 | −0.103 | −0.348 | 0.500 | 0.032 | −0.399 | −0.226 |
| 5 | 1.144 | 1.002 | 1.000 | 0.477 | −0.486 | −0.242 | −0.299 | 0.200 | 1.075 | 0.208 | 0.448 |
| 10 | 1.045 | 1.001 | 0.990 | 0.316 | −0.197 | −0.168 | −0.164 | 0.100 | 0.191 | 0.122 | 0.125 |
| 50 | 1.008 | 1.000 | 0.998 | 0.141 | −0.073 | −0.071 | −0.071 | 0.020 | 0.022 | 0.021 | 0.020 |
| 100 | 1.000 | 1.000 | 0.999 | 0.100 | −0.051 | 0.063 | −0.050 | 0.010 | 0.011 | 0.011 | 0.010 |
| $\infty$ | 1.000 | 1.000 | 1.000 | 0.000 | 0.000 | 0.000 | 0.000 | 0.000 | 0.000 | 0.000 | 0.000 |

| C: | The Classical transformation | $Y = 2\sqrt{X}$ |
| A: | Anscomb's transformation | $Y = 2\sqrt{X + 0.375}$ |
| F-T: | Freeman-Tukey transformation | $Y = \sqrt{X} + \sqrt{X + 1}$ |

suppose that $X/n$ is 0.5 of which the square root is 0.7071. Then referring to a table of common sines, we find that

$$\sin(45°) = 0.7071$$

whence the inverse relationship

$$\sin^{-1}(0.7071) = 45°$$

We then take 45 as the transform of the original data on which the statistical operations are to be performed. Assessing the usefulness of this transformation is not so easy as for the Poisson distribution. Its efficacy depends on the size of the sample.

In Figure 4.8 is shown, as a function of the binomial parameter, the variance for a sample of size 10 when the outcomes are untransformed (dotted line), and transformed (continuous line). Both curves are symmetrical about $p = 0.5$; so, in general, it will be unnecessary to present more than half the curve. In Figure 4.9 are shown values for $n = 5$, 10, and 40 represented as continuous lines and the relative variance for the untransformed data as an interrupted line. It must be noted that as a criterion

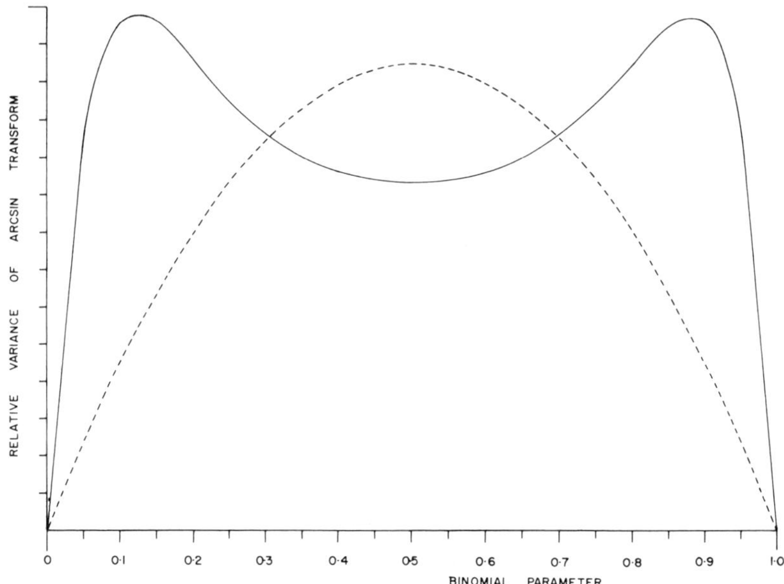


**Figure 4.8.** The variance of the arcsine–square root transform of a binomial variate of order 10 as a function of the parameter (*continuous curve*). For comparison the variance of the untransformed variate is shown (*interrupted line*). (The units on the vertical scale are 0.02; but this fact is irrelevant, as it is in Figure 4.9 also, to the *stability* of the variance.)

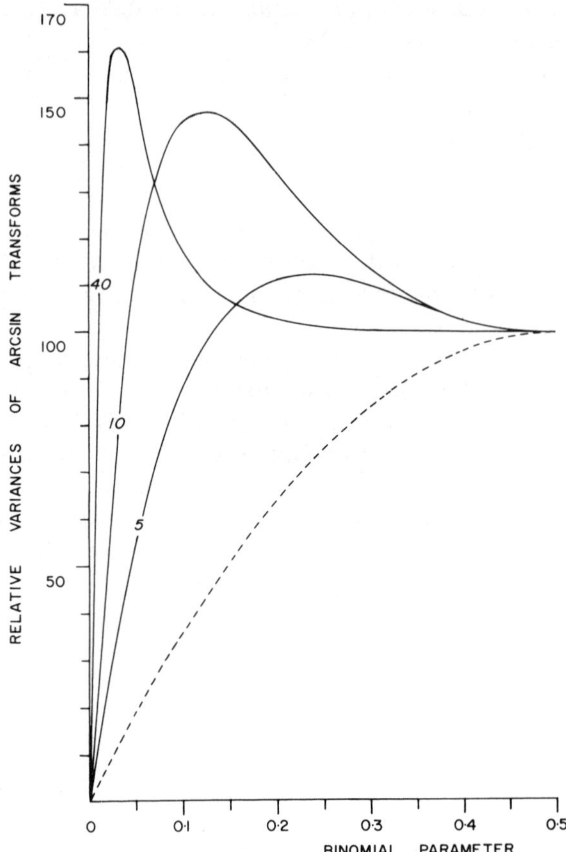



**Figure 4.9.** The variances of the arcsine–square root transform of binomial variates of orders 5, 10, and 40, as functions of the parameter. To make the comparison easy, variances are scaled so that at $p = 0.5$ all measure 100 units. For comparison, the variance of the untransformed variate (the shape of which is the same for all sizes of sample) is also shown (*interrupted line*).

of variance-stabilization it is the *relative* variance for different values of $p$ that matters. Accordingly they are each scaled so that the variance at $p = 0.5$ is taken as 100. The stability of the variance is not impressive, though perhaps it is adequate for larger values of $n$ for $p$ between 0.1 and 0.9. Calculating the variance is tedious and can probably not be done exactly by any analytic procedure. These findings seem to suggest that even for fairly large sample sizes, the transformation should be used with caution.

**The relationship between the mean and the variance is known empirically**

Commonly in practice there is no reason a priori for supposing that a random variable follows any particular distribution or what (if any) relationship there should be between mean and variance. The course of action is then largely dictated by the data. But there may be grounds intermediate between formal certainty (as in the Poisson) and empirical conviction that may suggest a course of action.

We may illustrate this intermediate state first, using as an example the class of random variables in which not the variance, but the standard deviation, is proportional to the mean. This property arises in a variety of circumstances, which we shall consider severally.

**1. Methodological.** In many chemical analyses, there is a limited range over which an apparatus is known to give accurate readings. Below this range the measurement may be obscured by "background noise"; above this range the system of measurement may be nonlinear.

**Example 4.6.** Suppose that the usable range is 20–100 mg/100 ml and that there is an experimental error with a standard deviation of 2 mg/100 ml. In three specimens the concentrations are 90, 180, and 270 respectively. The last two are outside the range of accuracy of the apparatus and it will be necessary to dilute them, say twofold and threefold respectively. The second value is now reduced to a concentration of 90 and will be read with a standard deviation of 2 mg. But when the correction is made for dilution by multiplying by two, the standard deviation must also be multiplied by two. Likewise the error of the third specimen is multiplied by three. Hence we get the results

| Mean | Standard Deviation |
|------|--------------------|
| 90 | 2 |
| 180 | 4 |
| 270 | 6 |

This relationship is due entirely to experimental method. It is a common type.

**2. Physiological.** It has long been known among physiologists that the impact of an increase in a stimulus depends on the present state of stimulation of a nerve. An increase of 40 candle power that would be painful to a person previously in darkness, would be imperceptible to one adapted to bright sunlight. One expects, with increasing intensity of stimulus, that a wider range of values will be indistinguishable, and hence an increasing error of measurement to occur. The claim that the response is proportional to the logarithm of the intensity of the stimulus (Fechner's law) is probably an oversimplification (15); but to have survived so long, the statement must be at least approximately true. Taking logarithms of the intensities of stimulation then should tend to make the ranges of "indistinguishable" values independent of the mean intensity. Reference to Table 4.5 shows that this transformation is appropriate where the standard deviation is a fixed multiple of the mean.

**3. Notional.** The next point is rather subtle and (like many subtle points) best explained by a joke. An astronomer during a lecture paused and asked the audience how old they thought the world is. A voice from the back said "About four billion and four years." The astronomer asked him how he arrived at such a figure and the student replied "Well, I heard you lecture on this subject four years ago and you said it was about four billion years old then." If the lecturer had been talking about the length of time for which Polish jokes had been in existence, there would have been no joke at all. Reflection suggests that the reason is that while the time of neither the exact origin of the earth nor of Polish jokes may be known, there is a big difference in the ranges of our uncertainty. Four more years would be swallowed up in the uncertainty of the one but not of the other. This illustration seems to suggest that scientists think of error as a percentage of the measurement. The eighteenth-century biblical scholar who, solely by analyzing the Old Testament, concluded that the world was created in 4004 B.C., giving the very month and day, at about half past nine in the morning (Greenwich mean time?), now seems ludicrous.

**4. Conceptual.** This point is more subtle still, but even more fundamental. It may explain the problem discussed at length by Raimi (31, 32) why, of measurements of all classes, the leading digit is more commonly 1 than 9. The way in which we group phenomena into classes and call them by a specific name is logarithmic or something like it. We distinguish between a "pony" and a "horse" or between a "cottage" and a "house," although the absolute differences in size are relatively small. We would group Venus, Mercury, and the Earth together as "medium-sized planets," although the absolute differences are measured in thousands of miles and trillions of tons. Conversely, we distinguish "shrimps" from "jumbo shrimps" (*Anglice* "prawns"), though the difference is a matter of a few grams. We called $H_2O$ a small molecule and collagen a "giant molecule," though both are so small as to tax physical methods of observation to the utmost.

The same phenomenon occurs on the time scale. By an "acute" illness we mean one lasting days; a "subacute" illness lasts months, a "chronic" illness, years. The ranges are obviously different, and the variances of duration must be different. But it is not evident a priori why diseases are grouped in this way, though I do not imply that the grouping is capricious or without value. Or again, a hematologist will make an important distinction between a bleeding time of 3 minutes and one of 20 minutes, but very little between 200 minutes and a thousand. Indeed most would not bother to time the process beyond 60 minutes and many give up interest much earlier. Again I do not imply that this attitude is unreasonable; my only point is that investigators do in fact impose their own concepts on their measurements, and we must expect the variance to reflect this fact.

**5. Interpretive.** The concepts discussed in the last section may be a blemish. On the other hand they may be important correctives for measurements inappropriately chosen in the first place. Thus the common practice of grading the degree of albuminuria as "+," "++," and so forth, meaning perhaps a bare trace, an easily detected trace, a heavy precipitate, and a solid coagulum, may be deplored by the biochemist on the grounds that the groupings have different ranges as judged by quantitation. Nevertheless, the clinician may not be interested in the amount of albumin but what its significance may be. The difference between "none" and "a trace" is minute chemically but vast prognostically; and conversely for the difference between daily outputs of 20 G and 25 G. Here, then, we might find a well-balanced grouping on the scale of "significance" which would give rise to serious statistical problems when exact quantitation is employed. I am sure, for example, that in physical dimensions the paintings of Delacroix show greater variance than those of Rembrandt; but this is no reason for believing that there is a greater variance of artistic merit. It is simply that Delacroix's paintings are on a larger scale, providing a greater opportunity for variation, and that physical dimensions are an incongruous measure of artistic merit.

**6. Ontogenic.** An understanding of mechanisms involved in the production of a characteristic may suggest that the variance should be relatable to the mean in some fashion. In considering the effects of a drug, digoxin for instance, we may expect all those treated to behave in the same way if dosage is very low (none exhibit any change) or very high (all are poisoned); it is for the doses in the "therapeutic range" that we may expect the greatest variation. And this is true regardless of what the distribution of individual outcomes may be—for instance whether binomial or not, whether there are genetic components in sensitivity, etc.

Or again, in discussing normalization we have seen a case (Example 4.2) in which lognormality may be expected, in the growth of a thrombus. The fact that taking logarithms converts a multiplicative, into an additive, process suggests that the variance might be stabilized by the same process, since in general

$$\text{Var}(XY) \neq \text{Var}(X) + \text{Var}(Y)$$

regardless of $E(X)$ and $E(Y)$ even if $E(X)$ is independent of var$(X)$ and $E(Y)$ of var$(Y)$; whereas if $X$ and $Y$ are independent

$$\text{Var}(\log X + \log Y) = \text{Var}(\log X) + \text{Var}(\log Y)$$

which will yield a stable variance if $E(\log X)$ is independent of $\text{Var}(\log X)$ and likewise for $Y$.

**The empirical criterion.** When sample sizes are small it may be difficult to detect small systematic differences in the variance from inspection. On the other hand it is likely that if the differences are small they hardly matter. We may apply two empirical criteria in looking for a relationship between mean and variance.

1. Are the sample variances from the different samples strikingly different? One may deal with this question by simple inspection, which is the most useful and the safest approach. There exist formal tests of homogeneity

of variance but they are highly sensitive to departures from Gaussian normality in the data. Thus if we reject the null hypothesis of homogeneous variance it is difficult in most cases to say whether it is normality or homogeneity of variance that the data violate. Furthermore if it is proposed to use a (nonlinear) transform of the data, the transform and the data cannot both be exactly Gaussian; hence the analysis will not allow us to decide which gives a more homogeneous variance and we are again driven back on inspection. I seldom do formal tests for heterogeneity of variance.

2. An easier criterion to use, a sounder and more sensitive one, is to see whether any *systematic* relationship exists between the sample means and the sample variances. If there is a significant correlation, then even if the test of heterogeneity is not significant, we may be sure that there is indeed heterogeneity. As will be pointed out in Chapter 8, regression analysis may detect heterogeneity of means where analysis of variance fails. The same point applies here. The type of test of correlation employed will depend on what is known about the data; but if need be we may use nonparametric tests (see Chapter 15).

**Example 4.7** By way of an example we shall consider the data of Table 4.8. These results come from a series of experiments on the amount of "thrombus" (i.e., solid deposit) formed in a flow chamber in 20 minutes. The method has previously been discussed (Example 4.2., Fig. 4.4) as an illustration of a grossly skewed distribution more or less normalized by logarithmic transformation. The five experiments deal with the effects of five treatments (the details of which are irrelevant to our purpose). Beneath each of the columns are shown the sample mean and standard deviation of the original values.

Two facts are evident: there are big differences among the means, and also among the standard deviations. An attempt to do conventional tests between the first and the last groups would be embarrassed by the differences in the standard deviations, the last sample value being 126.792 times the first, which makes the variance 16,076 times as great, far more that could be comfortably ascribed to chance.

But secondly there is a systematic relationship between the standard deviation and the mean: the experiments are arranged in ascending order of the means and it turns out that they are also in ascending order by standard deviations. Now if we arrange the means in ascending order there are 5! = 120 orderings for the five standard deviations and of these only two would be a perfect (monotonic) fit like this. We are tempted, then, to see whether there is proportionality. In fact, the ratio of standard deviation to mean ("the coefficient of variation") is rather similar in all five samples: remarkably so, considering the small sample sizes. As noted previously, we might have reason to hope that logarithmic transformation would stabilize the variance; and indeed it succeeds very well in this respect, as is evident from the values shown at the bottoms of the columns. The sample standard deviations are, of

**Table 4.8. Weights in mg of deposits formed in a standard flow chamber in five sets of experiments (arranged in ascending order of means)**

| | Experiment number | | | | |
| --- | --- | --- | --- | --- | --- |
| | 1 | 2 | 3 | 4 | 5 |
| | 0.01 | 0.12 | 0.10 | 0.50 | 0.30 |
| | 0.01 | 0.14 | 0.10 | 1.00 | 0.40 |
| | 0.05 | 0.14 | 0.90 | 2.52 | 0.40 |
| | 0.10 | 0.35 | 1.00 | 4.07 | 0.60 |
| | 0.20 | 0.41 | 1.40 | 5.94 | 0.80 |
| | 0.20 | 1.89 | 2.10 | 9.59 | 1.20 |
| | 0.40 | 2.46 | 6.60 | 9.63 | 1.40 |
| | 0.50 | 3.02 | 9.80 | 12.77 | 1.60 |
| | 0.90 | 3.39 | | 20.23 | 1.90 |
| | | | | | 2.50 |
| | | | | | 3.70 |
| | | | | | 5.00 |
| | | | | | 5.20 |
| | | | | | 7.80 |
| | | | | | 9.80 |
| | | | | | 19.00 |
| | | | | | 21.20 |
| | | | | | 34.00 |
| | | | | | 48.00 |
| | | | | | 82.20 |
| | | | | | 90.00 |
| | | | | | 141.50 |
| Natural data | | | | | |
| $m$ | 0.263 | 1.324 | 2.750 | 7.361 | 21.750 |
| $s$ | 0.293 | 1.360 | 3.531 | 6.411 | 37.150 |
| $\dfrac{s}{m}$ | 1.114 | 1.027 | 1.284 | 0.871 | 1.708 |
| Natural logarithms of data | | | | | |
| $m$ | $-2.160$ | $-0.459$ | 0.067 | 1.500 | 1.588 |
| $s$ | 1.628 | 1.421 | 1.691 | 1.225 | 1.898 |

course, not exactly equal—we could not expect that in small samples. But the *systematic* relationship to the mean is abolished; and the ratio of the largest to the smallest standard deviation is 1.549, which makes the ratio of the variances 2.400:1—small, considering that we have picked extreme values.

Note, incidentally, that the ranking of the coefficients of variation for the natural numbers is the same as the ranking for the standard deviations of the logarithms; in descending order experiments 5, 3, 1, 2, and 4. This will usually tend to be so, though there will be pathological exceptions.

As a fuller illustration of the relationship, in Fig. 4.10 are shown the relationships between sample deviations and sample means for 21 sets of flow-

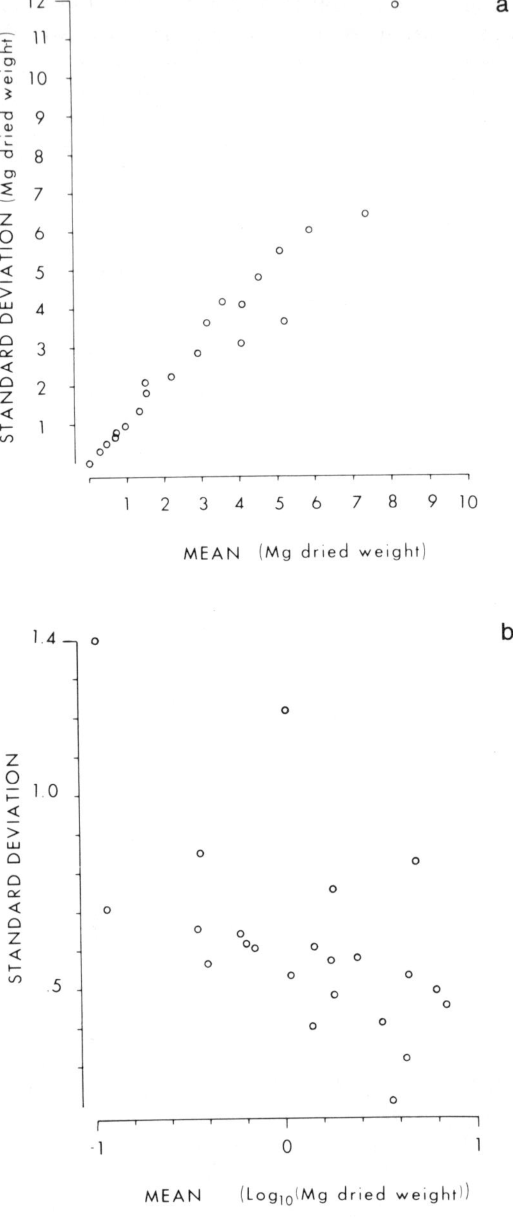



**Figure 4.10.** (*a*) The relationship between the sample means and standard deviations from 21 sets of studies on the formation of solid deposit in a flow chamber in an extracorporeal circulation. There is a clear trend with evidence of proportionality. (*b*) The relationship between the sample means and standard deviations of the logarithms of the data presented in (*a*). There is considerable scatter largely ascribable to the small sizes of the samples; but the systematic trend has been abolished.

chamber experiments in pigs. In Fig. 4.10a there is a striking trend to proportionality; one point actually lies at the origin (since heparin in full doses prevents deposition entirely, so there is a mean of zero and a variance of zero). In Fig. 4.10b are shown the sample standard deviations and means for the logarithms of the values (The point at the origin cannot be plotted at all, since the logarithms of all the values are $-\infty$.) There is no longer any systematic relationship evident. (If the variance were perfectly stabilized, the expected values would all lie on a horizontal line; but the samples are too small for us to expect the sample mean to be close to the true mean.)

**Example 4.8.** A final example is one in which (unlike the foregoing) there is no rational prior reason for believing either that the standard deviation should be heterogeneous or that it should be proportional. In Table 4.9 are shown some results from an experiment (33) on platelet survival in rabbits receiving either inert saline injections or injections of the drug sulfinpyrazone. (This drug is uricosuric, but since uric acid is not a metabolite in

**Table 4.9. Mean survival in days of blood platelets in rabbits receiving saline injections or sulfinpyrazone injections**

| Treated with saline | | Treated with sulfinpyrazone | |
|---|---|---|---|
| $x$ | $\log_e x$ | $x$ | $\log_e x$ |
| 0.988 | $-0.0121$ | 1.516 | 0.4161 |
| 0.989 | $-0.0111$ | 1.953 | 0.6694 |
| 1.075 | 0.0723 | 2.136 | 0.7589 |
| 1.116 | 0.1098 | 2.446 | 0.8945 |
| 1.126 | 0.1187 | 2.539 | 0.9318 |
| 1.132 | 0.1240 | 2.565 | 0.9420 |
| 1.255 | 0.2271 | 2.722 | 1.0014 |
| 1.438 | 0.3633 | 2.735 | 1.0061 |
| 1.473 | 0.3873 | 2.885 | 1.0595 |
| 1.534 | 0.4279 | 2.943 | 1.0794 |
| 1.639 | 0.4941 | 2.974 | 1.0899 |
| 1.681 | 0.5194 | 3.327 | 1.2021 |
| 1.874 | 0.6281 | 3.376 | 1.2167 |
| 1.890 | 0.6366 | 3.706 | 1.3100 |
| 1.925 | 0.6549 | 3.763 | 1.3252 |
| 2.089 | 0.7367 | 4.695 | 1.5465 |

*Natural data*

| | | | |
|---|---|---|---|
| $m$ | 1.4515 | 2.8926 | |
| $s$ | 0.3675 | 0.7718 | |
| $\dfrac{s}{m}$ | 0.2532 | 0.2668 | |

*Natural logarithms of data*

| | | | |
|---|---|---|---|
| $m$ | 0.3423 | 1.0281 | |
| $s$ | 0.2548 | 0.2734 | |
| $e^m$ | 1.408 | 2.796 | |

the rabbit, its action on platelets must be otherwise explained. The action was discovered fortuitously in studies on platelet survival in men with gout.)

There are thirty-two animals involved, assigned randomly to the two groups. The results are arranged in ascending order. In accordance with our custom, results are given to three decimal places.* (They have been estimated on the exponential model by unweighted least squares, which is appropriate for the rabbit [34].)

Although we believe, on massive evidence, that the true *mean* platelet survival follows closely a Gaussian distribution, so that there is nothing to be gained on this account from transformation, we have again an embarrassing difference in standard deviations: the $F$ test (see p. 119) shows that this is more than can be comfortably attributed to chance ($F = 4.410$ $p < 0.005$). But again the coefficients of variation are almost the same. Again the logarithmic transformation produces a more stable variance. Again the larger coefficient of variation is reflected in the larger standard deviation for the transformed values. When the mean of the logarithms is decoded by taking the antilogarithms, the resulting geometric mean is slightly less than the mean of the untransformed values.


## PROBLEMS

**4.1** Prove that the antilogarithms of the mean of the logarithms of any two positive numbers $x$ and $y$ is the square root of their product. (This is known as their geometric mean.) Can you generalize this statement?

**4.2.** Prove that the geometric mean of two unequal positive numbers is always less than their arithmetic (ordinary) mean. Could you guess at a generalization useful in statistics? (It is somewhat difficult to prove.)

**4.3.** Prove

**a.** if $x$ is positive and

$$y = \log x$$

any increase in $x$, however small, produces an increase in $y$.

**b.** If $x$ is positive and

$$y = 1/x$$

any increase in $x$ produces a decrease in $y$.

(The first is a monotonically increasing transformation, the second a monotonically decreasing one.)

**4.4** The standard method of measuring the activity of coagulation is by timing until a clot first appears; the greater the activity the shorter the time. As we have discussed elsewhere (7) Macfarlane's cascade model proposes that the speed of coagu-

---

*It has been mistakenly supposed that we claim three decimal places as the "accuracy of the method" (whatever that may mean). This is a common error, to confuse the criterion for the accuracy of an estimate with that for the resolving power of a measurement. The topic is discussed in considerable detail elsewhere (3).

lation depends on the product of the activities of many coagulant factors. How would you expect clotting time to be distributed?

**4.5** The strength of a lens may be measured by its focal length. The combined focal length, $F$, of two lenses with focal lengths $x$ and $y$ is

$$F = \frac{xy}{x + y}$$

which is not a linear relationship. Can you find a transform of focal length that will give a linear formula? (This problem should be easy for ophthalmologists!)

**4.6.** The following data (Table 4.10) have been furnished to me by Dr. T. E. Kelly on the number of nanomoles of phenyliduronide hydrolized per mg of protein in 18 hours at ambient temperature in normal subjects, in patients with Hurler's disease (mucopolysaccharidosis type I), and in obligatory heterozygotes.

**Table 4.10. Rate of phenyliduronide hydrolysis ng/mg protein in 18 hours at ambient temperature**

| Control subjects | Obligatory heteroxygotes for Hurler's disease | Homozygous affected with Hurler's disease |
|---|---|---|
| 115 | 39.5 | 0.0 |
| 120 | 54.8 | 0.0 |
| 122 | 59.7 | 0.0 |
| 125 | 66.6 | 0.0 |
| 129 | 71.9 | 4.2 |
| 132 | 72.0 | 4.4 |
| 132 | 78.6 | 4.8 |
| 132 | 86.0 | 4.8 |
| 139 | 86.3 | 7.1 |
| 143 | 93.0 | 8.4 |
| 144 | 102.0 | 12.9 |
| 146 | 115.0 | 17.0 |
| 146 | 124.0 | |
| 147 | | |
| 151 | | |
| 155 | | |
| 158 | | |
| 160 | | |
| 165 | | |
| 168 | | |
| 168 | | |
| 177 | | |
| 179 | | |
| 192 | | |
| 196 | | |
| 203 | | |
| 213 | | |

**Table 4.11. Percentage of lymphocytes bearing IgG globulin in 29 patients with primary immunodeficiency**

| | | |
|---|---|---|
| 0.1 | 3.5 | 8.4 |
| 0.2 | 4.4 | 9.3 |
| 0.6 | 4.5 | 10.0 |
| 0.8 | 4.6 | 11.2 |
| 1.0 | 4.8 | 14.5 |
| 1.3 | 5.4 | 15.5 |
| 1.4 | 5.5 | 17.0 |
| 2.2 | 6.0 | 18.7 |
| 2.2 | 8.2 | 32.9 |
| 3.2 | 8.4 | |

In an anaylsis to be based on the assumptions of normality and homoscedasticity, how would you handle these data?

**4.7.** In Table 4.11 are given some data from a study by Gajl-Peczalsa and associates (35) showing the percentages of lymphocytes bearing IgG globulin in twenty-nine patients with primary immunodeficiency. We shall assume that they came from a single population and are selected at random.

 **a.** Devise a suitable normalizing transformation.

 **b.** What is the variance-stabilizing transformation likely to be? Comment on its use here.

**4.8.** Devise an inverse for the Freeman-Tukey transform and verify it for $x = 2$.

# PART II
# THE STATISTICS OF THE GAUSSIAN ("NORMAL") DISTRIBUTION

# 5
# HYPOTHESIS TESTING: APPLICATIONS

In this chapter we shall discuss the applications of the principles laid out in Chapter 2 to certain more or less standard problems which are frequently encountered in connection with the Gaussian distribution and distributions derived from it. As a preliminary at least pages 10–13 should be mastered.

At the outset it is appropriate to recall (7) the distinction between definite values obtained in a sample, and variates or random variables, which are quantities not only not known, but unknowable, because they have not yet been realized. A sample value may be represented by a number such as 167 cm. Once selected, it is a constant and has no distribution, moments, modes, etc. If we wish to talk in general terms about such numbers, already selected, without specifying them, we use lowercase roman symbols. We might, for instance, code "$x$" for 167 cm and may, in principle, decode it at will. But "The height of the first male I shall see after breakfast tomorrow morning" is a random variable about which we may conjecture, viewing it as a chance outcome with a distribution, moments, and so forth. It is written with an uppercase roman letter, X. Once I have identified him, of course, his height becomes a definite quantity and hence represented by a lowercase letter.

## THE NORMAL RANGE

As in Chapter 2, hypotheses are tested on samples. We shall first deal with the simplest sample, which is that of size one.

More and more the clinician uses measurement in making decisions. Particularly since the multiple-channel analyzer has been introduced, laboratory results include a statement of "the normal range" for each test. Results outside this range are marked with asterisks, and are intended to raise questions in the physician's mind. It would be foolish to suppose that this system solves the clinician's problems, and nobody of experience would make this mistake. We shall have more to say about the "normal range" later (Chapter 11); but let us here discuss the problem on its merits.

An intelligent and responsible use of the normal range involves three questions: provenance, statistical meaning, and interpretation.

## Provenance

Where did the data come from and what are they supposed to represent? This is a common gap in quantitative medicine. It is seldom possible to find out what, if any, sampling scheme has been used and what the supposed population of reference comprises. A major problem is in defining what a "healthy" population is since, apart from the problem of defining health (3, 36), the supposed population may include patients with latent, larval, and subclinical disease. "Normal" levels of blood cholesterol may be based on people of whom some have coronary disease that cannot yet be recognized clinically. This aspect of the problem is so massive that I can only refer the reader to epidemiological textbooks (37, 38) and an excellent review by Feinstein (4, 39).

## Statistical meaning

The common criterion of the "normal range" is to pick the healthiest 95% of the population. Most often this is the central 95%; but in certain cases the remaining 5% are put in one tail. What are considered abnormal intelligence quotients are confined to one tail; so are levels of alpha fetoprotein—mistakenly so, Lau (40) believes; clinicians are seldom concerned with low sedimentation rates. It is not altogether clear on what grounds the choice between one-tailed and two-tailed is made. Presumably value judgments come into it; prognosis must be important—both sodium-losing and sodium-retaining states are harmful; but while mental deficiency is clearly deleterious, we have yet to show convincingly that genius is.

The responsible clinician must have some concern in every such case with how the 95% range was established. Broadly speaking there are two approaches. The safer, but less efficient, is the so-called empirical one: to determine from a large, carefully constructed sample that level below which $2\frac{1}{2}\%$ of values lie ("lower limit") and that above which $2\frac{1}{2}\%$ lie ("upper limit"). A corresponding modification would be made for a one-tailed rejection region. This method has been used where massive data are available such as in population surveys of lipids (41). Alternatively the variate may be found to follow some specific distribution, its parameters estimated, and the appropriate percentiles calculated. Either of these approaches is legitimate, but they must be soundly applied. One cannot simply compute the sample mean ($m$) and variance ($s^2$) and argue that 95% of the population will lie within the limits ($m - 2s$) and ($m + 2s$). This statement is approximately correct for a Gaussian distribution; but we have had many illustrations of the misuse of the argument in general, in discussing transformations (Chapter 4).

Once a "normal range" is established, it is treated as the acceptance region. Values inside are "within normal limits" and compatible with health. Values elsewhere are in the rejection region and abnormal: they call for further investigation. The whole process is a test of a hypothesis against a one- or two-tailed alternative. A minor peculiarity is that the "test statistic" is the mean of a sample of one; but that creates no logical problems; and the other notions discussed in Chapter 2—power, errors of the first kind ("false positives"), and of the second kind ("false negatives")—apply in the usual fashion. Since there may be many alternative hypotheses (the eosinophil count may be raised to differing degrees by many conditions), it may be useful to think even in terms of power functions. All this is straightforward.

**Interpretation**

The difficulty lies in interpreting the results. *The only guaranteed property of "the normal range" is that 5% of every normal population will be misdiagnosed as having a disease.* For rare and trivial conditions this criterion may seem outrageously radical. For common and serious diseases such as appendicitis, especially in childhood, it may be perilously conservative. Although this concept of the normal range has its champions even among respected statisticians such as Healey (42), there is little doubt that it is oversimplified and that in practice many other factors are, and should be, taken into consideration. Details are best deferred to Chapter 11.

It is easy to discredit this idea of the normal range (3, 43). Can anything be said in its favor?

**1. The choice of the tails as rejection regions.** There are many theoretical objections which may be raised, since the alternatives to health are so vaguely defined. But there are two general arguments in favor of the normal range. First, genetic and evolutionary reasoning suggests that the physically best endowed will tend to be either in the center of the distribution (as is so for blood pressure), which warrants a two-tailed rejection region; or that the system has not attained equilibrium and is moving to one extreme (e.g., intelligence), which calls for a one-tailed rejection region. Secondly, most physicians think much more readily in means than in any other moment; and the mere fact that the test is introduced into clinical practice assures us that the mean values in some diseases will differ from that in the healthy. Of course, as we have pointed out (43), the abnormal measurements may lie in both directions from the healthy mean, sometimes even in the same disease. The blood androgen level in the XYY syndrome, for instance, seems to have much the same *mean* level as in the normal male karyotype, but the variance is much larger, and thus extreme values are relatively more frequent in the abnormal state (44). Pyelonephritis may produce either sodium retention or sodium loss. Cholera may exhibit both febrile and "algid" (shocked) phases. Many such examples might be quoted. These arguments are plausible rather than rigorous.

**2. The size of the rejection region.** This is a more intricate matter to defend. There is some risk of confusion between two functions assigned to the normal range: as a *discriminant* of the normal; and as a *definition* of the normal. If the issue is the recognition of disease, clearly it makes no sense to decree that an unvarying 5% of the population is diseased: the prevalence of leprosy in the United States is much less than that of rheumatoid arthritis; and their individual frequencies vary greatly from one country to another. To set diagnostic criteria such that an unvarying 5% is "affected" would be gross overdiagnosis in some instances, and conversely. But when the range is being used as an *arbitrary* definition of the normal, its size helps us to contain the dimensions of our task. Granted the idea that the mentally retarded require special legal protection, we can scarcely afford to have 70% of the population so classified. Two hundred years ago, the "village idiot" was expected to do what work he could and to take responsibility for himself. Now, we can afford to treat a prearranged percentage as vulnerable; but not everybody outside an institution is intelligent enough to be immune to exploitation—and further economic prosperity may change the percentage. Even so the current proportion of the population treated is by no means uniform. Much more than 5% of the population wear glasses. Much less than 5% have plastic operations to correct facial abnormalities. By regulating levels of "normality," we can adjust the load of special care to the means available to deliver it.

Elvebach (45) has suggested that we should see the normal range, not as a basis for hard and fast decisions, but rather as a measure of the degree of unusualness of the value and hence as a general index that something may be the matter. Certainly I am sure that is the way in which all of us use new and unfamiliar tests; but it is rapidly modified by experience. Provided this use of the normal range produces neither complacency nor compulsive overinvestigation or overtreatment, it seems a sensible middle-of-the-road interpretation.


## TESTS ON THE MEAN

### One-sample tests. Variance known

Tests on individual readings such as we have just discussed may be regarded as tests on the sample means of samples of size 1. We will now generalize the method to those for samples of size $n$. If the true mean is $\mu$, the sample mean may (as in Chapter 3) be conveniently represented by the corresponding roman letter $m$, which is simply the sum of the observed values divided by the number of them. As stated elsewhere (7), if the values are random, independent samples with the same Gaussian distribution, $N(\mu, \sigma^2)$, then the sample mean, viewed as a random variable and hence denoted by an upper case letter, $M$, has the distribution

$$M = \frac{\Sigma X}{n} \sim N\left(\mu, \frac{\sigma^2}{n}\right) \tag{5.1}$$

i.e., it has the same mean and one-$n$th of the variance for a single reading. Since this is the first practical sampling problem discussed in this book, before going further it is advisable to make sure that the terms are all clearly understood.

1. The sample must be picked "at random." The term is discussed in detail elsewhere (7). Strictly, it means that a population of values must be defined and the probabilities of all possible subsets be specified in advance. In the present context I shall confine my comments to the simple random sample. It requires that values be not picked out because they are extreme, or close to the mean, but that all values of the class be equally likely to be picked. For instance, heights in a sample of ten basketball players may be a random sample of heights for basketball players but not for the population at large. The mean blood cholesterol in "all Americans" *cannot* be determined by excluding from the sample all those with myxedema, nephrosis, and familial hypercholesterolemia. Furthermore, in simple random sampling all combinations of cases must be equally likely. In certain instances a more general formulation is desirable, e.g., deliberately heavier sampling from certain groups (e.g., Negroes, females, professional men); or selection may be deliberately clustered, e.g., families rather than individual persons. In such cases the formal requirement is that (at least in principle) the probability of all possible subsets of the population being selected should be specified beforehand.

2. The sample is not haphazard, that is, collected in whatever way is convenient. In general a hospital corridor is not the place to find samples of the healthy any more than a morgue is. Nor can we expect the sample to be adequate if the investigator is to be allowed to make vague judgments in the individual case. It is not only conscious dishonesty that we have to be concerned with but unconscious biases, concealed factors that bear on the processes by which the cases are encountered, and so forth. For details of biases see (3).

3. Size of the sample has nothing whatsoever to do with bias unless the sample is so large as to include everyone in the population (which makes sampling unnecessary).

4. Haphazard mixtures of samples each of which has its own peculiar biases will not in general lead to unbiased samples.

The proper manner in which to collect a simple random sample is given briefly in (7) and for more subtle and elaborate schemes, the reader is referred to the classical discussions (46–48).

Let us now return to tests on a single sample of Gaussian values.

**Example 5.1.** Suppose that the data contained in Table 1.1 in the first chapter are drawn from a general population in which the standard deviation is known* to be exactly 0.35 Kg. Test at the 5% level the null hypothesis that the mean does not differ from 3.5 Kg, the mean for the population at large (a given).

We should now be fairly satisfied that these data follow a Gaussian distribution closely: whatever discrepancies there may be will be much reduced when the mean of the sample is used, in accordance with the central limit theorem (see Chapter 17 and reference 7). Then, according to the information, the variance of a single reading from this population is $(0.35)^2$ and the variance of the mean of a sample of size 54 will be

$$\frac{(0.35)^2}{54} = 0.002,268,518,5\ldots$$

The square root, $0.047,628,96\ldots$, is the standard deviation of the sample mean. These calculations are exact to as many decimal places as we care to take them.

The next issue is where we shall put the rejection region. Since the babies are from an inbred population, in part at least, general experience would suggest that the mean weight is likely to be depressed and hence that the rejection region should be in the lower tail. We know from Table 2.3 that the appropriate region will be "all values below 1.6449 standard deviations below the mean of the population at large," which is

$$3.5 - 1.6449 \times 0.047,628,96\ldots = 3.421\ldots \text{ or less}$$

The sample mean is

$$\frac{183.74}{54} = 3.402\ldots$$

clearly inside the rejection region; so we reject the hypothesis and conclude that the infants are on average smaller than the population mean.

The corresponding approach to the significance level would be that the sample mean is

$$\frac{3.5 - 3.402,59\ldots}{0.047,628,96\ldots} = 2.045,1\ldots$$

standard deviations from the population mean, and reference to appropriate tables of the standard Gaussian variate shows the corresponding (one-tailed) probability to be $0.0204\ldots$.

**Comment on accuracy in calculations.** It is wise to preserve as many significant figures in the intervening calculations as feasible; rounding may

---

*This assumption may seem implausible. It is motivated by two facts. First it is occasionally warranted: for instance, massive data from another source may be available, or the variance may be known because a transformation has been used. Second, the analysis is somewhat easier if the variance is known and the exposition that follows is somewhat easier to grasp.

lead to fair-sized errors by the end of the analysis. As I have pointed out elsewhere (3), in the statistical context *the precision of an estimate is not indicated by the number of figures carried but by the standard deviation of the estimate*. With the present result it scarcely matters whether the significance level is represented as 0.0204 or $1/50$. Retaining many figures is merely irrelevant, not misleading.

### Two-sample tests: variances known and equal

Let us suppose that two independent samples are taken, one from each of two subpopulations. For both, the variances are known and equal ($\sigma^2$). The first sample contains $n_1$ values and hence (under the usual assumption of independence) the sample mean has the variance

$$\frac{\sigma^2}{n_1}$$

The second sample comprises $n_2$ units and hence the sample mean has variance

$$\frac{\sigma^2}{n_2}$$

Now because of the independence of the samples, under the null hypothesis the variance of the difference of the means is the sum of their individual variances (7), i.e.,

$$\sigma^2 \left[ \frac{1}{n_1} + \frac{1}{n_2} \right] \tag{5.2}$$

The standard deviation of the difference of the means is the square root of this quantity.

**Example 5.2.** For the data of Table 1.1 test the null hypothesis that there is no difference between the mean birth weights of the two populations (inbred and control), against the hypothesis that the inbred babies are lighter. Assume the variances for both populations are as before $(0.35)^2$.

The variance of the difference between the sample means under the null hypothesis will be

$$(0.35)^2 \left[ \frac{1}{27} + \frac{1}{27} \right] = 0.009,074\ldots$$

giving a standard deviation of $0.095,257\ldots$. As before, the rejection region at the 5% level one-tailed is 1.6449-or-more times this quantity below 0 (the difference in means under the null hypothesis), i.e., any value less than

$$(0.095,257\ldots)(-1.6449) = -0.156,689,7\ldots$$

We now compute the means: for inbred children $91.35/27 = 3.3833\ldots$ and for controls $92.39/27 = 3.4218\ldots$ The difference is

$3.3833\ldots - 3.4218\ldots = -0.0385\ldots$, which is well inside the acceptance region. Thus we do not reject the null hypothesis at this level.

### Two-sample tests: variances known and unequal

Consider the case where the means of two independent samples are to be compared, the variance of both parent populations being known but unequal, $\sigma_1^2$ and $\sigma_2^2$, respectively; the sample size from the first is $n_1$, and from the second $n_2$. Then by a slight modification of (5.2), the variance of $D$ the difference of the means, viewed as random variables, $M_1$ and $M_2$, assuming independence is

$$\text{var}(M_1 - M_2) = \text{var}(D) = \frac{\sigma_1^2}{n_1} + \frac{\sigma_2^2}{n_2} \tag{5.3}$$

The individual means being normally distributed, their difference will be also. The test can be carried out in the usual way. It is not necessary to use any of the variance-stabilizing procedures of Chapter 4 where the data are Gaussian and the distribution of $D$ is known except for its mean.

### Two-sample tests: variances known; correlated data

A common pattern for two-sample tests is that in which the data in the two samples are not independent but have regular individual correspondences: for example, "before" and "after" results on the same patients; or comparisons between twins, one being on one treatment, one on the other; or "matched" controls and cases. Generally such a system is used because it eliminates some of the variance and makes for more powerful tests. But many investigators have forgotten that they matched (or even why they matched!) by the time analysis starts. However we may not use the formulae (5.2) or (5.3), which are based on the express assumptions of independence, which we may be almost certain are violated here. *Data that were paired by design must be analyzed as paired data.* * However, we must have a clear idea of what is independent. The members of any particular pair will not be independent over the sample space for the whole population: clearly so—they have been picked precisely because they are similar; but they will be supposed independent of the members of other pairs. It would be appropriate to include in the sample a set of twins (one being assigned to each group) but not two sets of twins from the same family.

Let us denote the typical pair of readings by $a_i$ and $b_i$ from a sample of size $k$ pairs. The assumptions are thus that these two readings are correlated,

---

*It may be argued that if we ignore nonindependence we overestimate the standard deviation of the difference and that if we then reject the null hypothesis at the 5% level, we are if anything understating our claims. But this argument seems objectionable. In the first place, it is gratuitously imprecise. It may bias the analysis in favor of accepting the null hypothesis. On the other hand, if the paired readings are negatively correlated, quite the opposite effects will result. Finally, if we do not know the variance and must estimate it, we have a different number of degrees of freedom for the $t$ test (see below).

and members of all other pairs are correlated, but the $a$'s are independent of each other and so are the $b$'s. So also are $a_i$ and $b_j$ for all values of $i \neq j$. The covariances (see 7; also glossary) are the same for each pair and the usual assumptions, normality and a constant variance, are used. If we define

$$d_i = a_i - b_i \qquad \text{for all } i$$

then it can be shown that

$$\text{var}(D) = \text{var}(A) + \text{var}(B) - 2\,\text{cov}(A, B) = \sigma_D^2$$

The test of whether the population mean of $a$'s and that of $b$'s differ is a test of the hypothesis that for the mean difference $\mu_D$

$$H_0: \quad \mu_D = 0$$

against some appropriate alternative. The test is thus based on the distribution under the null hypothesis of the sample mean, $M_D$

$$M_D \sim N\left(0, \frac{\sigma_D^2}{n}\right)$$

**Example 5.3.** Consider the following data (Table 5.1) from a study on dietary manipulations in rabbits (49). As we have seen in Chapter 4, using the logarithms of serum cholesterol normalizes the mean and stabilizes the variance. We shall reasonably assume it also stabilizes the covariance. Suppose the variance for the transformed value is known to be 0.03 and the covariance is 0.01. Then the variance of the difference of a pair is

$$\sigma_D^2 = 0.03 + 0.03 - (2)(0.01) = 0.04$$

**Table 5.1. Serum cholesterol levels (mg/100 ml) in 12 rabbits before and after 60 days of a diet containing margarine**

| | Before diet | | After diet | | |
| Rabbit | value ($x$) | $\log_{10}x$ | value ($y$) | $\log_{10}y$ | $\log_{10}(y/x)$ |
|---|---|---|---|---|---|
| 1 | 42 | 1.623 | 51 | 1.708 | 0.084 |
| 2 | 67 | 1.826 | 160 | 2.204 | 0.378 |
| 3 | 35 | 1.544 | 36 | 1.556 | 0.012 |
| 4 | 64 | 1.806 | 54 | 1.732 | −0.074 |
| 5 | 53 | 1.724 | 55 | 1.740 | 0.016 |
| 6 | 67 | 1.826 | 63 | 1.799 | −0.027 |
| 7 | 55 | 1.740 | 134 | 2.127 | 0.387 |
| 8 | 74 | 1.869 | 46 | 1.663 | −0.206 |
| 9 | 83 | 1.919 | 153 | 2.185 | 0.266 |
| 10 | 67 | 1.826 | 48 | 1.681 | −0.145 |
| 11 | 55 | 1.740 | 84 | 1.924 | 0.184 |
| 12 | 31 | 1.491 | 55 | 1.740 | 0.249 |

SOURCE: Data of Murphy et al. (49).

and since 12 values are being averaged, the standard deviation of the difference will be $\sqrt{(0.04/12)}$. Now we know that given that $\log(A_i)$ and $\log(B_i)$ are normally distributed, whether they are correlated or not, *any* linear combination of them, i.e., of the form

$$c_1 \log(A_i) + c_2 \log(B_i)$$

will follow a Gaussian distribution. Hence we know the form of the distribution, its variance and (under the null hypothesis) its mean. It remains to decide on the size of the test, and whether a one-tailed or a two-tailed rejection region would be appropriate. We note that rabbits are usually sensitive to dietary cholesterol and any result in which the serum level falls following the ingestion of it would be explained away as due to chance or intercurrent factors. So we set the size at 0.01 and make the test one-tailed. Referring to Table 2.3 we find the rejection region to be all sample values above

$$0 + (2.3263) \left( \sqrt{\frac{0.04}{12}} \right) = 0.134,3\ldots$$

The sample mean value $m_D$ is

$$m_D = 0.0936\ldots$$

which is within the acceptance region. Hence the hypothesis is not rejected at this level.


## TESTS ON VARIANCES

If $X$ is a random Gaussian variate with mean $\mu$ and variance $\sigma^2$ then the quantity

$$Z = \frac{X - \mu}{\sigma}$$

is the standard normal form; that is, it has a mean of zero and a variance of 1. The square of this quantity follows the distribution of a $\chi_1^2$ ("chi-square with one degree of freedom"). The sum of $k$ such squares, if they are independent, follows a $\chi_k^2$ distribution.* The form of the $\chi^2$ distribution varies according to the number of degrees of freedom, and particular sizes of test for each must be tabled separately. For large numbers, good approximations exist (7).


### Tests on variances: the mean is known

The average value of $(X - \mu)^2$ is, by definition, the variance of $X$ and the average value of $Z^2$ is thus

---

*The term *degrees of freedom* is one which will be somewhat more fully discussed in chapters 7 and 12. But roughly speaking it refers to the number of independent pieces of information or its equivalent on which an estimate is based.

$$E(Z^2) = \frac{\mathrm{var}(X)}{\sigma^2} = \frac{\sigma^2}{\sigma^2} = 1$$

If we denote the sum of the squares of $k$ such independent quantities of the form $(X - \mu)$ by $G$, then

$$\frac{G}{\sigma^2} = \sum \frac{(X - \mu)^2}{\sigma^2} \sim \chi_k^2$$

The appropriate estimate of the variance of $X$ where the mean is known is $G/k$ which, when divided by $\sigma^2$, follows a distribution of $\chi_k^2/k$. However, in tests of hypotheses it is usually more convenient to use the distribution of $G/k$, which should be a $\sigma^2\chi_k^2/k$.

**Example 5.4.** Mishkel (50), in a series of nearly three thousand children in Canada, found the mean total serum cholesterol to be 70 mg/100 ml and the standard deviation to be 17. (At this age we may treat the distribution as Gaussian and the sample is so large that we may assume the estimates to be exact.) Kwiterovich (51) on a sample of size 36 obtained the figures 74 and 11 mg/100 ml respectively. Test the hypothesis that Kwiterovich's result for the standard deviation does not differ from the true figure for Mishkel's population, assuming that the population from which his sample is taken has the same mean.

Since Kwiterovich's sample is rather small, we shall settle for a 5% test, and in the absence of any indications to the contrary we shall use a two-tailed test having half of the error of the first kind in each tail. Now since we do not have to estimate the mean, there are 36 degrees of freedom; and referring to appropriate $\chi^2$ tables we find the 2.5% and 97.5% values to be 21.336 and 54.437 respectively. Hence we require the limits for the acceptance region

$$21.386 \le \left[\frac{x - \mu}{17}\right]^2 \le 54.437$$

or, multiplying throughout by $17^2$,

$$6{,}166 \le \Sigma(x - \mu)^2 \le 15{,}732$$

Now we infer that Kwiterovich estimated his own mean from his sample so that in estimating the variance he will in the usual fashion have divided by $(n - 1) = 35$. Hence his residual sum of squares was

$$35(11)^2 = 4{,}235$$

This in turn was found by subtracting from the original sum of squares the correction factor $m\Sigma x = nm^2 = (36)(74)^2 = 197{,}136$. We thus infer the original sum of squares to be $4{,}235 + 197{,}136 = 201{,}371$. Now in general

$$\Sigma(x - \mu)^2 = \Sigma x^2 - 2\mu\Sigma x + n\mu^2$$

which in this case (using Mishkel's sample mean for $\mu$) is

$$201{,}371 - 2 \times 70(74 \times 36) + 36(70)^2 = 4{,}811$$

and this lies outside the acceptance region. We therefore conclude that the variance in Kwiterovich's data is not equal to that of Mishkel's.*

### Tests on variances: the mean is not known

Much more commonly, the mean is not known and must be estimated from the data. We know that the unbiased estimate of $\sigma^2$ is then found by dividing the residual sum of squares from the *sample estimate* of the mean by $(n - 1)$ rather than $n$. But there is a further effect: since the mean is obviously not independent of the values being averaged, the distribution of the sum of squares divided by the true variance proves to be equivalent to the sum not of $n$ chi-square values, but of $(n - 1)$, each with one degree of freedom. The limits of the acceptance region will be changed accordingly.

**Example 5.5.** Consider again the data of Example 5.4 but suppose the mean is not known. We shall test whether the variance in Kwiterovich's data differs significantly from that of Miskell's. As before we infer the sample sum of squares from the sample mean to be 4,235. The limits of the acceptance region are computed for 35 degrees of freedom:

$$(20.569)(17)^2 = 5{,}944.441$$

and

$$(53.203)(17)^2 = 15{,}375.667$$

The sample statistic does not fall within these limits, so we reject the hypothesis at the 5% level.


## TESTS WHERE NEITHER THE VARIANCE NOR THE MEAN IS KNOWN

### Comparison of the variances

Let us denote the sum of squares, not from the true mean, but from the sample mean by $T$. We have seen that

$$T = \Sigma(X - M)^2$$

has two important properties (provided the $X$'s are Gaussian, independent, and homoscedastic, the sample size being $n$ and the sample mean $M$)

$$\frac{T}{\sigma^2} \sim \chi^2_{n-1}$$

---

*The reader may be appalled at the complexity of my illustrative example. But in reviewing published data critically, it is often necessary to "reconstruct" the original sum and sum of squares so that the reader may do his own analysis. Simpler, more orthodox, examples will be worked out in problems at the end of the chapter.

and

$$\frac{T}{n-1} = S^2_{n-1}$$

is an unbiased estimator of $\sigma^2$.

Thus the ratio

$$\frac{S^2_{n-1}}{\sigma^2} \sim \frac{\chi^2_{n-1}}{n-1}$$

i.e., a chi-square statistic divided by its degrees of freedom.

Now let us suppose we have a second such quantity based on $w$ observations and denoted $S^2_{w-1}$, in a sample which, according to the null hypothesis, is drawn independently from the same Gaussian distribution.

Then it proves most convenient to test the null hypothesis of equal variances, not by testing whether $(S^2_{n-1} - S^2_{w-1})$ is zero, but whether their ratio is unity. Elsewhere (7) this ratio is defined as the $F$ distribution, the shape of which depends on both $(n-1)$ and $(w-1)$. Thus in using it we have to be concerned about three features: the number of degrees of freedom in the numerator, the number in the denominator, and the size of the test. It does not matter which estimate goes in the numerator, but in interpreting the result the number of degrees of freedom will in general matter.

**Example 5.6.** Consider the data from Example 4.8. We are assured that the data are reasonably close to Gaussian. The test of the equality of the variances might be handled by arguing a priori that the only effect of the drug could be to increase the variance and the only alternative to the null hypothesis of equal variance will be to accept that the variance is larger in those receiving sulfinpyrazone. This surmise is a matter of (scientific) taste. One would then put the estimate for the saline group in the denominator and reject the null hypothesis only if the ratio is significantly greater than unity. I would be happier with a two-tailed test, and it is then convenient to put the larger estimate in the numerator and do a one-tailed test. Consulting standard tables we find that with 15 and 15 degrees of freedom we have the following critical values.*

| Size of test | | Critical values of $F$ | |
| --- | --- | --- | --- |
| | One-tailed | Two-tailed | |
| | | Lower | Upper |
| 0.05 | 2.403,5 | 0.349,39 | 2.862,1 |
| 0.01 | 3.522,2 | 0.245,71 | 4.069,8 |

It will be seen readily that the lower limit for any particular size of test is the reciprocal of the upper limit. (Why?) Moreover, if we *always* put the larger mean square in the numerator, under the null hypothesis we are giving

---

*To be completely correct we should use a lower case for these particular values of the $F$ distribution. Unfortunately custom is against me. In practice this inconsistency causes little confusion.

ourselves two chances to reject the null hypothesis since either the control or the treated value may appear in the numerator. In such a test, then, we treat 2.862,1 as the upper critical value at the 5% (not the 2½%) level. If on the other hand we agreed *beforehand* to put the control estimate in the numerator, 2.862,1 would be the critical value for the 2½% level and 0.349,39 would be the lower critical level (which can never be attained if the larger mean square always becomes the numerator).

In this case we have the following estimates for $\sigma^2$:

For saline: 0.135,06

For sulfinpyrazone: 0.595,68

whence (putting the larger mean square in the numerator)

$$F = \frac{0.595,68}{0.135,06} = 4.410$$

which would lead us to reject the null hypothesis at the 1% level.

### Pooled estimate of the variance

If we are satisfied that the variances are equal we may devise a pooled estimate of the variance readily enough. Suppose we have two residual sums of squares which we denote by $T_1$, with $(n_1 - 1)$ degrees of freedom, and $T_2$ with $(n_2 - 1)$ degrees of freedom. Now

$$\frac{T_1}{\sigma^2} \sim \chi^2_{n_1 - 1}$$

$$\frac{T_2}{\sigma^2} \sim \chi^2_{n_2 - 1}$$

and if these quantities are independent, their sum, being the sum of two independent chi-square statistics, is a chi-square statistic with the sum of their degrees of freedom (7) with expectation equal to the sum of the degrees of freedom. Thus

$$\frac{T_1 + T_2}{\sigma^2} \sim \chi^2_{n_1 + n_2 - 2}$$

and thus

$$\mathrm{E}\left[\frac{T_1 + T_2}{n_1 + n_2 - 2}\right] = \sigma^2 \tag{5.4}$$

i.e., the sum of the sums of residual squares divided by the sum of the degrees of freedom is an unbiased estimator of the (common) variance; and, divided by the true variance, it follows a chi-square distribution divided by the degrees of freedom.

The foregoing development is quite sound; but perhaps we have glossed rather too easily over the assumption that the true variances are equal. The evidence (to my mind) should be positive rather than negative; the onus of proof should be on those who wish to invoke homoscedasticity. From inspection of the $F$ tables it will be evident that for small sizes of sample, the critical values are large. Thus for 4 and 4 degrees of freedom, the critical $F$ value at the 5% level is 6.383. But I would be very reluctant to assume that sample estimates of the variance differing by a factor of six were indeed derived from distributions with the same variance, although the $F$ value would be within the acceptance region. For this reason some practical statisticians demand sample values of $F$ closer to unity than a literal interpretation of Neyman-Pearson hypothesis testing would demand.

### The $t$ distribution

We have seen earlier, under the null hypothesis that the mean of the distribution is $\mu$, the variance $\sigma^2$, and the sample mean, $M$, based on $n$ observations, that the statistic

$$\frac{M - \mu}{\sqrt{\dfrac{\sigma^2}{n}}} = Z \sim N(0, 1)$$

If $\sigma^2$ is unknown and must be estimated by the usual estimator $S^2 = T/(n-1)$ from the data, reflecting the variation due to sampling, then the statistic

$$\frac{M - \mu}{\sqrt{\dfrac{S^2}{n}}}$$

follows not the Gaussian but the $t$ or Gosset* distribution. The critical values depend on the degrees of freedom on which the estimate $S$ is based, exactly as happens with the chi-square or the $F$ distribution. Indeed it is easy to show (3) that the square of the $t$ statistic with $(n-1)$ degrees of freedom is an $F$ statistic with 1 and $(n-1)$ degrees of freedom. As $n$ becomes large, $t$ becomes more nearly a standard Gaussian variate. An approximation is given in Appendix 2.

### One-sample tests: neither the mean nor the variance known

**Example 5.7.** Consider again Example 5.3 but suppose the variances and covariances between the paired readings are homogeneous but un-

---

*Because Gosset in 1908, with Edwardian diffidence, published under the *nom de guerre* "Student," it is common to refer to this variate as "Student's $t$ distribution." I propose to break with this stilted tradition. Note also that, as for the $F$ distribution, well-entrenched practice prevents us from adhering to the conventional distinction between uppercase and lowercase letters. The symbol $t$ is used indiscriminately to mean both a random variable and a particular outcome. The symbol $T$ has unfortunately been reserved to another context.

known. Test at the 5% level whether the mean before treatment differs significantly from a (geometric) mean of 55 mg/100 ml.

In the absence of any indicator as to whether the geometric mean is likely to be more or less than 55, we do a two-tailed test. One may solve this problem by setting up a rejection region either for the sample mean or for the $t$ value. The latter is usually less trouble. Logarithms to the base ten (expressed to three decimal places as before) give the following results:
Sample mean

$$m = \frac{20.934}{12} = 1.744,5$$

Sample variance

$$s^2 = \frac{36.708,708 - (20.934)^2/12}{12 - 1} = 0.017,213,181,8\ldots$$

The estimated standard deviation of the sample mean

$$s_M = \sqrt{\left(\frac{0.017,213,181,8\ldots}{12}\right)} = 0.037,873,893,6\ldots$$

The mean under the null hypothesis, 55, has the common logarithm 1.740,362,689. Hence

$$t = \frac{1.744,5 - 1.740,36..}{0.037,87\ldots} = 0.109,2$$

The critical value at the 5% level for 11 degrees of freedom, two-tailed is 2.009,85. . . . The null hypothesis is not rejected.

### Two-sample tests: correlated data

**Example 5.8.** Test at the 5% level whether the mean change on diet in the data in Table 5.1 differs from zero.

Here the appropriate rejection region seems to be in the upper tail. While most observers would accept that the treatment might increase the serum cholesterol, most would flatly dismiss any suggestion that it would reduce it. A one-tailed test will therefore be done. Note that the data are paired and must be analyzed as such. The test will be on the paired differences and will remain a one-sample test with 11 degrees of freedom, not 22 as would be appropriate for a two-sample test. Also note that even if the (unknown) variances in the two sets of measurements are not the same (for instance, if the treatment increases it) we may still do a paired test. This property in paired data avoids a great deal of trouble, as we shall see in Chapter 17. The paired differences may be found either by direct subtraction or by taking the logarithm of the ratio of the second, to the first, reading. Values are given in the last column of Table 5.1.

The sample mean is

$$\frac{1.124}{12} = 0.093,666\ldots$$

The sample variance is

$$\frac{0.536,388 - (1.124)^2/12}{11} = 0.039,191\ldots$$

The estimated standard deviation of the mean is

$$\sqrt{\left(\frac{0.039,191\ldots}{12}\right)} = 0.057,148\ldots$$

Hence

$$t = \frac{0.093,666\ldots}{0.057,148\ldots} = 1.639$$

The critical level, one-tailed for $t$ with 11 degrees of freedom and size 0.05, is 1.796. The hypothesis is therefore not rejected. (Note that with a two-tailed test it would have been further still from the rejection region.)

**Two-sample tests: uncorrelated data**

In this instance, the commonest type of two-sample analysis, we have neither sample variates that can be reduced to standard normal form, nor values that are tied together by correspondence. It may be an intricate problem to solve, or we may have to be content with an approximate solution. Details are best deferred to Chapter 17. We shall deal here with one particular type of case, where there is good evidence that the variances in the two groups (although unknown) are equal. Then under the usual conditions of normality and independence, we may use the pooled estimate of the variance discussed above. It will be independent of the difference between the sample means, which is normally distributed.

**Example 5.9.** Chern and Beutler (52) studied pyridoxine kinase levels in erythrocytes in black and white Americans. They present evidence that the sample values are Gaussian or close to it, plots of cumulative proportions on normal probability paper being virtually linear. The sample statistics are shown in Table 5.2. The ratio of the variances gives

$$\frac{(0.329)^2}{(0.305)^2} = 1.163\ldots$$

The critical value at the 5% level is 1.66; so the hypothesis of homoscedasticity is not rejected at this size. Furthermore, the ratio is reassuringly close to unity. So with confidence we proceed to compute a pooled estimate of the variance. We suppose that the estimates provided are found by the usual

**Table 5.2. The activity of pyridoxine kinase in mg/gm Hb in the erythrocytes of white and black Americans**

| Racial group | Sample size | Sample mean | Sample standard deviation |
|---|---|---|---|
| White | 51 | 1.305 | 0.329 |
| Black | 52 | 0.754 | 0.305 |

SOURCE: Data of Chern and Beutler (52).

method-of-moments formula. Hence we reconstruct the residual sums of squares as follows

$$\text{For whites } (0.329)^2 \times (51 - 1) = 5.412,050$$
$$\text{For blacks } (0.305)^2 \times (52 - 1) = 4.744,275$$

Adding these sums and dividing the sum of the degrees of freedom, 101, we get

$$S^2 = \frac{10.156,325}{101} = 0.100,557$$

The standard error of the difference is estimated by

$$\sqrt{\left[ 0.100,557 \left( \frac{1}{50} + \frac{1}{51} \right) \right]} = 0.063,110\ldots$$

Under the null hypothesis that the means are equal,

$$t = \frac{1.305 - 0.754}{0.063,110} = 8.730$$

which leads to firm rejection at the one in a thousand level of the null hypothesis that the means are equal.

## PROBLEMS

**5.1.** Benevolent societies in the United States campaign on the grounds that many people suffer from the condition on which they wish to promote research—defective vision, dyslexia, mental retardation, hearing loss, stunted growth, etc. The size of the population affected is with considerable regularity quoted at five million or ten million people. Comments?

**5.2.** In twelve young adult males with acute appendicitis, I found on admission the following temperatures (°C); 36.0, 36.8, 37.0, 37.4, 37.4, 37.4, 37.5, 37.7, 38.0, 38.1, 38.2, 38.3. On the basis of extensive data, body temperature in young, healthy, male adults is Gaussian with mean 37.0 and standard deviation of 0.5.

**a.** Choose a suitable acceptance region for the (null) hypothesis that in this disorder in such patients, the mean temperature is normal, using a test of size 0.05.

    **b.** Compute the power curve for suitable sample means less than $37.5°C$.

    **c.** Test the hypothesis.

**5.3.** The logarithm of the blood cholesterol level in the total population is close enough to being Gaussian for practical purposes. The mean of the logarithms to the base 10 in men aged 25 is 2.35 and the variance is 0.014. In a sample of 36 men of this age with familial hypercholesterolemia the mean is 2.40.

    **a.** Test the hypothesis that the mean level is greater than normal in the latter group.

    **b.** Compute the probability of an error of the second kind against the alternative hypothesis that the mean is 2.5, the variance being the same as under the null hypothesis.

**5.4.** The 576 male medical students enrolling at a particular medical school in 1975 had a mean height of 170 cms with a standard deviation of 2.4 cms. Test at the 5% level whether the mean height for this population differs from the national mean height of 168 cms.

# 6
# ESTIMATION ON GAUSSIAN DATA

In this chapter we shall confine ourselves to estimation of the parameters of the univariate Gaussian distribution: $\mu$ the centrality parameter which measures location; and $\sigma$ which is the measure of dispersion.

## THE CENTRALITY PARAMETER

We have seen that the Gaussian distribution is symmetrical about $\mu$. In consequence, $\mu$ is also the mean of the distribution. Our intuitive estimate of $\mu$ is thus that given directly by the method of moments. What can we say about its properties? Can this estimate be improved on? We may first give some thought to how we may approach actual data.

### Preliminaries

The prior convictions we may or may not have relate to two main issues: first, whether or not the data are indeed Gaussian, and second, whether the parameter is likely to have some particular value. These prior convictions are everything and nothing. They represent a great gulf between the mathematical statistician and the empirical scientist; they have had a tremendous divisiveness in theoretical statistics itself. They are at once the refuge of the bigot and the last bulwark against absurdity. It is no trivial matter to preserve a sensible balance between fanciful ornamentation and insouciant agnosticism. Here the reader might with advantage refer back to Chapter 2 where we set out to obtain an answer to one question (the probability of the hypothesis being true) and ended up with a very precise answer to something else (how readily the hypothesis would account for the data). A "know-nothing" policy of estimation, confining attention to what the present data contain, may yield a very precise answer to an irrelevant question and may even mislead (as we shall see in due course). Still, this peril must not lead us to mistreat our data.

First, then, why should a variate be Gaussian? (There is no reason why measurements on "normal" people—whatever that may mean—should be normal in the sense of Gaussian; that is merely the feeblest of puns.) We have three main prior reasons.

1. The conditions underlying the central limit theorem (7, and see Chapter 17) may be known, or thought, to apply. If our data consist of sums of means of many observations, even if they are not identically distributed, we may expect them to be Gaussian or close to it. If, for instance, we repeatedly time how long it takes for 500 emissions to occur from a stable radioactive source, we may expect the times to approximate a Gaussian distribution. (They actually follow a gamma process of order 500.) Rather less certainly, we may expect human adult height to be virtually Gaussian because height reflects the sum of many small contributions; but occasionally height is disrupted by one major factor, as in achondroplasia; and such patients would comprise a splinter population.

2. We may expect a variate to be Gaussian because it has been made so artificially. Such is the case for the intelligence quotient or the constructive "liability" that Falconer has used to provide an underpinning for his threshold model of a dichotomous genetic trait with a multifactorial orgin (53).

3. We may expect a variate to be Gaussian because extensive experience of the present population, or of populations similar to it, shows it to be Gaussian.

Second, what can we say about the true value which $\mu$ may be expected a priori to take? We may have powerful theoretical arguments; but this is a rare experience in medicine. Usually we hold prior ideas from the collective empirical experience. In practical situations this evidence is usually somewhat ambiguous. If there were secure data, there would rarely be interest in the individual estimate on another sample of data, especially if small. The main use in these well-characterized instances is as a monitoring device. If an investigator reports a mean birth weight in a human population of 5 kg, either he will earn discredit or it will be inferred that he has encountered an extraordinary population that merits much closer study.

At the other extreme, there may be many data in such blatant conflict that prior probabilities may be discounted. The levels in blood of creatine phosphokinase in muscular dystrophy vary over more than an order of magnitude from one report to another (54, 55). The test has diagnostic value only if each laboratory sets up its own control standards; and to dismiss a set of values as "unacceptable" is much more hazardous than would be so for many common measurements. Moreover, there are genuine differences among populations and subpopulations which neither can, nor should, be obscured by the prior convictions.

There exist also situations in which the prior probability represents not a purely subjective indeterminacy about an undoubted parameter but a distribution of a set of individual parameters, each corresponding to distinct distributions. This notion causes considerable mystification where the sample space is not clearly defined. Let us take the steps a little more slowly.

**Example 6.1.** Consider a single reading on blood cholesterol level taken at random in an individual patient picked at random. We shall suppose two things: that there is a true mean value for a patient; and that the true mean differs from one patient to another. Some patients have habitually high levels, others medium or low. But all of them vary from time to time. How may we estimate from this sample reading the mean *for that patient?* The evidence suggests that serum cholesterol has a lognormal distribution in populations and so do the readings within the individual patient about his true mean; so it is reasonable to apply Gaussian theory to the logarithm of the value. We have two sources of uncertainty here. How far is the patient's true mean value from the true mean for the population at large? (We shall show that this issue is relevant to our answer.) And, how far is the sample value on the patient, from his true mean? Let us suppose that on the transformed scale the mean $\mu$ for a subject $A$ in the population $\sim N(\nu, \zeta^2)$ and the sample reading, $X$, in the patient chosen $\sim N(\mu, \sigma^2)$. We suppose $(X - \mu)$ is independent of $\mu$ over the sample space of patients.

We note that in the sample value two events are jointly fulfilled: that $A$, the subject chosen, has a true mean value of $\mu$ and that the reading $x$ is obtained in him. Hence the (joint) probability density of such a reading is the product of a prior (marginal) probability density and conditional probability density.

$$f(\mu, x) = \left[ \frac{1}{\zeta\sqrt{2\pi}} \, e^{-\frac{1}{2}\left(\frac{\mu - \nu}{\zeta}\right)^2} \right] \left[ \frac{1}{\sigma\sqrt{2\pi}} \, e^{-\frac{1}{2}\left(\frac{x - \mu}{\sigma}\right)^2} \right]$$

$\qquad\qquad$ Likelihood the $\qquad\quad$ Likelihood of the
$\qquad\qquad$ patient's mean is $\mu$ $\quad$ sample value $x$
$\qquad\qquad\qquad\qquad\qquad\qquad\quad$ given patient's
$\qquad\qquad\qquad\qquad\qquad\qquad\quad$ mean is $\mu$

Now the result is anomalous. If the reading $x$ on the patient is truly taken at random, then *by definition* it is an unbiased estimate of the mean for that subject. But we do not know in what part of the distribution of means of subjects $A$'s true mean lies. It might for instance be $(x + c\sigma)$ and the sample value is $c$ standard deviations below it. The *conditional* probability density is exactly the same as the *conditional* probability of a reading $c$ standard deviations above the mean from a subject with a mean of $(x - c\sigma)$. But in the situation portrayed in Figure 6.1 the prior probability of the latter explanation is much greater than that of the former. Applying the argument for all possible values of $c$, incorporating the prior probability, we obtain the remarkable (but nonetheless perfectly reasonable) result that the best estimate of the patient's average value is not $x$, but closer to the value $\nu$. This implies that $X$ is *not* an unbiased estimator of $\mu$ over the joint sample space. Cast in other terms, if we take a group of subjects in whom the first sample value is $X$ their

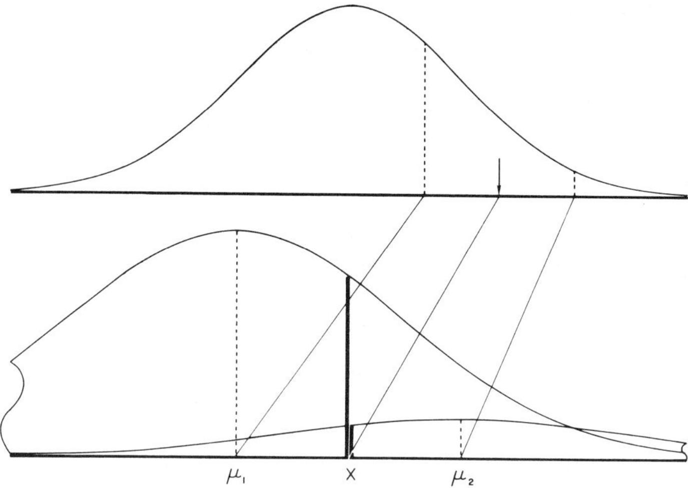


**Figure 6.1.** Regression towards the mean. The observed value, $x$, indicated by the arrow, is subject to some random Gaussian error of measurement. Given the true mean, x is as likely to be a value some specified distance above $\mu_1$ as the same distance below $\mu_2$. But the *upper* diagram (which represents the distribution of true mean values among patients) shows that $\mu_1$ is a priori the more probable value of the two, and the (conditional) distributions in the *lower* diagram are scaled proportionately. The relative joint densities are indicated by the two black lines on either side of $x$. The same argument for any pair of means symmetrically placed about $x$, lead us to a similar conclusion. Thus the most plausible estimate of the true value will lie to the left of $x$ (the side on which the mean of the prior density lies).

true means (and in particular the second readings on them) will on average be closer to the global mean, $v$.

This shift is known as a regression toward the mean. It is strictly a property of individual readings, but statistically it is discerned mainly in samples. It is important to stress that it is a logical consequence of error of measurement. If we were dealing with an almost error-free measurement of a random variable, such as height, we would expect little regression toward the mean. One must not be careless about this argument, however, which is why the details have been explained at some length. Reflection will make it clear that the regression toward the mean is due to the shape of the (prior) probability density function of mean values within individuals. Suppose, however, that the (prior) distribution is bimodal, such as may be expected if there are two distributions of individual means, for example if there is a polymorphism for one main gene, and complete dominance for the one type. Then the ratio of the priors of points symmetrically chosen about the observed value will be quite different. The details are somewhat complicated (56); the expected value for a second determination will tend to be "attracted" to the nearer mode in the prior probability. For a wide range of cases this will mean a regression *from* the mean and *to* the mode. In the example displayed in Figure 6.1 the mode and the mean of the prior distribution of means are the same. Indeed if we were starting anew, the expression *regression toward the mode* would be preferable.*

## MLE of the centrality parameter

Where we have no prior probability distribution, or for one reason or another elect not to use it, we are, in effect, using a uniform prior distribution, and resting our inference entirely on the likelihood curve for the present data. To be sure, the epistemological nature of the inference differs from that by the Bayesian method with an explicit uniform prior, in that it leads to no formal statement about the posterior distribution (any more than the Neyman-Pearson test of a hypothesis gives a posterior probability of the hypothesis). Nevertheless, in practice, the MLE is viewed as being an inference on an equal footing with the Bayesian estimator.

The likelihood of a set of independent observations from the same Gaussian distribution has the form

$$\prod_i \frac{e^{-\frac{1}{2}\left(\frac{x_i - \mu}{\sigma}\right)^2}}{\sigma\sqrt{2\pi}}$$

The logarithm of this product to the base $e$ is the sum of the logarithms which we shall denote by $L$:

$$L = -n \log \sigma - \frac{n}{2}\log(2\pi) - \frac{1}{2}\Sigma\left(\frac{x_1 - \mu}{\sigma}\right)^2 \tag{6.1}$$

Now the only part of this expression that depends on $\mu$ is the last term which we wish to maximize. (Nothing we can do to $\mu$ will have any effect on the values of the other terms, which may just as well be ignored.) Further, this last term is negative, and to maximize it, we wish its absolute value to be as small as possible. We thus perceive that the MLE involves minimizing

$$\frac{1}{2\sigma^2}\Sigma(x_i - \mu)^2$$

The term before the summation sign does not vary with $\mu$ and we have to find the minimum value of the summed terms. In other words, the MLE is simply the unweighted least squares estimate (LSE) which, as we have shown in Chapter 3, is the plain sample mean. Gauss himself discovered this fact,

---

*I note here from discussions with Dr. M. Feinlieb that he and his colleagues have independently reached the same conclusion.

although the MLE was not explicitly formulated as a method until Fisher's paper, some fifty years ago (57).

### The lower bound on the MLE of the $\mu$

Consider the logarithm of the likelihood for *one* sample value. Let us denote the first two terms of (6.1) above by $R$. At $M = \mu$ the last term is zero. If $x = \mu - \delta$ the last term becomes $-(\delta^2/2\sigma^2)$ and likewise for $x = \mu + \delta$ it is $-(\delta^2/2\sigma^2)$. Then using our approximate formula from 3.3, we find the reciprocal of the variance for a single datum to be

$$\frac{2(R - 0) - \left(R - \dfrac{\delta^2}{2\sigma^2}\right) - \left(R - \dfrac{\delta^2}{2\sigma^2}\right)}{\delta^2} = -\frac{1}{\sigma^2}$$

Note that the result (which is asymptotically exact) is the same whatever the value of $\delta$ so we may view the above as exact. The variance for $n$ values will be the sum of $n$ such terms. As usual we find the variance by taking the expectation of minus the reciprocal of this quantity; and since $\sigma^2$ and $n$ are constants

$$E\left(\frac{\sigma^2}{n}\right) = \frac{\sigma^2}{n}$$

which verifies here that the variance of the MLE of $\mu$ is exactly equal to the lower bound and the MLE is thus absolutely efficient. Also, obviously, as $n$ increases the variance goes to zero; and the estimator is unbiased. Hence the estimator is consistent. There is no point in trying to find a consistent estimator with lower variance.

### Interval estimation on $\mu$

Confidence limits on $\mu$ may be formulated in various ways. To my mind the easiest way is to think of a 95% confidence interval as comprising that set of values of the parameter which would leave the estimate inside their acceptance regions.* Indeed there is an analogy (which like all analogies must not be pushed too far) between a confidence region and an acceptance region. The main difference is that an acceptance region is set up about a value of the parameter whereas a confidence region is set up about an estimate.

**Example 6.2.** In a detailed study of the distribution of plasma clotting time (58) using certain standard sources of thromboplastin the results in 25 patients shown in Table 6.1 were obtained. Further studies show the distribution to be reasonably close to Gaussian, and invoking the central limit theorem (Chapter 17 and reference 7) further reassures us. Let us suppose for simplicity and in the first place that the standard deviation is 1.5

---

*This quick-and-easy treatment will be examined explicitly in Chapter 12.

**Table 6.1. Plasma clotting time in patients on phenylindanedione**

| | | | | |
|---|---|---|---|---|
| 30.6 | 30.0 | 29.0 | 29.6 | 30.6 |
| 31.0 | 29.8 | 28.6 | 29.6 | 30.6 |
| 30.8 | 29.8 | 29.2 | 29.2 | 29.6 |
| 32.8 | 29.6 | 29.0 | 29.8 | 34.4 |
| 32.6 | 28.2 | 28.6 | 31.4 | 34.0 |

SOURCE: Data of Murphy and Denson (58).

seconds, which we know to be approximately correct. We may set ourselves two problems with these data.

First, is the result compatible at the 5% level (two-tailed) with the hypothesis that the true mean for this population is 30 seconds? This is set up as a test of hypothesis. The variance of the sample mean is

$$\frac{(1.5)^2}{25} = 0.09$$

so that the standard deviation of the sample mean is 0.3. At the 5% level the acceptance region will be within 1.96 times this amount (or 0.588) on either side of the surmised mean, i.e.,

$$30.0 - (1.96)(0.3) \quad \text{to} \quad 30.0 + (1.96)(0.3)$$

i.e.,

$$29.412 \quad \text{to} \quad 30.588$$

The sample mean is 30.336 so we do not reject the hypothesis.

Second, what are the 95% confidence limits on the sample mean? As before, the standard deviation of the sample mean is 0.3 and the extreme values of $\mu$ for which the sample mean would just be inside the acceptance region are

$$30.336 - 0.588 \quad \text{and} \quad 30.336 + 0.588$$

or

$$29.748 \quad \text{and} \quad 30.924$$

All intervening values of $\mu$ would be less than 0.588 from the sample value and for any of them the null hypothesis would not be rejected at the 5% level. (We may note incidentally that the surmised mean, 30.0, falls inside this interval as it always will whenever the sample mean of a Gaussian variate falls inside the corresponding acceptance region. [Why?] Thus we have the opportunity to test the hypothesis as a byproduct of computing the confidence intervals. For my part, I find this a much more useful way of looking at data than concentrating attention on the formal test of the hypothesis.)

The important point to note here is that there is a considerable similarity between the two processes, testing the hypothesis and determining the confidence interval, and the calculations overlap; nevertheless they are two logically distinct operations which must not be confused.


## THE DISPERSION PARAMETER

We have noted that this quantity is (for the Gaussian variate) also the variance of the distribution. The method-of-moments estimator is then simply the sample variance:

$$\hat{\sigma^2} = \Sigma \; \frac{(X_i - \mu)^2}{n}$$

or if $\mu$ is unknown and must be estimated $(M)$ from the data

$$\hat{\sigma^2} = s^2 = \Sigma \; \frac{(X_i - M)^2}{n-1} = \frac{T}{n-1} \quad \text{(say)}$$

both of which are unbiased (7). However the MLE in the second situation has $n$ as the denominator and is consequently biased, although consistent. Because of Gaussian normality

$$\frac{\Sigma \, (X_i - M)^2}{\sigma^2} = \frac{(n-1)S^2}{\sigma^2}$$

is distributed as a $\chi^2_{n-1}$ the mean of which is $(n-1)$ and the variance $2(n-1)$. Thus if the true mean is unknown

$$\text{var}\left(\frac{S^2}{\sigma^2}\right) = \text{var}\left(\frac{\chi^2_{n-1}}{n-1}\right) = \frac{1}{(n-1)^2}\,(n-1)$$

$$\frac{1}{\sigma^4}\,\text{var}(S^2) = \frac{1}{n-1}$$

$$\text{var}(S^2) = \frac{\sigma^4}{n-1}$$

For large $n$ the distribution of $S^2$ approaches the Gaussian.

**Interval estimation on $\sigma^2$**

Since if the true mean is unknown

$$\frac{S^2}{\sigma^2} \sim \frac{\chi^2_{n-1}}{n-1}$$

then

$$S^2 \sim \frac{\sigma^2\chi^2_{n-1}}{n-1}$$

and confidence limits can be obtained by using the appropriate values from a $\chi^2$ table.

**Example 6.3.** Consider again the data of Table 6.1. We wish to estimate the variance and set 95% confidence limits on the result. Calculation gives us

$$\Sigma\, x_i^2 = 23{,}069.84$$

$$m\Sigma\, x_i = \underline{23{,}006.822{,}4}$$

$$\Sigma\, (x_i - m)^2 = 63.017{,}6 = g \quad \text{(say)}$$

To obtain the MLE we divide by 25 to get 2.520,704. This estimate, however, we know to be biased. The unbiased estimate is found by dividing by one less than the sample size, 24, which yields 2.625,733...

To find the confidence interval, we note that a priori for the residual sum of squares, $T$,

$$\frac{T}{\sigma^2} \sim \chi^2_{24}$$

The 2.5 and 97.5 percentile points for a $\chi^2_{24}$ distribution are 12.401 and 39.364 respectively. Then we set up the 95% confidence relationship.

$$12.401 \leq \frac{63.017{,}6}{\sigma^2} \leq 39.364$$

Dividing through by 63.0176 gives

$$0.196{,}78\ldots \leq \frac{1}{\sigma^2} \leq 0.624{,}65\ldots$$

Then taking the reciprocals (which inverts the inequality signs)

$$5.081{,}65\ldots \geq \sigma^2 \geq 1.600{,}89\ldots$$

which gives the desired confidence limits.

**Estimation of $\sigma$**

Since the standard deviation is by definition the positive square root of the variance, the relationship is monotonic, the variance being by its nature positive; and by the invariance properties discussed in Chapter 3, the MLE of $\sigma$ is simply the square root of the MLE of $\sigma^2$. This estimator enjoys the properties of consistency, efficiency and asymptotic normality. However, it is easily shown, for finite samples, to be negatively biased. This bias exists for all such estimates of the standard deviation based on an unbiased estimate of the variance, whatever the parent distribution. Granted that the variance of $S$ from a finite sample is not zero, by definition

$$\text{var}(S) = E(S^2) - [E(S)]^2 > 0$$

$$E(S^2) > [E(S)]^2$$

Now the left hand side is $\sigma^2$ because of unbiasedness. Hence taking positive square roots

$$\sigma > E(S)$$

Various methods have been proposed for obtaining an unbiased estimate of $\sigma$. An approximation which is simple and accurate for even small samples from Gaussian distributions was proposed by Gurland and Tripathi (59):

$$\hat{\sigma} = s\left[1 + \frac{1}{4(n-1)}\right]$$

where $s$ is the square root of the unbiased estimate of the variance and $n$ the sample size.

Note that confidence limits on $\sigma$ are easily found by taking the square root of the confidence limits on $\sigma^2$.

**Example 6.4.** From the data of Table 6.1 and Example 6.3 we have

$$s = \sqrt{2.625,733} = 1.620,4\ldots$$

The correction factor for bias is

$$1 + \frac{1}{(4)(24)} = 1.010,416$$

giving

$$\hat{\sigma} = 1.637,3\ldots$$

The confidence limits on $\sigma$ are found directly from those on $\sigma^2$:

$$\sqrt{5.081,6\ldots} = 2.254$$

and

$$\sqrt{1.600,89\ldots} = 1.265$$

**Estimation when neither the mean nor the variance is known**

This problem has again close affinities with testing of a hypothesis on the mean. We know that

$$\frac{M - \mu}{\sqrt{\dfrac{S^2}{n}}} \sim t_{n-1}$$

Thus we estimate the mean and the variance in the usual fashion. Then the 95% confidence limits on $\sigma$ will be given (using the appropriate critical values of $t$) by

$$m - \frac{ts}{\sqrt{n}} \quad \text{and} \quad m + \frac{ts}{\sqrt{n}}$$

**Example 6.5.** Using the data on Table 6.1, we wish to find the 95% confidence limits on $\mu$. We have estimated the mean as 30.336 and the variance as 2.625,733.... We divide the latter by 25 and extract the square root to get 0.324..., which is the estimated standard deviation of the sample mean. The limits on the acceptance region of size 0.05 for the $t$ distribution with 24 degrees of freedom are (2.063,9)(0.324) on either side of the sample mean, which give

$$30.336 - 0.668,87 \quad \text{and} \quad 30.336 + 0.668,87$$

or the confidence interval

$$29.667 \quad \text{to} \quad 31.005$$


## PROBLEMS

**6.1.** A colleague trying a new cholesterol-lowering drug, selected a group of patients with severe "hypercholesterolemia" (i.e., 300 mg/ml or over) for treatment. A paired $t$ test of measurements "before" and "after" treatment showed a reduction in serum level which was highly significant, both biologically and statistically. Discuss the therapeutic implications of these findings.

**6.2.** Reed and Anderson did a massive study on the progeny of mentally retarded parents (60). The issue was the extent to which their reproduction was dysgenic. Parents were grouped as having mental retardation (I.Q. $< 70\%$) or not. Where both parents were retarded the mean I.Q. of the progeny was 74%; where only one was, the mean I.Q. was 90%; where neither was retarded the mean was 107%. Comment on these results.

**6.3.** Compute the MLE of the population mean of the square root of the variate for control subjects in Problem 4.6 and find 95% confidence limits on it. What would the corresponding confidence limits on the square of the value be?

**6.4.** For the data on the other two groups in Problem 4.6 (heterozygous and affected) set 95% limits on the estimated difference between the means of the square roots. Would you accept the hypothesis that there is no difference between the means? What size of test is involved in this statement?

**6.5.** Compute 95% confidence limits on the geometric mean of the data in Table 4.11.

**6.6.** In Example 4.8 compute 99% confidence limits on the ratio of the geometric mean survival of platelets in the animals receiving sulfinpyrazone to those receiving saline.

# 7
# ANALYSIS OF VARIANCE

The reader may have noted a regular alternation in recent chapters between testing of hypotheses and estimation. My treatment of analysis of variance deals entirely with the former whereas regression, which will occupy the next two chapters, is concerned with both, but mainly with estimation. In analysis of variance we have the null hypothesis that none of the means of several groups differ.

## THE FUNDAMENTAL PROBLEM

The treatment of the *t* test in Chapter 5 required that each test be concerned with answering one question. Scientific method could consist of designing experiments each with the objective of answering one question.

Consider in greater detail the sitosterol experiment previously mentioned (49). Three questions were of interest. Does dietary cholesterol affect serum cholesterol level in rabbits? Does sitosterol affect serum cholesterol in these animals? Does it do so only in those receiving cholesterol? Each one could be tested by a separate experiment including a control group receiving no treatment; but this would mean that six groups would be required, three of them controls (which seems excessive). Furthermore it may not be at once apparent what coherence there is to the set of three questions. Are they merely three questions being explored? Or are they facets of a single idea? We can confront both these problems by analysis of variance, which is the subject of the present chapter.

First the matter of scientific coherence. In the first place there is little hope of demonstrating any inhibiting effect from sitosterol on the blood cholesterol unless dietary cholesterol influences blood cholesterol. Hence the first question and the need to test the relationship. Of course there already exists a good deal of evidence on the point, but that is merely a result of the history of research. There might very well not have been any such evidence. Moreover, it is always more or less questionable to compare one's results with data from other sources. There may be some peculiarity in the present animals, the drinking water, or what not which might invalidate the comparison.

Secondly, before exploring possible mechanisms of any such effect, we need to test whether there is any effect at all from sitosterol. Hence the second question.

Then thirdly, granted that sitosterol can be shown to lower blood cholesterol level, we might wonder whether it does so because it regulates endogenous synthesis of cholesterol; or whether its action is on cholesterol absorption. If the latter is the case, sitosterol should have an effect on blood cholesterol only if there is dietary cholesterol, the absorption of which it could block. Hence the need to do the last experiment.

At first sight it might seem that the obvious method is to compare three groups of animals each with the *same* control group; but on a little reflection it will be seen that this might mislead. We set up the null hypothesis in each case that some factor has no effect and there is a risk (represented, under the null hypothesis, by the size of the test or the significance level, as the case may be) of rejecting the null hypothesis falsely just because the samples happen to be unrepresentative. But suppose it is the control group which happened to have unrepresentative values; there would be some risk that not one, but all three, hypotheses would be rejected and all because of the one freak sample of control values. The three tests are clearly not independent. It looks, perhaps, as if we have no choice: either three experiments or the risks from nonindependence.

The problem can be solved by a suitably designed experiment; but to understand the reasoning we shall have to reexamine the $t$ test and construct the more general solution by analogy. Suppose that two values $X$ and $Y$ are taken at random and independently from the same population. The standard method of estimating the variance is used, i.e.,

$$S^2 = \frac{\left[ X^2 + Y^2 - \dfrac{(X+Y)^2}{2} \right]}{2-1} = \frac{X^2}{2} + \frac{Y^2}{2} - \frac{2XY}{2}$$

$$= \frac{(X-Y)^2}{2} \tag{7.1}$$

## PARTITION OF A SUM OF SQUARES

In the long run, it will be helpful to develop the ideas of partition of a sample space and of a sample sum of squares. Let us take the simplest non-trivial case (Fig. 7.1a). Consider two independent, identically distributed Gaussian variates, $X$ and $Y$. The domain of values for $X$ is represented on the horizontal axis, that for $Y$ on the vertical. Any particular outcome can be represented by a point Q with coordinates $x$ and $y$ respectively. Now draw a line OA at 45° passing through the origin. Since $X$ and $Y$ have the same expectation, the representation of our estimate of it must lie on this line; the two coordinates of any point on it will be equal. Drop a perpendicular from Q on OA intersecting OA at B and the $X$-axis at C. Drop a perpendicular B$m$ from B to OC. The perpendicular from Q to OC has height $y$ and intersects OC at a distance $x$ from the origin (by the definition of coordinates). Now the triangle Q$x$C being isosceles (having two angles of 45° each and a right angle), $x$C equals $y$ and OC is therefore the sum $(x + y)$. From similar triangles, $m$ is clearly the midpoint of OC and hence equal to the sample mean. Since OB$m$ is isosceles and right-angled, OB is $m\sqrt{2}$.

Now what exactly have we done with this maneuver? $(OQ)^2 = x^2 + y^2$; that is, $(OQ)^2$ corresponds to what we call the sum of squares of the sample values. Further, since the triangle OQB is right-angled, $(OQ)^2$ is also equal to $(OB)^2 + (BQ)^2$. As we have seen $(OB)^2$ equals twice the square of the sample mean. By subtraction and using the formula

$$\Sigma x^2 - \frac{(\Sigma x)^2}{n} = \Sigma(x - m)^2$$

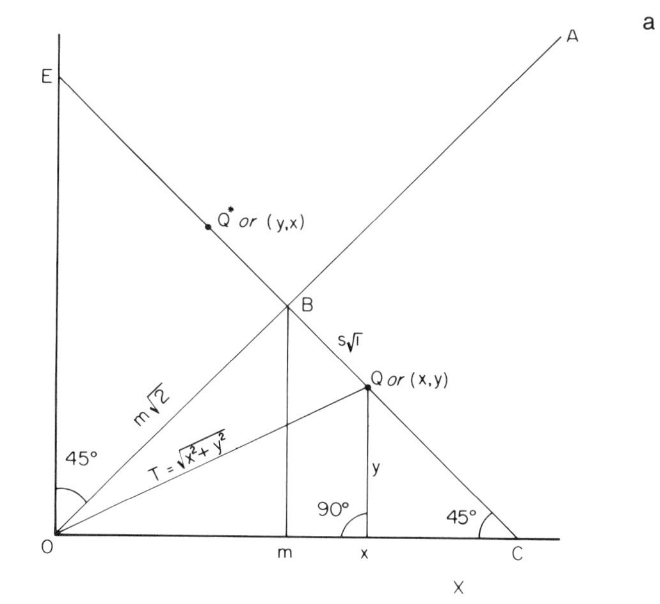


**Figure 7.1.** (*a*) Partition of the sum of squares for two independent random realizations of the same variate. The sample values, $x$ and $y$, are used as the coordinates of the point $Q$. The sum of their squares is the square of the distance $T$ $(=OQ)$ from the origin. It may be decomposed into a component, $(OB)^2$ corresponding to twice the square of the sample mean and $(BQ)^2$ corresponding to the sample estimate of the variance. (Note that, formally, $x$ and $y$ are being used to denote *both* points and distances from the origin. This is inelegant but should not be confusing, and it has the virtue of immediateness.) (*continued*)

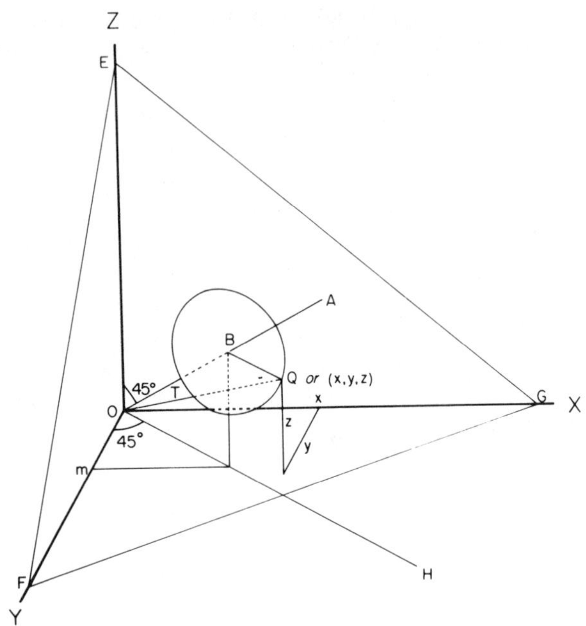


**Figure 7.1.**   (*continued*) (*b*) Partition of the sum of squares of three independent random realizations of the same variate. The values, *x, y,* and *z* are used as the coordinates of the point *Q* in the three-dimensional system. The sum of their squares is the square of the line *OQ*. It may be decomposed into (1) a component *OB* corresponding to $\sqrt{3}$ times the sample mean. The projection of this point on any axis (e.g., the *y*-axis as here) is at a distance from the origin equal to the sample mean, *m*; and (2) *BQ* corresponding to $\sqrt{2}$ times the sample standard deviation. The latter has two degrees of freedom associated with it, i.e., the two-dimensional sample space in which the circle lies.

$(BQ)^2$ equals the sum of the squares of *x* and *y* from the sample mean: when divided by $(n - 1)$ —in this case 1—it yields the sample estimate of the variance. In our usual analysis, we are partitioning the total sum of squares into a part concerned with the mean and a part concerned with the variance.

However, there is another property which must receive attention. Before we take the sample, the outcome is free to occupy any point in the plane: starting from any point it may be varied freely in any two directions at right angles to each other. We therefore say that it has two degrees of freedom. However, if we require the sample to have the particular sample mean *m*, the sample value is restricted to points on the line BC and hence it has only one degree of freedom. Likewise, if we apply the constraint that the total sum of squares be the quantity $T^2$, the sample point is confined to a circle of radius $T$ about the origin: it is not confined to a *straight* line, but there is no point in the plane at which it is free to move in two directions at right angles. Again if we require the sample standard deviation to be $s$, the point is free to move along one of two lines parallel to OA and at a distance of $s$ from it, but it is not free to move at right angles to these lines. If we apply two of these constraints, e.g., a given mean and variance, the sample point is not free to move at all and therefore has no degrees of freedom. If both are nonnegative, we could reconstruct both sample values, $x$ and $y$, if we knew both $s$ and $m$. In the example shown, there are only two possible points that satisfy both: Q and Q*. We recognize then, that one degree of freedom is used in estimating each of the two parameters. We also note that the lines representing (a fixed multiple of) the estimate of the mean (OB) and standard deviation (BQ) are at right angles to each other: their geometric relationship is represented by the Greek word orthogonal (which merely means "right-angled"). No amount of movement along one line has any influence on the value of the other. The property implies uncorrelatedness, and with normal distributions it guarantees statistical independence of the estimates.

This illustration can be extended to a sample of size three (Fig. 7.1b). The arguments are similar, merely somewhat more involved. The total sum of squares $T^2$, represented by $(OQ)^2$, is again partitioned into a sum of squares associated with the mean, $(OB)^2$, and one with the variance $(BQ)^2$. These components are again at right angles. Algebraically this assumes a more familiar form:

$$(OQ)^2 = (OB)^2 + (BQ)^2$$

$$\Sigma x_i^2 - \frac{(\Sigma x_i)^2}{3} = \Sigma(x_i - m)^2$$

As in the previous example we also examine the degrees of freedom. At the outset, the random point representing any sample is free to move in any three directions at right angles, and therefore has three degrees of freedom. For any given mean, the point is confined to the plane EFG (which is in fact not of finite extent as shown in the diagram but extends in all directions). On a plane it has (as before) two degrees of freedom. If we now fix the variance as well, the point is confined to the circle shown in this plane where it still has one degree of freedom. In practice we partition the degrees of freedom in the sample space into one for the mean and two for the variance.

It is clearly impossible to represent geometrically the relationships for samples of sizes greater than three, but more general statements can be made algebraically. We may readily accept that the total sample space is partitioned into two: one degree of freedom for the estimate of the mean $(n - 1)$ degrees of freedom for the variance. In general we decide whether a mean effect (e.g., a treatment) differs significantly from zero by comparing the length of the projection on to any (or all) the axes against the residual mean square error (that part projected on the subspace for variance).

Note further that if the mean were known then B would not have to be on the line EC in Figure 7.1a or the plane EFG in Figure 7.1b. The number of degrees of freedom for variance would be whatever the dimensionality of the sample space is. Note, however, that in both cases the mean must lie on the line OA. Finally, note that the perpendicular is the shortest possible path from Q to the line OA. That is, it is precisely the least-squares estimate of the mean!

Now we take two samples of equal size, $n$, and compute the sample means of each, $m_X$ and $m_Y$. These statistics are samples from a population of means of such samples which have the same mean, $\mu$ and (sample) variance $\sigma^2/n$. Thus we have the estimate, using (7.1) above

$$\frac{s^2}{n} = \frac{(m_X - m_Y)^2}{2}$$

or

$$s^2 = \frac{n(m_X - m_Y)^2}{2}$$

Now we have made $2n$ observations but the only information we have extracted from them is the two sample means: in effect, by this treatment the sample has been reduced to two quantities and from them the variance has been estimated. Thus the resulting estimate has associated with it one degree of freedom. This estimate is based on variation *between* means of two samples.

### The "among" estimate of the variance

Following this line of argument it is easy to see how from *several* sample means we could make an estimate of the variance. We refer to such an estimate based on three or more samples as the "among samples estimate" which we shall denote by $s_a^2$. The actual form of the estimate based on $k$ samples will not be so simple as in (7.1) but in general of the form

$$s_a^2 = [\Sigma m_i^2 - (\Sigma m_i)^2/k]\frac{n}{k - 1} \tag{7.2}$$

This estimate will, of course, have $(k - 1)$ degrees of freedom associated with it. In particular if the distribution from which the samples are taken is Gaussian then

$$\frac{S_a^2(k - 1)}{\sigma^2}$$

follows a chi-square distribution with $(k - 1)$ degrees of freedom.

### The "within" estimate of the variance

Now the variance could be estimated in quite a different way. From *each* sample of $n$ observations the variance could be estimated by the usual formula

$$s^2 = \frac{\Sigma x^2 - (\Sigma x)^2/n}{n - 1}$$

and since the variance is, after all, an average, the estimates obtained in this way may be combined by averaging the values furnished by the $k$ samples; thus:

$$s_w^2 = \frac{\Sigma s_i^2}{k} \tag{7.3}$$

Each estimate has $(n - 1)$ degrees of freedom associated with it. This pooled estimate of the variance is denoted by $s_w^2$ because it is obtained by examining how the values behave *within* groups. Given a Gaussian distribution each quantity $(n - 1)S_i^2/\sigma^2$ is distributed as a chi-square with $(n - 1)$ degrees of freedom. Their sum follows a chi-square with $k(n - 1)$ degrees of freedom (7).

Evidently, given that the samples are mutually independent

$$\frac{k(n - 1)S_w^2}{\sigma^2}$$

is distributed as a chi-square statistic with $k(n - 1)$ degrees of freedom.

### The pooled sample estimate of the variance

Finally, of course, we could pool all the $nk$ sample values, which under the null hypothesis are all independently and identically distributed, to obtain an estimate of the variance which would have $(nk - 1)$ degrees of freedom associated with it. We shall denote it by $S_t^2$.

### Relationships among the estimates

The reader may have painstakingly followed the foregoing with mounting perplexity. How can it solve our problem? Well, it can be shown with some tedious, but not particularly difficult, algebra:

1. That under the null hypothesis (i.e., all the samples are in fact drawn from identical populations), the two estimates $S_a^2$ and $S_w^2$ are uncorrelated, which, if the population is Gaussian, also implies that they are independent.

2. That in all cases the total degrees of freedom $nk - 1$ is the sum of the degrees of freedom within, and among, samples, i.e.,

$$k(n - 1) + k - 1 = nk - 1 \tag{7.4}$$

3. Each estimate of the variance is found by dividing a residual sum of squares from a mean by the appropriate number of degrees of freedom. If the

total, the within, and the among, residual sums of squares are represented by $R_t$, $R_w$, and $R_a$ respectively, i.e.,

$$R_t = S_t^2(nk - 1)$$

$$R_w = S_w^2 k(n - 1)$$

$$R_a = S_a^2(k - 1)$$

then there is an identity which has already been illustrated for two special cases in figures 7.1a and 7.1b

$$R_t = R_w + R_a \tag{7.5}$$

This relationship is useful as a check on the accuracy of the calculations.

4. The relationships in equations (7.4) and (7.5) are true even if the samples are not all of the same size. In such cases the simplest way to compute the sums of squares is to pool all values and find the sum of squares from the common mean, $R_t$; to find the sum of squares of each sample from its own mean and add these sums together to give $R_w$; and find $R_a$ by subtraction.

From the foregoing systems, where all values are taken at random and independently, from one Gaussian distribution or from several Gaussian populations which have the same parameters, then *ex hypothesi* they are estimating the same quantity $\sigma^2$. The only respect in which the estimates differ in general is in the number of degrees of freedom associated with them. Then the statistic with which we will in fact test the null hypothesis is the *ratio* between $S_a^2$ and $S_w^2$. The ratio, as is familiar from Chapter 5, follows the $F$ distribution with the appropriate numbers of degrees of freedom.

We have said previously that the test to be used is a generalization of the two-sample $t$ test between samples of size $n_1$ and $n_2$ respectively. This point we shall now clarify. It can be easily shown (see the appendix to this chapter) with a little rearrangement that the value, not of $t$ but of $t^2$, is the ratio of two estimates of a variance. The numerator is based on one degree of freedom and calculated from the difference between two means (it thus corresponds to $S_a^2$). The denominator is a within-groups estimate, $S_w$, based on $(n_1 + n_2 - 2)$ degrees of freedom. The estimates can be shown to be uncorrelated and hence (assuming normality) independent. Under the null hypothesis, then, the ratio follows an $F$ distribution with 1 and $(n_1 + n_2 - 2)$ degrees of freedom.

Thus, in general, the square of a $t$ statistic with $n$ degrees of freedom follows an $F$ with 1 and $n$ degrees of freedom. (The alert reader will at once recognize a special case. When $n$ is very large, the $t$ distribution is close to the Gaussian. In consequence, its square is more or less the square of a Gaussian variate which follows a chi-square with one degree of freedom.)

It is easy to arrive at a more general result. The $F$ statistic is the ratio of two estimates of the variance. Now these estimates are consistent. Hence when the number of degrees of freedom for the denominator is very large, we may expect that estimate to be for all practical purposes equal to the true variance. Thus the $F$ statistic approaches

$$F = \frac{S^2}{\sigma^2}$$

If it is multiplied by the number of degrees of freedom for the numerator it gives precisely the chi-square statistic with that number of degrees of freedom.

Multiply an $F$ statistic with $n_1$ and $n_2$ degrees of freedom respectively, by $n_1$. Then for large values of $n_2$ it assumes the form of a chi-square statistic with $n_1$ degrees of freedom. The number of degrees of freedom for the chi-square is that associated with the numerator.

The $t$ distribution was established somewhat earlier than the $F$ and tends to persist in its original form although it might with advantage be squared and treated as a special case of the $F$. However, one should recognize that in the $t^2$ distribution, as with the chi-square, the sign of the deviation is lost and hence one-tailed tests cannot be done. (It will be abundantly clear from what has gone before that I do not consider this any great loss.)

**Example 7.1.** Consider further the problem which was propounded early in this chapter. There were four groups of rabbits studied: one receiving basic diet with margarine *(M)*; one receiving cholesterol as well *(MC)*; one receiving margarine and sitosterol *(MS)*; and one receiving all three *(MCS)*. There are four groups, among which there are three degrees of freedom (Table 7.1). Logarithmic transformation of the blood cholesterol levels is both normalizing and variance stabilizing.

The first question which might be asked is whether *any* of these treatments makes any difference whatsoever. The matter can be tested by a simple $F$ test (Table 7.2). It is simplest to calculate the residual sum of squares within the individual groups and pool them; then to compute the total residual sum of squares from the common mean; then, invoking formula 7.5, to find the component among groups by subtraction. In each case the degrees of freedom are computed and divided into the residual sum of squares to give the estimate of the variance. In the fifth column is given the $F$ statistic and the significance level in the last column. The null hypothesis is dismissed as extremely implausible.

So far, so good. But there are two fundamental weaknesses in answering such a general question. If the result leads to rejection of the null hypothesis, it is still far from clear *which* of the treatments is having any effect. The question we have answered is, after all, not one of those which was propounded at the beginning; and the answer is a mere throat-clearing, an ill-formed statement that "there is gold in them thar hills."

But, to extend the analogy, a sufficiency of hill will obscure even quite a large amount of gold. If one, and only one, of the effects is large and the experiment is such that many degrees of freedom for treatments are involved,

**Table 7.1. Level of cholesterol in serum (mg/100 ml) in rabbits 190 days after the start of a special diet**
*Transformed Data**

| | Kind of lipid added to basic diet | | |
| --- | --- | --- | --- |
| Margarine (*M*) | Margarine and cholesterol (*MC*) | Margarine and sitosterol (*MS*) | Margarine, cholesterol and sitosterol (*MCS*) |
| 623 | 1888 | 41 | 1511 |
| 716 | 1702 | 924 | 1083 |
| 914 | 1348 | 1041 | 204 |
| 519 | 519 | 447 | 898 |
| 580 | 1549 | 230 | 792 |
| 771 | 1539 | 929 | 732 |
| 398 | 1270 | 255 | 826 |
| 820 | 1292 | 653 | 602 |
| 633 | 1297 | 505 | 301 |
| 380 | 1378 | 204 | 863 |
| 462 | 1330 | 279 | 556 |

Source: Data of Murphy et al. (49).

*The transform $y = 1000(\log_{10}x)$ is normalizing and variance stabilizing. Transformed values are given to the nearest integer.

so comprehensive a test may fail to detect it. This casual comment may be sustained by statistics, but it would perhaps be laboring the point to prove that needles may be lost in haystacks.

Thus we must honestly confront the three questions with which we started. The estimate of variance among treatments has three degrees of freedom, the same as the number of questions to be answered. This is no coincidence, and in a properly designed* experiment it will perhaps always be so. And in fact one degree of freedom may be used to test each question. This brings us to the notion of a contrast.

## Contrasts

Consider the first question: whether dietary cholesterol has any effect on blood cholesterol in rabbits. The null hypothesis says that there is none and hence that for equal numbers of rabbits the expected sum of the serum cholesterol levels (or in this case their logarithms) should be the same. Now 22 animals receive cholesterol in one of two combinations (*MC,* and *MCS*) and 22 do not (*M* and *MS*). The results are combined and under the null hypothesis the expectation of the difference is zero.

*We are not considering experiments of which the statistician is trying to make sense afterwards. Nor are we discussing artificial meaning derived not from the scientific content of the problem, but from statistical theory.

**Table 7.2. Analysis of the transformed data in Table 7.1**

|  | M | MC | MS | MCS | Total |
|---|---|---|---|---|---|
| $n$ | 11 | 11 | 11 | 11 | 44 |
| $\Sigma y$ | 6,816 | 15,112 | 5,508 | 8,368 | 35,804 |
| $m_y$ | 619.64 | 1,373.82 | 500.73 | 760.73 | 813.73 |
| $\Sigma y^2$ | 4,525,720 | 21,947,892 | 3,920,804 | 7,656,304 | 38,050,720 |
| $\Sigma(y - m_{Y_k})^2$ | 302,278.55 | 1,186,751.64 | 1,162,798.18 | 1,290,538.18 | 8,916,028.73 = within sum of squares, $R_t$ |

$$\Sigma\Sigma(y - m_{Y_k})^2 = 302,278.55 + \cdots + 1,290,538.18 = 3,942,366.55 = \text{pooled within sum of squares}$$

| Source of variation | Degrees of freedom | Sums of squares | Mean square | $F$ | $P$ |
|---|---|---|---|---|---|
| Treatments | $4 - 1 = 3$ | 4,973,662.18* | 1,657,887.39 | 16.821 | $<10^{-4}$ |
| Within | $4(11 - 1) = 40$ | 3,942,366.55 | 98,559.16 | | |
| Total | $44 - 1 = 43$ | 8,916,028.73 | | | |

*Calculated as the difference between the "total" and "within" sum of squares.

Thus if the totals are denoted by

$$T_M + T_{MS} = T_0$$

$$T_{MC} + T_{MCS} = T_C$$

Then

$$T_0 - T_C$$

should be distributed with a mean of zero. What is its variance? Each result has a variance of $\sigma^2$ and each group contains 11 animals. Thus the variance of the total is 44 $\sigma^2$. A suitable test is thus

$$t = \frac{T_0 - T_C}{s\sqrt{44}} \qquad\qquad (7.6)$$

or squaring it,

$$t^2 = F = \frac{(T_0 - T_C)^2}{44s^2}$$

where $s^2$ is estimated from within samples.

More formally we may set up this *contrast* (as it is called) by writing down a set of coefficients by which the individual group totals are multiplied. Thus the numerator of (7.6) consists of the sum of the totals we multiplied by the following coefficients.

| | |
|---|---|
| $M$ | 1 |
| $MC$ | $-1$ |
| $MS$ | 1 |
| $MCS$ | $-1$ |

The number by which the estimate of variance in the denominator is multiplied is the sum of squares of these numbers multiplied by the number in each group, i.e.,

$$11[(1)^2 + (-1)^2 + (1)^2 + (-1)^2] = 44$$

The second question is whether sitosterol has an effect on serum cholesterol. Grouping according to whether or not sitosterol is present is equivalent to using above the coefficients (1), (1), $(-1)$, and $(-1)$, i.e., combining $M$ and $MC$ and comparing them with $MS$ and $MCS$ pooled. The coefficient in the numerator is again 44.

The final question is whether the effect of sitosterol depends on the presence of cholesterol in the diet. Suppose it does not; then the effects of cholesterol and sitosterol are simply additive. In the grouping

$$(M + MCS) \quad \text{versus} \quad (MC + MS)$$

each group contains 22 rabbits receiving margarine, 11 cholesterol, and 11 sitosterol. Under the null hypothesis, the fact that in the first group the same rabbits receive both the cholesterol and the sitosterol and in the second group the animals receiving the sitosterol differ from those receiving the cholesterol does not matter; but if their effects interact (i.e., the null hypothesis is wrong) then the difference in grouping does matter. The appropriate contrast is thus (1), $(-1)$, $(-1)$, and (1).

We have generated three tests from four groups of animals. But this does not at once meet the basic criticism mentioned above about nonindependence of the tests. What has been gained by this maneuver? Would any other three contrasts do as well?

Under the null hypothesis, the estimates of variance from these particular three contrasts are uncorrelated and therefore (under normality) independent. But not any set of contrasts would have this property. The point is easily tested by looking at the coefficients. (We assume, as here, that *there is the same number of animals in each group.*)

The three contrasts will be calculated in problem 7.1. The resulting system has the following properties:

1. The algebraic sum of the coefficients in every contrast is zero. If this is not so, the expected value of the corresponding sum under the null hypothesis is not zero.

2. The *dot products* are all zero—that is, taking any two contrasts, multiplying corresponding terms and adding the results gives a total of zero. Consider, for example, the first two contrasts; we get the following products

```
M        (1)(1)   =    1
MC      (−1)(1)   =   −1
MS       (1)(−1)  =   −1
MCS (−1)(−1)  =    1
Total                 0
```

The reader may verify that the same is true for the other two dot products, i.e., the first and third, and the second and the third, contrasts. In general for $n$ contrasts there will be $n(n - 1)/2$ dot products to be checked. Dot products all zero ensure that the estimates of variance (obtained from each contrast by squaring it and dividing by the appropriate number) are uncorrelated and hence, under normality, independent.

A set of contrasts leading to independent estimates of variance is said to be *orthogonal.* The inner meaning of this term (see below), which comes from theory of vector spaces, need not concern us. But it may be helpful to revert to figures 7.1a and 7.1b. Here we saw that a total sum of squares could be broken down into components. Just so a complete set of orthogonal contrasts breaks the total sum of squares among treatments into components mutually at right angles, each consuming one degree of freedom. For instance the radius QB in Fig 7.1b can be made the hypoteneuse of a right-angled triangle, the other two sides representing an orthogonal partition, so that the sums of squares add up to the total treatment sum of squares. The term *orthogonal contrasts* should be known, since it is in common use among statisticians.

The individual effects may now be tested by dividing the corresponding estimate of variance by the estimate determined within groups. The resulting $F$ tests are interpreted with 1 degree of freedom for the numerator and however many degrees of freedom may be available for the within-groups estimate. Or equally, the square root may be taken and the result interpreted as a $t$ with the number of degrees of freedom within groups.

It is in this final division that a blemish is introduced, since the same denominator is used for all the $F$ tests. Thus, though the numerators are all independent, the $F$ tests are not. However, for a reasonably large number of degrees of freedom within groups, this is a small distortion only, with much smaller chances that freakish behavior of any one group will dominate the system (as is the case with multiple $t$ tests between the individual groups and one control group).


## MATRIX INTERPRETATION OF ORTHOGONAL CONTRASTS

Readers may, if necessary, bypass this section. It is included for completeness and because it reveals a little more of the nice theory underlying the use of orthogonal contrasts. To understand it, however, some knowledge of matrix algebra is necessary such as is discussed in Appendix 1 at the end of this book.

There are four treatment groups and it seems that there must be four degrees of freedom of which, however, only three are available for contrasts. What has happened to the fourth? At this stage the reader may suspect that it has been used up in estimating the overall mean. A little thought will make it clear that this mean is estimated by the equivalent of a contrast in which the coefficients are (1), (1), (1), and (1) and the divisor is, as before, $11(1^2 + 1^2 + 1^2 + 1^2) = 44$. Of course, its expected value, unlike that of the other contrasts, is not zero. Combining this contrast with the other three gives a $4 \times 4$ matrix in which (as the reader may verify for himself) all the dot products are zero and hence all the corresponding totals independent.

This matrix can be further modified by dividing each element by the square root of the sum of squares of the elements in that row. In the present instance, since all the terms are 1 (they differ only in sign), the appropriate divisor is

$$\sqrt{1^2 + 1^2 + 1^2 + 1^2} = 2$$

The resulting matrix has the form          Corresponding to

$$
\begin{bmatrix}
\tfrac{1}{2} & \tfrac{1}{2} & \tfrac{1}{2} & \tfrac{1}{2} \\
\tfrac{1}{2} & -\tfrac{1}{2} & \tfrac{1}{2} & -\tfrac{1}{2} \\
\tfrac{1}{2} & \tfrac{1}{2} & -\tfrac{1}{2} & -\tfrac{1}{2} \\
\tfrac{1}{2} & -\tfrac{1}{2} & -\tfrac{1}{2} & \tfrac{1}{2}
\end{bmatrix}
\qquad
\begin{array}{l}
\text{Mean} \\[1.2em]
\text{Cholesterol} \\[1.2em]
\text{Sitosterol} \\[1.2em]
\text{Interaction}
\end{array}
$$

As before the dot products are zero; but now also the sum of squares of the elements for any line is unity. It will then be evident that the matrix is orthogonal, i.e., the product of the matrix and its transpose is the identity matrix:

$$
\begin{bmatrix}
\tfrac{1}{2} & \tfrac{1}{2} & \tfrac{1}{2} & \tfrac{1}{2} \\
\tfrac{1}{2} & -\tfrac{1}{2} & \tfrac{1}{2} & -\tfrac{1}{2} \\
\tfrac{1}{2} & \tfrac{1}{2} & -\tfrac{1}{2} & -\tfrac{1}{2} \\
\tfrac{1}{2} & -\tfrac{1}{2} & -\tfrac{1}{2} & \tfrac{1}{2}
\end{bmatrix}
\begin{bmatrix}
\tfrac{1}{2} & \tfrac{1}{2} & \tfrac{1}{2} & \tfrac{1}{2} \\
\tfrac{1}{2} & -\tfrac{1}{2} & \tfrac{1}{2} & -\tfrac{1}{2} \\
\tfrac{1}{2} & \tfrac{1}{2} & -\tfrac{1}{2} & -\tfrac{1}{2} \\
\tfrac{1}{2} & -\tfrac{1}{2} & -\tfrac{1}{2} & \tfrac{1}{2}
\end{bmatrix}
=
\begin{bmatrix}
1 & 0 & 0 & 0 \\
0 & 1 & 0 & 0 \\
0 & 0 & 1 & 0 \\
0 & 0 & 0 & 1
\end{bmatrix}
$$

Finally, we may note that the effect of multiplying the vector of the four group totals by this orthogonal matrix is to conserve the total value of the "treatment sum of squares." If the matrix were not orthogonal—for instance if, in general, sample sizes were not equal—this property would not hold.

## OTHER SETS OF CONTRASTS

The foregoing arguments clearly show that for the estimates of a set of components of variance to be independent, they must bear a certain relationship to one another. My contention is that *the set of contrasts appropriate is dictated by the nature of the scientific problem being explored.* The latter is the responsibility of the scientist as such: I have a poor opinion of the experimenter who does not think out clearly and in detail exactly what it is he wants to find out. Too often he ends up by doing an experiment which is inefficient and does not lead naturally to a set of orthogonal tests. For my part I see little point in doing simultaneous studies on different treatment groups among which there are no logical connections. (I realize that there will be many dissident opinions.)

Where there are three degrees of freedom among treatments, once any two contrasts are chosen, there remains one, and only one, contrast orthogonal to both of them. The penalty of a carelessly designed experiment is that one, and sometimes more, of the contrasts may be *scientifically* uninterpretable or at best, of no interest.

However, the same set of data in the example we have just discussed might have been used to answer a different set of questions. Properly, *the investigator should make up his mind which questions he wants to answer before he does the experiment;* otherwise he may be tempted to explore all possible contrasts until he gets a contrast with the highest possible $F$ value and claim it as the "purpose" of the experiment. Where it is chosen *because it yields an extreme F value,* it is somewhat difficult to interpret.

For the sake of illustration we shall pretend that the same set of data had been collected to answer a different set of three questions:

1. Do one or more of the steroids (cholesterol and sitosterol) combined with neutral fat (margarine) have any effect which neutral fat alone does not have?

2. Does sitosterol with or without cholesterol have any effect which cholesterol alone does not have?

3. Does sitosterol alter the effect of cholesterol?

As will be evident, the questions, from a *scientific* standpoint, do not make nearly such good sense as in the first set of contrasts. However we can still interpret the results. Moreover, this new set of contrasts will illustrate the case where coefficients in a contrast assume a value other than 1.

The obvious way to answer question 1 is to compare the total of the $M$ group with that for the $MC$, $MS$, and $MCS$ groups. However, since there are only 11 animals in the first group but 33 in the second, to compensate, the former group must be multiplied by 3, which gives the contrast

$$\begin{array}{ll} M & 3 \\ MC & -1 \\ MS & -1 \\ MCS & -1 \end{array}$$

We verify that the total is zero; then we divide by the sum of the squares of the coefficients multiplied by the number of animals in each group, i.e.,

$$11[3^2 + (-1)^2 + (-1)^2 + (-1)^2] = 132$$

The argument here is that the variance of the first component is

$$\text{var}(3 \times X) = 9 \text{ var}(X)$$

which because of independence is

$$9(11\sigma^2) = 99\sigma^2$$

It is clear in general that the component of variance involves the square of the coefficient in the contrast. The other three components contribute $33\sigma^2$. The second question is answered by comparing $MC$ with $MS$ and $MCS$, ignoring the $M$ group altogether. The contrast is thus 0, 2, $-1$, $-1$ and the divisor is

$$11[0^2 + 2^2 + (-1)^2 + (-1)^2] = 66$$

The final question is answered by a simple comparison between $MS$ and $MCS$. It gives the contrast 0, 0, 1, $-1$ with a divisor of

$$11[0^2 + 0^2 + (1)^2 + (-1)^2] = 22$$

These three contrasts are orthogonal, mutually and to the contrast 1, 1, 1, 1 with divisor 44, which estimates the overall mean.

Note that unlike the previous one, this set of contrasts yields differing divisors; however the sum of the squares associated with the three contrasts is again equal to the total sum of squares among treatments. Detailed calculations are given in Table 7.3.


## EXPERIMENTS INVOLVING MULTIPLE GROUPS

At this stage it is of value to consider a little more closely the types of experiments involving three or more experimental groups. The groups are, in general, distinguishable because the members of each have in common some feature not shared by members of another group. The features are often referred to as *treatments,* a word which must not be construed in any narrow therapeutic sense. It seems to me worthwhile to distinguish between two types of treatment. This distinction is made not for any statistical reason (statistical manipulations are precisely the same for both types). The reason is scientific, mainly the relevancy that the results have to causality. The matter has been discussed in detail elsewhere (3). The two types of treatment are

### 1. Manipulations

Treatments that are imposed from without in accordance with some strategy. The treatment in no way arises intrinsically from the group but is imposed on it in a totally unrelated fashion.

**Example 7.2.** Three groups of hypertensive patients are set up by random allocation from a common population. One group receives 50 mg of hydrochlorothiazide twice a day; one receives 0.25 mg of reserpine twice a day; and one receives one scruple of extract of larks' tongues every time it rains.

**Example 7.3.** Patients with streptococcal sore throats are randomly allocated to three groups. One group receives a placebo; a second receives 125 mg penicillin V by mouth every 6 hours; and the third receives 250 mg penicillin V by mouth every 6 hours.

### 2. Characteristics

These are treatments that *spontaneously* distinguish groups. They are not imposed from without. Whereas manipulations are typical concerns of the experimental scientist such as the laboratory pharmacologist or the Drosophila geneticist, characteristics are the concern of the observing, non-experimental scientist such as the astronomer, the political theorist, the

**Table 7.3 A set of contrasts for example 7.1**

| Scientific question | Alignment of groups | Contrast mean square | $F(1, 40)$ | $t_{40}$ | $p$ |
|---|---|---|---|---|---|
| 1. Do dietary steroids change any effect of dietary margarine on serum cholesterol? | $(3T_M)$ and $(T_{MC} + T_{MS} + T_{MCS})$ | $\dfrac{[(3 \times 6816) - (15{,}112 + 5508 + 8368)]^2}{11[3^2 + 1^2 + 1^2 + 1^2]}$ $= 552{,}512.12$ | 5.606 | 2.368 | .0228 |
| 2. Does dietary sitosterol have any effect on serum cholesterol? | $(2T_{MC})$ and $(T_{MS} + T_{MCS})$ | $\dfrac{[(2 \times 15{,}112) - (5508 + 8368)]^2}{11[2^2 + 1^2 + 1^2]}$ $= 4{,}049{,}350.06$ | 41.085 | 6.410 | $10^{-6}$ |
| 3. Does dietary sitosterol alter the effect of dietary cholesterol on serum cholesterol? | $(T_{MS})$ and $(T_{MCS})$ | $\dfrac{(5508 - 8368)^2}{11[1^2 + 1^2]}$ $= 371{,}800$ | 3.772 | 1.942 | 0.0592 |
| 4. Are there *some* differences among the groups? | | $= 4{,}973{,}662.18$ | 16.821 | — | $< 10^{-4}$ |

epidemiologist and, in many cases (because of ethical problems) the clinician. (There are exceptions—for instance the epidemiologist may manipulate as in the fluoridation of drinking water—but I am talking about *typical* concerns.) Again two examples are appropriate.

**Example 7.4.** Random samples are taken of males aged fifty from Asia, Africa, Europe, the Americas, and Australia, and their blood glucose levels measured after twelve hours of fasting.

**Example 7.5.** Random samples of men aged thirty who are 60 inches, 65 inches, 70 inches and 75 inches tall respectively are interviewed and the number of times they have bumped their heads going through doorways in the last six months recorded for each subject.

The distinction should now be clear. The scientist in Examples 7.2 and 7.3 determines which group shall receive a particular treatment. He does not determine which men shall be Australians or 60 inches tall—these qualities he takes as he finds them. On the other hand he does decide which antihypertensive treatment or which dose of penicillin is given.

There are some treatments that may be rigidly applied in accordance with policy: for instance in broad public health programs, such as disinfesting patients of body lice or the treatment of gonorrhea. It also applies to radical surgery; or to prophylaxis (e.g., use of vitamins). Here it is easy to see in classical statistical terms which is the treatment and which the outcome.*

A close look at the examples quoted will reveal another respect in which there are differences.

In Examples 7.2 and 7.4 there is no natural ordering of the groups. Granted that hydrochlorothiazide and reserpine have more effect than larks' tongues, which should be regarded as the more powerful? Indeed the whole object of the experiment might be to find the answer to that question. But even the answer would tell us nothing about dosage effect, simply what the effects of these two isolated dosages are. In the same way there is no reasonable ordering a priori of the five groups of men whose blood sugars are being measured, unless there is some hypothesis about them which has not been specified. In a general sense what can we mean by rating Australians as

---

*For the sake of completeness we should mention a third group where response depends on treatment and treatment on response. This kind of situation is exemplified by "treatment to effect" which is common, even usual, in clinical medicine. In the management of diabetes with insulin the objective is to control it within limits. A trial dose is given and the blood glucose level falls. If it falls too much the dose is reduced. If it falls insufficiently the dose is increased. Thus both the dosage and the response are random variables and dependent on one another. The dosage largely determines the response, which in turn largely determines how the dose should be modified, which in turn largely determines the new response, and so indefinitely. It is not at all clear which is to be regarded as the treatment (in the statistical, not the therapeutic, sense) and which the effect. Of course in reality the management of diabetes is vastly more complicated than a mere regulation of blood sugar.

This is a miserably inadequate allocation of space to what is an extremely important aspect of clinical medicine, and in a textbook on medical biostatistics at that! It is an area which has been grossly neglected from a theoretical standpoint, and the brevity of this note reflects what is known rather than what ought to be known.

either higher or lower than Europeans? Conversely, when the outcome of the study is known, what does it tell us? The fact that the various groups differ in such-and-such a way that (perhaps) cannot be accounted for by chance may scientifically have only trivial implications.

In Examples 7.3 and 7.5, however, there is a natural ordering. In Example 7.3 three different dosage levels are tried (0, 125 and 250 mg). It would thus be possible from the three outcomes to make an attempt at constructing a dose-response curve. With a little luck or discernment it will be possible to guess what an optimal dosage is, or at least to gain insight as to what further dosages should be explored to find it. Likewise in Example 7.5 there is again a natural ordering (by height), and we might hope to find that height at which the risk of head-bumping exceeds some acceptable level and certain precautions about bumping their heads should be taken. It might conceivably transpire that at some intermediate height—say 71 inches—the risk is at a maximum: that very tall people always take precautions and do not need to be warned.

Now these distinctions are of considerable importance.

1. Where there is a natural *measured* ordering of the treatment, the problem is best viewed as one of regression (see Chapters 8 and 9). If this recommendation is not obvious, let the reader recognize that, whatever the ordering of the groups, analysis of variance will always give the same results, regression will not. This can only mean that regression is making more use of the data about treatment, and presumably it will be in general a more sensitive form of analysis.

2. Where there is no natural ordering or the natural ordering is likely to be trivial (such as arranging places of birth alphabetically), analysis of variance is appropriate.

3. Where there is natural ordering which is not explicitly measured, or the measure is of doubtful relevance, the analyst may take his choice. For example, suppose a psychologist wished to see whether in women high intelligence is partly due to beauty. He is prepared to interpret "intelligence" as meaning "intelligence quotient." How is "beauty" to be analyzed? He might order it as those who have been chosen "Miss America"; those who have not, but have won state titles; those who have won county titles only; and those who have not even won county titles. It would then be possible to do an analysis of variance to see whether there are any differences among the four groups. But the answer would not tell him whether there is a *systematic* relationship between the two characteristics. He might assign these individuals arbitrary scores, e.g., 4, 3, 2, and 1, according to their beauty league and treat the whole matter as a regression problem. This will certainly make his analysis more sensitive. It is a trifle difficult to imagine what the parameters of the regression curve may mean in real terms (see p. 46), but they would show whether the relationship is monotonic or not. It may be, for instance, that women of intermediate beauty are the most intelligent. This fact would

not emerge at all from analysis of variance in which ordering of beauty was totally ignored; but a quadratic regression line might demonstrate a significant negative estimate of the parameter for the quadratic term. But here the utility would end. It would be useless to try from the regression equation to make a precise estimate of the intelligence of a woman whose beauty has an intermediate grade, say the beauty queen for the Mohawk Valley or the panhandle of Texas, each of which comprises several counties.


## ONE-WAY AND TWO-WAY EXPERIMENTS

There is an evident difference between two experiments previously discussed, the antihypertensive study (Example 7.2) and the cholesterol-sitosterol experiment discussed in detail in the earlier part of this chapter. They are similar in that the treatments are essentially of the "nonmeasured" type—although doses are specified they are not really germane, and no attempt is being made to make any general inference about dosage levels. But they differ in respect to balance and symmetry. The cholesterol-sitosterol experiment is balanced. Clearly there are four possible combinations of treatments: with and without sitosterol and with and without cholesterol. In fact all possible combinations are studied and with the same number of animals in each group.

The design might be expressed thus:

|            | Cholesterol |           |            |
|------------|-------------|-----------|------------|
| Sitosterol | Present     | Absent    | Total      |
| Present    | 11 animals  | 11 animals | 22 animals |
| Absent     | 11 animals  | 11 animals | 22 animals |
| Total      | 22 animals  | 22 animals | 44 animals |

The symmetry* is nicely preserved if the sitosterol and the cholesterol are interchanged. We are studying the individual effects of two treatments and the interaction between them. It is thus referred to as a two-way design or more generally as a $2 \times 2$ factorial experiment. It might be generalized in a variety of ways. Cholesterol intake might be tried at three levels and sitosterol at four levels and the appropriate balanced experiment would be a $3 \times 4$ factorial experiment.

On the other hand the antihypertensive experiment does not have this pattern. There are three treatments each tried at one level only and not in combination with any other treatment. There is here no way of constructing a nontrivial two- or three-way display. The experiment is incorrigibly one-

*The symmetry is mathematical not scientific. It is not implied that dietary cholesterol would be an appropriate treatment for hypersitosterolemia. There is little interest in whether or not it would be.

dimensional. I can conceive of instances where such experiments represent a valuable approach; but in my experience they are rare. It is the kind of experiment performed where the investigator wishes to condense a great deal of work into one experiment and get it over quickly, like a cocktail party given to business acquaintances who are not friends. It is a technique used in screening for efficacy among a large number of compounds for which hopeful claims have been made on scanty evidence. It is characteristically the approach of the man on the "fishing expedition," substituting energy for thought. The perceptive reader will recognize my distaste for this kind of experimentation and I take no pains to disguise it.

Undoubtedly there are one-way multiple-group experiments that look like science and not like prospecting. It is commonly necessary to study the effect of several different doses of a drug, and this problem (as we have seen) is best treated as a problem in regression. It is a legitimate use of multiple groups. But I see little point in comparing several alternative treatments except economy, and, as I imply, this is often false economy. Surely if two drugs are to be compared for efficacy, this is best done by a simple experiment involving the two treatments alone, and a $t$ test? What need is there for a control group? If they are both worth testing for efficacy against a control group, is it not usually worth exploring their *combined* effect and making a $2 \times 2$ design? We encounter difficulty in interpreting the antihypertensive experiment as described above. The multiple comparison problem recurs, but what contrasts can we use? There exist orthogonal contrasts such as

(Larks' tongues) v. (Hydrochlorothiazide and reserpin)

(Hydrochlorothiazide) v. (Reserpin)

It is easy to interpret the second which could have been done by studying two groups only. But it is difficult to attach any meaning to the first *which is the only contrast orthogonal to the second,* even if we make the formally unjustified assumption that the larks' tongues are a control group. If we make the first contrast

(Larks' tongues) v. (Reserpin)

which does explore a useful question, the only contrast orthogonal is

(Larks' tongues and reserpin) v. (Hydrochlorothiazide)

which is, if possible, more meaningless than ever.


## ALTERNATIVE METHODS OF MULTIPLE COMPARISONS

To avoid such absurdities there have been several ingenious attempts to provide means of making multiple comparisons. Noteworthy among these are those of Tukey (61), Duncan (62), and Scheffé (63). The first two call for extensive special tables. An up-to-date review of the topic is given by Miller (64). However, it is pitched at a technical level.

Clearly if $n$ independent tests were done at $\alpha = 0.05$ the probability under the null hypothesis of any one test being not rejected is 0.95; and the probability that no hypothesis will be rejected will be $(0.95)^n$. Hence the alternative, that not all will be accepted or (in other words) that at least one will be rejected, i.e., considered "significant" and at least one error of the first kind made, is:

$$1 - (0.95)^n$$

which becomes embarrassingly large as $n$ increases. Of course within any experiment the problem is further complicated by nonindependence of the tests. The methods cited are designed to compensate for these distortions. They are methods I rarely use myself and about which I can write with little authority; my tendency is to regard them as diplomatic methods devised in their kindness by statisticians for answering questions from the results of experiments which are either of formally bad design, or are formally well-designed but with inadequate thought on the scientific plane. They do, of course, have value in analyzing data that are not the result of explicit design but (for instance) have been casually published, as in vital statistics. I am sure that there are instances where the most thoughtful and accomplished scientist may have need of them: for instance where a pharmacologist is studying chemically related drugs and wishes to capitalize on the fact that they will probably have related actions. One interesting consequence of these statistical methods is to throw into sharp focus the epistemological problems underlying testing hypotheses. Recent developments, far too complicated to be discussed here, are highly promising. But in such a text as this, written by a physician for the use of medical scientists, I adhere obstinately to my view that prevention is better than cure, and that elegance of analysis should not be called upon as a remedy for superficial scientific reasoning about the issues, or careless design.


## TWO-WAY ANALYSIS OF VARIANCE

In the discussion on paired $t$-tests it was pointed out that if beforehand an important source of heterogeneity exists, e.g., age, it may make for a more powerful experiment to pair the subjects, giving one of the treatments to one member of each pair. The effect of age (in which there may be little interest) can be effectively eliminated by taking the paired differences for the outcomes and doing the analysis on the differences.

Where three or more treatments are involved it may be profitable to use an analogous method.

**Example 7.6.** For instance, consider the following experiment which we have performed (65). We started with 32 piglets that were divided among four groups given four different diets. Flow-chamber experiments to measure the rate of formation of a "thrombus" (see p. 81) were to be done once on each. Now it was impracticable to do them on all animals at the same time: experiments could be done only on Thursdays and no more than four animals studied each day. Thus the studies had to be spread over a period of eight weeks. This period is sufficiently long to produce a considerable difference in the sizes of the animals. Since the effect of the experiments is to produce an acute arteriovenous fistula, the hemodynamic effects are likely to depend on the size of the animal, which may affect the rate of flow and hence (conceivably) the rate of thrombus formation.*

Thus it seemed reasonable each week to study one piglet from each group. The order in which the piglets in each group were to be studied was decided by randomization at the start. Thus we may think of the process being equivalent to the following.

1. The 32 piglets were randomly assigned to 8 *blocks* of four each, which denote the date on which the experiment was to be done.

2. A piglet from each block was randomly assigned to one of each of the four *treatments.*

Then the experiment was carried out. Thus there is a two-way randomization on two features: one (diet) being of major interest and the other (age at experiment) of minor importance only. This is known as two-way design and belongs to the type known as a completely randomized block design. It enables the following estimates of variance to be studied.

|  | Units | Degrees of freedom |
|---|---|---|
| Total variation | 32 | 31 |
| Variation among weeks | 8 | 7 |
| Variation among treatments | 4 | 3 |
| Other sources |  | 21 |

What are these other sources? Within each cell of the $8 \times 4$ table there is only one observation and hence no degrees of freedom for estimating the variance within (under the null hypothesis that neither diet nor age has an effect). Where, then, have the degrees of freedom gone? They are associated with estimates of variance based on *interaction* between diet and age. It is plausible, for example, to suppose that the effect of diet is cumulative: the longer the animal has been on the diet the greater the effect on its thrombogenic mechanisms. Thus a one-way analysis of variance of animals at the start might show no effects ascribable to differences among the treatments, whereas at two months there might be profound differences. It is this change with time that interaction is testing. The associated number of degrees of freedom is

(Number of treatments − 1)(Number of blocks − 1)

*In the event, this did not prove to be so; nevertheless it seemed to be a precaution worth taking.

Here it is

$$(8 - 1)(4 - 1) = 21$$

Thus the table for analysis of variance will be set up as follows:

| Source | Degrees of freedom |
|---|---|
| Treatments | 3 |
| Blocks | 7 |
| Interaction between treatments and blocks | 21 |
| Total | 31 |

Since there is no "within cells" estimate of the variance, there is no means of testing whether or not interaction is significant. It is then customary to test treatment and block effects against interaction.

*Replication within* is a more satisfactory design and one which we shall illustrate. Suppose there are $t$ treatments and $b$ blocks giving $bt$ cells in all. Suppose that there are $btr$ animals (or other units) so that each of the $bt$ cells contains $r$ values.* The new feature here is that there are $(r - 1)$ degrees of freedom within each cell to estimate the variance. When we pool them from the $bt$ cells there will be $bt(r - 1)$ such degrees of freedom. For definiteness let $t = 4$, $b = 8$, and $r = 5$, giving 160 observations in all. The analysis of variance now assumes the following form:

| Source | Formula | Degrees of freedom Present example |
|---|---|---|
| Treatments | $t - 1$ | 3 |
| Blocks | $b - 1$ | 7 |
| Interaction between blocks and treatments | $(t - 1)(b - 1)$ | 21 |
| Within cells | $bt(r - 1)$ | 128 |
| Total | $btr - 1$ | 159 |

**Example 7.7.** As part of a more elaborate exploration of the characteristics of the distribution of plasma clotting time (58) the following experiment was performed. In four patients samples of blood were taken once a week for six weeks.† On each sample three "blind" determinations of plasma clotting time were done. On other evidence, logarithmic transformation of the measurement is known to be approximately both normalizing and variance stabilizing.

The structure of the experiment is a randomized block "replicated

---

*In other cases we might suppose there is only one animal in each cell but $r$ observations were made on each. Replicate determinations were not available for the flow chamber experiments, since the procedure is lethal.

†Theoretically we have a time-series problem here (see p. 179). With an interval of a week between samples, it may safely be ignored.

within," that is, in effect, there are multiple values within each cell of the two-way table (Table 7.4a). The appropriate calculations are shown in the second part (Table 7.4b). The sample variance over weeks is calculated as before from the appropriate sum of squares obtained by calculating the sum of the squares of the weekly totals each divided by the number of readings it comprises ($3 \times 4$). From this is taken the square of the grand total divided by the total number of values it comprises ($3 \times 4 \times 6 = 72$). Variation among patients is computed analogously although the divisor of the first term is different, each total being based on $3 \times 6 = 18$ values. The sum of squares from which we estimate variation within cells (which is usually called "experimental error") is found by summing the sums of squares computed individually for each cell. The total residual sum of squares is found in the usual fashion and the sum of squares for interaction is found by subtraction (Table 7.4c).

Under the null hypothesis, the quantities obtained by dividing each sum of squares by the corresponding number of degrees of freedom are all estimating the same quantity $\sigma^2$. We may test each estimate against experimental error. All three components are greater than can be readily ascribed to it. In particular, the differences among patients are striking.

A more conservative test is obtained by comparing variations among patients and among weeks against, not error, but interaction. The reason is that, for example, variation among means for patients when estimated from marginal totals may reflect to some extent what is really due to interaction. Theoretically the denominator of the $F$ test should be the estimate based on error if the main treatments are random, and the estimate from interaction if they are not. In most practical situations it is difficult to decide whether choice of a treatment should be regarded as random or not.


## MORE ELABORATE DESIGNS

It will have been obvious enough that the notion of an $a \times b$ factorial experiment can be readily extended to an $a \times b \times c$, i.e., a three-way, experiment; or to a four-way experiment, etc. The subjects in a therapeutic trial of tuberculosis might for instance be treated or not with streptomycin, with para-amino salicylic acid, and with isonicotinic acid hydrazide, yielding a $2 \times 2 \times 2$ factorial experiment; or in a descriptive study subjects might be grouped two ways by sex, four by marital status (married, single, widowed, or divorced), and five ways by continent of birth, giving a $2 \times 4 \times 5$ experiment.

There are appropriate designs for all such experiments, in which special textbooks abound (66, 67). They include elaborate balancing mechanisms for various obtrusive but irrelevant external factors; methods of reducing the somewhat formidable numbers of subjects required, by sacrificing certain information not of particular interest; methods of combining data obtained from several populations, no one of which is large enough to accommodate a complete experiment.

An exposition of these ingenious and mathematically sleek methods would be out of place in this text. They were devised largely for agricultural experiments where, comparatively speaking at least, they are easy to execute. In animal experimentation they are less apposite. Since interpretation depends, for simplicity if not for cogency, on balance and orthogonality, the premature death of some animals from some accidental unrelated cause, with the resulting deficiencies in data, make analysis much more complicated. It is true there are computer methods of artificially balancing a defective structure, but they are elaborate and time-consuming and quite beyond the present enterprise.

When it comes to applying these methods to human beings the difficulties proliferate. The investigator cannot order four dozen unrelated persons with Henoch's purpura the way he could order four dozen C57Bl mice.

**Table 7.4. Variation in plasma clotting time in normal untreated subjects**

a. Basic data expressed as logarithms to the base 10

| Weeks | Subjects | | | |
| | A | B | C | D |
|---|---|---|---|---|
| 1 | 1.215 | 1.210 | 1.301 | 1.210 |
|   | 1.210 | 1.236 | 1.292 | 1.230 |
|   | 1.210 | 1.230 | 1.279 | 1.241 |
| 2 | 1.230 | 1.236 | 1.255 | 1.241 |
|   | 1.230 | 1.236 | 1.274 | 1.230 |
|   | 1.246 | 1.246 | 1.262 | 1.236 |
| 3 | 1.215 | 1.230 | 1.236 | 1.225 |
|   | 1.230 | 1.204 | 1.255 | 1.215 |
|   | 1.204 | 1.230 | 1.241 | 1.236 |
| 4 | 1.188 | 1.250 | 1.265 | 1.220 |
|   | 1.193 | 1.230 | 1.265 | 1.193 |
|   | 1.193 | 1.210 | 1.270 | 1.215 |
| 5 | 1.188 | 1.274 | 1.279 | 1.230 |
|   | 1.182 | 1.270 | 1.301 | 1.236 |
|   | 1.215 | 1.250 | 1.279 | 1.210 |
| 6 | 1.225 | 1.288 | 1.260 | 1.220 |
|   | 1.236 | 1.283 | 1.283 | 1.225 |
|   | 1.204 | 1.270 | 1.274 | 1.199 |

SOURCE: Data of Murphy and Denson (58).

**Table 7.4.** (*continued*)

b. Computations

| Weeks | | A | B | C | D | Total |
|---|---|---|---|---|---|---|
| | | | | Subjects | | |
| 1 | $\Sigma x$ | 3.635 | 3.676 | 3.872 | 3.681 | 14.864 |
| | $\Sigma x^2$ | 4.404,425 | 4.504,696 | 4.997,706 | 4.517,081 | 18.423,908 |
| | $\Sigma(x - m)^2$ | 0.000,016,6... | 0.000,370,6... | 0.000,244.6... | 0.000,494,0... | — |
| 2 | $\Sigma x$ | 3.706 | 3.718 | 3.791 | 3.707 | 14.922 |
| | $\Sigma x^2$ | 4.578,316 | 4.607,980 | 4.790,745 | 4.580,677 | 18.557,646 |
| | $\Sigma(x - m)^2$ | 0.000,170,6... | 0.000,066,6... | 0.000,184,6... | 0.000,060,6... | — |
| 3 | $\Sigma x$ | 3.649 | 3.664 | 3.732 | 3.676 | 14.721 |
| | $\Sigma x^2$ | 4.438,741 | 4.475,416 | 4.642,802 | 4.504,546 | 18.061,505 |
| | $\Sigma(x - m)^2$ | 0.000,340,6... | 0.000,450,6... | 0.000,194,0... | 0.000,220,6... | — |
| 4 | $\Sigma x$ | 3.574 | 3.690 | 3.800 | 3.628 | 14.692 |
| | $\Sigma x^2$ | 4.257,842 | 4.539,500 | 4.813,350 | 4.387,874 | 17.998,566 |
| | $\Sigma(x - m)^2$ | 0.000,016,6... | 0.000,800 | 0.000,016,6... | 0.000,412,6... | — |
| 5 | $\Sigma x$ | 3.585 | 3.794 | 3.859 | 3.676 | 14.914 |
| | $\Sigma x^2$ | 4.284,693 | 4.798,476 | 4.964,283 | 4.504,696 | 18.552,148 |
| | $\Sigma(x - m)^2$ | 0.000,618,0... | 0.000,330,6... | 0.000,322,6... | 0.000,370,6... | — |

| 6 | $\Sigma x$ | 3.665 | 3.841 | 3.817 | 3.644 | 14.967 |
|---|---|---|---|---|---|---|
| | $\Sigma x^2$ | 4.477,937 | 4.917,933 | 4.856,765 | 4.426,626 | 18.679,261 |
| | $\Sigma(x - m)^2$ | 0.000,528,6... | 0.000,172,6... | 0.000,268,6... | 0.000,380,6... | — |
| Total | $\Sigma x$ | 21.814 | 22.383 | 22.871 | 22.012 | 89.080 |
| | $\Sigma x^2$ | 26.441,954 | 27.843,929 | 29.065,651 | 26.921,500 | 110.273,034 |
| | $\Sigma(x - m)^2$ | | | | | 0.061,278,4... |

c. Analysis of variance

The estimates of variance among weeks ($\sigma_w^2$), among persons ($\sigma_p^2$), and among replicates ($\sigma_e^2$) are given by:

$$(4 - 1)\hat{\sigma}_p^2 = \frac{(21.814)^2}{18} + \frac{(22.383)^2}{18} + \frac{(22.871)^2}{18} + \frac{(22.012)^2}{18} - \frac{(89.080)^2}{72} = 0.036,026,11\ldots$$

$$(6 - 1)\hat{\sigma}_w^2 = \frac{(14.864)^2}{12} + \frac{(14.922)^2}{12} + \frac{(14.721)^2}{12} + \frac{(14.692)^2}{12} + \frac{(14.914)^2}{12} + \frac{(14.967)^2}{12}$$

$$- \frac{(89.080)^2}{72} = 0.005,391,94\ldots$$

$$6 \times 4 \times (3 - 1)\hat{\sigma}_e^2 = 0.000,016,6\ldots + 0.000,370,6\ldots + \cdots + 0.000,380,6\ldots = 0.007,053,33$$

| Source of variation | Degrees of freedom | Sum of squares | Estimate of variance | $F$ | $P$ |
|---|---|---|---|---|---|
| Weeks | $6 - 1 = 5$ | 0.005,391,94... | 0.001,078,38 | 7.339 | $<0.001$ |
| Persons | $4 - 1 = 3$ | 0.036,026,11... | 0.012,008,70 | 81.723 | $\ll 0.001$ |
| Interaction between weeks and persons | $5 \times 3 = 15$ | 0.012,807,055 | 0.000,853,80 | 5.810 | $<0.001$ |
| Replicates ("Experimental error") | $6 \times 4(3 - 1) = 48$ | 0.007,053,33 | 0.000,146,94... | | |
| Total | $6 \times 4 \times 3 - 1 = 71$ | 0.061,278,44 | | | |

Once in the study, subjects may not merely die prematurely; they may be lured to a great distance by employment, politics, or love; they may conceive a distaste for the investigator or the treatment and refuse to participate further; they may meddle illegally with the dosage they are supposed to take, try supplementary medication, or pry into a blinding procedure. (I have had all these misfortunes overtake me in clinical studies.)

In my opinion attempts to use in man anything more complicated than, or perhaps even as complicated as, a completely randomized block design, is to court disaster. I have rarely used them myself and I try to discourage other clinicians from doing so.

But I have an even more radical suspicion of complex designs, namely, in the interpretation of the analysis. It represents the divarication between a scientist, to whom mathematics is a method to explore ideas, and the statistician, who is preoccupied with an exhaustive partition of the components of variance. The focus of my concern here is interaction, which is sufficiently important to merit a section on its own.


## INTERACTION

In the cholesterol-sitosterol experiment, interaction emerged as the measure of the extent to which treatment effects are not additive. If the expected effect of cholesterol, $[C]$, in a particular dosage is to raise the serum cholesterol level by $k$ mg per 100 ml and that of sitosterol, $[S]$, to raise it by $z$ mg per 100 ml ($z$ may be either a positive or a negative quantity, i.e., sitosterol might lower serum cholesterol level), then interaction implies that the expected combined effect, $[CS]$, is not the sum of the individual effects, i.e.,

$$[CS] \neq [C] + [S]$$

If $[CS] > [C] + [S]$ the interaction is positive or, as a pharmacologist would say, these treatments are *synergistic*; and if $[CS] < [C] + [S]$ interaction is negative (competitive inhibition, etc.).

In a three-way experiment the effects $[A]$, $[B]$, and $[C]$ of three treatments may exhibit no interaction when taken in pairs, i.e.,

$$[AB] = [A] + [B]$$

$$[AC] = [A] + [C]$$

$$[BC] = [B] + [C]$$

but the total combined effect may exhibit interaction, i.e.,

$$[ABC] \neq [A] + [B] + [C]$$

or

$$2[ABC] \neq [AB] + [AC] + [BC]$$

Such an effect is called a second-order interaction. Enthusiasts for the more elaborate experiments will happily deal with sixth and seventh order interactions; and the resulting analysis of variance will exhibit a precise additivity of the appropriate components (degrees of freedom and sums of squares).

### Scientific interpretation

Interaction, however, is difficult to interpret satisfactorily from a scientific standpoint.

1. There is some peril that "interaction," really a device of accountancy—a statement of the extent to which such-and-such effects are nonadditive—will be reified. A perplexed accountant can always "balance his books" by ascribing discrepancies between credit and debit to "petty cash" or "miscellaneous small donations." If he is stupid as well as perplexed he may eventually come to believe that the terms refer to real phenomena, that is, he may reify them. Eddington (writing about the "reality" of the FitzGerald contraction) puts it admirably: "Apart from deliberate use of the balance-sheet to conceal the actual situation, it is not well adapted for exhibiting realities, because the main function of the balance-sheet is to balance and everything else has to be subordinated to that end" (68). It is, as Chesterton has pointed out (69), rather like the way the more careless biologists used to talk about the "missing link," originally an embarrassing hiatus in the evolutionary theory, but ultimately treated as if it were a real being.

Thus "interaction" is in its *scientific* aspect simply a confession of the inadequacy of our scientific model: a recognition that the addition of effects does not describe what happens. In the cholesterol-sitosterol experiment the presence of interaction would make it necessary to abandon any simple idea that both substances were simply diffusing from the lumen of the gut into the blood stream. Beyond that, it tells us nothing about underlying processes of the interaction; it is a term without interpretable *scientific* meaning. I do not regard a measurement as a meaning.

2. It is, indeed, arguable whether an interaction is even a useful measurement. In the cholesterol-sitosterol experiment, for instance, the treatments were used at two levels only, "present" and "absent." While interaction was substantial at these dosages (13.5% of the total treatment sum of squares), at a different dosage it would more than likely have quite different importance. It is easy to conceive why this should be so, for chemical or biological reasons. But a priori there was little reason, other than convenience, for picking these particular dosages. Hence, the actual measurement has little general utility unless there is a clear scientific theory for interpreting it: for example if the mathematical form of the departure from additivity is known; or at least that values are obtained that relate importance of interaction to the levels of the treatment concerned.

3. But conversely, absence of interaction, even if securely demonstrated by powerful experiments, is not at all the same thing as saying the mecha-

nisms of action are unrelated. An additivity or otherwise of the effects of cholesterol and sitosterol is merely the composite result of what may be numerous types of interactions, the effects of which may cancel each other out. Simplicity is an ideal of description and proof. But we have no right to demand that nature be simple. To sum up all in one term is a "black-box" treatment with a vengeance. It is description without enlightenment.

4. Interaction may be a pure artifact of measurement. Suppose that the effects of cholesterol and sitosterol really are additive, i.e.,

$$[CS] = [C] + [S]$$

In our experiment, however, it was convenient to work with the logarithms of the values because they normalized the data and stabilized the variance. It is evident, if

$$a = b + c$$

and all three exceed 2, that

$$\log a < \log b + \log c = \log bc$$

Thus

$$10 = 6 + 4$$

Taking common logarithms

$$\log_{10}(10) = 1 < 0.7781 + 0.6021 = 1.3802 = \log_{10}24$$

Thus on the natural scale the effects might be additive, on the transformed scale they would exhibit "negative interaction": indeed this is precisely what happened in the cholesterol-sitosterol experiment. Had there been no outside evidence it would have been perilous to conclude that absorption of cholesterol and of sitosterol bore any relationship to each other. The use of logarithms is justified by the fact that, for cholesterol at least, the effect of dietary cholesterol appears to be proportional to the initial cholesterol level.

But what a scientist would mean by an "interaction," which is a mechanism of mutual impact, is a *logical* relationship which must clearly be independent of the scale of measurement and how readings are formally manipulated.* But "interaction" in the statistical sense does not have this invariance.

It will be obvious that if choice of a wrong scale may generate a spurious "interaction" it may just as well obscure a true one. The objectives of transformations (Chapter 4) may include the abolition of inconvenient interactions. While this has computational advantages, it has no scientific merit and, if misinterpreted, may be grossly misleading.

5. All these problems raised with $2 \times 2$ experiments are compounded

---

*Arsenic is not the less poisonous because the amount given is measured in grams rather than ounces.

for more elaborate studies. I can explain what a seventh order interaction is mathematically; but it is beyond my imagination what scientific meaning it could have. This is hardly surprising: black boxes do not become more illuminating by becoming larger or developing more knobs. It is barely possible that a significant interaction in a two-way experiment would suggest to me the next point to be explored; for a three-way or higher experiment, my understanding would falter. As much as anything else it is this lack of illumination which leaves me disinclined to do anything more complex than two-way experiments.


## ILLUMINATION VERSUS DESCRIPTION

The foregoing must not be interpreted as an outright condemnation of complex designs: it represents the viewpoint of a scientist, rather than of a statistician. There is a distinction between their two objectives which is important. Scientists and statisticians are both concerned with testing hypotheses but, in their narrower roles, they have different perspectives. The pure scientist is concerned with hypotheses about mechanisms, the pure statistician with hypotheses about quantities. The distinction has been illustrated previously for new mutations (Chapter 2).

The scientist really wishes to understand the biological relationship between dietary cholesterol and dietary sitosterol. As a step to this he may seek help from the statistician to see whether the measurable predictions of some theory he may have about the interaction are verified.

The statistician, as such, knows nothing about biological mechanisms; the only pertinent currency in which he can deal is how the measurable quantities behave on formal analysis. If he is wise, he will resist the temptation to reify his results or to try to interpret them to the scientist in conceptual terms. The most he can furnish is a prescription for making, in some sense, *the best possible description of how the outcomes relate to the treatments, severally and jointly*. The description will, of course, include point estimates, confidence limits, and the like.

Fortunately, in practice, this gap in interests is bridged by other people less pure in their specialization. Not all observers are narrowly concerned with mechanisms. The analyst is concerned with the Framingham Study because of the light it throws on the mechanism of coronary disease; but the clinician may use the same data to improve the accuracy of his prognoses, that is, he is concerned, as the statistician is, with "black-box" descriptions. On the other hand, unlike the statistician the true, and not merely nominal, *bio*statistician has a genuine interest in biological mechanisms and is prepared, if given the opportunity, to discuss with the scientist what the next step should be—a conceptual rather than a mathematical matter.

The specialist is concerned with his own problem and from his own standpoint, and it is hardly to be wondered at that his solutions are expressed primarily in terms which he himself can understand and manipulate. But as we have seen in the too brief allusion to "treatment to effect," a heavy price may have to be paid when the commendable rigor imposed within a science merely becomes rigidity when put to use elsewhere.

## ANALYSIS OF VARIANCE AND ESTIMATION

Throughout this chapter we have treated analysis of variance as if its objective is the testing of hypotheses. Most textbooks on experimental design (66, 67) also lay a fair amount of stress on using it to estimate the absolute or relative sizes of components of variance. This extension only makes sense if the null hypothesis—that all apparent differences among subgroups are due to the sampling error—is rejected. One may write down the variance of the mean of any particular subtotal in terms of the various components. (We shall have some illustrations of this method in Chapter 11.)

I confess to major misgivings about estimation in this context, for reasons too complicated to deal with here (70, 71). Suffice it to say that while I can readily enough interpret a difference between means (which, should the system ultimately prove deterministic, can be interpreted as a difference in nonrandom magnitudes), I get into difficulties, mainly ontological, in doing the same for components of variances (which in a deterministic system would seem to have no meaning at all). The most profound problems arise in the classical theory of quantitative genetics (71). Until there is some agreement as to what we may mean *empirically* by "random," I would rather abstain from attempting to describe something with no clear meaning and (in the views of many) not even existence.

Those who do not accept this view will have plenty of guidance from standard texts.

## APPENDIX

### The relationship of the $t$ and $F$ distributions

Let two samples of independent values, $x_1, \ldots, x_u$ and $y_1, \ldots, y_v$ be taken from the same Gaussian population with variance $\sigma^2$. The $t$ statistic has the form

$$t = \frac{m_X - m_Y}{s\sqrt{\dfrac{1}{u} + \dfrac{1}{v}}}$$

Where

$$s^2 = \frac{\Sigma(x_i - m_x)^2 + \Sigma(y_j - m_Y)^2}{u + v - 2}$$

Then squaring $t$ and dividing above and below by $(1/u + 1/v)$ gives

$$t^2 = \frac{\dfrac{uv}{u + v}(m_X - m_Y)^2}{s^2} \tag{1}$$

The denominator is clearly an estimate of the variance based on $(u + v - 2)$ degrees of freedom.

Further, since for any random variable $R$

$$\mathrm{Var}(R) = E(R^2) - [E(R)]^2$$

then

$$E(R^2) = \mathrm{var}(R) + [E(R)]^2$$

Now considering the numerator

$$E(M_X - M_Y)^2 = \mathrm{var}(M_X - M_Y) + [E(M_X - M_Y)]^2$$

But $X$ and $Y$ have the same mean and are independent. Hence

$$E(M_X - M_Y)^2 = \mathrm{var}(M_X) + \mathrm{var}(M_Y) + 0^2$$

$$= \frac{\sigma^2}{u} + \frac{\sigma^2}{v} = \frac{\sigma^2(u + v)}{uv}$$

Thus for the entire numerator of (1)

$$E\left[\frac{uv}{u + v}(M_X - M_Y)^2\right] = \frac{uv\sigma^2}{u + v}\left(\frac{u + v}{uv}\right) = \sigma^2$$

Hence the numerator is an estimate (based on one degree of freedom) of the variance. Finally, since the population is Gaussian the numerator and the denominator are independently distributed.

Thus $t^2$ is distributed as an $F$ with 1 and $(u + v - 2)$ degrees of freedom respectively.


## PROBLEMS

**7.1.** Carry through the calculations for the first set of contrasts in Example 7.1 and verify that the sum of squares for the three contrasts equals the total for treatments. Do an appropriate statistical test on each contrast.

**7.2.** Verify that the second set of contrasts for Example 7.1 is orthogonal and construct the appropriate orthogonal matrix.

**7.3.** In an experiment, 60 guinea pigs were randomly assigned to five groups of equal size. The first group received no treatment (—). All members of the other four groups received an injection of heat-killed streptococci into the right hip joint (*S*). As well, the several groups received respectively, systemic butazolidine (*SB*); systemic penicillin (*SP*); both (*SBP*); and systemic cortisone (*SC*). Comment on the design of the experiment. Propose, with reasons, a suitable set of contrasts.

**7.4.** In Table 7.5, are blood urea-nitrogen levels in four groups of subjects with malignant hypertension (21), half of them women, and half with renal disease. Do an appropriate analysis of variance, including a set of contrasts, and comment on the results. (The sizes of the four groups have been artificially balanced by random selection of equal subsets of cases.)

**Table 7.5. Blood urea–nitrogen level in malignant hypertension (mg/100 ml)**

| Men, essential | Women, essential | Men, renal | Women, renal |
|---|---|---|---|
| 14 | 13 | 6 | 20 |
| 16 | 13 | 41 | 20 |
| 20 | 15 | 42 | 20 |
| 22 | 15 | 70 | 62 |
| 23 | 20 | 87 | 64 |
| 25 | 40 | 140 | 79 |
| 28 | 67 | 143 | 84 |
| 33 | 71 | 143 | 100 |
| 36 | 93 | 143 | 103 |
| 56 | 95 | 149 | 121 |
| 68 | 100 | 155 | 170 |
| 93 | 111 | 161 | 173 |
| 131 | 147 | 162 | 183 |
| 155 | 189 | 215 | 250 |

**7.5.** In the course of a study (72) on the effect of dietary phospholipid, the following control values were separately obtained on different occasions in six subjects. (Table 7.6. Measurements are in mg/100 ml.) Test whether

**Table 7.6. Variation in the blood level of serum phospholipid**

| | | Subject | | | |
|---|---|---|---|---|---|
| 1 | 2 | 3 | 4 | 5 | 6 |
| 347 | 204 | 203 | 164 | 185 | 238 |
| 321 | 198 | 218 | 191 | 203 | 253 |
| 331 | 194 | 214 | 179 | 210 | 257 |
| 347 | 205 | 200 | 158 | 208 | 270 |
| 328 | 200 | 198 | 181 | 225 | 257 |

there is heterogeneity among the readings (transformed) in the subjects. Comment on the result.

**7.6.** Platelet adhesiveness (a ratio close to unity) is approximately Gaussian. In the experiment (65) discussed in Example 7.6 the values shown in Table 7.7 resulted. Do an analysis of variance.

**Table 7.7. The adhesive index of platelets in relationship to dietary fat in pigs**

| Age at death (weeks) | Treatment group | | | |
| --- | --- | --- | --- | --- |
| | Control (C) | Neutral fat alone (N) | Neutral fat and cholesterol (NC) | Egg yolk (EY) |
| 10 | 1.04 | 1.36 | 0.99 | 1.34 |
| 11 | 1.08 | 1.04 | 1.10 | 1.16 |
| 12 | 1.05 | 1.38 | 1.18 | 1.21 |
| 13 | 0.91 | 1.08 | 1.02 | 1.23 |
| 14 | 1.02 | 1.16 | 1.50 | 1.34 |
| 15 | 0.71 | 1.49 | 1.04 | 1.16 |
| 16 | 1.01 | 0.96 | 1.26 | 1.27 |
| 17 | 1.12 | 1.10 | 0.75 | 0.94 |

**7.7.** Do an analysis of variance on the data from Table 4.10.

**7.8.** Many investigators of platelet survival plot their results on linear graph paper, draw a line through the points by eye and take the intercept of the line with the abscissa as a measure of mean platelet survival in accordance with a distribution-free property discussed elsewhere (7). Some sets of

**Table 7.8. Duplicate graphical estimates of platelet survival in three sets of data by ten expert observers (in days)**

| | Data set | | | | | |
| --- | --- | --- | --- | --- | --- | --- |
| | 1 Reading | | 2 Reading | | 3 Reading | |
| Observer | First | Second | First | Second | First | Second |
| 2 | 11.3 | 11.7 | 11.3 | 11.8 | 11.1 | 11.2 |
| 3 | 10.2 | 10.1 | 9.8 | 9.8 | 10.4 | 9.6 |
| 4 | 12.0 | 11.5 | 12.0 | 11.0 | 11.5 | 11.3 |
| 5 | 11.4 | 11.2 | 11.6 | 11.2 | 11.4 | 10.8 |
| 6 | 11.6 | 11.2 | 11.2 | 11.3 | 11.2 | 11.2 |
| 8 | 8.8 | 8.9 | 9.4 | 9.6 | 9.5 | 9.2 |
| 9 | 10.9 | 11.7 | 11.3 | 11.2 | 11.4 | 11.3 |
| 10 | 11.4 | 11.4 | 11.6 | 11.5 | 11.3 | 11.2 |
| 12 | 12.0 | 11.0 | 11.5 | 11.0 | 11.0 | 10.8 |
| 13 | 11.0 | 10.5 | 11.0 | 11.0 | 11.0 | 10.8 |

points with known mathematical properties were plotted on such graph paper, photocopied, and distributed to several world experts (73). Each expert unwittingly read the same set of data twice. A random subsample of the results is given in Table 7.8. Analyze the results, assuming that the errors are Gaussian and have homogeneous variances.

# 8
# SIMPLE REGRESSION ANALYSIS

In the analysis of the relationship between two (or more) variables, the statistician may treat them symmetrically or asymmetrically. For definiteness suppose the variables are the amount of exercise a man takes in ergs per day ($E$) and his weight in kilograms ($W$). A symmetrical analysis would explore such questions as whether or not these quantities are statistically independent; and, if not, how closely and in what way they are related and whether the relationship is direct or inverse. In such an approach there is no interest in cause and effect. Logically one might reasonably hold, for example, that—

1. being overweight interferes with taking exercise;
2. lack of exercise leads to increasing weight;
3. lack of exercise and being overweight are attributable to common causes such as watching sports programs on television and following the intercalated recommendations of the manufacturers of beer, rather than swimming or playing football.

The common statistical methods employed in such instances involve measurements of association, of which by far the most widely used are the correlation coefficients.

In an asymmetrical analysis, the statistician wishes to explore how much one variable *depends* on another. The word *depend* here encompasses at least two meanings: dependence in a causal sense, and dependence in a mathematical sense. The former meaning is a commonplace. How far does exercise reduce weight? To be able to answer this question presupposes that one of the variables (exercise) will be manipulated and the other (weight) observed.* In the mathematical sense we are accustomed to exploring such matters as, given the radius, what the area of a circle is. The radius is called by the mathematician the independent variable and the area, the dependent variable. These two designations cannot be used in statistics because of confusion with independence in yet a third sense, the statistical. If I call $X$ an in-

---

*It must not be supposed that the response of one variable to manipulation of the other is used as an empirical criterion of cause. One is obliged, for instance, to keep continually in mind the impact of confounding (3). The exercise itself may have little impact on weight, it may merely interfere with drinking beer. My point is that the notion of causality would demand the relationship. The absence of any discernible response would exclude empirical causality.

dependent variable, there is some risk of the confused supposition that $X$ is independent of $Y$ and therefore unrelated to it. But the idea of the relationship is pertinent. The statistician may ask the question "Given the amount of exercise a man takes what would one estimate his weight to be?" Or equally, "Given his weight what would we estimate his amount of exercise to be?" Here no manipulation is involved; and which one may be treated as the primary variable and which the secondary one may be arbitrarily decided. The method of exploring such problems, regression analysis, is the subject of the next two chapters.

### Terminology

In the simplest case there is a regressor variable which is viewed as either the causal or the predictor variable, as the case may be; and a response, or outcome, variable which results from it or is predicted by it. Simple linear regression is simply dealt with in chapters 8 and 9. Regression is of many different types, some of which call for methods far beyond the scope of this book. For those who are interested in pursuing them and have the necessary mathematical knowledge to do so, an appropriate reference is given (74). In other cases, the general reader may make do with the account of matrix algebra given in Appendix 1. But he should be warned that it will be of little value to him merely to leaf through the text: he should master this brief account thoroughly, and memorize and practice both the operations and the rules.

## CLASSIFICATION OF REGRESSION MODELS

The following classification must not be viewed as a bill of fare. The reader need neither hope nor fear (according to temperament) that all of these methods will be discussed; any more than he believes, in these days of ostentatious catering, that most of the wines which appear on a wine list are actually in the cellar! Many of these models—if one may extend the metaphor—are a La Tâche or a Château Margaux for the grand occasion, to be handled by experts and appreciated by the discerning. The matters discussed in this book are the *ordinaires*. But for reasons to be discussed later, it seems desirable to lay the subject out in generality and to deal with the more treatable ones in this and the following chapters.

A classification can be applied to the regressor and the response variables and to several aspects of each. All possible combinations may be encountered.

### The regressor variable

The regressor variable may be classified according to the following characteristics:

**1. Number.** Weight might be predicted from one variable such as height; or jointly from height and chest circumference; or from three: height, chest circumference, and age; and so on. Where two regressors or more are involved, we speak of multiple regression analysis. (See Chapter 9.)

**2. Degree.** If $Y$ is to be predicted from $x$, it may be profitable to consider the effect not only of $x$ but of $x^2$, of $x^3$, and so on.

Thus the model

$Y = a + bx + U$   is linear,

$Y = a + bx + cx^2 + U$   is quadratic,

and generally

$Y = a + bx + cx^2 + \ldots + U$   is a polynomial,

where, in each case, $U$ is the random component of $Y$. (See Chapter 9: polynomial regression.)

**3. Stochastic structure.** An experimenter may decide before doing his study to use six equally spaced dosage levels on each of four subjects, the levels being multiples of the dosage dispensed by the pharmaceutical company. Here the regressor, dosage, is clearly fixed. This is the so-called Type 1 regression model.

"Random" dosage may be understood in two senses.

a. The dosage may be truly selected by a random process for each subject; the so-called Type 2 regression model.

b. The *intended* dosage may be fixed but the actual dose given *measured* with some error. This Berksonian* model fortunately behaves (with a few minor adjustments) like a Type 1 model.

There is a third (and statistically rather disreputable) situation in which, viewed under some aspects, the regressor is a random variable but the statistician finds it convenient to treat it as fixed. Thus height and weight are jointly distributed random variables, but they may be treated as if the predictor variable (height) is fixed and the weight is a random variable. This supposition could be more readily justified if from a large sample of natural pairs of values, the statistician were to pick out at random four people who were 65 inches tall, five at 66 inches, ten at 67 inches and so on, where the numbers to be selected are decided nonrandomly in advance. In fact this method of reconstruction of a Type 1 model from "Type 2 data" is rarely formally used. Use of the procedure is not usually warranted.

**The response variable**

The response variable also may be classified, according to the following characteristics:

**1. Continuity.** First, it may be some characteristic that either is, or is not, present. The subject may be living or dead; married or single; male or

---

*Medical readers may take comfort from the fact that Berkson is not only a proper statistician but also a proper doctor. There are not many such.

female. (These alternatives admit of no gradations. Except in political invective, a person cannot be half dead. It takes more than the new morality to admit of an intermediate between the single and married states. Some may be of ambiguous sex, but the state still cannot be measured.) Variables of this kind are said to be quantal. The field of statistics that deals with them is called theory of bioassay, since it was largely stimulated by the problems of the estimation of effective dosage where the outcome must be either positive or negative.

Secondly, the outcomes may be discrete but multiple: for example, number of progeny, number of college degrees earned, and so forth. In some situations it may be appropriate to treat diseased states as a succession of steps through which the patient goes on his way from health to death, e.g., health—atherosclerosis—angina  pectoris—coronary  thrombosis—death.  I will leave it to my readers to decide whether in any particular case this is biologically a realistic model.

Thirdly, the outcome may be continuous, as weight, age at death, time until the graft is rejected, blood sugar level, and so on. This third type is the usual subject of regression analysis.

Although it is not within the corpus of classical statistics, we should consider briefly yet a fourth type of regressor which may have a considerable interest in the future. The idea has been discussed elsewhere (1). Suppose that we wished to relate expectation of further survival to the blood uric acid level in a population of men. The higher the level, the more damage to the kidneys we may infer, and hence the worse the prognosis. However, the uric acid level may itself be deleterious: because of its low solubility, it will crystallize at a critical concentration. It then may cause gout, further kidney damage and, it is argued by some, atherosclerosis. The prognosis in years plotted against concentration will show a continuous decline with increasing level, a sudden drop at the critical concentration at which crystallization occurs, and then a further steady decline. The fact that the critical level of crystallization may vary from one patient to another is not the point, merely a matter that may obscure the existence of the discontinuity. Again, the relationship between cardiac function and the size of the heart may yield several discontinuities that represent episodes of coronary occlusion. A heedless use of polynomial regression may merely conceal these highly informative jumps. The reader may encounter references to "splines" or angularities in the regressor curve, which must not be confused with the foregoing (Fig. 8.1).

**2. Number.** Sometimes—though rarely—the outcome may refer to the joint values of two or more characteristics. One might wish to predict adult height and weight jointly from the mid-parental height, the family income, and the age at onset of puberty.

One such study we have done was on the frequencies of the A, B, and O genes for the ABO blood group in the Indian province of Andhra Pradesh (75). The regressor variable was time. The outcomes were, naturally, quantal, but in three, rather than two categories.

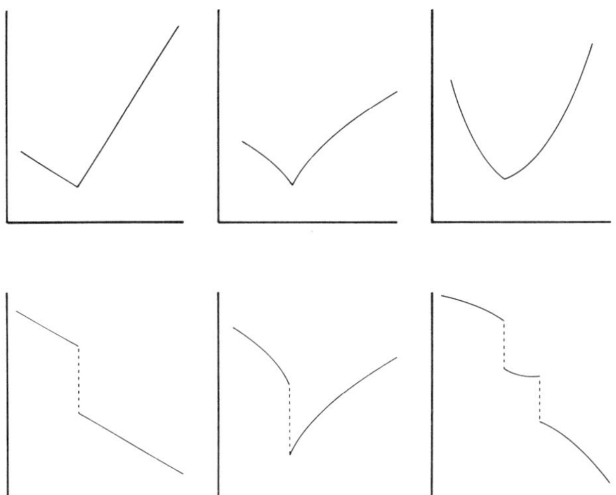

**Figure 8.1.** Splines and discontinuities. In the *upper* three diagrams, the curves are continuous, that is, they could be drawn without lifting the pencil; but on each there is a point at which there is an abrupt change of direction. The "slope" of the curve near that point will depend on which side we are approaching from. Such a curve is known as a spline. In the *lower* three diagrams there are one or more discontinuities, points at which there is an abrupt jump in the function, represented by the dashed lines. There may or may not also be an abrupt change in slope. The curve at the *bottom left* consists of two parallel segments, while the other two show changes in slope on the two sides of the discontinuities.

**3. Serial correlation.** This is a more substantial matter. The experimentor must distinguish between regression made on cross-sectional and on longitudinal data.

Thus if pairs of observations are available—say "age at last birthday" and "height"—orthodox regression analysis of height on age is quite appropriate: model 1 can be used, provided the numbers from each age group are not randomly selected.

On the other hand what happens if yearly readings are available for ten years on each of 50 children? The ordinary supposition in regression analysis is that the errors in the readings (i.e., the departures from conditional expectation given age) are uncorrelated and hence, if the random components are Gaussian, independent. But ten annual readings on a boy are hardly independent: a boy who begins the study unusually tall for his age will at least *tend* to remain unusually tall for his age one year later and, though the influence is gradually decaying, two years later as well. Such a set of readings on one person that are expected to show serial correlation constitutes a *time series*. This subject has given rise to much bloodthirsty algebra but depressingly little useful methodology unless a great many observations per subject are available. The statistician, like the physician, is often much better at rec-ognizng illness than treating it!

Nevertheless, on this subject I have reservations. The problem of serial correlation is at least sometimes an artifact generated by carelessness in defining the sample space and inappropriateness of the model. While it is obvious that deviations from the common regression line will be serially correlated within any one boy when the sample space consists of "all readings on all boys," it is much less obvious that they will be correlated if the sample space is "all readings on this boy." But more, serial "correlation" between adjacent points may merely reflect the fact that the curve fitted is of so incongruous a form as to produce large systematic *biases* which contribute in similar degree to adjacent points. If these biases are not recognized they will be ascribed to the "random" error. The serial "correlation" in the foregoing example would clearly be enhanced if at each measurement the tall boy was wearing high-heeled boots and the small boy was barefoot. This effect would be due entirely to a persistent bias.

I am not unsympathetic to the notion of a time serial analysis;* but most of the examples I have ever heard produced for the enlightenment of us biologists lend themselves to such interpretations. The subject is mentioned here to put the reader on his guard about the problems of analyzing longitudinal data, and because sooner or later he is sure to hear the problem discussed and will be the better for understanding it.

**4. Linearity and nonlinearity in the participation of the parameters.** In the examples quoted above, most generally in the polynomial, the parameters appear in linear form, that is, as constants or simple coefficients of a power of the regressor variable. A common instance in which they are *not* linear is encountered in nuclear medicine. For example the half-life of $^{32}$P is 14.3 days, and mean survival of a radionuclide can always be computed from the half-life by dividing by $\log_e 2 = 0.693$, giving in this case 20.6 days. Then if at time zero the radioactivity emits on average 200 particles per minute, at time $x$ days, it will emit $Y$ particles per minute where

$$E(Y) = 200e^{-x/20.6}$$

Each time 14.3 days is added to $x$, the expected value of $Y$ is halved. This process is *exponential decay*. More generally with an initial count of $c$ and mean survival of $\mu$, the appropriate model is

$$E(Y) = ce^{-x/\mu}$$

If the $y$ and $x$ are known but $c$ and $\mu$ are not, there are special problems in estimation. For example, least squares estimation might be used; but the reader will find no straightforward method of using it from Appendix 1 on matrix algebra. There is, of course, the added problem that at least one com-

---

*The term is used improperly where, as here, there is no true *stochastic* association between readings but merely a recurrent bias; but in practice it is rarely possible to make the distinction.

ponent of $Y$ is Poission error (q.v.); with many systems of radioactive counting, it will depend on $E(Y_i)$. The fundamental distinction is this: that since the value of the regressor is known, any specific function of it is known and may be replaced by a constant. Thus, if $x$ is 2 then $ax^3 = 8a$ which is merely a linear function of $a$. But $e^{-ax} = e^{-2a}$, which is not a linear function of $a$. Matrix methods are designed to solve simultaneous *linear* equations.

This problem of nonlinearity may be tackled by taking logarithms of the $y$:

$$\log y_i = \log c - \frac{x_i}{m} + W_i$$

The method of least squares provides us with unbiased estimates of $\log c$ and of $1/m$. From them estimates of $c$ and $m$ are easily obtained. However, they are not necessarily unbiased. That is, that

$$E(\log C) = \log c$$

does not, in general imply

$$E(e^{\log C}) = c$$

nor does

$$E\left(\frac{1}{M}\right) = \frac{1}{\mu}$$

imply

$$E(M) = \mu$$

The other problem lies over what the effect of transforming the $y$ will be on the characteristics of $U_i$. For this reason errors on the transformed scale have been written with a different symbol $W_i$. The fact that the former has a mean zero and a stable variance will not guarantee that the latter does.

Most people in nuclear medicine ignore these refinements, commonly rendering them unimportant by doing large numbers of counts. Where the extent of labeling is small, however, serious problems exist and then elaborate methods of analysis may be needed, which go far beyond the scope of this book. We ourselves have encountered this problem in estimating mean survival of blood platelets labeled in vivo (to minimize the disturbance of platelet function) and where the size of dose was limited by both physical and pharmacological factors. The theoretical details are considerable. The interested reader who is not overawed by calculus may wish to read the matter in detail (76).

## Motivation

This full classification is presented for the following reasons:

1. For the sake of completeness and in an attempt to unify what may at

first appear to be quite unrelated ideas. For instance, quantal bioassay methods are a type of regression in which the expected value of the response variable is a probability; but they are commonly treated as if they are totally unrelated to regression.

2. Understanding of the scope of a model is sometimes possible only when an example, not covered by it, is explicitly presented. Thus, it is rather difficult, except by example, to explain the difference between a model nonlinear in the regressors, such as a polynominal, which may be handled by matrix methods; and a model that is nonlinear in the parameters, such as the exponential, which cannot.

3. The reader should understand clearly the assumptions underlying a model and not confuse a Type 1 regression model with a Type 2, or a time-series model with either. Many textbooks do not stress these points sufficiently, with the result that regression methods are commonly misapplied.

4. The reader may have brought to his attention the existence of special models that may apply to some problem in which he is involved. He may not find his solution here; but he may find a reference or at least be enabled to approach a statistician with a clear idea of what kind of help he needs.

## SIMPLE REGRESSION ANALYSIS: MODEL 1

The relationship here is of the form

$$Y_i = \alpha + \beta x + U_i$$

Here $Y_i$, the response variable, is in part random ($U_i$) and in part a deterministic function of $x_i$, where $x_i$ is a realization of a quantity which is a mathematical, but not a random, variable. The common method to be used will be the method of least squares (*LSE*) which, as we have seen, corresponds to MLE if the errors are Gaussian.

### Assumptions and properties

These characteristics are intimately related.

1. Least square estimates may be obtained in all cases provided that both the regressor and response variables are measured. Estimates so obtained will always be unbiased (provided the data are unbiased).

2. If we know that the variance of the outcome variable is the same, whatever the value of the regressor, and if the covariances between all pairs of points are zero over the sample space of the data, then three properties follow (from the Gauss-Markov theorem):

The estimates of all the parameters are unbiased.

The estimates have a smaller variance than any other estimate obtained by a linear function of the data.

The variances and covariances of the estimates are known, or at least can be estimated simply from the analysis.

3. If we may further assume that the errors are normally distributed, then

     a. *LSE* is also the maximum likelihood estimate and therefore fully efficient.

     b. The distributions of the estimators are known and may be used in the testing of hypotheses, the setting of confidence limits, and the construction of conditional distributions.


## THE ONE-PARAMETER MODEL

Suppose that it is known that two quantities must be proportionate, but the proportion is unknown. For instance, $y$, the amount of a standard acid solution required to neutralize an amount ($x$) of a given alkaline solution properly mixed, is proportional to $x$. One might find the value of $y$ for various values of $x$ and plot the results. Then, if $x$ is zero, $y$ is zero, that is, the regression line must pass through the origin. Thus the fitted line should be forced through the origin, and the appropriate model is

$$Y = \alpha x_i + U_i \tag{8.1}$$

where the $x_i$ are known; $\alpha$ is a fixed, but unknown, parameter; $U_i$ is a random variable, unknown and only indirectly of importance. We assume that the mean value of $U_i$ is zero. It is readily shown* that the *LSE* of $\alpha$ is given by $a$ where

$$a = \frac{\Sigma x_i y_i}{\Sigma x_i^2}$$

and the variance of $A$, the random variable of which $a$ is a realization is

$$\mathrm{var}(A) = \frac{\sigma^2}{\Sigma x_i^2}$$

where $\sigma^2$ may be estimated by

$$\frac{\Sigma y_i^2 \, \Sigma x_i^2 - \Sigma (x_i y_i)^2}{(n - 1) \, \Sigma x_i^2}$$

**Example 8.1.** The following experiment was performed ad hoc. A radioactive source, $^{137}\mathrm{Cs}$, was counted under identical conditions for various periods of time. We expect the mean count $E(Y)$ to be some multiple ($\alpha$) of the time of counting ($t$). Background radiation was, by comparison, negligible. The results are displayed in Table 8.1. One obvious composite estimate of radioactivity is simply to pool the times and the numbers of counts and divide the former sum into the latter. It will be illuminating, however, to treat the problem as one in regression.

*For example by applying the theory in the Appendix on matrix algebra.

**Table 8.1. The relationship of radioactive count to duration of counting**

| Time ($t$) (seconds) | Counts ($y$) | $\sqrt{yt}$ |
|---|---|---|
| 3 | 3,511 | 102.630 |
| 4 | 4,701 | 137.128 |
| 5 | 5,681 | 168.538 |
| 10 | 11,690 | 341.906 |
| 20 | 23,205 | 681.249 |
| 30 | 34,776 | 1,021.411 |
| 40 | 46,378 | 1,362.028 |
| 50 | 58,085 | 1,704.186 |
| 162 | 188,027 | 5,519.076 |

Direct estimate: 188,027 counts in 162 seconds = 1,160.66 counts per second.
Indirect estimate using the square root transform:

$$\sqrt{\hat{\alpha}} = \frac{\Sigma\sqrt{y}\sqrt{t}}{\Sigma(\sqrt{t})^2} = \frac{\Sigma\sqrt{yt}}{\Sigma t}$$

$$\sqrt{\hat{\alpha}} = \frac{5,519.076}{162} = 34.068,4$$

$$(\sqrt{\hat{\alpha}})^2 = 34.068,4^2 = 1,160.66$$

Regression sum of squares on the transformed scale is

$$\frac{(5,519.076)^2}{162} = 188,025.918,247$$

Residual sum of squares from regression is

$$\frac{188,027 - 188,025.918}{8 - 1} = 0.154$$

compared with the theoretical figure of 0.25

Since the count should be proportional to the mean, the best straight line must go through the origin. The problem is how to determine it. There is a complication because the variances are not homogeneous—to the contrary, since the number of counts in a fixed time follows a Poisson distribution,* the variance equals the expected number of counts for each reading. As discussed in Chapter 4, the transformation

$$Y = \sqrt{X}$$

has a variance close to ¼ if $E(X)$ is 15 or greater. Furthermore, the distribution of the outcome cannot be Gaussian, being discrete. This anomaly matters little because the Poisson with a fairly large mean is approximated well by the Gaussian; however it always has slight positive skewness, which the

*See Chapter 14.

transformation tends to correct (see Chapter 4). A further advantage of the transform is that the value of $\sigma^2$ does not have to be estimated, since, at least for large $\mu$, it is known to be $1/4$.

But, as the reader can readily verify by graphing, if $y$ is *exactly* proportional to $t$, $\sqrt{y}$ is not proportional to $t$. However $\sqrt{y}$ is proportional to $\sqrt{t}$. Thus to preserve linearity in the relationship it is necessary to take the square root of the times also. Note that while $\sqrt{A}$ is an unbiased estimator of $\sqrt{\alpha}$ its square is not an unbiased estimator of $\alpha$ (although it is a consistent one). This is the price we pay for ensuring uniform variance in the errors. However, for such large counts, the negative bias is trivial.

In the analysis, then, we shall not attempt to estimate the slope, but its square root. The appropriate calculations are shown in the bottom of Table 8.1. The quantity of interest, finally, is estimated by squaring the result.


## THE TWO-PARAMETER MODEL

More commonly, there is no reason a priori to suppose that the regression line goes through the origin: that is, there is no reason to believe that the response variable should be proportional to the regressor variable.

For example $Y$, the weight in grams of a tube containing blood, will comprise a fixed part, $\alpha$, which is the weight of the tube and a (mathematically) variable part $\beta x$, the weight of the blood where $x$ is the volume of blood in ml and $\beta$ its (unknown) specific gravity. Now if there is an error of measurement, $U$ with mean zero and a variance of $\sigma^2$, then the model is

$$Y_i = \alpha + \beta x_i + U_i \tag{8.2}$$

The investigator knows neither of the two parameters and proposes to determine the specific gravity of blood by adding further known volumes of blood, thus obtaining $n$ values of $x_i$ with the corresponding values of $y_i$.

Now it may be objected that this example is contrived; and that what an investigator would actually do would be to weigh the dry tube first, then add a known volume of blood and reweigh. The difference in weights divided by the volume is the specific gravity.

Well, of course, this is what would be done in a system in which there is no error of measurement. If, however, there is error, it is clearly safer to rely on multiple observations; but one cannot do multiple weighings on the same dry tube independently: unconscious bias on the part of the observer will always tend to make them agree with each other. Finally, one cannot simply subtract the dry weight $(y_o)$ from the other weights and reduce the problem to the one-parameter case by plotting the differences against the volumes. The differences $(Y_1 - Y_0)$ and $Y_2 - Y_0)$ will not be independent, since both involve the term $Y_0$ which contains its unknown random error. The same applies to all such differences. One cannot reasonably treat the readings for the empty tube as if, unlike the others, it involves no error.

**Note.** A mistake commonly made in analyzing data* is to treat *one* of a number of measurements, *all* of which involve random errors, as if it alone were an exact quantity which may without distortion be subtracted from all the others. The form that this mistake commonly takes is in the analysis of serial observations in a patient after a drug has been administered: the investigator subtracts the initial reading $(Y_0)$ from each of the subsequent ones and fits his regression line through the differences. But the errors of the differences are correlated since they all involve $Y_0$. In consequence simple regression analysis cannot be applied. A better way to handle such a system is to fit a line through all the points which will very likely show the first reading also to be in error. (Even so one must remember that this constitutes a time series, see above.)

### Fitting a regression line

The method in Appendix 1 shows how to fit a regression line for this two-parameter model. The estimates of $\alpha$ and $\beta$ are not independent so that a complete description of the estimators involves not only point and interval estimates of $\alpha$ and $\beta$ and their variances, but the covariance between them as well. Let the mean value of the $x$'s be denoted by $m_x$, and that of the $y$'s by $m_Y$. The least squares estimators yield a line that must go through a point of which the coordinates are $(m_x, m_Y)$. Thus the first step is to find these quantities. Then $\beta$ is estimated by the ratio of the sample estimate of the "covariance" between $x$ and $Y$ and that of the "variance" of $x$. (Strictly speaking, since $x$ is not a random variable it can have neither a variance nor a covariance; but the calculations are carried through as if it were random.) Thus the method-of-moments estimate of the "covariance" of $x$ and $Y$ is given by the identity

$$\text{``cov}(x, y)\text{''} = \frac{\Sigma(y_i - m_Y)(x_i - m_x)}{n - 1} = \frac{\Sigma x_i y_i - (\Sigma x_i)(\Sigma y_i)/n}{n - 1}$$

In words: compute the sum of the products of each $x$ with its own $y$ and subtract the sum of the $x$'s multiplied by the sum of the $y$'s divided by the number of pairs of observations. Next compute the "variance" of $x$ by the usual identity

$$\text{``var}(x)\text{''} = \frac{\Sigma x_i^2 - (\Sigma x_i)^2/n}{n - 1}$$

Then the least squares estimate of the regression slope, $\beta$, is given by $\hat{\beta}$ where

$$\hat{\beta} = \frac{\text{``cov}(x, Y)\text{''}}{\text{``var}(x)\text{''}}$$

---

*I could cite a case in an august medical journal, in which such an error was left unchanged even when I pointed it out in the editorial review of the paper; but I have no wish to make enemies. The mistake is common enough. There are further problems involving time-series analyses (see p. 179).

At this stage we have estimates of both the slope of the line and a point through which it passes, and the rest is easy. Set

$$m_Y = \hat{\alpha} + \hat{\beta}m_x$$

which can be solved for $\hat{\alpha}$, the least squares estimate of the intercept:

$$\hat{\alpha} = m_Y - \hat{\beta}m_x$$

## CONFIDENCE INTERVALS

It is necessary at the start to recognize two distinct kinds of confidence statements which are readily confused.

1. On the expected value of $Y$ given $x_i$. Here there is interest simply in the regression line; that is, the investigator wishes to make statements about the conditional expectation of $Y$ given $x$. This quantity is written, in accordance with the notation of probability algebra (7), as: $E(Y|x_i)$. Here $Y$ is a random variable, $x_i$ is not. For example, he may wish to state what the *average* of the fall in blood pressure ($Y$) is after ingestion of a particular dose ($x_i$) of a drug. From (8.2) its expectation is

$$E(\alpha + \beta x_i + U_i) = \alpha + \beta x_i$$

Clearly, if he had an unlimited amount of information, he would know the answer to this question exactly. This, indeed, would be true even if he had no observation at that particular level of dosage, provided that the mathematical *form* of the curve being fitted is correct.

The conditional expectation of $Y$ given $x$ is in fact the precise regression line of $E(Y)$ on $x$. The sample data have yielded an estimate of this line.

Likewise, if an unlimited amount of data were available, the variance of individual values would be known exactly: hence under the assumption of Guassian normality, the characteristics of the response variable would be precisely known. More generally, the statistician wishes to be able to make statements of the form "I am 95% confident that for $x = 20$ the mean value of $Y$ lies between 120 and 135."

2. On $Y(x_i)$, i.e., a future value of $Y$ given $x = x_i$. Even if the parameters of the regression line were known *exactly* there would necessarily be uncertainty about predicting any *individual* outcome, because of course the whole basis of the model is that $Y(x_i)$ is a random variable. Hence there is need to make confidence statements about individual readings as well. The objective of the statistician is to say "I am 95% confident that given $x = 20$ in a future *individual* result the response variable taken at random will be between 55 and 200."

The distinction is sufficiently important to merit some elaboration. An excellent illustration of the difference is encountered in genetic counseling for multifactorial traits. If two parents are both of somewhat short stature because of "constitution" (i.e., not from any single defective gene), it is known that the best prediction one can make about the *average* height of any future child, $E(Y)$, is the sample average of the two parental heights $(x_i)$. This is the figure that is usually quoted to them. However, the parents may wish to know on the one hand what the prospects are that their child will be tall enough to be a policeman; and on the other hand what the chance will be that the child's height is so much lower than their's that he will need special schooling. An insurance company underwriting the liabilities of such parents might be content with $E(Y|x)$, relying on the fact that with hundreds of thousands of children on average their calculations will come out right and the gains cancel the losses. But the parents are not planning to have hundreds of thousands of children; and what children they do have, they must look after themselves. Hence they will have some concern with the distribution of possible outcomes, not merely the average.

The argument is as follows. Suppose that the true mean value for $Y$, when the regressor variable assumes the value $x_i$, is denoted by $E(Y|x_i)$. Then the observed value of $Y$ for any future value would be distributed about its mean, $E(Y|x_f)$ with a variance of $\sigma^2$. But, of course $E(Y|x_f)$ is not known exactly: we merely have an estimate of it, $\widehat{E(Y|x_f)}$, which will have its own distribution about the true expectation. Its variance will in general be some function of $\sigma^2$, and the values of the regressor variable from which the regression line was constructed. Let us denote this variance by $\sigma^2 k(x)$ where $x$ is used to denote the $x_1, x_2, \ldots x_n$ data values. Finally, suppose that the future point about which we want to make predictions is uncorrelated with the present data. Then the variance of $Y(x_f)$ about $E(Y|x_f)$ would have two uncorrelated parts:

1. The variance, $\sigma^2$, of $Y(x_f)$ about $E(Y|x_f)$

2. The variance, $\sigma^2 k(x)$ of $E(Y|x_f)$ about $E(Y|x_f)$ where $k(x) = 1/n + (x_f - m_X)^2 / \Sigma(x_i - m_X)$. The variance of the sum of any two uncorrelated variables is the sum of the variances (see 7). Hence

$$\text{var}[(Y|x_f) - \widehat{E(Y|x_f)}] = \sigma^2 + \sigma^2 k(x)$$

which is a measure of how far off we may expect to be in predicting a future value of $Y$ for the particular value $x_f$.

Finally, then, the confidence limits on the future value would be quoted in the fashion appropriate for any Guassian random variable as

$$E(Y|x_f) - z\sigma\sqrt{[1 + k(x)]}$$

and

$$E(Y|x_f) + z\sigma\sqrt{[1 + k(x)]}$$

where $z$ is the value of the standard normal deviate appropriate for the required confidence level.

If $\sigma^2$ is unknown and must be estimated from the same data as those from which the intercept and regression line have been estimated, a $t$ value is substituted for $z$ with degrees of freedom two less than the number of pairs of observations. We "lose" two degrees of freedom, one for each of the two parameters estimated. (This adjustment is in accord with common sense. If there were only two points, a line could be fitted that would go exactly through both without any error at all. If two degrees of freedom are subtracted, the distribution of the point estimate of the regression slope minus its expectation and divided by its standard error follows the $t$ distribution with no degrees of freedom, which is not a distribution at all. Furthermore, of course, there would be no estimate of $\sigma^2$ since there is no residual sum of squares.) Thus the unbiased estimate of $\sigma^2$ for the two-parameter model is in general

$$\hat{\sigma}^2 = \frac{\Sigma e_i^2}{n-2} = \frac{\Sigma(y_i - \hat{\alpha} - \hat{\beta}x_i)^2}{n-2}$$

Since the direct computation of the residual errors is tedious, we use a general result in Appendix 1 to obtain the formula

$$\Sigma e_i^2 = \Sigma(y_i - m_y)^2 - \frac{[\Sigma(x_i - m_x)(y_i - m_y)]^2}{\Sigma(x_i - m_x)^2}$$

The foregoing algebra may look rather awesome; but the method is really quite simple. All that is required is a little practice in going through the steps.

**Example 8.2.** My colleague, Dr. Juan Baertl, has kindly furnished the data in Table 8.2. from a survey of nutrition that he conducted in Peru (77). The argument was that if nutrition is marginal, body growth may outstrip its supply of protein; thus as body size increases, the amount of serum protein may be diluted and hence the concentration of protein in serum may fall. Several measures of this relationship were made; those in Table 8.2. are body length (as a measure of size) and the level of serum protein (as a measure of nutrition). We shall condition on body size and estimate the regression of serum protein on it. The plot of the results in Figure 8.2. suggests that it might not be too inaccurate to assume a linear relationship (though we shall have more to say about the matter later). Then the appropriate calculations are carried out in the latter part of Table 8.2. There seems little point in going through them step by step. The entries in the table are self-explanatory and follow what has already been discussed in the text.

## LINEAR COVARIANCE ANALYSIS

Consider a trait that, we have reason to believe, is linearly age-dependent, such as the logarithm of the blood level of cholesterol. We wish to test whether there is a difference in mean values in two (or more) groups, such as men and women or those with, and those without, a particular chromosomal anomaly. The simplest strategy would be to pick random samples from the two groups for a particular age and do a *t* test or whatever might be appropriate. So far, so good. But what age of patient shall we choose? For instance, there is evidence that men with, and without, coronary disease have differing mean blood cholesterol levels but that this difference, conspicuous in patients in their forties, is less in older patients and absent in those in their seventies (78). Thus we will get different answers according to which group we study. In other cases, the age trends may be parallel and the difference would be the same, whatever age we looked at, provided that we confine our attentions to patients of one age. We could legitimately compare the groups if they had the same age distributions. Ignoring age and making our comparison on the marginal distribution, however, will smudge the distributions and reduce the power of the test. The former restriction is irksome in many practical cases where the investigator must be content with what he can get and not discard cases for no better reason than that he wants the groups to balance. Moreover, it will be clear that in grouping by age (in the interests of age-homogeneity), he will be discarding the information arising from the relationship between age and the variate; and as a result the confidence limits may be very much wider than they need be. The point is illustrated in an example presented in detail elsewhere (3).

The question arises, then, as to whether we may attempt to compare the regressions in the various groups. Assuming that the true relationship is linear in both groups (this point must not simply be taken for granted but must be verified, at least graphically), the regressions may differ as to (*a*) position and (*b*) slope. If the regression lines are parallel, it does not matter where we take the difference; it is then customary to compare the intercepts. If the regression lines are not parallel, then they will intersect. Should the intersection occur at a reasonable age, it would be meaningless to ask, without qualification, which group had the higher level. For instance, boys are on average taller than girls; but reference to a standard source (79) shows that girls undergo their adolescent growth spurt earlier than boys and that between the ages of eleven and thirteen years are on average taller. Hence, categorical statements are warranted only if the age is taken into consideration. With a complicated relationship such as the latter where, moreover, there is little difficulty in obtaining good data, it is wisest to keep to statements about the age-specific means. But where data are scanty and the relationship is simple, or can be made so by transformation, we are advised to use covariance analysis.

**Example 8.3.** It will be illuminating to reanalyze the data on blood urea nitrogen (BUN) in various subgroups of patients with malignant hypertension discussed in problem 7.3. In the interests of balance, a random subsample only of the data were analyzed there. In Table 8.3a the data are shown in full, as logarithms to the base 10, together with the ages of the patients. Preliminary calculations (in part b of Table 8.3.) shows some considerable dif-

**Table 8.2. Relationship of height to total serum protein in a severely malnourished population of children**

(a) The Data

| Body length (cms) $(x)$ | Total serum protein (gm/100 ml) $(y)$ | Ratio* | Code* |
|---|---|---|---|
| 49.5 | 5.32 | 0.1075 | + − − |
| 51.5 | 6.96 | 0.1351 | + − + |
| 54.5 | 7.87 | 0.1444 | + + + |
| 55.0 | 6.15 | 0.1118 | + − − |
| 55.5 | 6.20 | 0.1117 | + − − |
| 57.0 | 7.30 | 0.1281 | + + + |
| 57.0 | 6.21 | 0.1089 | + − − |
| 57.3 | 7.33 | 0.1279 | + + + |
| 59.0 | 7.66 | 0.1298 | + + + |
| 59.5 | 7.84 | 0.1318 | + + + |
| 62.0 | 7.43 | 0.1198 | + + + |
| 63.0 | 7.43 | 0.1179 | + + + |
| 63.5 | 6.12 | 0.0964 | + + − |
| 65.0 | 7.63 | 0.1174 | + + + |
| 65.0 | 6.54 | 0.1006 | + + + |
| 66.0 | 7.06 | 0.1070 | + + + |
| 66.0 | 4.70 | 0.0712 | − − − |
| 66.0 | 4.82 | 0.0730 | − − − |
| 66.5 | 6.44 | 0.0968 | + + + |
| 67.5 | 7.20 | 0.1067 | + + + |
| 68.3 | 6.66 | 0.0975 | + + + |
| 69.8 | 3.82 | 0.0547 | − − − |
| 70.0 | 3.69 | 0.0527 | − − − |
| 70.5 | 5.74 | 0.0814 | + + + |
| 71.0 | 4.76 | 0.0670 | − − − |
| 71.5 | 2.43 | 0.0340 | − − − |
| 72.0 | 6.82 | 0.0947 | + + + |
| 72.5 | 4.58 | 0.0632 | − − − |
| 73.5 | 3.93 | 0.0535 | − − − |
| 74.0 | 3.86 | 0.0522 | − − − |
| 75.0 | 4.15 | 0.0553 | − − − |
| 75.0 | 4.30 | 0.0573 | − − + |
| 75.0 | 4.15 | 0.0553 | − − − |
| 77.0 | 3.10 | 0.0403 | − − − |
| 80.0 | 4.65 | 0.0581 | − + + |
| 80.5 | 4.29 | 0.0533 | − + + |
| 80.5 | 3.29 | 0.0409 | − − − |
| 81.0 | 4.65 | 0.0574 | − + + |
| 81.5 | 4.67 | 0.0573 | − + + |
| 84.0 | 5.51 | 0.0656 | − + + |
| 85.5 | 3.71 | 0.0434 | − + − |

SOURCE: Data of Baertl (77).
*The significance of these columns will emerge in the discussion of problem 11.1.

**Table 8.2.** (*continued*)

(b) Analysis

$n = 41$

$$\Sigma x = 2{,}794.4 \qquad \Sigma y = 226.97$$

$$m_x = 68.156 \qquad m_Y = 5.535{,}85$$

$$\Sigma x^2 = 193{,}935.22 \qquad \Sigma y^2 = 1{,}350.858{,}7 \qquad \Sigma xy = 15{,}085.468$$

$$\frac{(\Sigma x)^2}{n} = 190{,}455.399 \qquad \frac{(\Sigma y)^2}{n} = 1{,}256.472{,}705 \qquad \frac{(\Sigma x)(\Sigma y)}{n} = 15{,}469.389{,}46$$

$$\Sigma(x - m_x)^2 = 3{,}479.821 \qquad \Sigma(y - m_Y)^2 = 94.385{,}955 \qquad \Sigma(x_i - m_x)(y_i - m_Y) = -383.921{,}46$$

$$\hat{\beta} = \frac{-383.921{,}46}{3{,}479.821} = -0.110{,}327{,}933{,}5$$

$$\hat{\alpha} = 5.535{,}8 - (0.110{,}327{,}933{,}5)(68.156) = 13.055{,}4$$

Residual sum of squares from the regression line is $\Sigma d^2$ where

$$\Sigma d^2 = \Sigma(y - m_Y)^2 - \hat{\beta}\Sigma(x_i - m_x)(y_i - m_Y)$$

$$= 94.385{,}995 - (-0.110{,}327{,}933{,}5)(-383.921{,}46)$$

$$= 94.385{,}995 - 42.357{,}261{,}31 = 52.028{,}733{,}69$$

$$\hat{\sigma}^2 = \frac{52.028{,}733{,}69}{41 - 2} = 1.334{,}070{,}0$$

$$\hat{\sigma} = 1.155{,}02$$

(c) 95% confidence limits

$t_{39} = 2.0227 \qquad ts = 2.3363$

| $x$ | $k(x) = \dfrac{1}{n} + \dfrac{(x - m_x)^2}{3{,}479.821}$ | $ts\sqrt{k(x)}$ | $ts\sqrt{1 + k(x)}$ | $\widehat{E(Y|x)}$ | 95% confidence limits | | | |
| | | | | | On $E(Y|x)$ | | On $Y(x)$ | |
| | | | | | Lower | Upper | Lower | Upper |
|---|---|---|---|---|---|---|---|---|
| 49 | 0.129,842 | 0.841,85 | 2.483,35 | 7.6493 | 6.8074 | 8.4911 | 5.1659 | 10.1326 |
| 50 | 0.119,119 | 0.806,34 | 2.471,54 | 7.5390 | 6.7326 | 8.3453 | 5.0674 | 10.0105 |
| 55 | 0.074,129 | 0.636,09 | 2.421,35 | 6.9873 | 6.3512 | 7.6234 | 4.5660 | 9.4087 |
| 60 | 0.043,506 | 0.487,31 | 2.386,58 | 6.4357 | 5.9484 | 6.9230 | 4.0491 | 8.8223 |
| 65 | 0.027,253 | 0.385,68 | 2.367,92 | 5.8841 | 5.4984 | 6.2697 | 3.5161 | 8.2520 |
| 68.156 | 0.024,390 | 0.364,87 | 2.364,62 | 5.5359 | 5.1710 | 5.9007 | 3.1712 | 7.9005 |
| 70 | 0.025,367 | 0.372,11 | 2.365,75 | 5.3324 | 4.9603 | 5.7045 | 2.9667 | 7.6982 |
| 75 | 0.037,851 | 0.454,53 | 2.380,10 | 4.7808 | 4.3262 | 5.2353 | 2.4007 | 7.1609 |
| 80 | 0.064,703 | 0.594,28 | 2.410,70 | 4.2291 | 3.6348 | 4.8234 | 1.8184 | 6.6398 |
| 85 | 0.105,923 | 0.760,37 | 2.456,92 | 3.6775 | 2.9171 | 4.4379 | 1.2206 | 6.1344 |
| 86 | 0.115,892 | 0.795,34 | 2.467,97 | 3.5672 | 2.7718 | 4.3625 | 1.0992 | 6.0351 |

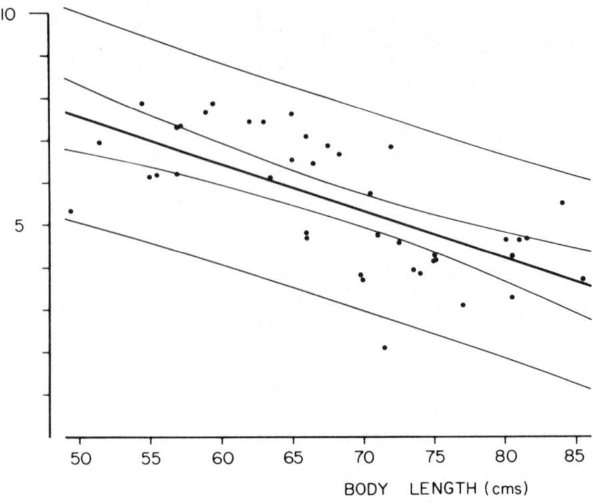


**Figure 8.2.** Linear regression of the serum level of protein in Gm per dL on body length in Peruvian children. The *heavy* line is the regression; the *inner* pair of curves are the 95% confidence limits on the mean; and the *outer* pair are the 95% confidence limits on a future value $Y_f$ given that the length of the child is $x_f$. (Data of Dr. Juan Baertl, 77.)

ferences in the mean ages $(m_x)$ and in the BUN levels $(m_Y)$; the renal patients are younger and have higher mean levels. The signs of the regression slopes are disconcertingly discrepant.

Our detailed analysis of covariance (in part c of Table 8.3.) comprises three steps.

i. Comparison of the mean residual squared error in the four subgroups each from its own regression line. The values shown for the mean squares are reasonably similar and the use of a pooled estimate of the residual mean square is warranted. (A purist would demand that we do a formal test of homoscedasticity. Classical tests do exist; but they are exquisitely sensitive to departures from Gaussian normality. I doubt that for samples of sizes such as those in our table, this degree of heterogeneity would prove statistically significant.) We pool the four sets of residuals and divide by the total degrees of freedom for error to get an estimate of the residual mean square, 0.150,510.

ii. Regression coefficients. The idea underlying the test is similar to that already familiar from analysis of variance. We argue that if the differences among the slopes are entirely due to random error, the variance among the four actual estimates of the regression slopes has an associated sum of squares with three degrees of freedom, which is estimating the same mean square error as the residual sum of squares within subgroups. The most convenient way in which to compute the sums of squares for differences in sample regression slopes is first to pool the residual sums of squares from regression within groups (Table 8.3b). Then we fit a "common regression line (adjusted for means)" by pooling within groups the sums of squares and cross-products from their means. This is equivalent to shifting bodily each set of data points so that the points representing their means are superimposed. I have tried to represent this process in Figure 8.3. In the top row of Figure 8.3. the scatter diagrams are shown for each of the four subgroups with a pair of intersecting lines representing the mean age and the mean logarithm of the BUN. In the *lower left-hand* diagram the four *top* diagrams are superimposed with hairlines aligned. The reader may best get a feel for it by making transparencies of four scatter diagrams and superimposing the hairlines. Thus any differences among the intercepts will be canceled. We then operate as if these shifted values are the true ones and find the residual sum of squares from the best-fitting straight line. Because only one line is being fitted, instead of four (as in *i* above), the residual sum of squares has three more degrees of freedom associated with it. The mean squares are computed; the $F$ test gives a ratio close to unity and is clearly not significant. We may then confidently ignore the differences in regression slopes and pool the residuals from the common regression line to get an estimate of the variance of the residual error 0.150,075, this time with 67 degrees of freedom.

iii. Finally we get to the main problem in which we are interested. Granted that the four groups have differing means for both regressor and response variables, can the latter be accounted for by the former? We may fit a single regression line to the uncorrected data: that is, we make no correction for means. In Figure 8.3. this maneuver is displayed in the lower right diagram. Here it is the coordinate axes that are superimposed, not the axes representing the means (which are all shown). Thus instead of four intercepts we estimate only one; the resulting difference in the number of degrees of freedom (3) can be matched against the difference in the sum of the squares of the residuals. The significance level of the $F$ test is between 0.25 and 0.1. Evidently we could ignore differences among groups altogether and argue as if the BUN is merely a function of age. One might argue, for instance, that the capacity of the patient to withstand severe uremia in malignant hypertension decreases with age and the lower level with increasing age is due to preferential death of those with higher values.

**Comment.** The reader will perceive that such an analysis is not the last word and must be interpreted with caution. Here linear effects of age and sampling error "explain away" differences in means: there is no need to invent an ad hoc explanation for the difference in the means. On the other hand, that does not prove that the effect of age is an adequate explanation. I might argue that the damage to the kidney occurs more slowly in renal disease and the boy has more time to adapt to it. In general, investigators would require stronger evidence to accept covariance analysis when the covariate uncovers a difference rather than explaining one away. The analysis always

**Table 8.3. Relationship of the blood urea nitrogen level ($Y$) to age ($x$) in four subgroups of patients with malignant hypertension**

(a) Raw Data

| Men, essential (ME) | | Women, essential (WE) | | Men, renal (MR) | | Women, renal (WR) | |
|---|---|---|---|---|---|---|---|
| $x$ | $\log_{10} y$ | $x$ | $\log_{10} y$ | $x$ | $\log_{10} y$ | $x$ | $\log_{10} y$ |
| 36 | 1.322 | 30 | 2.276 | 13 | 1.940 | 7 | 1.792 |
| 38 | 1.362 | 42 | 1.602 | 18 | 2.210 | 13 | 2.013 |
| 39 | 1.146 | 43 | 1.301 | 19 | 2.207 | 21 | 1.462 |
| 41 | 1.833 | 48 | 1.826 | 25 | 2.155 | 22 | 1.924 |
| 50 | 1.362 | 48 | 1.176 | 27 | 2.146 | 25 | 2.279 |
| 50 | 1.146 | 49 | 1.176 | 29 | 2.332 | 28 | 1.613 |
| 51 | 2.117 | 50 | 1.114 | 40 | 2.210 | 29 | 2.238 |
| 51 | 1.342 | 51 | 2.167 | 45 | 1.623 | 34 | 1.380 |
| 52 | 2.093 | 52 | 1.968 | 45 | 1.771 | 39 | 1.301 |
| 52 | 1.556 | 54 | 2.045 | 51 | 0.778 | 39 | 2.238 |
| 54 | 2.190 | 54 | 1.978 | 52 | 2.155 | 41 | 2.083 |
| 56 | 1.398 | 55 | 2.000 | 53 | 2.173 | 42 | 2.230 |
| 56 | 1.748 | 57 | 1.851 | 53 | 1.613 | 43 | 1.301 |
| 57 | 1.204 | 63 | 1.114 | 54 | 1.845 | 44 | 2.000 |
| 58 | 1.968 | | | 55 | 2.190 | 45 | 1.556 |
| 59 | 1.301 | | | | | 46 | 2.061 |
| 60 | 1.519 | | | | | 46 | 1.898 |
| 61 | 2.212 | | | | | 50 | 1.806 |
| 62 | 2.207 | | | | | 52 | 2.262 |
| 63 | 1.380 | | | | | 53 | 2.398 |
| 65 | 1.447 | | | | | 59 | 1.301 |
| 76 | 1.380 | | | | | | |

(b) Preliminary calculations

| | ME | WE | MR | WR | Total |
|---|---|---|---|---|---|
| $n$ | 22 | 14 | 15 | 21 | 72 |
| $\Sigma x$ | 1,187 | 696 | 579 | 778 | 3,240 |
| $m_x$ | 53.95 | 49.71 | 38.60 | 37.05 | 45.00 |
| $\Sigma x^2$ | 65,969 | 35,402 | 25,563 | 32,572 | 159,506 |
| $\Sigma(x - m_x)^2$ | 1,924.954 | 800.857 | 3,213.600 | 3,748.952 | 13,706 |
| $\Sigma y$ | 35.233 | 23.594 | 29.348 | 39.136 | 127.311 |
| $m_Y$ | 1.6015 | 1.685,3 | 1.956,5 | 1.863,6 | 1.768 |
| $\Sigma y^2$ | 59.323,743 | 42.111,024 | 59.633,916 | 75.609,488 | 236.678,171 |
| $\Sigma(y - m_Y)^2$ | 2.898,094 | 2.348,393 | 2.213,576 | 2.674,893 | 11.565,800 |
| $\Sigma xy$ | 1,913.500 | 1,164.711 | 1,099.984 | 1,450.642 | 5,628.837 |
| $\Sigma(x - m_x)(y - m_Y)$ | 12.519,500 | $-8.247,857$ | $-32.848,800$ | 0.746,380 | $-100.158$ |
| $b_{Y \cdot x}$ | 0.006,504 | $-0.010,299$ | $-0.010,222$ | 0.000,199 | $-0.007,308$ |
| $\Sigma d^2*$ | 2.816,669 | 2.263,450 | 1.877,802 | 2.674,744 | 10.833,885 |
| $s^2$ | 0.140,833 | 0.188,621 | 0.144,446 | 0.140,776 | 0.154,770 |

*$\Sigma d^2 =$ Sum of squares from the regression line.

Pooled products from individual means

$$\Sigma\Sigma(x - m_x)^2 = 1,924.954 + 800.857 + 3,213.600 + 3,748.952 = 9,688.363$$

$$\Sigma\Sigma(y - m_Y)^2 = 2.898,094 + 2.348,393 + 2.213,576 + 2.674,893 = 10.134,956$$

$$\Sigma\Sigma(x - m_x)(y - m_Y) = 12.519,500 - 8.247,857 - 32.848,800 + 0.746,380$$
$$= -27.830,776$$
$$b = -0.002,875$$

Sum of squares from common regression line (adjusted for means)

$$= 10.134,956 - \frac{(-27.830,776)^2}{9,688.363}$$

$$= 10.055,009$$

---

(c) Analysis of covariance

i. Analysis of residual errors from individual regression lines

| Group | Sum of Squares | Degrees of Freedom | Residual Mean Square |
|---|---|---|---|
| ME | 2.816,669 | 20 | 0.140,833 |
| WE | 2.263,450 | 12 | 0.188,621 |
| MR | 1.877,802 | 13 | 0.144,446 |
| WR | 2.674,744 | 19 | 0.140,776 |
| Pooled residuals | 9.632,665 | 64 | 0.150,510 |

ii. Analysis of differences in regression slopes

| | Sum of Squares | Degrees of Freedom | Mean Square | F |
|---|---|---|---|---|
| Sum of square of residuals from common regression line (adjusted for means) | 10.055,009 | 67 | 0.150,075 | |
| Pooled residuals | 9.632,665 | 64 | 0.150,510 | |
| Variation in regression slopes (by difference) | 0.422,344 | 3 | 0.140,781 | 0.935 |

---

iii. Analysis of difference in intercepts

| | Sum of Squares | Degrees of Freedom | Mean Squares | F |
|---|---|---|---|---|
| Sum of squares of residuals from common regression line (unadjusted) | 10.833,885 | 70 | | |
| Sum of squares of residuals from common regression line (adjusted for means) | 10.055,009 | 67 | 0.150,075 | |
| Variation in intercepts (by difference) | 0.778,876 | 3 | 0.259,625 | 1.730 |

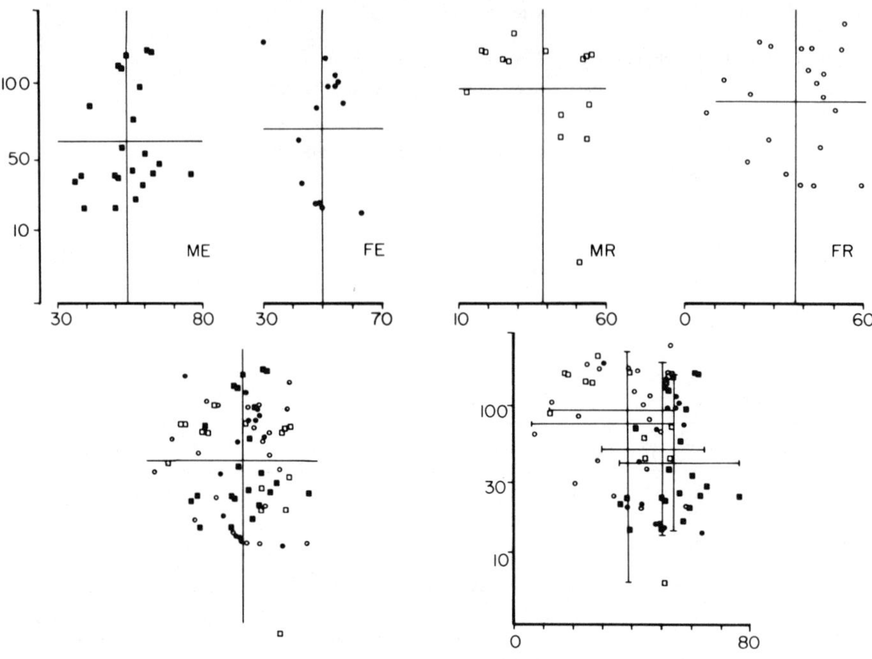

**Figure 8.3.** Linear covariance analysis. In the four *top* diagrams are shown the relationship between blood urea nitrogen and age. The former is plotted on a logarithmic scale. Males (*M*) are represented by squares, females (*F*) by circles. "Essential" malignant hypertension (*E*) is shown in black symbols, renal malignant hypertension (*R*) in open symbols. The four hairlines are superimposed in the *lower left* diagram which "adjusts" for differences in means. In the *lower right* diagram the coordinate axes are superimposed.

involves faith in extrapolation, which is notoriously treacherous. It is always comforting where (as here) there is considerable overlap in the values of the covariate (age).

## PROBLEMS

**8.1.** If the investigator is free to choose values of the regressor variable but the size of sample is limited, what values should he choose in order to get the best estimate of the regression slope?

**8.2.** The following data were obtained in an (unpublished) experiment.

| Subject | Total serum cholesterol ($x$) (mg/dL) | Total serum phospholipid ($Y$) (mg/dL) |
|---------|---------------------------------------|----------------------------------------|
| 1 | 300 | 362 |
| 2 | 278 | 242 |
| 3 | 250 | 227 |
| 4 | 209 | 164 |
| 5 | 240 | 185 |
| 6 | 298 | 250 |

Assume both variables have a lognormal distribution.

    **a.** Fit the best straight regression line of phospholipid on serum cholesterol and find 95% confidence limits on $E(Y|x)$ and on $Y(x)$.

    **b.** What confidence limits would be appropriate if the regression of $Y$ on $x$ does not differ significantly from zero?

    **8.3.** Prove the statements in Example 8.1. that

    **a.** $\sqrt{A}$ is an unbiased estimator of $\sqrt{\alpha}$.

    **b.** $(\sqrt{A})^2$ is a negatively biased estimator of $\alpha$.

    **8.4.** A claim was made that the increase in blood pressure when an arm is plunged in iced water ("the cold pressor response") is a useful predictor of high blood pressure. A counter claim was made that the response is merely a fixed proportion of the blood pressure before the inversion. (See [19] for details.) How would you test the latter claim?

# 9
# POLYNOMIAL AND MULTIPLE REGRESSION

It seems appropriate to deal with these methods in a separate chapter. For one thing, though commonly appropriate, they are used much less than the simpler regression models; for another, while the ideas involved are not difficult, it is tedious and confusing to deal with problems of any complexity without having recourse to matrix algebra. The reader who is not already well up on applied mathematics is advised to master at least the material covered in Appendix 1.

At the start, we distinguish between

1. *Polynomial regression*, which is appropriate where the response variable is a function of a weighted sum of various powers of one regressor variable; and

2. *Multiple regression*, in which the response variable is such a function of two or more distinct regressor variables.

## POLYNOMIAL REGRESSION

To start the development of this idea, we take a nonmedical example. Suppose a man wishes to devise empirically a formula for buying enough paint to cover containers in the form of a cylinder which is 150 cm tall. His formula is to be based on his experience of painting a number of cylinders of the same height but of various diameters. Let—

$y$ = volume of paint in ml;
$x$ = radius of the cylinder in cms.

The cylinder has a curved surface with area $(2\pi x)(150)$, which is a function of $x$; and two ends, each of area $\pi x^2$, which is a function of $x^2$. He has found that a constant amount of paint $(\alpha)$ sticks to the side of the can in which he buys it and cannot be used; further, the techniques for painting ends and the curved surface of the cylinder are different and require different amounts of paint per unit area. Then the appropriate relationship will be of the form

$$y_i = \alpha + \beta x_i + \gamma x_i^2 + u_i$$

which is a quadratic regression containing three parameters ($\alpha$, $\beta$, and $\gamma$) to be estimated, and $u_i$ is an error term. We could generalize this further by supposing the container has a flat bottom, but the top is to be built up with paint into the form of a hemisphere, the volume of which is $2\pi x^3/3$. Allowing yet a third technique for building up volumes we would obtain a cubic regression formula

$$y_i = \alpha + \beta x_i + \gamma x_i^2 + \delta x_i^3 + u_i$$

which contains four parameters ($\alpha$, $\beta$, $\gamma$, and $\delta$).

These basic ideas prepare us for the model that the surface area of the human body (which is geometrically a much more complex and irregular figure) may have a far from simple relationship to height or to chest circumference. It is altogether too much to hope that the simple one-parameter model—the idea that surface area should be proportional to height—will be anything like adequate.

At this stage the reader does well to read again the comments in Chapter 2 about the distinction between modeling and description. The mathematical structures of the formulae for painting the cylinder are clearly models; that for estimating the surface area of the body much less so. Many uses of polynomial models are even less articulate. The scientist who is secure about the structure of the model will, of course, wish to estimate the appropriate parameters and to explore the characteristics of the residual errors—whether they are indeed normally distributed, and so forth. Testing the goodness of fit will be of only minor importance. At the other extreme, the purely descriptive study, the scientist may be groping for a formula with no idea beforehand even about what its degree may be. He may be simultaneously trying to find the mathematical form of the regression equation, estimate its parameters, and explore the characteristics of the residual errors. I know of no theoretically sound basis for doing so, though certain traditions have been built up in practice. I suspect that the lack of theory reflects the fact that mathematics cannot really be expected to discharge what is properly the function of the scientist.

## THE QUADRATIC MODEL

Suppose that the relationship

$$Y_i = \beta_0 + \beta_1 x_i + \beta_2 x_i^2 + U_i$$

holds under the usual conditions (the random components, $U_i$, being Gaussian, independent, and homoscedastic; the $x$'s not random variables; none of the parameters known).

Then the model may be written in the matrix equation

$$
\begin{bmatrix} y_1 \\ y_2 \\ \cdot \\ \cdot \\ y_n \end{bmatrix} = \begin{bmatrix} 1 & x_1 & x_1^2 \\ 1 & x_2 & x_2^2 \\ \cdot & \cdot & \cdot \\ \cdot & \cdot & \cdot \\ 1 & x_n & x_n^2 \end{bmatrix} \begin{bmatrix} \beta_0 \\ \beta_1 \\ \beta_2 \end{bmatrix} + \begin{bmatrix} u_1 \\ u_2 \\ \cdot \\ \cdot \\ u_n \end{bmatrix}
$$

$$n \times 1 \qquad\qquad n \times 3 \qquad\quad 3 \times 1 \qquad\quad n \times 1$$

or more compactly, in matrix notation*

$$\mathbf{y} = \mathbf{X}\boldsymbol{\beta} + \mathbf{u}$$

Using the equation given to us by the Gauss-Markov theorem (q.v.) that the least squares estimate of the vector $\beta$ is the vector $\hat{\beta}$ where

$$\hat{\beta} = (\mathbf{X}'\mathbf{X})^{-1}\mathbf{X}'\mathbf{y}$$

we first compute $\mathbf{X}'\mathbf{y}$

$$
\begin{bmatrix} 1 & 1 & \cdot & \cdot & 1 \\ x_1 & x_2 & \cdot & \cdot & x_n \\ x_1^2 & x_2^2 & \cdot & \cdot & x_n^2 \end{bmatrix} \begin{bmatrix} y_1 \\ y_2 \\ \cdot \\ \cdot \\ y_n \end{bmatrix} = \begin{bmatrix} \Sigma y_i \\ \Sigma x_i y_i \\ \Sigma x_i^2 y_i \end{bmatrix}
$$

$$3 \times n \qquad\qquad n \times 1 \qquad\qquad 3 \times 1$$

Likewise, by writing $\mathbf{X}$ above in place of $\mathbf{y}$, the reader may assure himself

$$
\mathbf{X}'\mathbf{X} = \begin{bmatrix} n & \Sigma x & \Sigma x^2 \\ \Sigma x & \Sigma x^2 & \Sigma x^3 \\ \Sigma x^2 & \Sigma x^3 & \Sigma x^4 \end{bmatrix}
$$

$$(3 \times n)\,(n \times 3) = \qquad\qquad 3 \times 3$$

*The convention here is that boldface type is used for multiple measurements, lowercase for a vector, uppercase for a matrix. This clashes with the more general convention that uppercase letters indicate random variables; since most of our concerns are statistical this creates little confusion in practice. It helps to remember that (in this treatment) the $x$'s are never random variables, the $y$'s and $u$'s are random variables or realizations of them.

(Once the pattern is perceived, the form of the matrix can be written at sight.) The inverse of this matrix may then be computed. Its algebraic form is messy, but in practice this is simply a $3 \times 3$ matrix, all the elements of which are known numbers. The inversion of such a matrix is illustrated in Appendix 1 but the experimenter may be content to leave this task to the computer. Then finally the universe matrix is postmultiplied by the vector $\mathbf{X'y}$.

**Example 9.1.** The volume in relationship to age of the ventral cochlear nucleus in the brain stem (medulla oblongata) in man is governed by two processes: the increase in number of cells that it contains, a process most active early in development; and loss in cell volume, which is a major cause for the physiological decrease in brain mass that occurs with increasing age. Thus we might guess that a description of changes in this volume with age requires at least two components, the *relative* importances of which change with age. The easiest such process to deal with (not necessarily the correct one) is a quadratic regression of the type just discussed. This analysis is thus a kind of "weak" model in which the broad processes are represented but the degrees in which age is represented have no precise meaning.

The data I shall use come from a study* previously published (80). The volumes were estimated by measuring serial sections of the brain stem, which is naturally possible only after death. The data obtained must therefore be interpreted with caution, since those who died early or who survived to advanced ages may well be unrepresentative. There are, of course, further perils (3) in interpreting the relationship between synchronic studies (i.e., studies done at the same time on several people of diverse ages) and diachronic studies (multiple studies on the same patient over time), which have been discussed under "time series analysis" elsewhere (q.v.). In the present instance, of course, it is impossible to get diachronic data; but the impossibility of obtaining data must never be used as a pretext for laxity of interpretation. The pertinence of this synchronic study to the temporal evolution of the nucleus must remain conjectural, and await evidence from other sources, perhaps by noninvasive vital procedures.

The data then: age at death and estimated volume of the nucleus are given for each of the 23 persons in Table 9.1. The observations, being on different and unrelated persons, may be reasonably supposed independent. We shall suppose (without any justification here) that they are also normally distributed with homogeneous variance. The details of the analysis are shown in the lower part of Table 9.1. They turn out much as we might expect: a positive coefficient for the term in $x$ that dominates the early part of the curve and a smaller negative one that gradually takes command as $x^2$ grows in relation to $x$.

---

*The principal mover was Bruce Konigsmark, whose untimely death has deprived neuropathology of a patient and self-effacing scholar of high standing.

**Table 9.1.** The volume of the ventral cochlear nucleus (in cubic millimeters) in relationship to age in years (to the nearest year) in 23 subjects

| age $(x)$* | $x^2$ | $x^3$ | $x^4$ | volume $(y)$ | $y^2$ | $xy$ | $x^2y$ |
|---|---|---|---|---|---|---|---|
| 0 | 0 | 0 | 0 | 2.71 | 7.3441 | 0 | 0 |
| 0 | 0 | 0 | 0 | 2.52 | 6.3504 | 0 | 0 |
| 0 | 0 | 0 | 0 | 3.62 | 13.1044 | 0 | 0 |
| 0 | 0 | 0 | 0 | 7.16 | 51.2656 | 0 | 0 |
| 0 | 0 | 0 | 0 | 5.26 | 27.6676 | 0 | 0 |
| 2 | 4 | 8 | 16 | 6.86 | 47.0596 | 13.72 | 27.44 |
| 7 | 49 | 343 | 2,401 | 8.49 | 72.0801 | 59.43 | 416.01 |
| 19 | 361 | 6,859 | 130,321 | 10.42 | 108.5764 | 197.98 | 3,761.62 |
| 20 | 400 | 8,000 | 160,000 | 7.51 | 56.4001 | 150.20 | 3,004.00 |
| 27 | 729 | 19,683 | 531,441 | 9.59 | 91.9681 | 258.93 | 6,991.11 |
| 27 | 729 | 19,683 | 531,441 | 10.17 | 103.4289 | 274.59 | 7,413.93 |
| 31 | 961 | 29,791 | 923,521 | 11.02 | 121.4404 | 341.62 | 10,590.22 |
| 35 | 1,225 | 42,875 | 1,500,625 | 13.20 | 174.2400 | 462.00 | 16,170.00 |
| 40 | 1,600 | 64,000 | 2,560,000 | 14.20 | 201.6400 | 568.00 | 22,720.00 |
| 43 | 1,849 | 79,507 | 3,418,801 | 11.00 | 121.0000 | 473.000 | 20,339.00 |
| 48 | 2,304 | 110,592 | 5,308,416 | 10.07 | 101.4049 | 483.36 | 23,201.28 |
| 60 | 3,600 | 216,000 | 12,960,000 | 13.32 | 177.4224 | 799.20 | 47,952.00 |
| 64 | 4,096 | 262,144 | 16,777,216 | 12.70 | 161.2900 | 812.80 | 52,019.20 |
| 68 | 4,624 | 314,432 | 21,381,376 | 8.60 | 73.9600 | 584.80 | 39,766.40 |
| 80 | 6,400 | 512,000 | 40,960,000 | 9.28 | 86.1184 | 742.40 | 59,392.00 |
| 82 | 6.724 | 551,368 | 45,212,176 | 10.88 | 118.3744 | 892.16 | 73,157.12 |
| 82 | 6,724 | 551,368 | 45,212,176 | 8.12 | 65.9344 | 665.84 | 54,598.88 |
| 90 | 8,100 | 729,000 | 65,610,000 | 9.75 | 95.0625 | 877.50 | 78,975.00 |
| Total | 825 | 50,479 | 3,517,653 | 263,179,927 | 206.45 | 2,083.1327 | 8,657.53 | 520,495.21 |

SOURCE: Data of Konigsmark and Murphy (80).

*These figures have been slightly simplified from the original (by rounding to the nearest integer). In consequence the estimates are slightly changed.

$$(\mathbf{X}'\mathbf{X})^{-1} = \begin{bmatrix} 23 & 825 & 50{,}479 \\ 825 & 50{,}479 & 3{,}517{,}653 \\ 50{,}479 & 3{,}517{,}653 & 263{,}179{,}927 \end{bmatrix}^{-1}$$

$$= \begin{bmatrix} 0.147{,}207{,}28 & -0.006{,}390{,}533{,}5 & 0.000{,}057{,}180{,}666 \\ -0.006{,}390{,}533{,}5 & 0.000{,}566{,}259{,}754 & -0.000{,}006{,}342{,}875{,}8 \\ 0.000{,}057{,}180{,}666 & -0.000{,}006{,}342{,}875{,}8 & 0.000{,}000{,}077{,}610{,}832 \end{bmatrix}$$

Also

$$\mathbf{X}'\mathbf{y} = \begin{bmatrix} 206.45 \\ 8{,}657.53 \\ 520{,}495.21 \end{bmatrix}$$

Premultiplying $\mathbf{X}'\mathbf{y}$ by $(\mathbf{X}'\mathbf{X})^{-1}$ gives $\mathbf{b}$ the least-squares estimates of $\beta$

$$\mathbf{b} = \begin{bmatrix} 4.826{,}968{,}86 \\ 0.281{,}648{,}71 \\ -0.002{,}712{,}622{,}4 \end{bmatrix}$$

Hence the regression equation:

$$y = 4.827 + 0.2816x - 0.002{,}713x^2$$

The regression sum of squares is

$$\mathbf{b}'\mathbf{X}'\mathbf{y} = 206.45 \times 4.826{,}968{,}86 + 8.657.53 \times 0.281{,}648{,}71 + 520{,}495.21 \times (-0.002{,}712{,}622{,}4)$$
$$= 2{,}023.002{,}9$$
$$\mathbf{y}'\mathbf{y} = 2{,}083.132{,}7$$

By subtraction the residual (error) sum of squares is 60.129,8 which is associated with 20 degrees of freedom. Hence

$$s^2 = 3.006{,}5$$

## COMPUTATION OF CONFIDENCE LIMITS

Suppose that we were to take any future person of age $x_f$ and from the same population. The best prediction for the mean volume of his ventral cochlear nucleus would, in accordance with the regression analysis, be $E(Y|x_f)$ where

$$E(Y_f|x_f) = \hat{\beta}_0 + \hat{\beta}_1 x_f + \hat{\beta}_2 x_f^2$$

or, in matrix notation, if we let

$$\mathbf{h}' = [1 \quad x_f \quad x_f^2]$$

then

$$E(Y_f|x_f) = \mathbf{h}'\hat{\boldsymbol{\beta}}$$

Now we know from Appendix 1

$$\text{var}(\hat{\boldsymbol{\beta}}) = (\mathbf{X}'\mathbf{X})^{-1}\sigma^2$$

Hence

$$\text{var}(\mathbf{h}'\beta) = \mathbf{h}'\text{var}(\beta)\mathbf{h} = \mathbf{h}'(\mathbf{X}'\mathbf{X})^{-1}\mathbf{h}\sigma^2$$

Of course $\sigma^2$ is not known, but is estimated by $s^2$ from the residual (error) sum of squares ($ESS$) which is the total sum of squares ($\mathbf{y}'\mathbf{y}$) minus the estimated regression sum of squares ($\hat{\beta}'\mathbf{X}'\mathbf{y}$). Thus

$$\sigma^2 = \frac{\mathbf{y}'\mathbf{y} - \hat{\beta}'\mathbf{X}'\mathbf{y}}{23 - 3}$$

The denominator is the number of data points (23) less 1 for each of the three parameters estimated from them.

The form of the confidence interval is interesting. Denoting the typical term of the inverse matrix $(\mathbf{X}'\mathbf{X})$ by $c_{ij}$ and multiplying out $\mathbf{h}'(\mathbf{X}'\mathbf{X})^{-1}\mathbf{h}$ we get as the coefficient of $\sigma^2$ a quartic $\mathbf{k}'\mathbf{x}$ the coefficients of which are the diagonal sums of the elements.

$$\mathbf{k}'\mathbf{x} = c_{11} + x(c_{12} + c_{21}) + x^2(c_{31} + c_{22} + c_{13}) + x^3(c_{32} + c_{23}) + x^4(c_{33})$$

Making use of the symmetry of the inverse matrix (i.e., $c_{ij} = c_{ji}$ for all $i$ and $j$) this can be reduced somewhat in complexity. Finally, then, using the appropriate $t$ value with 20 degrees of freedom we have the confidence limits on $E(Y|x_f)$

$$\hat{\beta}_0 + \beta_1 x_f + \beta_2 x_f^2 \pm t\hat{\sigma}\sqrt{\mathbf{k}'\mathbf{x}}$$

And of course for the future individual reading itself, $(Y|x_f)$

$$\hat{\beta}_0 + \hat{\beta}_2 x_f + \hat{\beta}_2 x_f^2 \pm t\hat{\sigma}\sqrt{1 + \mathbf{k}'\mathbf{x}}$$

What is curious about these limits is that in general the square root of a quartic cannot be supposed to have a single "waist" as occurs for linear regression (where the confidence interval is narrowest at the sample mean, $m_x$). Here there will be two locally narrowest intervals with a locally widest interval in between; and of course the interval expands to infinity beyond both ends. The pattern is illustrated in Table 9.2. This curious pattern does not seem to have been widely recognized. (Although this general method of computing confidence intervals for polynomial regression is well known and given in the textbooks, I have never known any statistician, however senior, who has ever actually used it. The only publications I know of in which it is encountered are ours.)

In the present instance we have the following coefficients:

$$
\begin{aligned}
k_0 &= c_{11} & &= 0.147,207,28 \\
k_1 &= c_{21} + c_{12} & &= 2(-0.006,390,533,5) = -0.012,781,067,0 \\
k_2 &= c_{31} + c_{22} + c_{13} & &= 2(0.000,057,180,666) + .000,566,259,75 \\
& & &= 0.000,680,621,082 \\
k_3 &= c_{32} + c_{23} & &= 2(-0.000,006,342,875,8) = \\
& & &\quad -0.000,012,685,751,6 \\
k_4 &= c_{33} & &= .000,000,077,610,832
\end{aligned}
$$

Since three parameters are estimated, the appropriate $t$ value has 20 degrees of freedom associated with it. At the 5% level of significance, this gives the value 2.0860 and thus $t\hat{\sigma}$ is 3.6170. The appropriate calculations are presented in some detail in Table 9.2. It is evident that as age increases, the width of the confidence intervals decreases to a minimum at about 14.7 years; it then increases to a maximum, almost twice as wide, at 43.35 years of age or so and then decreases to a second minimum at 64.5 years of age and then increases without limit.

The behavior of the regression line and the confidence limits are shown in Figure 9.1.


## TESTING HYPOTHESES

It is perhaps in polynomial and multiple regression that the conceptual defects of the classical theory of hypothesis testing are most obtrusive. If it is a matter of asking "Is the relationship between $E(Y_i)$ and $x_i$ adequately represented by a polynomial of the fifth degree?" little harm is done; but the statistician continually finds himself forced to answer the question "What order of polynomial best fits the data?" In other words he is not being asked merely to estimate a specified number of parameters, or which of them is significantly different from zero; he is being asked to determine the *mathematical form* of the relationship. The latter problem, however, falls within the domain of the biomathematician who starts from the data and the notions of science, not the statistician who is supposed to work with the errors inside a given model. Moreover it seems absurd that what the true relationship may be should depend on how the sample was obtained. Yet if the true relation-

**Table 9.2. The quadratic regression function on the data from Table 9.1, with appropriate confidence limits**

| Age in years | $k(x) = \Sigma k_i x^i$ | $t\hat{\sigma}\sqrt{k(x)}$ | $t\hat{\sigma}\sqrt{1 + k(x)}$ | $E(Y\|x)$ | 95% confidence limits | | | |
|---|---|---|---|---|---|---|---|---|
| | | | | | On $E(Y\|x)$ | | On $Y(x)$ | |
| | | | | | Lower | Upper | Lower | Upper |
| 0 | 0.147,207 | 1.387,7 | 3.874,0 | 4.827,0 | 3.439,3 | 6.214,7 | 0.953,0 | 8.701,0 |
| 5 | 0.098,780 | 1.136,8 | 3.791,4 | 6.167,4 | 5.030,6 | 7.304,2 | 2.376,0 | 9.958,8 |
| 10 | 0.075,549 | 0.994,2 | 3.751,1 | 7.372,2 | 6.378,0 | 8.366,4 | 3.621,1 | 11.123,3 |
| 14.7195 | 0.069,728 | 0.955,1 | 3.741,0 | 8.385,0 | 7.429,9 | 9.340,1 | 4.644,0 | 12.126,0 |
| 15 | 0.069,745 | 0.955,2 | 3.741,0 | 8.441,4 | 7.486,2 | 9.396,6 | 4.700,4 | 12.182,4 |
| 20 | 0.074,765 | 0.989,0 | 3.749,7 | 9.374,9 | 8.385,9 | 10.363,9 | 5.625,2 | 13.124,6 |
| 25 | 0.085,170 | 1.055,6 | 3.767,8 | 10.172,8 | 9.117,2 | 11.228,4 | 6.405,0 | 13.940,6 |
| 30 | 0.096,682 | 1.124,6 | 3.787,8 | 10.835,1 | 9.710,5 | 11.959,7 | 7.047,3 | 14.622,9 |
| 35 | 0.106,192 | 1.178,7 | 3.804,2 | 11.361,7 | 10.183,0 | 12.540,4 | 7.557,5 | 15,165,9 |
| 40 | 0.111,751 | 1.209,1 | 3.813,7 | 11.752,7 | 10.543,6 | 12.961,8 | 7.939,0 | 15.566,4 |
| 43.350 | 0.112,828 | 1.214,9 | 3.815,6 | 11.938,8 | 10.723,9 | 13.153,7 | 8.123,2 | 15.754,4 |
| 45 | 0.112,577 | 1.213,6 | 3.815,1 | 12.008,1 | 10.794,5 | 13.221,7 | 8.193,0 | 15.823,2 |
| 50 | 0.109,051 | 1.194,4 | 3.809,1 | 12.127,8 | 10.933,4 | 13.322,2 | 8.318,7 | 15.936,9 |
| 55 | 0.102,718 | 1.159,2 | 3.798,2 | 12.112,0 | 10.952,8 | 13.271,2 | 8.313,8 | 15.910,2 |
| 60 | 0.096,287 | 1.122,3 | 3.787,1 | 11.960,5 | 10.838,2 | 13.082,8 | 8.173,4 | 15.747,6 |
| 64.510 | 0.093,593 | 1.106,5 | 3.782,4 | 11.707,4 | 10.600,9 | 12.813,9 | 7.925,0 | 15.489,8 |
| 65 | 0.093,632 | 1.106,8 | 3.782,5 | 11.673,5 | 10.566,5 | 12.780,1 | 7.890,8 | 15.455,8 |
| 70 | 0.099,790 | 1.142,5 | 3.793,1 | 11.250,5 | 10.107,9 | 12.393,1 | 7.457,4 | 15.043,6 |
| 75 | 0.120,964 | 1.258,0 | 3.829,5 | 10.692,1 | 9.434,1 | 11.950,1 | 6.862,6 | 14.521,6 |
| 80 | 0.164,520 | 1.467,1 | 3.903,2 | 9.998,1 | 8.531,0 | 11.465,2 | 6.094,0 | 13.901,3 |
| 85 | 0.238,987 | 1.768,2 | 4.026,0 | 9.168,4 | 7.400,2 | 10.936,6 | 5.142,4 | 13.194,4 |
| 90 | 0.354,061 | 2.152,2 | 4.208,8 | 8.203,1 | 6.050,9 | 10.355,3 | 3.994,3 | 12.411,9 |

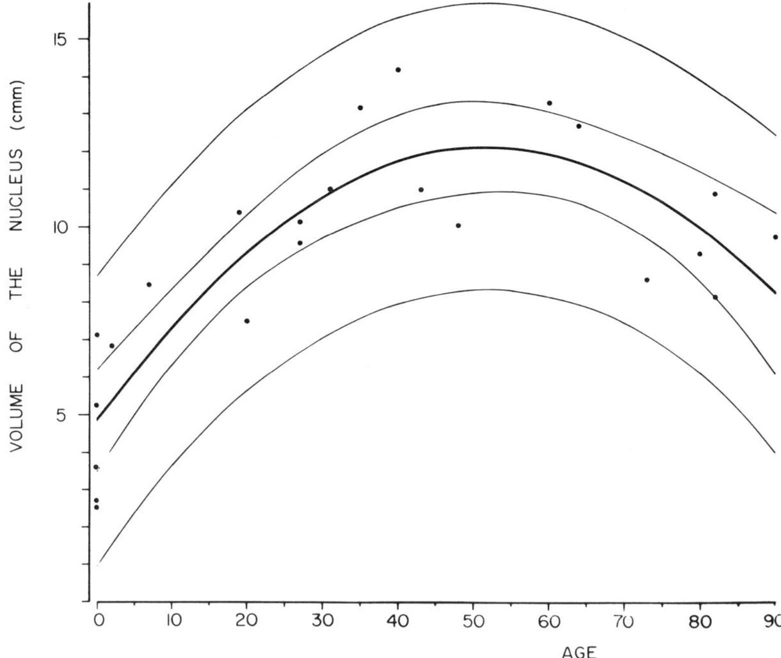


**Figure 9.1.** Quadratic regression curve of the volume of the ventral cochlear nucleus on age in man. The *heavy* line is the regression, the *inner* pair of lines the 95% confidence limits on the mean, and the *outer* pair the 95% confidence limits on any single future value.

ship is a polynomial of order ten, and if there are only nine data points, there is no unambiguous method even of estimating eleven parameters from nine data, to say nothing of testing for significance. Hence it would logically be necessary to specify in advance the minimum size of sample; but if the form of the curve is to be determined from analysis of the data, we are driven to the position that we must have the sample before we know what the size of sample should be!

There is another problem. Suppose it is shown in a particular instance that the contribution to the regression sum of squares from the term $x^4$ is highly significant *provided* the term in $x^2$ is not included, e.g., if we use the model

$$y_i = \alpha + \beta x_i + \delta x_i^3 + \zeta x_i^4 + u_i$$

Suppose also that the term $x^2$ contributes significantly if the term in $x^4$ is ignored and we use the model

$$y_i = \alpha^* + \beta^* x_i + \gamma^* x_i^2 + \delta^* x_i^3 + u_i^*$$

Which model should be used? If the model is intended merely as a computational device, then over the range of sample values it probably matters little which is used. But if the model is to be used for extrapolating beyond experience (always a hazardous matter) or if the biologist attaches *meaning* to the individual powers of $x$, it matters enormously. Yet using statistical tests of hypotheses as a means of making such decisions, here as elsewhere, leads to the preposterous conclusion that the relative importances of the various terms *depend on the order in which the statistician thinks of them*. But surely it will not seriously be argued that the statistician's mode of operation tells us anything about the metaphysical truth of the model?

Thus, I approach the whole problem of hypothesis testing in this context with even more misgiving than usual. If (as a statistician) I had my choice, I would insist that the biologist specify his model in advance, regard the statistical problem as one in estimation using the methods previously discussed, with perhaps some goodness-of-fit test as a criterion of the plausibility or otherwise of the biological model. As it is, I can be persuaded to discuss the topic from a statistical standpoint only on three conditions:

1. That I attach no biological (or more generally scientific) meaning to the terms in the polynomial or to the magnitudes of the coefficients of the individual terms. The expression is not a model, but a computational device. In just the same way, the elaborate formulas for interpolating given in Appendix 2 are designed for arithmetical accuracy and not in the least to throw light on the nature of the true relationship between $y$ and $x$. If the scientist wishes to give meaning to the terms, he should do so in advance and not look to the statistician to find will-o'-the-wisps on which he will confer a spurious identity.

2. The regression line is to be used for interpolation only, and not extrapolation. The accuracies of these two processes may be in conflict.

3. Other things being equal, the simpler the formula the more useful. The simplicity depends on both the numbers of terms and the complexity of each.

Then our strategy will involve two principles:

a. We start with terms of the lowest power of $x$.

b. When adding further terms produces no *disproportionate* decrease in the residual (error) sum of squares, further elaboration is stopped. However, because empirical experience suggests that, in general, even and odd powers of $x$ tend to make different contributions, the practice is to desist from further elaboration when *two* consecutive powers of $x$ have produced no significant improvement.

The method that is usually applied is to fit successively increasing numbers of terms in $x$, estimating the best-fit coefficients at each step. Thus given a sample size $n$, starting with a linear fit we minimize

$$S_a = \Sigma(Y_i - \alpha_0 - \alpha_1 x_i)^2$$

Of the $n$ degrees of freedom, two have been allocated to the parameter space, giving the estimates $\hat{\alpha}_0$ and $\hat{\alpha}_1$, while $(n - 2)$ are assigned to error. As with analysis of variance in general, the total sum of squares $\mathbf{y}'\mathbf{y}$ can be split into $S_a$, and $R_a$, the regression sum of squares, which is the estimate of the amount of the total variability that can be accounted for by introducing these two parameters. If the mean square for regression is significantly greater than for error, i.e., if

$$F_{2,n-2} = \frac{R_a/2}{S_a/(n-2)}$$

is significantly greater than 1, we may conclude that linear regression has improved the precision of the description. In the next step we minimize

$$S_b = \Sigma(Y_i - \beta_0 - \beta_1 x_i - \beta_2 x_i^2)^2$$

in the process transferring a third degree of freedom from the error to the parameter space and effecting a corresponding transfer in the sum of squares to give a new regression sum of squares, $R_b$. The question now is whether the process has been worthwhile: has the reduction in the error sum of squares "gained" on the reduction in degrees of freedom, or is it no greater than would occur readily by chance for any degree of freedom? The test is

$$F_{1,n-3} = \frac{R_b - R_a}{S_b/(n-3)}$$

The denominator, $(R_b - R_a)$ is commonly referred to as the "sum of squares for the quadratic component adjusted for the intercept and the linear component." Note that because of the change in the complexity of the regression equation we cannot expect $\hat{\alpha}_0$ and $\hat{\beta}_0$ to be equal or $\hat{\alpha}_1$ and $\hat{\beta}_1$.

In the next step we find a new set of coefficients that minimize

$$S_c = \Sigma(Y_i - \gamma_0 - \gamma_1 x_i - \gamma_2 x_i^2 - \gamma_3 x_i^3)^2$$

and test the improvement with

$$F_{1,n-4} = \frac{R_c - R_b}{S_c/(n-4)}$$

and so on.

When he sets out to apply this, the reader will find that the procedure is simple enough but the calculations rapidly become tedious. Thus when the $\mathbf{X}'\mathbf{X}$ matrix is computed to fit a quartic equation (i.e., one involving a term in $x^4$) he is obliged to compute the sum of all powers of $x$ up to the eighth, i.e.,

$$\mathbf{X}'\mathbf{X} = \begin{bmatrix} n & \Sigma x & \Sigma x^2 & \Sigma x^3 & \Sigma x^4 \\ \Sigma x & \Sigma x^2 & \Sigma x^3 & \Sigma x^4 & \Sigma x^5 \\ \Sigma x^2 & \Sigma x^3 & \Sigma x^4 & \Sigma x^5 & \Sigma x^6 \\ \Sigma x^3 & \Sigma x^4 & \Sigma x^5 & \Sigma x^6 & \Sigma x^7 \\ \Sigma x^4 & \Sigma x^5 & \Sigma x^6 & \Sigma x^7 & \Sigma x^8 \end{bmatrix}$$

If a quintic equation is proposed, another row and column will be introduced, which involves computing all powers up to $x^{10}$. Simply inverting these matrices is a heavy burden; and until the computer became available very few such problems were ever explicitly solved. Programs now exist which will do most of the calculations, including computing the $F$ values.

**Example 9.2.** Let us reconsider the data of Baertl discussed in Example 8.2. We were content at that time to fit a straight line; but inspection suggests that something a little more elaborate may be suitable. For the smaller children there is rather too large a proportion of points above the regression line, whereas for the taller children the converse is the case. These departures from prediction suggest that a sigmoid curve might be more appropriate, falling in the intermediate range and then leveling out. If this is to be expressed as a polynomial it might require a negative quadratic term and a positive cubic one.

In Table 9.3 are given the $\mathbf{X'X}$ matrix and the $\mathbf{X'y}$ vector for a polynomial regression function of order five. To fit a curve of any lesser degree, the appropriate submatrix and subvector may be used. For example, to fit a quartic we delete the last row and column of the matrix and the last element of the vector. We could likewise eliminate any other power by deleting the appropriate row and column. For instance, we might fit a quintic curve but without a cubic term by deleting the fourth row and column. Detailed calculations are somewhat tedious—in this case they were carried out on a standard computer program—and I shall merely present results. For each order of polynomial equation the regression sum of squares and the error mean square were computed and compared (Table 9.4). As each higher power of $x$ is added, one more parameter is estimated and hence one degree of freedom is transferred from the error to the parameter space. The usefulness of each step is tested by comparing how much the regression sum of squares has increased over that for the previous model, and tested against the error mean square under the new model. We may think of the $h$ degrees of freedom and the sum of squares left over after the polynomial of order $k$ has been fitted as being partitioned into:

1. A component corresponding to the parameter associated with the $(k + 1)^{\text{th}}$ power of $x$. It corresponds to the increment in the regression sum of squares when the least squares polynomial of order $(k + 1)$ is substituted for the polynomial of order $k$. It has one degree of freedom associated with it.

2. A sum of squares for error from the polynomial of order $(k + 1)$ having $(h - 1)$ degrees of freedom associated with it.

The test of the contribution of the term in $x^{k+1}$ is based on a comparison of the mean squares for the subdivisions produced by the partition.

The results in Table 9.4 show no significant contributions from the terms in $x^4$ and $x^5$ so in accordance with custom we stop at the last contributing power and fit a cubic equation. Note that the error mean square for the cubic regression line is 0.977,7 compared with 1.334,1 for the linear, an ap-

**Table 9.3. The $X'X$ matrix and the $X'y$ vector for a quintic regression equation**

$$X'X = \begin{bmatrix}
41 & 2.794{,}400 \times 10^3 & 1.939{,}352 \times 10^5 & 1.368{,}954 \times 10^7 & 9.816{,}177 \times 10^8 & 7.140{,}970 \times 10^{10} \\
2.794{,}400 \times 10^3 & 1.939{,}352 \times 10^5 & 1.368{,}954 \times 10^7 & 9.816{,}177 \times 10^8 & 7.140{,}970 \times 10^{10} & 5.263{,}669 \times 10^{12} \\
1.939{,}352 \times 10^5 & 1.368{,}954 \times 10^7 & 9.816{,}177 \times 10^8 & 7.140{,}998 \times 10^{10} & 5.263{,}669 \times 10^{12} & 3.926{,}500 \times 10^{14} \\
1.368{,}954 \times 10^7 & 9.816{,}177 \times 10^8 & 7.140{,}998 \times 10^{10} & 5.263{,}69 \times 10^{12} & 3.926{,}500 \times 10^{14} & 2.960{,}861 \times 10^{16} \\
9.816{,}177 \times 10^8 & 7.140{,}998 \times 10^{10} & 5.263{,}669 \times 10^{12} & 3.926{,}500 \times 10^{14} & 2.960{,}861 \times 10^{16} & 2.254{,}620 \times 10^{18} \\
7.140{,}998 \times 10^{10} & 5.263{,}670 \times 10^{12} & 3.926{,}500 \times 10^{14} & 2.960{,}861 \times 10^{16} & 2.254{,}620 \times 10^{18} & 1.732{,}045 \times 10^{20}
\end{bmatrix}$$

$$X'y = \begin{bmatrix}
2.269{,}700 \times 10^2 \\
1.508{,}547 \times 10^4 \\
1.021{,}223 \times 10^6 \\
7.037{,}581 \times 10^7 \\
4.933{,}377 \times 10^9 \\
3.514{,}683 \times 10^{11}
\end{bmatrix}$$

SOURCE: Data of Baertl (77).

**Table 9.4. Serial tests of the contributions of various powers of $x$ to goodness of fit of a polynomial regression for the data of Table 9.3**

| Highest power of $x$ | Parameters estimated | Regression sum of squares (RSS) | Increase in RSS | Residual sum of squares | Residual mean square | $t = \sqrt{F}$ | d.f. for error | P. |
|---|---|---|---|---|---|---|---|---|
| — | 0 | — | — | 1,350.858,7 | 32.947,8 | — | — | — |
| 0 | 1 | 1,256.472,7 | 1,256,472,7 | 94.386,0 | 2.359,6 | 23.076 | 40 | $<10^{-10}$ |
| 1 | 2 | 1,298.830,0 | 42.357,3 | 52.028,7 | 1.334,1 | 5.635 | 39 | 0.000,001,666 |
| 2 | 3 | 1,299.155,4 | 0.325,4 | 51.703,3 | 1.360,6 | 0.489 | 38 | 0.627,6 |
| 3 | 4 | 1,314,683,8 | 15.528,4 | 36.174,9 | 0.977,7 | 3.985 | 37 | 0.000,304,5 |
| 4 | 5 | 1,314.756,0 | 0.072,2 | 36.102,7 | 1.002,9 | 0.268 | 36 | 0.790,0 |
| 5 | 6 | 1,317.594,2 | 2.838,2 | 33.264,5 | 0.950,4 | 1.728 | 35 | 0.092,79 |

preciable reduction in uncertainty. Coefficients for polynomial regressions up to the quintic are given in Table 9.5.

**Confidence limits**

Denoting, as before, the typical term of the $4 \times 4$ inverse matrix $(\mathbf{X}'\mathbf{X})^{-1}$ by $c_{ij}$, premultiplying it by $\mathbf{h}' = [1\ x\ x^2\ x^3]$ and postmultiplying it by $\mathbf{h}$ and then collecting terms we get

$$\mathbf{k}'\mathbf{x} = \sum_{i=0}^{6} k_i x^i$$

where

$$
\begin{aligned}
k_0 &= c_{11} &&= 3.002{,}764{,}1 \times 10^3 \\
k_1 &= c_{12} + c_{21} &&= -2.713{,}188{,}8 \times 10^2 \\
k_2 &= c_{13} + c_{22} + c_{31} &&= 1.017{,}319{,}3 \times 10 \\
k_3 &= c_{14} + c_{23} + c_{32} + c_{41} &&= -2.025{,}945{,}622 \times 10^{-1} \\
k_4 &= c_{24} + c_{33} + c_{42} &&= 2.260{,}002{,}46 \times 10^{-3} \\
k_5 &= c_{34} + c_{43} &&= -1.339{,}024{,}86 \times 10^{-5} \\
k_6 &= c_{44} &&= 3.292{,}188{,}8 \times 10^{-8}
\end{aligned}
$$

Since there are 37 degrees of freedom for error, the appropriate $t$ value at the 95% level is 2.026,2 and the standard deviation, $s$, estimated by $\sqrt{0.977{,}700{,}12} = 0.988{,}78\ldots$ Thus $ts$ is 2.003,48, and the 95% confidence limits on $E(Y|x_f)$ are

$$\hat{E}(Y|x_f) \pm 2.003{,}48\sqrt{\mathbf{k}'\mathbf{x_f}}$$

while the 95% confidence limits on $Y(x_f)$ are

$$\hat{E}(Y|x_f) \pm 2.003{,}48\sqrt{(1 + \mathbf{k}'\mathbf{x_f}}$$

Values for the regression function and the two sets of confidence limits for several different values of $x$ are shown in Table 9.6 and the results displayed in Figure 9.2. In this instance because of the values of the coefficients there is only one minimum for $\mathbf{k}'\mathbf{x}$, attained when $x = 68.163$.

Note the following points:

1. The curve certainly fits the data better than the straight line (Fig. 8.3). There is no conspicuous clustering of the points above or below the regression line. This example, then, illustrates how in the linear fit an illusion might be created of correlated errors (such as may be encountered in principle in a time series) when perhaps it is simply an erroneous form of regression model. (See Chapter 8.)

2. The widths of the confidence intervals are somewhat less than with the linear fit. What is more comforting, their widths are less sensitive to $x$ so that they are more uniform over the range of observations. Note, however, how ominously the widths increase at both ends. Extrapolation in either direction leads to predictions which rapidly became unusably vague.

**Table 9.5. Best-fitting polynomial curves of various powers to the data of Table 9.3**

| Order of polynomial | Constant term | Coefficient of $x$ | Coefficient of $x^2$ | Coefficient of $x^3$ | Coefficient of $x^4$ | Coefficient of $x^5$ |
|---|---|---|---|---|---|---|
| Linear | 13.055,4 | $-0.110,327,94$ | | | | |
| Quadratic | 8.680,6 | 0.021,208,07 | $-0.000,970,415,36$ | | | |
| Cubic | $-205.080,4$ | 9.746,149,7 | $-0.146,375,69$ | 0.000,715,000,3 | | |
| Quartic | $-299.026,35$ | 15.495,336 | 9.276,925,35 | 0.002,018,980,4 | $-0.000,004,835,345,2$ | |
| Quintic | 3,973.296,6 | $-311.798,69$ | 9.663,989,6 | $-0.147,640,99$ | 0.001,112,195,6 | 0.000,003,307,506,1 |

**Table 9.6. The fitting of a cubic regression equation of serum protein level on body length**

| Length $(x)$ | $k(x) = \Sigma k_i \cdot x^i$ | $\mathrm{ts}\,\sqrt{k(s)}$ | $\mathrm{ts}\,\sqrt{1 + k(x)}$ | $E(Y\|x)$ | 95% confidence limits | | | |
| | | | | | On $E(Y\|x)$ | | On $Y(x)$ | |
| | | | | | Lower | Upper | Lower | Upper |
|---|---|---|---|---|---|---|---|---|
| 49 | 0.659,668 | 1.627 | 2.581 | 5.192 | 3.525 | 6.779 | 2.571 | 7.773 |
| 50 | 0.449,515 | 1.343 | 2.412 | 5.663 | 4.320 | 7.006 | 3.251 | 8.075 |
| 52.5 | 0.175,730 | 0.840 | 2.172 | 6.607 | 5.767 | 7.447 | 4.435 | 8.780 |
| 55 | 0.095,305 | 0.619 | 2.097 | 7.130 | 6.511 | 7.748 | 5.033 | 9.226 |
| 57.5 | 0.082,312 | 0.575 | 2.084 | 7.297 | 6.722 | 7.872 | 5.213 | 9.381 |
| 60 | 0.078,843 | 0.563 | 2.081 | 7.176 | 6.614 | 7.739 | 5.095 | 9.257 |
| 62.5 | 0.068,296 | 0.524 | 2.071 | 6.835 | 6.311 | 7.358 | 4.764 | 8.905 |
| 65 | 0.054,448 | 0.467 | 2.057 | 6.339 | 5.872 | 6.806 | 4.282 | 8.396 |
| 67.5 | 0.046,315 | 0.431 | 2.049 | 5.757 | 5.325 | 6.188 | 3.707 | 7.806 |
| 70 | 0.048,808 | 0.441 | 2.052 | 5.154 | 4.712 | 5.597 | 3.103 | 7.206 |
| 72.5 | 0.059,196 | 0.487 | 2.062 | 4.599 | 4.111 | 5.087 | 2.537 | 6.661 |
| 75 | 0.069,292 | 0.527 | 2.072 | 4.158 | 3.631 | 4.686 | 2.087 | 6.230 |
| 77.5 | 0.073,490 | 0.543 | 2.076 | 3.899 | 3.356 | 4.442 | 1.823 | 5.974 |
| 80 | 0.082,608 | 0.575 | 2.085 | 3.887 | 3.312 | 4.463 | 1.803 | 5.972 |
| 82.5 | 0.143,330 | 0.758 | 2.142 | 4.191 | 3.433 | 4.950 | 2.049 | 6.334 |
| 85 | 0.363,840 | 1.208 | 2.340 | 4.878 | 3.669 | 6.086 | 2.538 | 7.217 |
| 86 | 0.537,750 | 1.469 | 2.484 | 5.274 | 3.805 | 6.743 | 2.790 | 7.759 |

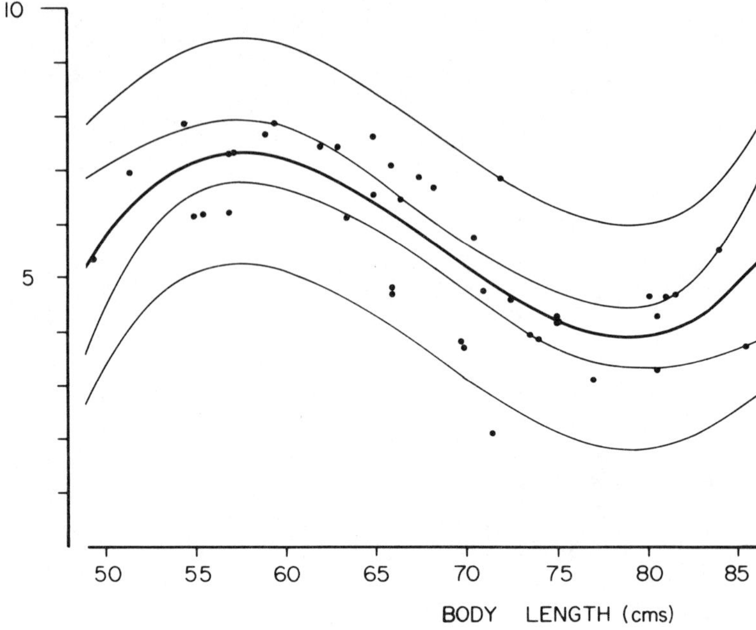


**Figure 9.2.** Cubic regression of serum protein on body length in Peruvian children. Compare with Figure 8.2.

**Interpretation of the curve.** As I have stressed above, such a regression equation should be treated as a purely descriptive device not necessarily having any scientific meaning at all. Note for example (Table 9.5) how drastically the estimates of the intercept change as the order of the polynomial increases.

However, while it would be foolhardy to make any claims for what the analysis has shown about the mechanisms involved, it would be a pity not to see what ideas the data might inspire. Intelligently and imaginatively done, surmise might suggest some useful line of investigation. But these speculations should be "intramural." To pursue them and describe the results (whether positive or negative) is science. To discuss them in print without experimentation or further analysis is science fiction. I regard my mission territory as lying in the former field. That I consider this matter at all and commit it to paper here is, if you will, a breach of my own policy; I have no intention of pursuing it further and would certainly not like to have it supposed that what follows is my considered interpretation of the results. It is merely an exercise in imagination which might lead to fruitful exploration.

What prompted the analysis in the first place was recognition that a child may outgrow his food supply and that in consequence his reserves of protein (which serum protein largely represents) are depleted and the concentration falls. This fall is adequately demonstrated in either the linear model or the quadratic model. But the coefficient of the cubic term is positive and while, as I have said, it may have no scientific meaning, what process *might* it represent?

The reversal might be due to a variety of mechanisms. It may be related to a change in diet as the children grow older, for instance when they are weaned. This is a matter which might be investigated. But it must be borne in mind that $x$ is not age but *height*; and while we invoked growth to explain the depletion, it seems clear that what we really had in mind was not amount, but rate, of growth. Depletion presumably reflects the discrepancy between needs for growth, and intake; and if growth were slowed down or arrested, the deficit would be reduced and perhaps even be reversed. Hence the next step would be to regress "fall of serum protein" against "rate of growth." It might turn out to be a much simpler relationship, perhaps even a simple ratio. But our analysis has not taken *time* into consideratiion at all. Hence it seems perilous to embark on any very elaborate policy until the matter has been much more clearly interpreted.

**On precipitate action.** At this stage, I may express my condemnation of the practice of basing massive, expensive, drastic and, for all we know, dangerous, campaigns in preventive medicine on an analysis of empirical relationships derived from observational data, with little attempt at anything more perceptive than a black-box inference. Manipulations of serum cholesterol in the hope that arterial disease will be prevented thereby is one such scheme. It may be brilliantly successful. Serum cholesterol may prove to be merely a straw in the wind—guilt by association.* It may be that high serum cholesterol in people with arterial disease is a protective device and the worst thing possible is to attempt to reduce it.† The wise might recall the fiasco of the oral hypoglycemic agents which lower blood glucose levels but increase the frequency of arterial complications. The critical discussion by Mann of the relationship of cholesterol to atherosclerosis (81) is worth study.

Unfortunately preventive medicine also has its Gresham's law: Bad research drives out good. For all the world loves an activist, and many a soldier would rather be led to his death by a fool than kept in barracks by a man of decent caution.

## MULTIPLE REGRESSION ANALYSIS

The methods involved in relating an outcome variable to a function of several regressor variables are substantially the same as for polynomial re-

---

*Because wisdom is commonly associated with gray hair, does this mean that the man who aspires after wisdom should dye his hair gray?
†Who, after all, treats pneumonia with cytotoxic drugs because there is an associated leucocytosis?

gression. Algebraically the only difference lies in the forms of the $\mathbf{X'X}$ matrix and the $\mathbf{X'y}$ vector. Geometrically the difference lies in the dimensionality of the display: for a function of a single regressor we need a line, for two regressors, a surface; and for three or more a hypersurface (which lies beyond the power of conventional illustration) is necessary.

Suppose, to take the simplest instance, there are two nonrandom regressor variables, $x$ and $z$, and a response variable $Y$ and that

$$y_i = \beta_0 + \beta_1 x_i + \beta_2 z_i + u_i$$

As usual we assume the error terms have a mean of zero, are uncorrelated, and have a homogeneous variance: that is, in matrix notation

$$\mathbf{E(U)} = \mathbf{0}$$

$$\mathrm{Var}(\mathbf{U}) = \mathbf{I}\sigma^2$$

The $\mathbf{X'X}$ matrix and the $\mathbf{X'y}$ vector for a sample of $n$ values can now be written down almost by inspection

$$\mathbf{X'X} = \begin{bmatrix} n & \Sigma x & \Sigma z \\ \Sigma x & \Sigma x^2 & \Sigma xz \\ \Sigma z & \Sigma xz & \Sigma z^2 \end{bmatrix} \quad \text{and} \quad \mathbf{X'y} = \begin{bmatrix} \Sigma y \\ \Sigma xy \\ \Sigma zy \end{bmatrix}$$

Then the whole process goes through exactly as before. The elements $c_{ij}$ of the inverse matrix $(\mathbf{X'X})^{-1}$ multiplied by $\sigma^2$ are the variances and covariances of the estimates of the parameters. Thus

$$\begin{aligned}
\mathrm{var}(\hat{\beta}_0) &= c_{11}\sigma^2 & \mathrm{var}(\hat{\beta}_1) &= c_{22}\sigma^2 \\
\mathrm{cov}(\hat{\beta}_0, \hat{\beta}_1) &= c_{12}\sigma^2 = c_{21}\sigma^2 & \mathrm{cov}(\hat{\beta}_1, \hat{\beta}_2) &= c_{23}\sigma^2 = c_{32}\sigma^2 \\
\mathrm{cov}(\hat{\beta}_0, \hat{\beta}_2) &= c_{13}\sigma^2 = c_{31}\sigma^2 & \mathrm{var}(\hat{\beta}_2) &= c_{33}\sigma^2
\end{aligned}$$

Estimates of these variances and covariances are found by substituting $s^2$ for $\sigma^2$ where

$$s^2 = \frac{ESS}{n - 3}$$

*ESS* is the error sum of squares found by subtracting from the total sum of squares $(\mathbf{y'y})$ the regression sum of squares *RSS* given by

$$[\hat{\beta}_0 \quad \hat{\beta}_1 \quad \hat{\beta}_2] \begin{bmatrix} \Sigma y \\ \Sigma xy \\ \Sigma zy \end{bmatrix} = \hat{\beta}_0 \Sigma y + \hat{\beta}_1 \Sigma xy + \hat{\beta}_2 \Sigma zy$$

From the structure of the regression equation, which is linear in $x$ and $z$, the surface will be simply a plane. The confidence limits are computed as before.

$$\mathrm{Var}[\widehat{E(Y}|x,z)] = [1 \quad x \quad z] \begin{bmatrix} c_{11} & c_{12} & c_{13} \\ c_{21} & c_{22} & c_{23} \\ c_{31} & c_{32} & c_{33} \end{bmatrix} \begin{bmatrix} 1 \\ x \\ z \end{bmatrix} \sigma^2 = \sigma^2 \mathbf{k(x, z)}$$

$$= [c_{11} + x(c_{21} + c_{12}) + z(c_{31} + c_{13}) + x^2(c_{22})$$
$$+ z^2(c_{33}) + xz(c_{32} + c_{23})]\sigma^2$$

as the reader can easily verify by multiplying out and collecting terms. Then by the same arguments as before, the confidence limits are

$$\text{On} \quad E(Y): \hat{\beta}_0 + \hat{\beta}_1 x + \hat{\beta}_2 z \pm t_{n-3} s\sqrt{\mathbf{k(x, z)}}$$

$$\text{On} \quad Y: x\hat{\beta}_0 + \hat{\beta}_1 x + \hat{\beta}_2 z \pm t_{n-3} s\sqrt{1 + \mathbf{k(x, z)}}$$

where the $t$ value at the appropriate confidence level is used. Of course, if $\sigma^2$ is known, it may be used with the standard normal deviate in place of $t$.

The foregoing treatment is condensed and will be hard to follow unless the reader has both mastered the matrix algebra in the appendix and worked through the first part of this chapter. If he has done so, then the foregoing should be easy enough and any difficulties may be put right by an example.

**Example 9.3.** Little and Shanoff have published in considerable detail the findings from a retrospective study in a group of about a hundred men with coronary disease and a like number of controls. Their ages are from 30 to 69 (about equally distributed over the decades). The details need not concern us here, though the interested reader may wish to consult some of their original papers. A great many aspects of these subjects were studied. For the purposes of this illustration we shall consider the coronary group only and the following characteristics considered in one of their papers (78): serum levels in mg per 100 ml of total cholesterol ($x$), total phospholipid ($z$), and the $S_f$0-12 fraction obtained by ultracentrifugation of lipoprotein ($y$). The first two measurements are comparatively easy to make; the last is more difficult and requires complicated apparatus which might not be available in many of the hospitals at which coronary disease would be a common disorder. The realistic question might arise, then, how necessary it would be to measure $y$ in order to discriminate between patients with, and those without, coronary disease. It may be that there is a high degree of redundancy in the data and it is not necessary to know all three measurements. How can we best infer $y$ given $x$ and $z$? This is a problem of multiple regression.*

Since the numbers are rather unwieldy—our purpose here is illustration, not to find a final answer—eight values have been chosen at random (by use of a table of random numbers) from each of the four decades. For those interested, the actual subjects chosen are identified according to the number-

---

*For purposes of analysis, they may be treated as nonrandom.

ing used by the authors. The 32 sets of values are listed in Table 9.7 with a few elementary calculations. In Table 9.8 we consider various models with appropriate tests of significance, which will be discussed. All three variables depart somewhat from normality and give rather better fits to the lognormal distribution. In the interests of simplicity and with little loss of accuracy we shall ignore this refinement in what follows. However, we shall interpret tests of significance cautiously.

First, however, the $\mathbf{X'X}$ matrix and $\mathbf{X'y}$ vector are set up and the total sum of squares ($\mathbf{y'y}$) computed. To begin with we do not consider any more complicated function of $x$ and $z$ than the first degree of either. Various models may now be fitted, by the usual matrix methods, deleting whatever rows and columns may be irrelevant to a particular model.

For instance, the simplest model (and the first one listed in Table 9.8) is that $x$ and $z$ are totally irrelevant to $y$ and that the only appropriate model is that $y$ consists of a mean (of which $\hat{\alpha}$ will be the estimator) from which there is a random deviation ($e_i$). Then the $\mathbf{X'X}$ matrix reduces to the scalar 32, and the $\mathbf{X'y}$ vector to the scalar 11,681. Thus the estimator of the only parameter, is $(\mathbf{X'X})^{-1}\mathbf{X'y}$, that is,

$$\hat{\alpha} = (32)^{-1}(11,681) = 365.03\ldots$$

readily identified as the commonplace sample mean or average. The regression sum of squares is as usual the product of the estimate and the $\mathbf{X'y}$ vector which is here

$$(365.03\ldots)(11,681) = 4,263,930.031\ldots$$

which is the same as $(11,681)^2/32$, the usual "correction term" for converting the total sum of squares to the sum of squares from the sample mean. Finally, the variance may be estimated by dividing the residual (error) sum of squares by the number of degrees of freedom, $32 - 1 = 31$, to give the usual sample estimate:

$$\hat{\sigma}^2 = \frac{4,594,441 - 4,263,930.031}{31} = 10,661.644\ldots$$

We can test the trivial question of whether the mean is significantly different from zero by dividing it by its standard error, which is $\sqrt{(\mathbf{X'X})^{-1}\hat{\sigma}^2}$

$$\sqrt{\frac{10,661.644\ldots}{32}} = 18.253$$

The test, with 31 degrees of freedom is

$$t = \frac{365.03\ldots}{18.253\ldots} = 19.998$$

which would of course lead to firm rejection of the hypothesis that the value that $\hat{\alpha}$ is estimating is zero.

**Table 9.7. Regression of $S_fO$-12 lipoprotein on cholesterol and phospholipid levels in serum**

| Identification number | Serum cholesterol level ($x$) (mg/100 ml) | Serum phospholipid ($z$) (mg/100 ml) | Lipoprotein level ($y$) (mg/100 ml) |
|---|---|---|---|
| 51 | 235 | 217 | 344 |
| 89 | 239 | 244 | 398 |
| 99 | 253 | 293 | 383 |
| 100 | 271 | 283 | 404 |
| 115 | 421 | 352 | 774 |
| 117 | 255 | 279 | 306 |
| 135 | 240 | 299 | 357 |
| 137 | 292 | 299 | 426 |
| 34 | 259 | 288 | 368 |
| 91 | 304 | 284 | 594 |
| 101 | 257 | 275 | 376 |
| 108 | 254 | 283 | 383 |
| 111 | 226 | 251 | 404 |
| 126 | 231 | 255 | 376 |
| 130 | 200 | 268 | 208 |
| 134 | 248 | 245 | 303 |
| 28 | 344 | 325 | 377 |
| 40 | 248 | 250 | 309 |
| 55 | 232 | 304 | 280 |
| 58 | 204 | 252 | 350 |
| 69 | 308 | 308 | 407 |
| 78 | 284 | 285 | 347 |
| 80 | 330 | 311 | 301 |
| 90 | 250 | 246 | 348 |
| 3 | 210 | 299 | 353 |
| 7 | 236 | 257 | 381 |
| 9 | 271 | 307 | 328 |
| 65 | 275 | 280 | 353 |
| 67 | 264 | 265 | 413 |
| 71 | 191 | 228 | 248 |
| 77 | 169 | 206 | 198 |
| 82 | 210 | 238 | 284 |
| $\Sigma i$ | 8,211 | 8,776 | 11,681 |
| $\Sigma i^2$ | 2,180,829 | 2,439,058 | 4,594,441 |

$$\mathbf{X'X} = \begin{bmatrix} 32 & 8,211 & 8,776 \\ 8,211 & 2,180,829 & 2,289,625 \\ 8,776 & 2,289,625 & 2,439,058 \end{bmatrix}$$

$$\mathbf{X'y} = \begin{bmatrix} 11,681 \\ 3,113,432 \\ 3,257,916 \end{bmatrix}$$

$$\mathbf{y'y} = 4,594,441$$

Source: Data of Little and Shanoff (78).

**Table 9.8. Analysis of the data from Little and Shanoff in Table 9.7**

| Total sum of squares $(\mathbf{y}'\mathbf{y}) = 4{,}594{,}441$<br>Model | Regression<br>sum of squares |
|---|---|
| $y = a_0 + e_0$ | 4,263,930.0 |
| $y = a_1 + b_1x + e_1$ | 4,446,424.7 |
| $y = a_2 + c_2z + e_2$ | 4,355,737.4 |
| $y = a_3 + b_3x + c_3z + e_3$ | 4,448,291.8 |
| $y = a_4 + b_4x + c_4z + d_4x^2 + e_4$ | 4,457,442.3 |
| $y = a_5 + b_5x + c_5z + f_5z^2 + e_5$ | 4,448,611.1 |
| $y = a_6 + b_6x + c_6x + d_6x^2 + f_6z^2 + e_6$ | 4,472,476.8 |

| Term | Adjusted<br>regression<br>sum of<br>squares | Error<br>mean<br>square | $F$ | $P$ |
|---|---|---|---|---|
| $x$ (adj for mean) | 182,494.68 | 4933.88 | 36.988 | $1.110 \times 10^{-6}$ |
| $z$ (adj for mean) | 91,797.41 | 7957.12 | 11.537 | $1.941 \times 10^{-3}$ |
| $x$ and $z$ (adj for mean) | 184,361.77 | 5039.63 | 36.582 | $<0.001$ |
| $x$ (adj for mean and $z$) | 92,564.36 | 5039.63 | 18.367 | $1.883 \times 10^{-4}$ |
| $z$ (adj for mean and $x$) | 1,867.09 | 5039.63 | 0.370 | 0.547 |
| $x^2$ (adj for mean, $x$ and $z$) | 9,150.50 | 4892.81 | 1.870 | |
| $z^2$ (adj for mean, $x$ and $z$) | 319.30 | 5208.21 | 0.061 | |
| $x^2$ and $z^2$ (adj for mean, $x$ and $z$) | 24,185.00 | 4517.19 | 5.354 | |

The next simplest model is to suppose that $y$ is a function of one of the other variables, say $x$. Of course strictly, both $x$ and $z$, like $y$, are realizations of a random variable; but for the purposes of this illustration, we shall suppose that they were fixed in advance. The appropriate submatrix and vector obtained by deleting terms involving $z$ (the last row and column) are

$$\mathbf{X}'\mathbf{X} = \begin{bmatrix} 32 & 8{,}211 \\ 8{,}211 & 2{,}180{,}829 \end{bmatrix} \quad \text{and} \quad \mathbf{X}'\mathbf{y} = \begin{bmatrix} 11{,}681 \\ 3{,}113{,}432 \end{bmatrix}$$

The determinant of the matrix is

$$32 \times 2{,}180{,}829 - (8211)^2 = 2{,}366{,}007$$

and the inverse of the matrix is easily obtained by the method of cofactors (which in this instance consists of interchange of diagonal terms with a suitable adjustment of sign). Thus

$$(\mathbf{X}'\mathbf{X})^{-1} = \begin{bmatrix} \dfrac{2{,}180{,}829}{2{,}366{,}007} & -\dfrac{8{,}211}{2{,}366{,}007} \\ -\dfrac{8{,}211}{2{,}366{,}007} & \dfrac{32}{2{,}366{,}007} \end{bmatrix}$$

$$= \begin{bmatrix} 0.921,733,959,4 & -0.003,470,404,039 \\ 0.003,470,404,039 & 0.000,013,524,896,59 \end{bmatrix}$$

From these results, multiplying by $\mathbf{X}'\mathbf{y}$ we get the intercept

$$0.921,733,959,4 \times 11,681 - 0.003,470,404,039 \times 3,113,432$$
$$= -38.092,28$$

and the gradient

$$-0.003,470,404,039 \times 11,681 + 0.000,013,524,896,59 \times 3,113,432$$
$$= 1.571,057,480,0$$

and the regression sum of squares $\hat{\beta}'\mathbf{X}'\mathbf{y}$ is

$$-38.092,28 \times 11,681 + 1.571,057,48 \times 3,113,432 = 4,446,424.709$$

These results can be found by the more familiar method laid out in Chapter 8 where linear regression problems are solved without the use of matrices.

We can go through the same process for the regression of $y$ on $z$, for which the appropriate inverse matrix is

$$(\mathbf{X}'\mathbf{X})^{-1} = \begin{bmatrix} 2.364,161,368 & -0.008,506,513,648 \\ -0.008,506,513,648 & 0.000,031,017,369,73 \end{bmatrix}$$

giving the estimates for the intercept of $-97.737,98$ and the gradient $1.687,399,1$. The regression sum of squares is given in Table 9.8.

Finally, we may make $y$ a linear function of both $x$ and $z$, which involves the use of the entire matrix, of which the inverse is

$$(\mathbf{X}'\mathbf{X})^{-1} = \begin{bmatrix} 2.504,581,1 & 0.002,173,771,1 & -0.011,052,351 \\ 0.002,173,771,1 & 0.000,033,651,104 & -0.000,039,410,881 \\ -0.011,052,351 & -0.000,039,410,881 & 0.000,077,173,388 \end{bmatrix}$$

which leads to the regression equation

$$\hat{E}(Y) = 16.270,158 + 1.764,905,1x - 0.379,589,9z$$

and the regression sum of squares shown in Table 9.8.

The findings are highly revealing. First, it will be noted from the last regression equation we have quoted, increases in $z$ are associated with a fall in $y$. However, when $z$ is considered alone without reference to $x$, the regression is a positive one, and $y$ increases $1.687\ldots$ mg/100 ml for every increase of 1 mg/100 ml in $z$. This paradox reflects the degree of redundancy in the information about serum $S_f$0-12 furnished by cholesterol and phospholipid levels in serum. Another way of looking at this redundancy is to compute the correlation coefficient* between these two measurements. This quantity is

---

*See Chapter 10 for details.

0.773 which must be reckoned on a scale between zero (which implies that the two pieces of information were uncorrelated) and unity (which implies complete redundancy). Note that if there were complete redundancy the matrix would be singular and thus could not be inverted.

When we assess the contribution to prediction from $x$ and $z$ taken jointly, we take the regression sum of squares less the sum of squares associated with the mean alone. The resulting sum of squares ("adjusted for the mean"), $R$, has two degrees of freedom associated with it, and under the null hypothesis that neither has anything to contribute to prediction, the ratio $R$ bears to the estimated residual mean square follows an $F$ distribution with 2 and 29 degrees of freedom. The null hypothesis is emphatically rejected at conventional levels of significance.

However, as in the instance of analysis of variance, we may not rest content with a "black-box" statement about their joint relevance; it seems clearly desirable to find out what they mean individually. This question is answered by the first and second tests in Table 9.8 respectively, and both are highly significant. However, the scientist may have an uneasy feeling that $x$ merely appears to be relevant because $z$ is relevant and $x$ reflects $z$. Scars on the tongue do not in themselves tell us anything about the fitness of a person to drive a car; but they are commonly a result of epilepsy, which does. Then, it will be argued, it is on the basis of the epilepsy and not on the scars, that the decision is made whether or not the person should have a driving license.*

The appropriate method of testing the proper contribution of a factor is individually. But the anomalous conclusion now emerges that the importances that we attach to the predictors depend on the order in which we take them. For instance we may ask the following two sets of questions (Fig. 9.3):
*Set 1.*

a. What is the importance of $z$ as a predictor? Table 9.8 shows that under the null hypothesis $F = 11.5$ and we might conclude that it is a significant predictor.

b. What is the importance of $x$ as a predictor over and above the value of $z$? This is found by subtracting the regression sum of squares associated with $z$ alone from that associated with $x$ and $z$ taken jointly. As with polynomial regression, we now test this quantity against the residual sum of squares and conclude, since $F = 18.4$, that it is useful additional information.

Now what if we did the analysis the other way around?
*Set 2.*

a. What is the importance of $x$ as a predictor of $y$? $F = 36.99$ and again we would conclude there is a highly significant effect. However

b. What is the importance of $z$ as a predictor over and above the value of $x$? Table 9.8 shows little evidence that it adds anything ($F = 0.37$).

If we slavishly supposed that "statistical significance implies biological significance" and vice versa, we would have arrived at different answers

*I take this excellent illustration from Sir Thomas Lewis.

about $z$ depending on the order in which the terms are analyzed. Now this conclusion is manifestly absurd, though there is no flaw in the mathematical reasoning. It embodies much of my dissatisfaction with this kind of method of interpreting the results from the analysis. If our entire concern is with

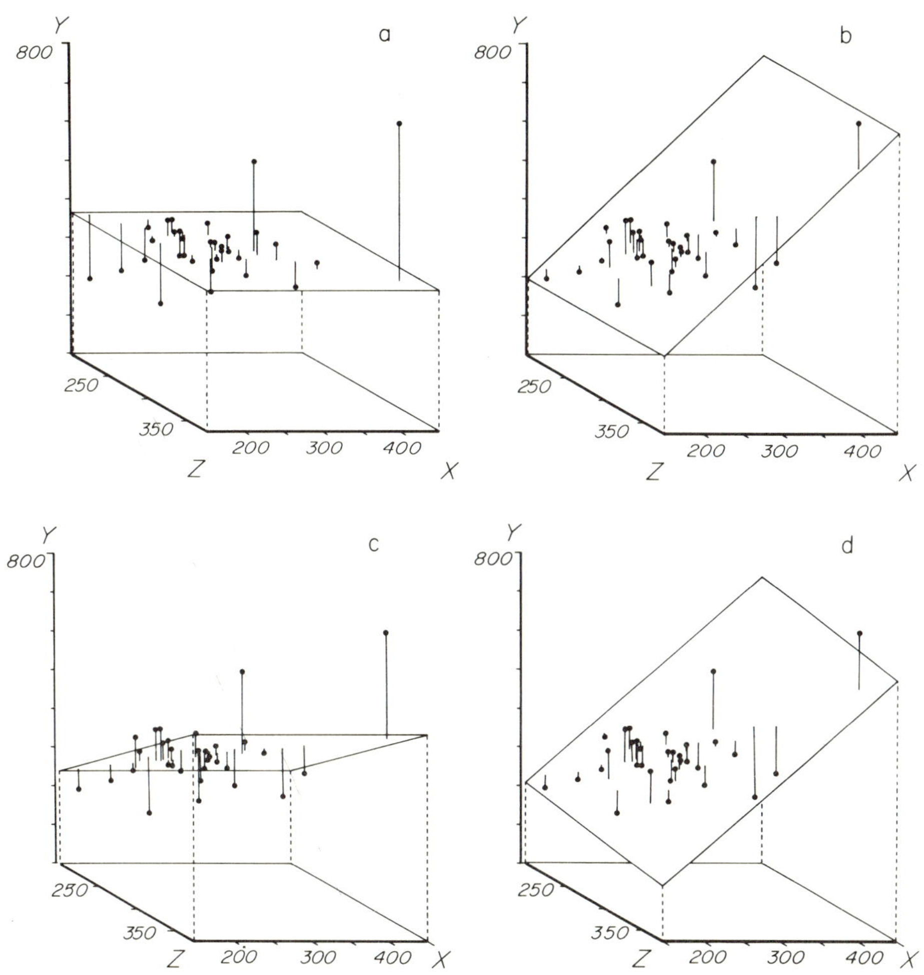

**Figure 9.3.** The regression of $S_f0$–12 lipoprotein fraction ($y$) on total blood cholesterol ($x$) and total phospholipid ($z$) in blood. All measurements are in milligrams per deciliter. Four models are shown. (*a*) The common mean only is considered significant. (*b*) A first degree regression on $x$ is introduced. (*c*) A first degree regression on $z$ is introduced. (*d*) First degree regression on both $y$ and $x$ is used. In each case the residual error for each point is represented by the length of line attached to it. If the line lies below the point, the residual error is positive, and conversely.

*prediction* then there is little harm done. But the investigator must resolutely resist the temptation to try to make the analysis into a method of inferring what the biologically significant relationships are between these three measurements.

To some readers, the manipulations will be easier to grasp if laid out in geometric terms. In Figure 9.3 the values of $y$ (vertical scale) are plotted against those of $x$ and $z$ jointly (horizontal scales). The plots are identical in all four diagrams. To these data are fitted four predictor surfaces, each datum point being connected by a straight line to the point on the surface that predicts it. The line may extend up from the point (the prediction is too high) or conversely. The lengths of these lines represent the residual errors, which are an index of goodness of fit: perfect prediction would mean that the points actually lie on the surface. Four surfaces are shown as functions of various combinations of variables.

a. The overall mean—i.e., $x$ and $z$ are ignored. This consumes one degree of freedom.

b. The slope of the plane is allowed to vary on the $x$ axis—i.e., $x$, but not $z$, is considered. This relaxation consumes a second degree of freedom.

c. The slope of the plane is allowed to vary on the $z$ axis but $x$ is ignored. (This alternative relaxation also consumes a second degree of freedom.)

d. The slope is allowed to vary on both axes. (Three degrees of freedom are used.)

The reader may note how on average the (sum of the squares of the) residual errors decrease with the amount of freedom allowed to the surface. Further degrees of freedom could be used to allow the plane to curve in one or other direction, and so forth.

## PROBLEMS

**9.1.** One definition of evolutionary fitness (82) is the probability that a particular genetic line will not become extinct. The sample estimate is obtained by solving a somewhat tedious polynomial equation which calls for iteration. However, the mean and variance of family size are easily computed. Clearly all three quantities are random variables and there is good reason to believe them closely related. The results in 15 Amish genealogies are shown in Table 9.9. Compute the means, variances, and covariances of the three variates.

**9.2.** What would be a reasonable multiple regression model in problem 9.1? Carry through the appropriate calculations. What assumptions underlie the interpretation of this result? Are they reasonable? Complete the multiple regression analysis.

**9.3.** In Table 9.10 are shown the age, the logarithm of the blood urea, and the logarithm of length of survival from diagnosis, in 22 men with malignant essential hypertension. Do a multiple regression analysis of the third variable on the first powers of the other two.

**Table 9.9. Empirical relationship of evolutionary fitnes of a male line to mean and variance of the number of line**

| Mean number of sons | Variance of number of sons | Probability of survival of line |
|---|---|---|
| 1.887 | 3.829 | 0.534 |
| 1.462 | 2.296 | 0.439 |
| 0.500 | 0.500 | 0.000 |
| 2.100 | 3.071 | 0.676 |
| 2.465 | 4.851 | 0.725 |
| 1.342 | 2.308 | 0.325 |
| 1.188 | 1.754 | 0.230 |
| 1.681 | 3.961 | 0.396 |
| 2.141 | 3.170 | 0.695+ |
| 1.711 | 1.671 | 0.668 |
| 2.233 | 3.876 | 0.700 |
| 1.814 | 3.715 | 0.510 |
| 2.431 | 4.847 | 0.695− |
| 2.150 | 4.776 | 0.587 |
| 1.214 | 2.514 | 0.187 |

SOURCE: Data of Murphy (82).

**Table 9.10. Survival in men with malignant essential hypertension**

| Age | $\log_{10}$BUN (mg/ml) | $\log_{10}$[Survival] (in months) |
|---|---|---|
| 76 | 1.380 | −1.000 |
| 51 | 2.127 | −1.000 |
| 61 | 2.212 | −1.000 |
| 56 | 1.748 | −0.875 |
| 58 | 1.968 | −0.778 |
| 41 | 1.833 | −0.778 |
| 62 | 2.207 | −0.632 |
| 54 | 2.190 | −0.574 |
| 50 | 1.362 | −0.398 |
| 52 | 2.093 | −0.331 |
| 52 | 1.556 | −0.062 |
| 51 | 1.342 | 0.000 |
| 59 | 1.301 | 0.097 |
| 57 | 1.204 | 0.176 |
| 39 | 1.146 | 0.352 |
| 65 | 1.447 | 0.477 |
| 63 | 1.380 | 0.477 |
| 38 | 1.362 | 0.544 |
| 56 | 1.415 | 0.699 |
| 60 | 1.519 | 0.740 |
| 36 | 1.322 | 0.778 |
| 50 | 1.146 | 1.146 |

SOURCE: Data of Kincaid-Smith et al. (21).

# 10
# CORRELATION

If it were quite certain that medical scientists knew nothing of correlation, the wisest course might be to deal lightly or not at all with the topic, the importance of which has certainly been overrated. It is a fact however that this frail statistic has been both much used and much abused. Hence, the statistical teacher must devote some space to putting the matter to rights. For my part, I would much sooner deal in regression; but correlation is still used in special fields (e.g., quantitative genetics). Moreover, it will be of some value in discrimination analysis (see Chapter 11).

In Chapter 8 we have considered the relationship between two variables, $x$ and $Y$. The former is a mathematical, but not a random, variable (hence the lowercase letter). The latter, $Y$, we view as in part a mathematical variable that is a deterministic function of $x$, and in part a random variable, $U$.

Let us now consider the case where both variables are random. For instance, we might be sampling adult males at random from a population and measuring both height and weight. Either might be denoted by $X$. This is a matter of regression, though of a slightly different form (Type 2) from that in Chapter 8 where the regressor variable is not random. (It is only reasonable to point out that, in practice, biostatisticians do not usually distinguish between the two types of regression.)

The theoretical underpinning for this new type of relationship is sketched elsewhere (7). It may be cast in very general terms. Here we shall confine our discussion to Gaussian data, though other types will be dealt with in Chapter 15.

## THE BIVARIATE GAUSSIAN DISTRIBUTION

Strictly, in the development that follows, we assume that $X$ and $Y$ jointly follow a bivariate Gaussian distribution. This condition is fulfilled if every linear combination of them with all possible constants $a$ and $b$

$$W = aX + bY \tag{10.1}$$

is Gaussian. An extensive testing of this property is tedious. I suggest the following relatively simple tests:

1. A scatter diagram should be constructed. In this context we use the weight as the coordinate of a point on one axis and height for that on the other. The scatter should be roughly elliptical, with the densest clustering of the points in the center. If the points clearly fall into separate clusters, the distribution is not Gaussian; but of course, one should not overinterpret the evidence of a small number of points, or weak evidence of clustering.

2. The separate (marginal) distributions of $X$ and $Y$ should conform to the normal distribution. These are special cases of (10.1) with $a$ and $b$ equal to (1 and 0) and to (0 and 1) respectively.

In principle, these conditions do not guarantee bivariate normality, as the theorists have demonstrated. But any such exceptions as I have seen are highly contrived, and I for one have little concern about encountering any of them in practice. Nevertheless, the scatter diagram remains a necessary precaution.

As we might surmise, each of the variates is characterized by a mean and a variance, which are estimated in the usual way, i.e.,

$$m_X = \frac{\Sigma x_i}{n}$$

$$s_X^2 = \frac{\Sigma(x_i - m_X)^2}{n - 1}$$

and analogously for $Y$.

We also need to determine a fifth parameter, the correlation coefficient. It is denoted by the Greek letter $\rho$ ("rho"); and the method-of-moments estimate of it is represented, in accordance with custom,* by the corresponding Roman letter $r$. By definition $\rho_{XY}$ is the ratio of the mean value of the expression

$$(X - \mu_X)(Y - \mu_Y) \tag{10.2}$$

termed the covariance of $X$ and $Y$ to the geometric mean of the variance of $X$ and $Y$, i.e.,

$$\sqrt{\text{var}(X)\,\text{var}(Y)} \tag{10.3}$$

---

*Yet once again, tradition violates our convention about uppercase letters. Since we are discussing the sample correlation coefficient as a random variable we should use an uppercase letter. However, "$R$" has been reserved to denote the multiple correlation coefficient (see below). The reader must infer from context whether $r$ signifies a random variable or a particular value of it.

## ESTIMATION

The common, method-of-moments estimate of the correlation coefficient requires that we compute the sample analog of (10.2). First we find

$$\Sigma(x_i - m_X)(y_i - m_Y) = \Sigma x_i y_i - \frac{\Sigma x_i \Sigma y_i}{n} \tag{10.4}$$

Note that

$$\frac{\Sigma x_i \, \Sigma y_i}{n} = m_X \Sigma y_i = m_Y \Sigma x_i$$

It is shown in (7) that dividing (10.4) by $(n - 1)$ gives an unbiased estimate of the covariance.* Thus the same divisors will be involved in all three components of $r$, the variances and covariances. So they cancel, and we may simply divide (10.4) by the geometric mean of $\Sigma(x_i - m_X)^2$ and $\Sigma(y_i - m_Y)^2$. Thus

$$r_{XY} = \frac{\Sigma(x_i - m_X)(y_i - m_Y)}{\sqrt{\Sigma(x_i - m_X)^2 \Sigma(y_i - m_Y)^2}}$$

**Example 10.1.** In Figure 10.1 are shown some results from the experiment discussed in Chapter 7 on the interaction between cholesterol and sitosterol (49). The data are for animals receiving no added dietary lipids. They comprise serum cholesterol levels at the beginning of the experiment ($x$) and six months later ($y$), and the estimated mean survival of the blood

**Table 10.1. Correlation between serum level of cholesterol in rabbits before and six months after starting a control diet (in mg/100 ml of blood)**

| Animal | Before ($x$) | 1000 × $\log_{10}(x/10)$ | After ($y$) | 1000 × $\log_{10}(y/10)$ |
|:---:|:---:|:---:|:---:|:---:|
| 1 | 40 | 602 | 13 | 114 |
| 2 | 61 | 785 | 48 | 681 |
| 3 | 50 | 699 | 22 | 342 |
| 4 | 60 | 778 | 22 | 342 |
| 5 | 38 | 580 | 32 | 505 |
| 6 | 86 | 934 | 25 | 398 |
| 7 | 60 | 778 | 42 | 623 |
| 8 | 33 | 519 | 23 | 362 |
| 9 | 58 | 763 | 48 | 681 |
| 10 | 62 | 792 | 32 | 505 |
| 11 | 112 | 1049 | 34 | 531 |
| 12 | 66 | 820 | 31 | 491 |

*Strictly speaking, it gives a realization (i.e., a particular example) of an unbiased estimator. The statement would be exact if one substituted uppercase letters for $x$, $y$, and $m$ throughout.

platelets measured in the last days of the experiment ($z$). We shall draw freely on these data. In the present illustration we shall consider the cholesterol readings only (Table 10.1). We obtain the following results.

$$\Sigma x_i = 726 \qquad\qquad \Sigma y_i = 372 \qquad\qquad \Sigma x_i y_i = 23,135$$
$$\Sigma x_i^2 = 49,058 \qquad\quad \Sigma y_i^2 = 12,828 \qquad\quad m_X \Sigma y_i = 22,506$$
$$m_X \Sigma x_i = 43,923 \qquad m_Y \Sigma y_i = 11,532 \qquad \Sigma(x_i - m_X)(y_i - m_Y) = 629$$
$$\Sigma(x_i - m_X)^2 = 5,135 \quad \Sigma(y_i - m_Y)^2 = 1,296$$

Then the sample correlation coefficient, $r_{XY}$, is

$$r_{XY} = \frac{629}{\sqrt{5,135 \times 1,296}} = 0.243,8$$

Note that nothing in the foregoing *calculations* depends on the marginal distributions of $X$ or $Y$ or their joint distribution. *Interpreting* the result is another matter, which we will discuss more fully later. But just as we can for-

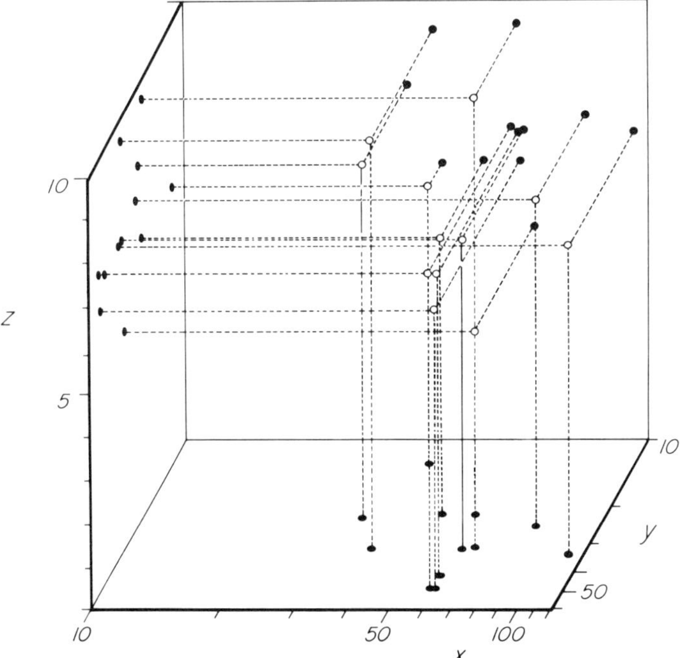


**Figure 10.1.** A three-way display of measurements in rabbits on serum cholesterol in mg/100 ml before ($x$) and after ($y$) six months on a control diet, and of platelet survival in days ($z$). The white dots have coordinates given by the three measurements, the black dots represent the projection of the white dots on the three surfaces corresponding to the three possible scatter diagrams. Projections are connected to the corresponding points by interrupted lines.

mally compute a standard deviation in any reasonable case, so we can compute a correlation coefficient. The reader should not fall into the trap of refusing to compute such a statistic at one extreme, and the trap of misinterpreting it at the other.


## PROPERTIES OF THE CORRELATION COEFFICIENT

1. Both $\rho$ and $r$ may assume any value between $-1$ and $+1$. The extreme values signify a perfect correspondence between $X$ and $Y$; that is, a scatter diagram would be simply a straight line, sloping upwards if the sign is positive, downwards if it is negative.

2. The correlation coefficient is unchanged by linear transformation of either or both variables. The correlation between weight in pounds and height in inches is the same as that between weight in kilograms and height in centimeters. But note that it would not be true of nonlinear transformations (e.g., logarithms or square roots). We shall have an illustration presently (Example 10.2).

3. For bivariate Gaussian data (which is what we are considering here) uncorrelatedness means statistical independence, and conversely. A correlation coefficient of 1 means complete statistical dependence and a value of 0 complete stochastic independence. This interchangeability between the terms *correlation* and *dependence* does not apply to distributions in general, and exceptions are easy to devise (7).

4. The correlation coefficient and regression coefficients are closely related; nevertheless they are distinct and must not be confused. The most important distinction is that the correlation coefficient is a symmetrical and mutual relationship between two random variables, from which symmetrical inferences are made. $X$ is correlated with $Y$ precisely to the degree that $Y$ is correlated with $X$. Indeed it might be argued that it is better even to *word* the relationship reflexively and say $X$ and $Y$ are correlated (with each other). It is disastrously easy to get into a habit of loosely using the words *regression* and *correlation* as if the terms were interchangeable.


## THE DISTRIBUTION OF THE SAMPLE CORRELATION COEFFICIENT $r$

This distribution is too complex to be of much interest here. It is extensively tabulated by David (83). Intuition rightly suggests that when $\rho$ is close to $-1$, the distribution of $r$ must be positively skewed since $r$ cannot be less than $-1$ and the distribution is constrained more on the lower side. The converse is true for $\rho$ close to $+1$. For $\rho$ close to 0 we infer it is nearly symmetrical (Fig. 10.2). It is shown with some difficulty that whatever the value of $\rho$, the transform

$$H = \frac{1}{2}\log_e\left[\frac{1+r}{1-r}\right] \tag{10.5}$$

derived by Fisher has three valuable properties. First, it is close to Gaussian even for small sizes of sample. Second, unlike $r$, it has a variance which does not depend on $\rho$, or at least is very insensitive to it. This variance is to a good approximation $1/(n-3)$. Third, it assumes the value 0 when $r = 0$ and is close to $r$ in the neighborhood of 0. The transform, $H$, is a monotonic increasing function of $r$, but it has infinite limits.

In the special case when $\rho = 0$, we have no need of such a transformation since the quantity

$$r\left|\sqrt{\frac{n-2}{1-r^2}}\right| \tag{10.6}$$

follows exactly the $t$ distribution with $(n-2)$ degrees of freedom and, for large $n$, at least, will be close to Gaussian. This function takes whatever sign $r$ has. We may note that (10.6) is a monotonic increasing function of $r$ over the interval $-1$ to $+1$. Hence, simple rejection regions for (10.6) furnish

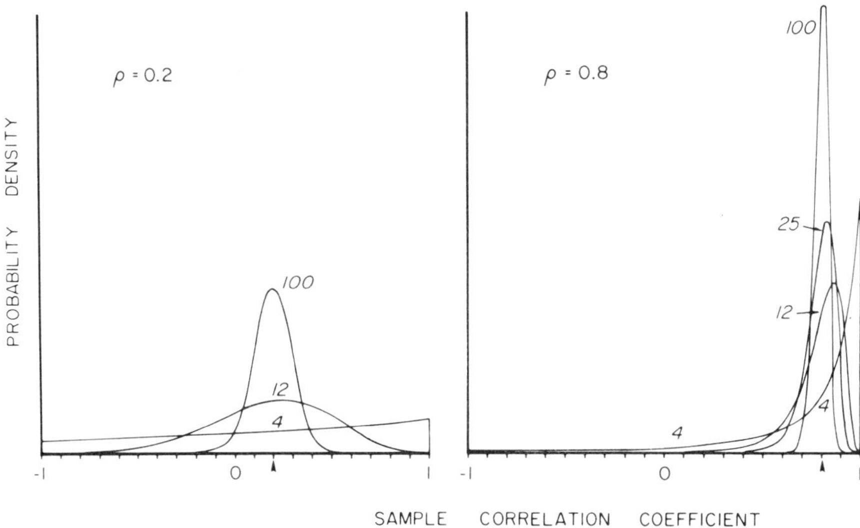


**Figure 10.2.** Distribution of the maximum likelihood (also the method-of-moments) estimate of the correlation coefficient of a bivariate Gaussian distribution. In each case the true correlation coefficient ($\rho$) is indicated at the *top left* and by an arrow on the abscissa. The italic number associated with each curve indicates the sample size. Where $\rho$ is close to O, with increasing sample size the distribution rapidly becomes Gaussian. However, if $\rho$ is close to the boundary of the parameter space, there is a long tail toward the center, which shrinks more slowly with increasing sample.

simple rejection regions for $r$. Also, while the domain of $r$ is finite, that of (10.6) is not.


## TESTS OF HYPOTHESES

Capitalizing on the above properties, we can deal with two standard cases.

### Null hypothesis: $\rho = 0$

This is by far the more commonly encountered case. From (10.6) we know the distribution under the null hypothesis and the rest is straightforward.

**Example 10.2.** Consider the data in Table 10.1. We will recompute the correlation coefficient, this time with a view to interpretation. Because the logarithm of the blood cholesterol level is nearly Gaussian, we surmise that the joint distribution will be bivariate Gaussian. Accordingly we shall operate on the logarithms of the values. To keep the calculations tidy we take logarithms to the base 10, retaining three decimal places, and multiply the result by 1,000. For convenience we may subtract 1,000 from the result (which is equivalent to dividing the original figure by 10). These last two maneuvers constitute a simple linear transformation, denoted $t(x)$ or $t(y)$, which, as I have remarked above, has no impact on the correlation coefficient. The calculations shown in Table 10.2 result.

Note not only that the correlation coefficient between the logarithms is not the same as that between the simple values. It is in fact somewhat, although not strikingly, larger. The exact test of the null hypothesis that the true correlation coefficient is zero is the $t$ test, which here has $12 - 2 = 10$ degrees of freedom.

From the $t$ table we find that the critical (two-tailed) value at a size of 5% is 2.228. The sample statistic is

$$t_{10} = 0.367{,}3 \left| \sqrt{\frac{10}{1 - (0.367{,}3)^2}} \right| = 1.249$$

Thus, the null hypothesis is not rejected.

**Example 10.3.** Test the hypothesis that for the data in Example 10.2 the true correlation coefficient is 0.7.

On the transformed scale this conjectured value is

$$\frac{1}{2}\log_e\left[\frac{1 + 0.7}{1 - 0.7}\right] = 0.867{,}3$$

The acceptance region will be within 1.96 times the standard deviation ($\frac{1}{3}$) on either side of this value, i.e.,

$$0.867{,}3 \pm 1.96/3 = 0.214{,}0 \quad \text{and} \quad 1.520{,}6$$

We do not reject the null hypothesis since $\frac{1}{2}\log_e[(1 + 0.3673)/(1 - 0.3673)]$ = 0.385,3.

Note that the data are consistent with both null hypotheses. The small sample size has resulted in little power, a point to bear in mind in interpreting such data.

**Test of the equality of two independent sample estimates of the correlation coefficient**

This procedure is straightforward, with the help of Fisher's transformation. The variance of the difference of the transforms is simply the sum of the variances.

**Test of the equality of several sample estimates**

This type of problem has analogies with analysis of variance, with this difference, that the variance of the transformed values does not need to be estimated from the data.

**Example 10.4.** In the study on malignant hypertension (21) the survival from the time of diagnosis was correlated with the age of the patient at diagnosis. After suitable normalization the results shown in Table 10.3 were obtained. Four groups (by sex and diagnosis) are involved. Since the approximate variances are known, we may weight the individual estimates by their reciprocals which is the same as multiplying by $(n - 3)$. The result, divided by the sum of these weights (in accordance with the appropriate section of Chapter 4) gives the weighted least squares estimate of the mean. The (weighted) sum of squares from the true mean would follow a $\chi_4^2$; since the mean has to be estimated, we lose a degree of freedom. The value of the statistic is less than 3 (the mean for the $\chi_3^2$ statistic), hence nowhere near significant.

**Table 10.2. Analysis of the correlation between the logarithms of the blood cholesterol levels from Table 10.1**

|  | Before treatment | After treatment | Cross-products |
|---|---|---|---|
| $\Sigma x_i$ | 9,099 | 5,575 |  |
| $m$ | 758.25 | 464.583 |  |
| $\Sigma x_i^2$ | 7,138,149 | 2,885,115 | 4,324,753 |
| $m\Sigma x_i$ | 6,899,316.75 | 2,590,052.083 | 4,227,243.75 |
| $\Sigma(x - m)^2$ | 238,832.25 | 295,062.916 | 97,509.25 |

$$r = \frac{97,509.25}{\sqrt{238,832.25 \times 295,062.916}} = 0.3673$$

**Table 10.3. Correlation between the age of onset and survivorship in malignant hypertension**

| Group | Number of subjects ($n$) | $n - 3$ | Sample correlation coefficient | Fisher's transform ($h$) | Weighted $h$ $= (n - 3)h$ | Weighted $h^2$ $= (n - 3)h^2$ |
|---|---|---|---|---|---|---|
| Men, renal | 15 | 12 | 0.2043 | 0.2072 | 2.4866 | 0.515,260 |
| Women, renal | 21 | 18 | −0.1951 | −0.1976 | −3.5574 | 0.703,062 |
| Men, essential | 22 | 19 | −0.2910 | −0.2997 | −5.6935 | 1.706,108 |
| Women, essential | 14 | 11 | −0.1420 | −0.1430 | −1.5726 | 0.224,833 |
| Total | | 60 | | | −8.3370 | 3.149,264 |

$$\Sigma(n_i - 3)h_i = -8.3370$$
$$\text{Sum of weights} = 60$$
$$\text{Weighted mean } (m_h) = -0.1389$$
$$\Sigma(n_i - 3)h^2 = 3.149,264$$
$$m_h\Sigma(n_i - 3)h_i = 1.158,414$$
$$\chi_3^2 = 1.990,851$$

SOURCE: Data of Kincaid-Smith et al. (21).

## ESTIMATION

The method-of-moments estimator for Gaussian data is the same as the MLE.* It happens that the absolute value of this estimate is a biased underestimate of the absolute value of $\rho$. A somewhat less biased estimate, $r^*$, is given by

$$r^* = r\left[1 + \frac{1 - r^2}{2(n - 4)}\right]$$

assuming $n \geq 9$ or so (84). Clearly as $n$ becomes large the term in brackets approaches 1 and $r$ becomes unbiased. In this section we will have to discuss two issues: interval estimation, and combining independent sample estimates.

### Interval estimates

The whole notion of estimation (as opposed to testing hypotheses) implies that we do not know the true value a priori. We could find two-tailed 95% confidence limits in the traditional way by finding those two values of $\rho$ such that the sample value just falls on the lower 2.5 percentile and the upper 97.5 percentile respectively. But this would be a tedious procedure. It seems much more sensible to use Fisher's transform and then treat the problem as a simple one in a Gaussian variate with known variance.

**Example 10.5.** Consider the estimate $\hat\rho = 0.367,3$ obtained in Example 10.2 from a sample of size 12. The transform, as we have seen above, is 0.385,3 and the standard deviation is $\frac{1}{3}$. So we may find the confidence limits on the transform as $0.385,3 \pm 1.96/3 = -0.268,0$ and 1.038,6

We must now convert these confidence limits back on to the scale of $\rho$. With a little manipulation of (10.5) we find

$$r = \frac{e^{2h} - 1}{e^{2h} + 1} \tag{10.7}$$

The confidence limits are found to be

$$-0.261,8 \quad \text{and} \quad 0.777,3$$

### Pooled estimates of the correlation coefficient

Let us suppose that we have several sets of data from each of which a sample correlation coefficient can be calculated. We have three major available strategies which are analogous to those discussed under "Linear Covariance Analysis" in Chapter 8.

1. The results may, of course, simply be pooled and a common correlation coefficient computed. But there may be reasons for not doing so. The means for the various samples may differ—for instance because different age

---

*I am assuming here that the variance of the two variates must also be estimated from the data.

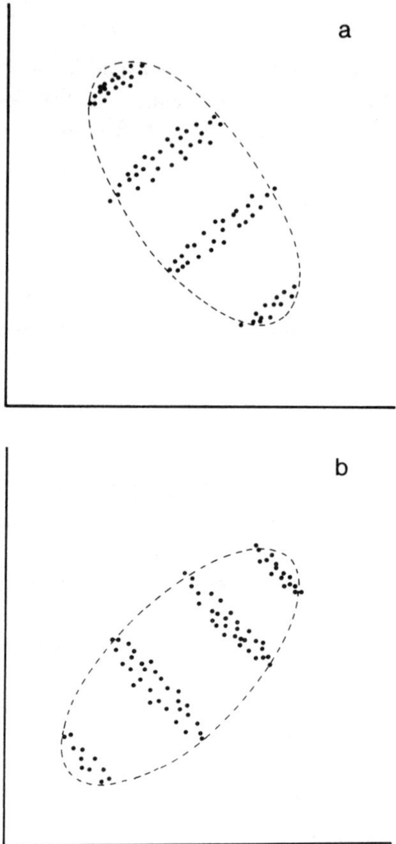

**Figure 10.3.** Spurious correlation because of heterogeneous means among subgroups (figmented data). In (*a*) scatter plots from four sets of data are shown. Taken separately, each shows a clear positive correlation as evidenced by the direction of the long axis of the scatter. When they are pooled, the overall trend (suggested by the dashed ellipse) is a negative correlation. Were the sample means for all four groups superimposed, the sample correlation coefficient would be positive. In (*b*) the converse artifact is shown. This incongruous result will be readily detected if a scatter diagram is done; it may readily be missed in the mindless use of the computer.

groups are to be combined; or several laboratories may be involved, each with its own standard values. To pool the results might give a spurious correlation. As extreme (and manifestly artificial) examples, in Figure 10.3 are shown two figures. In Figure 10.3a are shown several sets of data within each one of which there is a clear positive correlation; yet because of strikingly different means, the correlation for the data pooled without adjustment for differences in means is actually negative. In Figure 10.3b the converse is shown: within groups the correlations are negative, but after promiscuous pooling the correlation is positive.

2. We may compute the usual five raw moments for each group, compute the residual sums of square and the cross-products in the usual fashion (these calculations automatically adjust for differences in the mean); then we pool the residuals and compute the correlation coefficient as before. This approach is simple and reasonable. It will avoid the fallacy illustrated in Figure 10.3; but it will not adjust for differences in the true variance and covariance due, for instance, to variation in calibration from one laboratory to another. This problem could be circumvented if we had a rational basis for scaling without appeal to empirical data (e.g., if A's apparatus is known to read systematically 10% higher than B's). But scaling according to the empirical variances automatically converts the covariances into a sample correlation coefficient which cannot be handled by simple pooling.

3. We may obtain a pooled estimate of the correlation coefficient by using Fisher's transform and obtaining a pooled average by weighting inversely as the variance in accordance with the method discussed in Chapter 3. There are no special problems for samples of *large* size.

However, the sample estimate of Fisher's transform

$$H(r) = \frac{1}{2}\log_e\left[\frac{1 + r}{1 - r}\right]$$

is a biased overestimate of $H(\rho)$, the corresponding function of $\rho$. In fact, to a first approximation

$$E[H(r)] - H(\rho) = \frac{\rho}{2(n - 1)} \tag{10.8}$$

This formula has the undesirable property that the bias is a function of $\rho$, which is, of course, what we set out to find. The argument then, while reasonable, becomes somewhat untidy. To adjust for bias, we substitute for $\rho$ an initial estimate based on means of the individual sample values weighted inversely as their variances. Thus if the first sample is based on $n_1$ pairs of observations, we compute the corresponding transform $H(r_1)$ and weight by $(n_1 - 3)$ which is the approximate reciprocal of the variance. The sum of these weighted values divided by the sum of the weights gives us a first estimate of $H(\rho)$ which we shall denote $H_1(\rho)$:

$$H_1(\rho) = \frac{\Sigma(n_i - 3)H(r_i)}{\Sigma(n_i - 3)}$$

The inverse transform, $\hat{\rho}_1$, of this estimate (see 10.6 above) gives as the approximate bias for the $i^{\text{th}}$ set of data

$$\frac{\hat{\rho}_1}{2(n_i - 3)}$$

which we subtract from $H(r_i)$. The weighted mean of these corrected transforms is then computed to give a second estimate of the transform, which we denote $H_2(\rho)$. From it a second estimate of the bias is made and subtracted from the $H(r)$; a third pooled estimate $H_3(\rho)$ is now obtained, and so on. The enthusiastic should note, however, that excruciating exactitude of calculation is out of place since the correction for bias is of first order only and the inverse transformation creates its own biases.

**Example 10.6.** The data of Example 10.4 are reanalyzed by this method in Table 10.4.


## MULTIPLE CORRELATION*

Consider three random variables, $X$, $Y$, and $Z$, all defined on the same population. Can we construct an index of relatedness among them? We might do so in many ways. For instance we might pair the random variables in all possible ways and compute simple correlation coefficients for each pair: in this case $r_{XY}$, $r_{XZ}$, and $r_{YZ}$. Since the correlation coefficient of any random variable with itself is unity, and the correlation coefficient of $X$ and $Y$ is the same as that of $Y$ and $X$, we may set up a correlation matrix

$$\begin{bmatrix} r_{XX} = 1 & r_{XY} & r_{XZ} \\ r_{XY} & r_{YY} = 1 & r_{YZ} \\ r_{XZ} & r_{YZ} & r_{ZZ} = 1 \end{bmatrix}$$

In the first row and column, $X$ appears in every term; in the second row and column, $Y$; and in the third, $Z$.

Now suppose we were to devise a multiple regression equation of any of these three variables in terms of the other two: say

$$Y = \alpha + \beta X + \gamma Z$$

We apply this equation (using the least squares estimates of the parameters) to each of our sample values in turn, i.e., we compute and sum

$$\hat{y}_i = \hat{\alpha} + \hat{\beta} x_i + \hat{\gamma} z_i$$

We now have a value of $Y$ actually observed, $y_i$, and a "predicted" value $\hat{y}_i$, which is a realization of a random variable $\hat{Y}_i$. Then the correlation coefficient between $\hat{Y}$ and $Y$ is referred to as the multiple correlation coefficient. Clearly, since $\hat{Y}$ is the best predictor of $Y$, the two terms must in most cases have the same sign, and the sample correlation coefficient can never be negative.

Another way of representing the multiple correlation coefficient would be as the complement of the (positive) square root of the ratio

$$J = \frac{\text{Variance of the residual errors of the regression of } Y \text{ on } X \text{ and } Z}{\text{Variance of } Y}$$

*Some knowledge of matrix algebra (see Appendix 1) will be necessary to understand this section.

**Table 10.4. Iterative correction of bias in a pooled estimate of the correlation coefficient**

| | | Fisher's transform $h$ | | | | | Pooled estimate | |
| --- | --- | --- | --- | --- | --- | --- | --- | --- |
| | | Men renal | Women renal | Men essential | Women essential | Total | of $h(\rho)$ | of $\rho$ |
| First iteration | | | | | | | | |
|   Fisher's transform | $h(r_i)$ | 0.2072 | −0.1976 | −0.2997 | −0.1430 | | | |
|   Weight | $(n-3)$ | 12 | 18 | 19 | 11 | 60 | | |
|   Weighted sum | | 2.4866 | −3.5574 | −5.6935 | −1.5726 | −8.3370 | −0.1389 | −0.1380 |
| Second iteration | | | | | | | | |
|   Estimated bias | $\dfrac{-0.1380}{2(n-1)}$ | −0.0049 | −0.0035 | −0.0033 | −0.0053 | −0.0170 | | |
|   Corrected weighted sum | | | | | | −8.3200 | −0.1387 | −0.1378 |
| Third iteration | | | | | | | | |
|   Estimated bias | $\dfrac{-0.1378}{2(n-1)}$ | −0.0049 | −0.0034 | −0.0033 | −0.0053 | −0.169 | | |
|   Corrected weighted sum | | | | | | −8.3201 | −0.1387 | −0.1378 |
| Fourth iteration | $\dfrac{-0.1378}{2(n-1)}$ | Unchanged | | | | | | |

That is, if we denote the multiple correlation of $Y$ with $X$ and $Z$ by $R_{Y.XZ}$ then

$$R_{Y.XZ} = 1 - |\sqrt{J}|$$

This formulation may be extended to any number of variables, the "regressor" components being written after the period in the subscript of $R$.

The distribution of $R$ is in general very complicated. Under the null hypothesis of no correlation, however, if there are $n$ sample values and $k$ variates, the critical value of $R$ at a significance level, $\alpha$, is found from the $F$ distribution with $(k - 1, n - k)$ degrees of freedom

$$\sqrt{\frac{(k - 1)F}{(n - k) + (k - 1)F}}$$

## PROBLEMS

**10.1. a.** If in a scatter diagram, all the points fall on a straight line, parallel to one axis, what would this correlation coefficient be?

**b.** What would it be if they lay on a straight line with a gradient of 1 in 1,000,000?

**c.** How can you account for the difference logically?

**10.2.** If $X$ and $Y$ jointly follow a bivariate normal distribution

**a.** Write down the conventional test of the hypothesis that the coefficient of (linear) regression of $Y$ on $X$ is zero when the sample comprises $n$ pairs of values.

**b.** Hence prove the statement in the text that if $\rho = 0$

$$r \left| \sqrt{\frac{n - 2}{1 - r^2}} \right| \sim t_{n-2} \tag{10.9}$$

**10.3.** Show that (10.5) is a monotonic increasing function of $r$. (This can be done without appealing to the calculus.)

**10.4.** Find the approximate power curve of a test at the 5% level, based on a sample of size 28, of the hypothesis $\rho = 0.2$ against the alternate hypothesis $\rho > 0.2$.

# 11
# DISCRIMINATION
# AND DECISION

The scientist who is convinced that he is pursuing a fruitful line of inquiry and that some particular treatment does indeed have an effect, by conducting his experiment on a sufficiently large scale can achieve high power. He can give himself a good chance of rejecting the null hypothesis and of establishing the alternative conclusion in which he is interested.

If the samples are large then he confidently assumes that their sample means are Gaussian (see Chapter 17). The standard deviation of the differences in the means of two samples of the same size, $n$, with variances $\sigma_1^2$ and $\sigma_2^2$ is

$$\sqrt{\frac{\sigma_1^2 + \sigma_2^2}{n}}$$

Thus, even if the true difference between the means, $d$, is not zero but very small—far too small to have any *biological* significance—by choosing $n$ large enough he can make its standard deviation as small as he likes, and the ratio of the one fixed small value $d$ to the other arbitrarily small value can be made large enough to attain any desired level of *statistical* significance.

There is no theoretical limitation to the scale of the experiment and in matters of major importance there may be little practical limitation either. Basically this unlimited scope is attributable to the fact that the scientist is concerned with universals and that he may be content simply to show some difference in mean values between two groups—for example, between treated and untreated samples from a common population.

The primary concerns of the physician and many others who practice professions, however, are not with universals but with particulars, not with populations but with individuals. A physician who is confronted with the problem of deciding whether or not a particular patient has Paget's disease derives only limited help from the fact that the *mean* values a proposed diagnostic variate assumes in the normal and diseased populations differ. A very low value in his patient may be sufficient to rule out the disease, or a very high value sufficient to establish that he is not normal. But there remains the problem of what to do when the value lies in an equivocal range; and the size of this equivocal range will depend not merely on the means of the populations but on other factors as well, as we shall see presently. The problem for the physician is not one of what differences there may be in the *means* or other characteristics of the populations, but rather of how useful the test is in distinguishing between diseased and normal states in *individual* persons; that is, he is concerned with *discrimination*.

At first sight, this distinction may seem very puzzling; and it is because of the possibility of confusion that I have put so much emphasis here and earlier (7) on the difference between a distribution and a histogram, terms which many practical biostatisticians are content to treat as much the same. The random variable (e.g., alkaline phosphatase level) has a distribution; that is, we can make up a set of values that the result of a single measurement (yet to be taken at random) may assume, and the probability, or probability density, associated with each. In the real world, we usually do not know precisely what this distribution may be; but the larger our sample, the more detailed a histogram we can construct; and an indefinitely large sample displayed with sufficient precision and rescaled so that the area under it would equal unity would agree as closely as we please with the true distribution. We would then know the mean and the standard deviation of the distribution, properties which attach to predictive statements about any further *single* value to be picked at random. But even if we knew the distribution exactly (for instance, from theory), the next result to be randomly chosen would retain a priori all the uncertainty associated with the true distribution, although now that the true distribution is well characterized the degree of uncertainty is known exactly. In the course of finding the exact distribution, we have found the true mean precisely because the standard deviation (of the distribution *of the sample mean* but *not* that of the variate itself) has approached zero and a distribution with a standard deviation of zero is not a distribution at all but an exact quantity. If we now take a further sample of 100 random independent individual values with this distribution, the new sample mean a priori will have a standard deviation 1/10th of that of the parent distribution (i.e., of the distribution of any future single value that we select from it). This standard deviation of the sample mean, $\sigma/\sqrt{n}$ (or the sample estimate of it, $s/\sqrt{n}$), has little to do with the properties of the sample *estimate* of the standard deviation for the parent population based on the 100 values:

$$s = \left[ \frac{\Sigma(x_i - m)^2}{n - 1} \right]^{1/2}$$

which (as the sample size increases) does not approach zero but $\sigma$, the true standard deviation of the parent population. What the standard deviation of the mean, $\sigma/10$, describes is the sample standard deviation of the distribution of a large number of values *each of which is the sample mean of 100 independent readings.* Of course after any particular sample has been taken the sample mean has no distribution: it is what it is. Any probability statement we try to make at that stage describes what we *believe;* that is, it is a subjective probability statement, or (in somewhat less controversial terms) it is a probability statement about subjective indeterminacy (7). It will be clear, then, that as the sample size gets larger and larger, the distribution of the sample mean gets more and more compact. Thus with increasing sample size the degree of overlap of the distribution *of sample means* from distributions with different means becomes less and less, so that eventually a quite trivial difference in *means* will guarantee that the sample means are confidently distinguishable (Fig. 11.1 *right*). This result is quite readily reconciled with a large overlap in the parent distribution of (single) values (Fig. 11.1 *left*). But obviously in clinical practice the true reading for the individual patient is a single value, not the mean of a sample of many patients. Hence in making individual statements for this one patient, it is a distribution of the type in Figure 11.1 *left,* not one of the type in Figure 11.1 *right,* that is pertinent. That blood pressure levels are on average higher in Negro males than in White males in the United States, as has been shown repeatedly in large samples of readings, does not mean that blood pressure is a useful test for distinguishing to which of the racial groups an individual male belongs. With this background, how do we proceed in practice?

The uncertainty that the physician experiences in some of his cases, arising from the data of a particular test, is ultimately attributable to two sources: uncertainty about the value in the patient and, given the *true* value for that patient, uncertainty as to which class he belongs to. The distinction is an important one and worthy of elaboration.

### Uncertainty about the value in the patient

Some data on the patient may be easily obtained and highly reproducible: for example, height. Some are easily obtained and highly variable, such as blood pressure. Some are highly reproducible but difficult or dangerous or expensive to obtain (e.g., X-rays of bone). Some are both difficult and variable, such as measurements of pulmonary artery pressure by cardiac catheterization. There is little point in repeating tests that are highly reproducible:* we know of course that for independent readings the error of the mean has a standard deviation that goes to zero inversely as the square root of the number of observations. But if the standard deviation is very small to start with, the actual reduction produced by averaging a great many readings may be trivially small compared with other sources of variation.

---

*Of course a measurement may be in error, not because it is hard to make precisely, but because of gross carelessness: a height of 150 cm is read as 170 cm. In such cases, repeating the measurement to check the result is worthwhile.

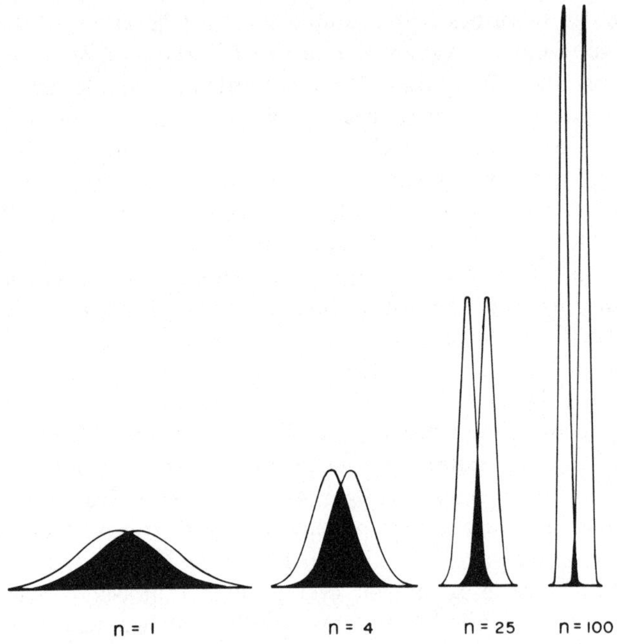



**Figure 11.1.** The degree of overlap of two Gaussian distributions is shown in the *left* diagram. The total proportion misclassified is shown by the area in black divided by the sum of the areas of the two distributions, which furnishes a measure of the *discriminating power* of the variate. When means of four independent measurements are taken the standard deviation of the distribution of such means is half that for individual measurements (*second* diagram) and the probability density curves require to be twice as tall. The overlap is correspondingly decreased. Likewise for samples of size 25 (*third* diagram) and 100 (*last* diagram) the overlap is further reduced. In the limit, it would be zero. But the discriminating value of the test (for single values) remains unchanged.

Where the result is highly variable, such as blood pressure, any competent physician would hesitate to make final pronouncements on the basis of one reading, unless the result is either extraordinarily high or low. In such cases, then, it is advantageous to deal with the average of several readings. However, two facts should be noted in this connection.

First, doing repeated tests may be impracticable, as for example measuring intracardiac oxygen levels or making measurements of pressure in the cisterna magna.

Second, the standard deviation of a mean need go to zero only if the measures are independent, and independence may not usually be assumed if several readings are taken close together. For example in determining what the patient's usual blood pressure is, the average of 20 readings taken during the first visit of the patient to the clinic will not necessarily provide a good measure—the average of one reading taken on each of 20 visits one week apart each would be much more informative. I am not proposing this latter as a routine diagnostic procedure; but I personally would hesitate to make any serious pronouncements on a patient's habitual blood pressure on the basis of any number of readings at one clinic visit only.

## Uncertainty of classification

But even if the physician knows exactly what the true mean value for the patient is, he may still be in serious doubt about classifying the patient, because the distribution of true readings for patients in the two populations of interest may exhibit a large overlap.

## Components of variation

Why is not all the uncertainty for the individual patient resolved by taking more and more measurements? To answer this question coherently, we shall have to make more explicit what the sources of variation are. As a talking point, consider blood pressure. We shall define the sample space as the blood pressure in one person yet to be picked at random from a particular population, that of people who constitute the reference set. It is in the mean for this corporate population that we are interested, not individual members of it as such. Analysis of error in detail would be an elaborate undertaking. Let us be content to say that there are three broad sources of variance:

1. Those due to permanent or semipermanent differences among subjects' means $(\sigma_a^2)$. They would reflect genetic factors, kidney damage, elasticity of the aorta, etc. Some of them change in the same patient with age, but relatively slowly.

2. Those due to relatively transient states $(\sigma_t^2)$: emotion, the effects of current drugs, diet, electrolytes, etc.

3. Those due to errors of measurement $(\sigma_e^2)$.

For simplicity suppose that these components operate independently so that the variances may be added to find the overall variance. Then if we denote any *single* reading by $Y$

$$\text{var}(Y) = \sigma_a^2 + \sigma_t^2 + \sigma_e^2$$

Where multiple measurements are taken the variance of the mean will depend on how the sample is composed, and what components may safely be regarded as independent. For instance, daily readings for a month on the same patient will do virtually nothing to reduce the first component of variance. A continuous tracing for a month with a perfectly accurate measuring device might come very close to telling us exactly what the mean reading in that particular patient is; but even this exact information tells us little about the true mean for the population at large. Even the *exact* mean for a patient yet to be picked at random would still have a variance of $\sigma_a^2$ about the *population* mean (which is what we want to determine). It would be far more informative to have one reading only on each member of a random sample from the population. In general, suppose we have $r$ replicate readings on a patient during a single visit. We shall also suppose them independent over the sample space of readings on that particular patient at that particular visit. Each patient is seen at $v$ independent visits. In all, $s$ independent subjects are studied. We average the results. Then the variance of their mean *(M)* about the true population mean would be

$$\text{var}(M) = \frac{\sigma_a^2}{s} + \frac{\sigma_t^2}{sv} + \frac{\sigma_e^2}{svr}$$

The reader may readily verify this formula by calculating the number of times each component is estimated and appealing to the independence of the components, which implies that all the covariances are zero.

**Example 11.1.** The distinction perhaps may be made clear by considering the diagnosis of the carrier state of hemophilia. The genotype of a woman may be known with virtual certainty, either from the phenotype of her father or those of her sons. Suppose, however, that there is no conclusive evidence one way or the other and we attempt to supplement meager genealogical data with examination of the woman's own blood. The standard method of assaying Factor VIII is crude—in expert hands blind duplicate determinations have an experimental error of perhaps 8 to 10% (85). Averaging blind replicate determinations will reduce the size of this error and, with a reasonable amount of laboratory work, a fairly good estimate of the true reading for that woman results. But reducing this source of error achieves only so much. True values in carriers vary from one to another for several reasons. Some mutant forms are more severe than others. Carrier sisters (whom we suppose to bear the same mutant gene) will vary in phenotype because of the (random) effects of Lyonization. Modifier genes may be variably present; and so forth. Even if, with some of the newer techniques, it were possible to measure the individual woman's Factor VIII level exactly (86), perfect separation of carriers and normal women into unambiguous groups is not to be expected. Of course the *mean* value for normal women may still be twice as high as that in carrier women; but the diagnostic hematologist is not concerned with women, he is concerned with each particular woman. The clinician must accept that certainty in diagnosis is rarely possible. Of course, this statement is not to make a virtue of vagueness, indecision, or incorrect diagnosis. There are undoubtedly diseases in which diagnosis can be made with complete certainty—for example in sickle cell disease. But at present such cases are exceptional. They may be used as an ideal of resolution. They can hardly be taken as representative of what the physician may reasonably expect to achieve in practice. If the clinician always demanded this kind of certainty before making a diagnosis he would seldom make any diagnosis at all.

### Practical steps

The clinician may attempt to resolve uncertainty by several maneuvers.

**1. Multiple determinations of the measurement.** The limitations of this approach have been discussed above.

**2. Other pertinent tests.** For preference these tests should not merely be discriminating, each in its own right, but mutually independent. More generally, the less the *redundancy* among tests, the better their combined discrimination.

**Example 11.2.** In Marfan syndrome, leg length has some diagnostic value; but once the length of the right leg is known, in most cases the length of the left leg furnishes little further information. On the other hand, trunk length as a second measurement is highly informative. How several pieces of information are to be best used for discrimination will be discussed later.

**3. Provocation tests.** If the variance can be reduced only with difficulty, or perhaps not at all, overlap between the distributions attached to competing hypotheses may still be reduced if the means can be further separated. The underlying idea is that defects in the structure or operation of a system may be heightened by taxing it. The murmur of mitral stenosis is made clearer by exercise. The pain of esophagitis is increased by an acid meal. A loading dose of glucose exposes mild degrees of glucose intolerance. The success of such measures depends on a *differential* effect: large effect in the diseased group, little in the normal. That such tests are not more successful and more widely used, is due to their general tendency to increase not only the difference in the means, but also the variances, with little consequent gain in discrimination. (As an analogy of the extreme case, measuring blood pressure in millimeters gives means that differ by a greater number of basic units than measuring it in centimeters. But the discrimination is unchanged.)

If it be granted that in general some degree of uncertainty is inescapable and some risk of misdiagnosis must be faced, what factors must be weighed and what is the best strategy to adopt in making decisions? There are two issues here: how misdiagnosis is related to the decision point(s); and what "values" (i.e., costs) are, and how they bear on choosing these points.


## FACTORS ON WHICH DISCRIMINATION DEPENDS

The error that will occur, whatever diagnostic criteria may be set up, depends on several factors.

### 1. The distributions of values in the healthy and diseased populations

Should there be no overlap at all in the distributions of interest, then no misclassification need occur. The dividing line between the two groups may equally well be drawn anywhere in the "no man's land" that lies between the two distributions.

**Example 11.3.** Bussey (11) found from his extensive data of polyposis coli that familial cases have 100 polyps or more, nonfamilial cases 50 or less (*his* Figure 59). Here there is no overlap. However, those data were collected mostly before colonoscopy was in common use. Thus the counts are based on less sensitive methods, which are not specified: presumably X-ray examinations, and study of specimens after operation, and at necropsy, where there will be a bias toward subjects with more advanced disease. Bussey is careful not to make any unwarranted claims.

Such a degree of separation is exceptional. Usually there is some overlap between the two distributions. For purposes of discussion, denote by the functions $f(x|H)$ the (conditional) probability density function of the measurement in healthy people and $f(x|D)$, that in diseased people. If these distributions are known exactly, either parametrically (e.g., if it is known that they are Gaussian, and the parameters are specified) or nonparametrically (if for example we have simply detailed histograms of sound samples from the population), then wherever the dividing line is drawn it is easy to determine what proportion of cases will be misclassified. For example, if all subjects in whom the serum alkaline phosphatase level is above 40 units are to be labeled Paget's disease, then any normal people above this level would be misclassified *(false positives);* for that matter people with other diseases that give rise to an increase in this variable, such as obstructive jaundice, may be misclassified also; and if the number of false positives from all causes is very high, the test, with this decision point, has low *specificity.* Conversely, if there are few false positives, the specificity is high; and if there are no false positives at all the test would be *pathognomonic.* Now of course the specificity of the test can in many cases be made very high by using only the most stringent standards for diagnosis. There is little doubt that the number of false positives in the last example would be greatly reduced if the dividing line between healthy subjects and those with Paget's disease were at a level of 150 units. The price for this high specificity of course would be that all those people with Paget's disease in whom the alkaline phosphatase levels are lower than 150 units would be misclassified in the opposite direction—that is, they would be *false negatives.* If this proportion is large, the test is then said to have low *sensitivity.* It is self-evident that the greater the specificity is made the lower the sensitivity, and conversely. If the serological diagnosis of syphilis is to be made at a titer of 1 in 50 or above, a patient would seldom be incorrectly labelled as syphilitic; but a great many cases of syphilis in whom the reactivity was at a lower titer (including most latent and late cases) would be missed.

## 2. The relative sizes of the distributions

The area under any probability density curve is equal to unity. Thus the foregoing argument is really about *proportions* of patients of a particular group misclassified one way or the other. It would be eminently reasonable to use these unmodified distribution functions where a priori the subject is as likely to be affected as not, e.g., in determining the carrier status of a woman whose father is normal and mother is an obligatory carrier of hemophilia. But outside of genetics, the physician can rarely make such a symmetrical assumption. He may then be concerned, in setting up his policy of diagnosis, to do the minimum amount of harm to the totality of his patients. While his concern is to do the best he can for his patient as an individual, nevertheless the occasional brilliant success in diagnosis is hardly a compensation for a great many dangerous failures. Seeing the problem in this light, he will want to take into consideration the prior probabilities of encountering patients who are normal and of those who are affected. Thus he will weigh not merely what the (conditional) probability distributions are but also what the prior odds are of the patient having the condition or not. The appropriate adjustment consists of multiplying each conditional probability by the prior probability of encountering the disease. This maneuver may be represented graphically as magnifying the vertical scale of the different distributions by the prior probabilities of encountering somebody from the several classes.

**Example 11.4.** For instance suppose a hospital is setting up a system of detecting the disorder phenylketonuria by quantitating the phenylalanine content of urine. The authorities might decide that the dividing line should be drawn in such a way that only 1% of healthy people will be false positives. Because of the configuration of the distribution curves this will allow (say) 90% of cases with this disorder to be picked up. Stated in these terms one might be pleased with both the sensitivity and specificity of the test. However, this disease is rare: it occurs with perhaps one ten-thousandth of the frequency of the healthy state. What is the composition of the cases that are diagnosed as positive? The results are laid out in Table 11.1. The vast majority of children who are singled out for further investigation would prove healthy. That is, the *yield* of cases (the proportion of true positives in all

**Table 11.1. The relationship of prior probability to yield of disease**

| | Competing interpretations | |
| --- | --- | --- |
| | Phenylketonuria | Normal |
| Prevalence | 0.000,1 | 0.999,9 |
| Frequency of positive test | 0.9 | 0.01 |
| Proportion of subjects with positive test and belonging to class | 0.000,09 | 0.009,999 |
| Yield of cases | $\dfrac{0.000,09}{0.000,09 + 0.009,999} =$ | 0.008,92 |

(i.e., approximately 1 in 112 positive cases has phenylketonuria)

positives) is low. Clearly, then, there is more to discrimination than the probability distributions for the two cases to be considered.

Such a low yield has certain serious consequences in practice. There may be a limit to the load of detailed confirmatory investigations the laboratory can undertake. Those with experience of such systems know how difficult it is to maintain interest, alertness, or even high standards of technical competence, when the yield of true cases is small. For many other disorders, there may not be better diagnostic criteria. Thus all "positives" (true or false) will be subjected to some procedure which is perhaps hazardous, expensive, or inconvenient, and inappropriate in over 99% of cases (although *which* 99% remains unknown). In the face of these defects it may not be practical to maintain a sensitivity of 90%.

### 3. The types of subjects that a particular kind of doctor encounters

They may be different from those met by a doctor of another kind. To some extent of course this will be a datum in diagnosis. A pediatrician rarely sees gout (except that due to specific genetic disorders, notably the Lesch-Nyhan syndrome). The geriatric specialist sees predominantly the common variety of gout. The genetic disorders in Negroes, in Jews, in Armenians, in Italians differ (87). Such illustrations could be multipled. There may be at work some case-selection that is in the nature of things, and does not reflect any conscious policy on the part of the referring doctor. For example the patient with coronary disease of trivial degree, or so severe that it causes sudden death, may never get to, still less beyond, the general practitioner. The hospital resident will tend to see patients of intermediate degrees of severity. Thus the picture of coronary disease that the general practitioner has in mind will be rather different from that of the hospital resident, and in consequence it may be necessary, and desirable, for each to set up his own standards as to where dividing lines for informative data should be drawn. This third factor is the phenomenon of ascertainment with which the genetical analyst is familiar.

The practical implications of this point are unexpected and even disturbing. By the same argument, how readily a physician would think of *kala-azar* in a patient with a big spleen would differ in Baltimore and in Teheran. But it should perhaps also differ at The Johns Hopkins Hospital from what it would in a small local hospital, partly because of the catchment of patients and partly because certain diagnostic problems tend to be sent to university hospitals. But even within a hospital these ascertainment factors differ. In our genetics clinic, for instance, the major causes of aortic regurgitation are such disorders as Marfan syndrome, mucopolysaccharidoses, and osteogenesis imperfecta; in this it differs widely from the cardiac clinic. So we are led inexorably to the unsettling conclusion that even within a single hospital diagnostic criteria should perhaps not be uniform!

**Where is the dividing line to be drawn?**

In the analysis so far, no system of values has been introduced. The actual drawing of the dividing line, however, must involve values, whether or not the practitioner likes to acknowledge them explicitly.

For instance, it can be shown (3, 29) that if the distribution curves for the two classes are suitably scaled to take account of the prior probabilities of the two distributions, then the point at which the two curves intersect is that which gives the minimum misclassification of results—that is, the expected *number* of false positives plus that of false negatives will be a minimum. But to use this criterion implies that the practitioner believes that the price paid for a false positive is exactly the same as that for a false negative. The human, as distinct from the medical, geneticist may be quite content with this policy since commonly he is interested merely in fact and not in the practical implications of the fact. But in many clinical situations such a detached attitude would be not merely inadequate but repellent.

**Example 11.5.** In secondary syphilis the price for a false negative is a serious one: the infected patient stands an excellent chance of being cured permanently if treated at this stage, whereas if his condition is neglected he may suffer dire tertiary and quaternary consequences; moreover, this is the stage at which the disease is most infectious and therefore the risk of spread to other contacts at its highest. Compared with these dangers the penalty for a false positive diagnosis is slight. Although the risks of treatment are not altogether negligible, they are comparatively small, and the only other price is the stigma of the diagnosis and the need for a lumbar puncture some time later to insure that neurosyphilis has not occurred. Not all the considerations lie in the one direction; one should attempt as accurate a diagnosis as possible, and should not treat patients for secondary syphilis on trivial evidence. But I would regard the cost of a false negative as being perhaps twenty or more times higher than that of a false positive and would shade my dividing line accordingly. Conversely, there are diseases in which (so far at least) nothing is to be gained by early diagnosis, so that a false negative imposes little cost, whereas a false positive diagnosis may lead to unnecessary disruption of the patient's life and certainly destruction of his peace of mind. Again the argument does not all lie on one side; but the dividing line should be shaded in such a way as to make standards of diagnosis rather stringent. There are of course other diseases in which the issues are much more complex. For example if in a patient with pulmonary symptoms the X-ray shows what appears to be a cancer, the criteria for diagnosis should be sensitive rather than specific, since early treatment provides the best hope of cure. But if a patient presents with bone pain and an X-ray shows an osteolytic lesion that is thought to be a secondary from cancer of the lung, then it is evident that if the conjecture is correct the disease is already past the stage of radical cure, by surgery at least, and the criteria of diagnosis should then be specific rather than sensitive. These and other aspects of rational diagnosis are considered in greater detail elsewhere (7).

## Comment

It is clear that, in principle at least, purely probabilistic aspects of making decisions are both unambiguous and firmly within the competence of the statistician. Incorporating values, once they have been decided on, is also simple enough. In practice, however, the statistician, as such, cannot assume the responsibility of assigning values. Although the physician is better informed about intermediate details, it can be argued that he should not be the arbiter either. The ultimate issues—what qualities do we really want? and what are we prepared to pay for them?—cannot be resolved by appeal to brute fact. Whether such-and-such a price is worth paying to reduce mortality or morbidity in a particular group, is a matter of judgment to be made by the responsible authority (perhaps society itself). I cannot deal with the subject adequately here. However, it is important that the question should at least be raised. To ignore it altogether would in effect impose a uniform cost for mistakes, which would be still more offensive than an arrogation of authority by the statistician. There is the added danger that the principle of uniform costs is an *implicit* policy, the intrusion of which might be overlooked entirely.

It seems, then, that the cost associated with a false positive or a false negative depends on whatever the value of the particular measurement is; in suitable circumstances the cost could even be represented as a mathematical function. The criterion (that, to me, seems most rational) for the best dividing line is that which minimizes the expected cost due to misclassification. It transpires, as might have been guessed, that this step is very simply dealt with. For each value of the discriminating variable, $x$, we compute a cost-benefit function under each of the hypotheses (or diagnoses). The function will be positive in proportion to the benefit derived and negative in proportion to the cost (or harm). If cost exceeds benefit, the function will be negative and conversely. This function is then multiplied by the density, the prior probability, and the ascertainment factor. The best dividing point is the value at which two such curves intersect.

The foregoing ideas may be summarized as follows. Let us represent the probability densities of the variate under the two hypotheses by $f(x|H)$ and $f(x|D)$ respectively; the relative sizes of the two populations by $S_H$ and $S_D$; the (conditional) ascertainment functions by $A(x|H)$ and $A(x|D)$; and the cost-benefit functions by $K(x|H)$ and $K(x|D)$. Then the best dividing point is that value of $x$ at which the curves representing the products of the four quantities intersect, i.e.,

$$S_H f(x|H) A(x|H) K(x|H) = S_D f(x|D) A(x|D) K(x|D).$$

There may be several places at which the curves intersect and hence this equation satisfied. In that eventuality either the policy-maker may settle for that dividing point which gives the minimum cost, or accept a division of the axis into several regions each labeled "abnormal" or "normal," according to which function has the higher value.

Complete illustrations of these notions from the clinical field are difficult to find because, despite the obvious nature of what has been discussed and the fact that any good clinician employs these ideas implicitly in making decisions, very little attention has been paid to the subject formally. One example that has been worked out in some detail is the diagnostic value of serum lipoprotein levels in coronary disease (43).

**Example 11.6.** Borgaonkar and his associates (88) have elaborated an index (which we shall denote by $X$) of dermatoglyphic patterns which may be used in distinguishing people with Down syndrome (trisomy 21, mongolism) from euploid (healthy) people. The diagnosis can often be made clinically with little difficulty and a definitive diagnosis can always be established by karyotyping done with sufficient care. However, clinical diagnosis is far from infallible; on the other hand karyotyping calls for considerable laboratory skill, is time-consuming (so that prompt answers are not available) and expensive. In such circumstances it is useful to have a discriminant such as the dermatoglyphic test, not so much to give the final answer, but as a coarse method of deciding which cases should, and which should not, be submitted to the more elaborate and expensive procedure. In the upper part of Figure 11.2 are shown the probability density functions for this dermatoglyphic index in healthy subjects (on the *left*) and those with Down syndrome (on the *right*). False negative results are shown to the left of the dividing line in solid black and false positives in the hatched area to the right. Something like 6 to 7% of all cases are misclassified by this criterion.

However, this analysis ignores the relative sizes of the two populations. Accordingly, in the lower part of the diagram are plotted what the actual frequencies would be in the population, assuming that healthy people are 200 times as common as those with Down syndrome. (This is certainly an underestimate of the true ratio in the population at large.) The "best" dividing line is now shifted considerably to the right, and in consequence the number of false positives has fallen to about one in 1,900 whereas the number of false negatives has increased to something over 56%. We may complete the analysis by supposing there is no ascertainment bias—that is, people with Down syndrome who have higher scores are no more likely than those with low scores to be seen by clinical geneticists; and we may suppose—quite arbitrarily—that for all values of $x$ the cost-benefit function of a false positive is only one-fifth as great as that of a false negative. Then the net result of incorporating these two further components is to increase the size of the frequency curve for persons with Down syndrome by a factor of 5 so that the final ratio

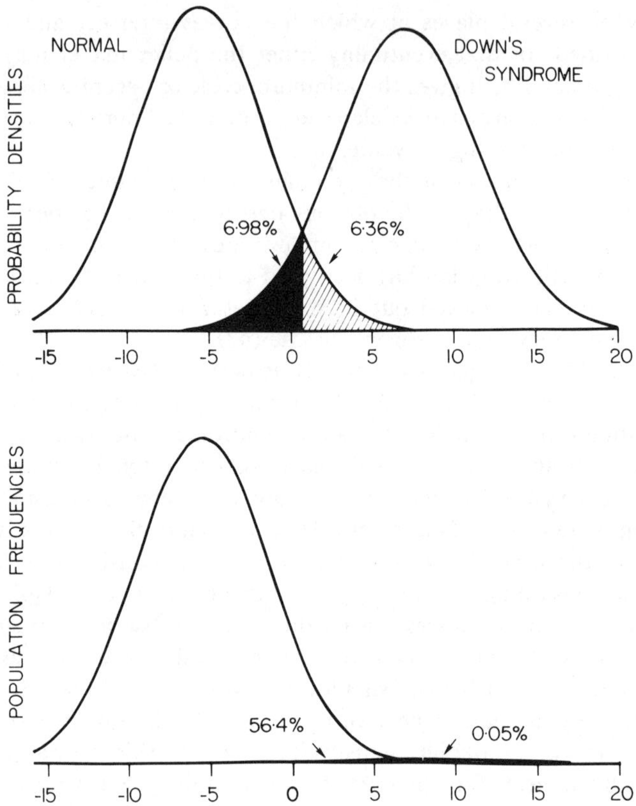


**Figure 11.2.** The discriminating power of a dermatoglyphic index in Down syndrome. False positives are *hatched,* false negatives in *black. Above* is shown the utility of the index when 50% of the population is affected, and *below* when normal persons are 200 times as common. Note the dramatic increase in the false-negative rate (when the criterion is the minimum number of misclassifications).

of the areas under the curves will be 40 to 1, a position intermediate between those shown in the two diagrams. This now leads to the best dividing point as 5.7085 which results in 0.29% of false positives and 37.23% of false negatives respectively.

The clinician is apt to be rather impatient with the foregoing formulation, which argues as if it were a common matter for a diagnosis to stand or fall by one particular finding. It is obvious that the making of decisions in clinical practice is a much more complex matter. However, there are at least three points to be made in defense of our formulation.

1. The cases that we have discussed, have been presented to illustrate the principles of making decisions, rather than because they are typical or

particularly plausible. The univariate case has been discussed quite simply because the inherent reasonableness of the decision is more easily grasped.

2. There are examples, rare but encountered from time to time, in which a clinical diagnosis does indeed depend on findings in one test. One such example is the serological test in latent syphilis. The test is obviously evaluated against the general background of the patient; but to rely on clinical diagnosis as a basis for making a decision would leave much to be desired.

3. Commonly, in procedures in public health unsuspected cases are to be detected by mass screening. A decision may, for logistic reasons, have to be based on a single measurement even if the decision is merely to sort out those cases that merit further investigation. We have already mentioned screening tests for phenylketonuria, which in most parts of the United States are compulsory by law even where there is no basis whatsoever for suspecting the child affected. The same kind of phenomenon is encountered in screening for Tay-Sachs disease which has come to be a very common practice among Ashkenazi Jews.

## BIVARIATE DISCRIMINANT ANALYSIS

As a first step to a generalization, consider discrimination with the aid of two tests. While they may not be entirely independent, nevertheless, a suitable combination of them may be more informative than either alone, even though one may appear to have little to do with the disease.

**Example 11.7.** Consider the diagnosis of almost any heart disease, say pulmonary stenosis. One relevant feature is ventricular hypertrophy, to be diagnosed on X-ray. If all subjects were pooled, the discriminating power of right ventricular size would be poor, because even among healthy people there is considerable variation. But much of this variation can be eliminated by taking into consideration the patient's size. In healthy and diseased subjects alike, heart size is related to body size; a heart size that would be healthy in a professional football player would cause consternation if found in a child of two years. Indeed, the point is so obvious that the clinician may be scarcely aware of using two data in making such an assessment.

Because of frequent mistakes in making such adjustments, it will be worthwhile to discuss explicitly the combination of two data.

### The use of ratios

There is commonly a loose argument something along the following lines. If $Y$ increases with $X$ and we are primarily interested in $Y$, we may adjust for variation in $X$ (which is beyond our control but is measurable) by considering the ratio of $Y$ to $X$. Thus, naive experimentalists will express drug dosage or dietary intake as so much "per kg body weight." Heart size is assessed as the ratio of its greatest diameter to the greatest diameter of the chest ("cardio-thoracic ratio"). Dieticians discuss atherogenic diets in terms of the P(olyunsaturated)/S(aturated) ratio of fatty acids. Psychometrists measure intellectual development as the ratio of observed to expected intelligence score ("the intelligence quotient"). And countless other examples. The *aficionado* of these contortions will occasionally see diagrams showing a two-way display of ratios ("Left-to-right shunt in liters per minutes per square meter body surface against dose of drug per kg body weight"). Even the toxicologists discuss mortality per man-year of exposure to some industrial toxin.

There is a proper use for ratios between two quantities which is justified *when each particular ratio between them has a homogeneous meaning;* that is, where the same notional inference can always be made from it. Logically clear illustrations in biology are not easy to find. For random variables the issue will be the behavior of mean values. We apply three criteria.

1. Is the relationship between the quantities linear? Dietary needs obviously are related to age, but not in a linear fashion. Needs at fourteen are greater than they are at seven; but needs at seventy are less than they are at twenty-five. It would be absurd to express dietary need as so much per year of age (Fig. 11.3, *top*).

2. Does the line of best relationship go through the origin? We expect the rate of exchange from dollars to francs to be the same for all sums; we do *not* expect the ratio of the temperature in degrees Fahrenheit to that in centigrade to be constant, as witness the values:

| Centigrade (C) | Fahrenheit (F) | F/C |
|:---:|:---:|:---:|
| 0 | 32 | ∞ |
| 10 | 50 | 5.00 |
| 20 | 68 | 3.40 |
| 30 | 86 | 2.87 |

Whether or not this condition is fulfilled may sometimes be decided from first principles; but in biology the relationship must usually be determined empirically. For instance, average weight is *not* proportional to average height; nor to the cube of it; nor is the cube root of weight proportional to height (26, 27). Thus the use of such ratios (e.g., "ponderal index") is to be condemned. Why, is clear from Figure 11.3 (*middle*). The part of the distribution in black is above the linear regression line and hence, if anything, overweight; but the corresponding ratios are below average and suggest that such people are underweight. The converse applies to the hatched part.

3. Are deviations from the best regression line proportionate? The points on any straight line through the origin will give the same ratios. But a ratio 1.1 times the mean ratio will mean a smaller *absolute* deviation from average near the origin than far from it (Fig. 11.3 *bottom*). There are plenty of instances where this condition of proportionality is fulfilled. We have seen

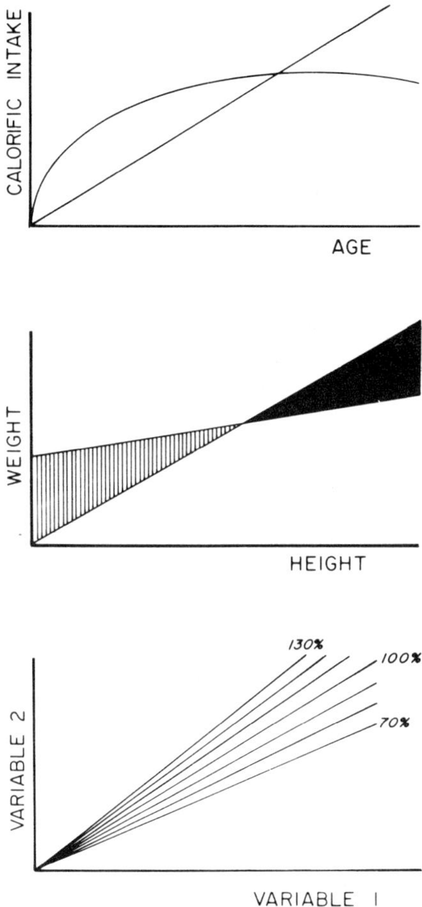


**Figure 11.3.** The use and abuse of ratios. From above down are shown: *Top*: An absurd use of ratios in a relationship that is clearly not linear. The aged would all appear undernourished and the young overfed, if the best straight line is fitted through the origin. *Middle*: An inappropriate use of a ratio when the regression is linear but does not go through the origin. Assuming that the true regression line is the ideal (a reasonable but by no means unexceptionable assumption) those with heights above the mean and weights in the segment in *black* are actually overweight but by the criterion of ratio of weight to height they will be below average. The converse applies to readings falling in the *hatched* area. These anomalies arise because the line of true relationship does not go through the origin, as the ratio implicitly supposes. *Bottom*: The assumptions where the ratio is appropriate. The line of relationship is not merely linear and through the origin, the deviations from the regression line are proportionate (i.e., the ratio of the standard deviation to the mean is constant).

several instances in Chapter 4. Under standard conditions the magnification in pelvimetry due to the distance from the pelvis to the plate will be proportionate, and any ratio on the pelvis will be the same as in the plate. But this property must, like the others, be empirically verified wherever it cannot be logically defended.

### The use of adjusted scores

To my knowledge this method originated in the writings of Roberts on the genetics of human intelligence (89) and blood pressure (90). Suppose that we wish to compare blood pressures in children in two districts in which the lithium content of the drinking water is high and low respectively. A properly designed study would avoid confounding with (for instance) age and sex, by proper matching, or stratification, and a corresponding analysis. But what if the study was not so designed and the data are too few to allow matching after the event? The common solution here is covariance analysis (cf. Chapter 8), which, in effect, determines the relationship of the observed values to those predicted for all subjects in each of the groups, and compares the results (although the discrepancies for the several subjects are not actually individually computed). The latter step is not necessary for a comparison of means; and theoretically we could calculate the overlap of the distributions after adjustment for age and sex, provided that all the slopes were parallel and known exactly, and also the differences in intercepts and the distribution of the errors from regression. However, Roberts wished to examine the individual deviations themselves, to test elaborate genetic hypotheses—e.g., that they comprise a mixture of distributions. Moreover, empirical study showed that the variance of the residual errors differs with age, and he proposed the construction of an age-specific $z$ variate, i.e.,

$$z = \frac{y_x - E(Y|x)}{\sqrt{\mathrm{Var}(Y|x)}} \tag{11.1}$$

where $x$ represents age and $y_x$ the reading at the age.

While I can see both the arguments for these steps and the intuitive appeal that they have, I have profound quarrels with the details (91). These criticisms, I may add, also apply in some degree to all systems of adjustment such as are widely used in population studies.

First, it makes strong assumptions about the relationship of the variable to the covariable. Just as a spurious serial correlation may be generated by an inappropriate model (see Chapter 8), so also may an anomalous distribution of residuals.

Second, adjusting for the covariable at all may distort a fundamental difference. If (to take an extreme example) we were to adjust intelligence for age, we might well demonstrate that the housefly is as intelligent as man, because the fly never attains an age of two years which would allow a direct comparison. It seems clear that we must have a theory for adjusting and not merely try to discern the theory after the adjustment has been made.

Third, the adjustment (11.1) involves concealing any heterogeneity of the variance that may exist. But it may be precisely in this heterogeneity that the information lies.

Fourth, the least squares (or other) procedure used to determine the expected value of the variable is based on certain minimum assumptions which, *after adjustment,* the analyst may proceed to show are untrue. But if they are untrue, then the adjustment is unsound and it is difficult to see what sound inferences can then be made from the residuals.

**Example 11.8.** Lest these arguments appear captious, consider the claim made by Platt (92) that there are two phenotypes—those with, and those without, the gene for a putative form of high blood pressure, essential hypertension; that they are indistinguishable in early life, but that with age they separate, those affected showing an increase in the mean blood pressure, the others not. Now *if* this hypothesis is true and we suppose that *within groups* the variance is constant, we foresee two consequences. The results, pooled by age, will show:

1. An increasing mean with age (being a mixture of groups, one that increases and one that stays constant).

2. An increase in variance because to the true random scatter within groups there is being added a systematic trend between groups.

These outcomes are the crucial predictions of the model, and the age-adjusted score is designed to conceal both! Of course it would be grossly unfair to Roberts to suppose that he had overlooked this fallacy and fallen into a logical trap. He was not developing Platt's hypothesis, but his own—that of multifactorial etiology. My only point is that those who have supposed his analysis a fair test of Platt's hypothesis have been grossly misled. Platt's hypothesis must be tested by algebra arising from *his* model. Nevertheless, Robert's analysis is to be seen as a description rather than a true model: he has no formal proposal as to *why* either the mean or variance changes with age—to him, these features are merely a nuisance obscuring the fundamental pattern.


## DISCRIMINANT ANALYSIS

### The basic ideas

The simple principle underlying the choice between two interpretations of the same set of measurements on a particular member of the population is to pick whichever has the greater (joint) likelihood. As we have noted above, this ensures the minimum misclassification both ways. If we know the mathematical form of the multivariate distributions under the two (or perhaps more) hypotheses, in principle there is no difficulty in choosing the dividing point. In practice, we may find the calculations tedious. Fortunately under certain circumstances (to be discussed later) the problem is dealt with by finding a simple linear combination of the measurements. Before we get down to details, a geometrical illustration of what we are about may help.

Consider, first, Figure 11.4. In the upper right hand corner we have represented two Gaussian distributions, identical except for their means ($\mu_1$ and $\mu_2$ respectively). We are looking at them "from above," so we cannot gauge their densities; but the percentiles are marked off, those for the first distribution above the line, those for the second, below it. It is evident that they are most crowded near the means and most widely spaced in the tails. Near the means, they are so thickly clustered that, with the thickness of line used, they fuse, and one can only identify every fifth percentile, which is marked slightly longer.

Now what would we see if we looked, not from above, but from the side? The answer would depend on our point of vantage. From immediately opposite (*bottom right*) we would see partially overlapping curves, the probabilities of misclassification about the point of intersection being shown black, and hatched, respectively. If we look more obliquely (*bottom left*) the distribution curves will appear more gaunt, thinner because of the obliquity, and hence (since the areas of both must be unity) taller. But since the two dis-

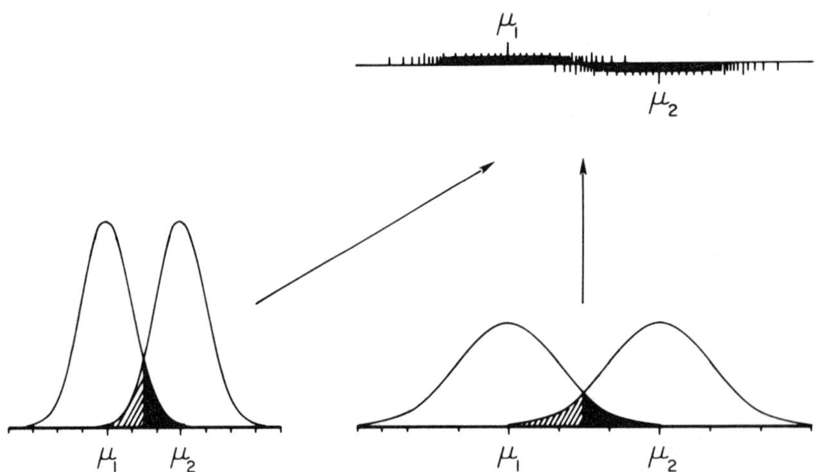



**Figure 11.4.** Univariate discriminating power. The percentiles for two distributions are marked at the *top right* in a "plan." Below are shown two frontal projections, full face (on the *right*) and oblique (on the *left*). The height of the probability density curve is increased where the domain is foreshortened, since area must be conserved. But the discrimination is exactly the same (except when the system is viewed end-on and the domain of the distribution is reduced to a point with an infinite height).

tributions are on the same plane, the proportions of overlap and misclassification will be the same. And this is true whether (as here) the distributions are Gaussian or of some other form. Simple scaling, as we have noted, does not improve discrimination. If we wish to identify to which of two groups a specimen belongs, it does not matter whether height is measured in inches or in centimeters. Only if we took the extreme case and looked actually along the line so that we would be unable to discriminate at all, would the vantage point matter.

But consider now a slightly more complicated relationship. In Figure 11.5 we represent two populations with bivariate distributions. We suppose

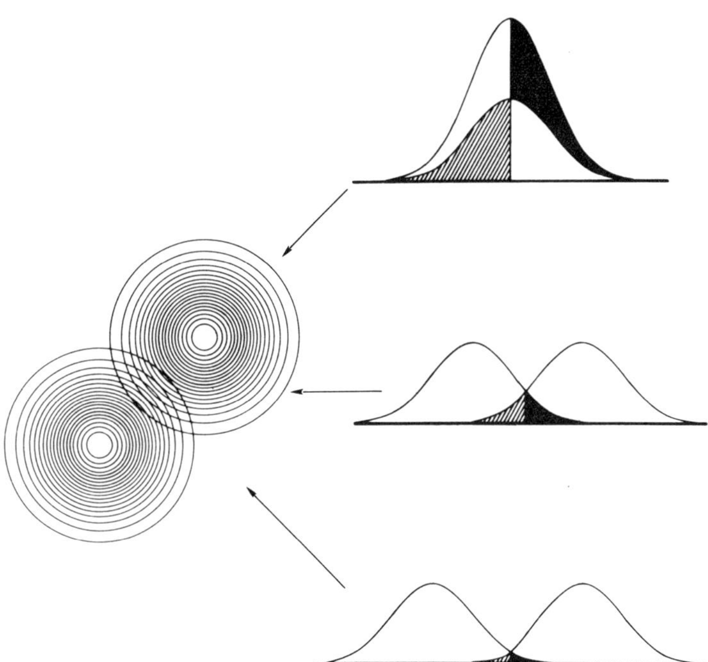

**Figure 11.5.** Bivariate discriminating power. Two groups with bivariate distributions for two uncorrelated Gaussian variates, differing only in their means, are to be distinguished. Every fifth percentile (from the center out) is displayed for both distributions, yielding two sets of concentric circles. From above downward are shown the side projections when viewed at *top*, end on. Here there is no discrimination and the two distributions can only be represented one on top of the other. The proportions misclassified (shown *hatched* and *black* respectively) are in each case 50%. *Middle*, viewed from an intermediate position in which there is "daylight" between the curves and the proportions misclassified greatly reduced. *Bottom*, viewed from the point of minimum misclassification. The objective of linear discriminant analysis is to find this latter optimal point.

that the two measurements are independent and have equal variances, that apart from means the populations are identical, and that their distributions are Gaussian. In the diagram they are represented by sets of concentric circles. The circles actually represent percentiles (at intervals of 5%) from the center outward; but that detail is of no importance here. What matters is that the circles are somewhat like contour lines which represent the locations of the distributions exactly as they do on the top line of Figure 11.4. Again what we see from differing vantage points—represented by the several arrows and accompanying diagrams—varies; but here both the form and the overlap change. In the *top* diagram the two distributions will overlap completely, and discrimination reduces to mere guessing with indeterminate error rates, which I have arbitrarily supposed 50% each way. To display this at all, I have had to put one distribution on top of the other. The other extreme, represented by the *lowest* diagram, shows the maximum amount of "daylight" between the distributions and the minimum misclassification. The *middle* diagram shows an intermediate position. The analogy with walking round the foot of two mountain peaks (or for those less acquainted with the outdoors, any two domes of St. Mark's Cathedral in Venice), looking for the maximum amount of sky between them, will prove helpful in this and other respects that we shall discuss later.

It is precisely the function of linear discriminant analysis to find the vantage point from which the separation is maximized. Why linear? In our analogy, it is because light travels in straight lines. While the floor of the valley between the peaks may in fact be curved and the "best" dividing line would then be curved, finding the best dividing point *by sighting* limits us to the use of straight lines. There are instances in which a curvilinear discriminant may be more sensitive; for instance in the diagnosis of Tay-Sachs disease, Gold and his associates have used such a method (93). However, in the interests of simplicity we shall confine our attentions here to the linear method.

Before translating our geometrical analogy into algebraic terms we may usefully analyze Figure 11.5 somewhat more closely.

First, we know that the percentile lines are circular, because the variables are uncorrelated and have equal variances. If they were positively correlated, the appropriate lines would be elliptical and the major axis inclined to both coordinate axes; and the greater the degree of correlation, the more eccentric the ellipses would be (see Chapter 10). If they were perfectly correlated the ellipses would degenerate into straight lines and the distributions into univariate. Thus this extreme case would become that shown in Figure 11.4, in which the vantage point does not affect the discrimination at all (provided it is not perfectly end on). These arguments suggest (what algebra confirms) that the more correlation there is between discriminating variates—the more redundancy there is—the less their joint discriminating power will improve their individual discriminating powers. For those who prefer a more statistical, less probabilistic, formulation, in Figure 11.6 are

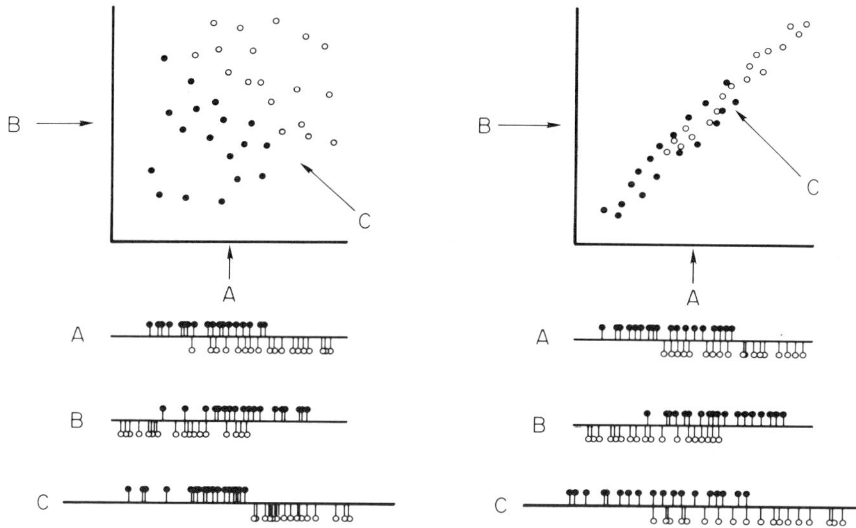

**Figure 11.6.** Discrimination in two (figmented) sets of data. In each case, the discrimination between the black and white points is shown: (*A*) when the variable on the abscissa only is used; (*B*) when the variable on the ordinate only is used; (*C*) when they are both used with optimal weights. The degree of misclassification can be read from the "head counts" opposite each. In the example on the *left*, the two variables have a low correlation, there is little redundancy, and the bivariate discrimination is more effective than either univariate. In the example on the *right*, the correlation is so high that the system is almost as extreme as that shown in Figure 11.4.

shown some (fictitious) data on two groups of points, black and white. On the *left* the variates represented by the two coordinates are almost uncorrelated and the clusters almost circular; on the *right*, the clusters are elongated ellipses. When these clusters are viewed from a point at an infinite distance looking in the direction *A* we are ignoring the variable on the vertical scale. The black points are shown above the line, the white below; and the degrees of overlap for both diagrams are about the same. Likewise from direction *B*, the variable on the horizontal is being ignored and again the overlaps are comparable. (Of course, the black dots will here tend to lie to the right of the white dots.) However, from the most discriminating viewpoint (direction *C*) the overlap is drastically reduced in the *first* diagram, but in the *second*, because of the redundancy, there is virtually no enhancement.

Second, what is the algebraic counterpart to "viewpoint"? When we look at a mountain from the southwest, what we see represents a composite of what we see from the south (*s*) and from the west (*w*); a linear combination of the two having the form

$$as + (1 - a)w$$

would give us something of this aspect: $a = 1$ would give us a pure southerly view, and $a = 0$ a pure westerly. As we may infer from this analogy and from the more detailed illustrations in Appendix 1, a linear combination is equivalent to a rotation of coordinate axes.

### The formal approach

Let us suppose that we construct a weighted linear function, $y$, of the discriminating covariables, $x_1, x_2, \ldots, x_k$. That is for the $i^{\text{th}}$ subject we compute

$$y_i = a_1 x_{1i} + a_2 x_{2i} + a_3 x_{3i} + \ldots + a_k x_{ki} \tag{11.2}$$

We thus have reduced a vector of measurements (say height, weight, intelligence, refractive error, etc.) to a single (scalar) quantity, which we shall call the *discriminant score* for that subject.

Now we are trying to discriminate between two groups (or perhaps more) and there will be a comparable number of sets of scores. Let us denote one sample of size $n$ by $y_1 \ldots y_n$ and the other (of size $r$) by $z_1 \ldots z_r$. Then we have two related problems: to find the misclassification for any particular set of $a$'s in (11.2) and to minimize this misclassification by a suitable choice of $a$'s. These problems may become complicated; but they become much simpler if we are prepared to assume, or can find a method (such as transformation) that ensures

1. That the $Y$ and $Z$ are Gaussian. It is not required that the $X$'s individually be Gaussian, and if $k$ is large we will be able to capitalize on the central limit theorem (see Chapter 17).

2. That $Y$ and $Z$ have equal variances. This is assured if the $X$'s have the same variances and covariances for the groups being compared.

Then the solutions to our problems become comparatively simple. It is not surprising that the difference between the means $d_j = m_{Y_j} - m_{Z_j}$ in the two groups of a particular discriminant is related to the size of the contribution of the $j^{\text{th}}$ variable to the score. Obviously if $d_j = 0$, the $j^{\text{th}}$ contributor does not help. Also if there is much redundancy in two discriminants, our weighting should reflect the fact. So the steps we take are

1. Compute the sum of squares for the data values for each variable from its group mean and pool the results for groups. Thus for the $j^{\text{th}}$ variable we have for the first group

$$\sum_i (x_{ji} - m_{X_j})^2$$

and we have a similar sum of squares for the other group. We add these two sums together. We thus have a pooled residual ("within") sum of squares for each of the $k$ contributing factors.

2. Compute for all possible pairs of factors the corrected cross products which we would use for the numerator of the correlation coefficient, i.e.,

$$\Sigma(x_j - m_{X_j})(x_h - m_{X_h})$$

and pool the sums for the two groups.

3. The results of (1) and (2) are now arranged in a symmetrical matrix. The rows and columns are as usual arranged in order, the first factor being in the first row and column, and so forth. This matrix would in fact become an estimated variance-covariance matrix if each term were to be divided by $(n + r - 2)$. We will denote this matrix by $\mathbf{M}$.

4. We set up two vectors, a $k \times 1$ vector $\mathbf{d}$ of differences between the means, the elements being arranged in order; and a vector of weights $\mathbf{a}$ represented in (11.2) above.

5. We then solve the matrix equation

$$\mathbf{Ma} = \mathbf{d}$$

to get the most discriminating set of weights

$$\mathbf{a} = \mathbf{M}^{-1}\mathbf{d}$$

**Example 11.9.** Consider again the diagnosis of malignant hypertension. In Table 11.2 are shown the sums of squares and products for men with renal hypertension and men with essential hypertension. We consider three discriminants: age at diagnosis $(a)$, the logarithm to the base 10 of the blood urea nitrogen $(u)$, and the survival from diagnosis $(s)$. The calculations are given in full and should be self-explanatory.

**Test of hypothesis**

Granted that we have found a best linear combination, has it been worth while? To test whether anything useful has been achieved, we set up an analysis of variance. There are two components of the sum of squares.

1. That within groups $(W)$ which turns out to be

$$W = \sum_{i=1}^{k} a_i d_i$$

It has $(n + r - k - 1)$ degrees of freedom associated with it.

2. That between groups which is

$$\frac{nr}{n + r} W^2$$

which has $k$ degrees of freedom.

The appropriate test is based on the $F$ statistic since the data are normally distributed.

**Interpreting the results**

It is a wise precaution to compute the discriminant score for each individual in the population, and to plot them out as a histogram. We may compare the results attained by various combinations of the discriminant variables by deleting or adding appropriate rows and columns of **M** and elements in **d**. The degree of separation will be evident from the histograms.

There are formal methods of assessing misclassification. However, we do well to recognize an inherent bias in the method. The method has consisted of finding those values of the $a$'s that minimize misclassification. But these estimates are random variables, depending as they do on data from random samples; and the true values, i.e., those we would get from samples of infinite size, will in general be somewhat different. But if we were in a position to use these somewhat different values we would no longer be at the point of minimum misclassification for this sample, and the amount of mis-

**Table 11.2. Discriminant analysis of data on essential, and renal, malignant hypertension in males**

| | Essential | | | | Renal | | |
|---|---|---|---|---|---|---|---|
| | Age | log BUN* | log survival from diagnosis | | Age | log BUN* | log survival from diagnosis |
| | 76 | 1.380 | −1.000 | | 19 | 2.207 | −2.079 |
| | 51 | 2.127 | −1.000 | | 55 | 2.190 | −0.699 |
| | 61 | 2.212 | −1.000 | | 40 | 2.211 | −0.632 |
| | 56 | 1.748 | −0.875 | | 52 | 2.155 | −0.301 |
| | 58 | 1.968 | −0.778 | | 27 | 2.146 | −0.301 |
| | 41 | 1.833 | −0.778 | | 45 | 1.623 | −0.273 |
| | 62 | 2.207 | −0.632 | | 25 | 2.155 | −0.247 |
| | 54 | 2.190 | −0.574 | | 53 | 2.174 | −0.247 |
| | 50 | 1.362 | −0.398 | | 51 | 0.799 | 0.097 |
| | 52 | 2.093 | −0.331 | | 13 | 1.941 | 0.176 |
| | 52 | 1.556 | −0.062 | | 29 | 2.332 | 0.477 |
| | 51 | 1.342 | 0.000 | | 18 | 2.211 | 0.544 |
| | 59 | 1.301 | 0.097 | | 53 | 1.609 | 0.544 |
| | 57 | 1.204 | 0.176 | | 54 | 1.845 | 0.699 |
| | 39 | 1.146 | 0.352 | | 45 | 1.769 | 1.439 |
| | 65 | 1.447 | 0.477 | | | | |
| | 63 | 1.380 | 0.477 | | | | |
| | 38 | 1.362 | 0.544 | | | | |
| | 56 | 1.415 | 0.699 | | | | |
| | 60 | 1.519 | 0.740 | | | | |
| | 36 | 1.322 | 0.778 | | | | |
| | 50 | 1.146 | 1.146 | | | | |
| $\Sigma x$ | 1187 | 35.260 | −1.942 | | 579 | 29.367 | −0.803 |
| $m_x$ | 53.95 | 1.602,72 | −0.088,27 | | 38.60 | 1.957,8 | −0.053,53 |
| $\Sigma x^2$ | 65,969 | 59.414,004 | 9.846,986 | | 25,563 | 59.664,135 | 9.007,123 |
| $s^2$ | 91.66 | 0.138,182 | 0.460,741 | | 229.54 | 0.154,959 | 0.640,295 |

$$\Sigma au = 1100.877 \qquad \Sigma as = 3.637 \qquad \Sigma us = -2.878,649$$

*BUN = Blood urea nitrogen in mm/100 ml blood.

**Table 11.2.** *(continued)*

---

$$\mathbf{M} = \begin{bmatrix} 5,138.554,545,5 & -20.164,472,72 & -\ 5.062,472,728 \\ -20.164,472,72 & 5.071,262,764 & -\ 5.097,055,236,37 \\ -\ 5.062,472,728 & -\ 5.097,055,236,37 & 18.639,696,097 \end{bmatrix}$$

$$\mathbf{d} = \begin{bmatrix} 15.350 \\ -\ 0.355,072,727,28 \\ -\ 0.034,739,393,939,4 \end{bmatrix}$$

$$\mathbf{M}^{-1} = \begin{bmatrix} 0.000,199,564,543,7 & 0.001,169,386,953,70 & 0.000,373,971,760,002 \\ 0.001,169,386,953,70 & 0.278,778,719,031 & 0.076,550,095,654,2 \\ 0.000,373,971,760,002 & 0.076,550,095,654,2 & 0.074,683,260,954,8 \end{bmatrix}$$

$$\mathbf{a} = \mathbf{M}^{-1}\mathbf{d} = \begin{bmatrix} 0.002,635,106,778,61 \\ -0.083,695,934,263,7 \\ -0.024,034,835,944 \end{bmatrix}$$

$$\mathbf{W} = \mathbf{a}'\mathbf{d} = 0.071,001,988,326,9$$

*Analysis of Variance*

| Source | d.f. | Sum of Squares | Mean Square | F |
|---|---|---|---|---|
| Between | 2 | .044,962,79 | 0.022,481,39 | 10.76 |
| Within | 34 | .071,001,99 | 0.002,088,29 | |

---

classification would be greater. Thus our sample estimate of misclassification always exaggerates somewhat the degree of discrimination, especially if small samples only are available. I know of no straightforward remedy for this defect that does not involve sacrifice of efficiency. A method commonly used is to divide the set of data randomly into two parts, estimating the $a$'s from the one part, and the degrees of misclassification by applying these estimates to the other part. But clearly the method is not efficient, since not all the data are being used to estimate the $a$'s.

## CLUSTER ANALYSIS

In discriminant analysis, we have assumed implicitly not only that we know the number of groups involved, but also which elements fall into each particular group. If the latter information is not known, we cannot compute either the difference between the means or the deviations from them. Discriminant analysis is above all a method to make best use of simple tests where the classification *can* be validated, but only by methods that are expensive, dangerous, or elaborate. For instance noninvasive methods of diagnosis are preferable to invasive, and one might wonder how a combination of (say) the blood pressure, auscultation, echocardiography, and chest X-ray (which are noninvasive) might be used to distinguish those with aortic stenosis from those without, as judged by cardiac catheterization (which is invasive and not without danger).

But what if there is no basis for validation? What if we do not know how many classes there are? These are problems that arise in connection with taxonomies. The methods of approach—which are controversial—include cluster analysis. They are highly technical in scope and far beyond my ambitions in this book. Nevertheless, the reader should know of their existence and seek help where, and when, appropriate.

## PROBLEMS

**11.1.** For the data of Table 8.3 find the ratio of serum protein to body length. Compute the mean and find which children are "better nourished" than average and which less. Compare the results with Figures 8.2 and 9.2.

**11.2.** For the data of Table 4.10 estimate the rate of misclassification when the test is used to distinguish carriers from wildtype (control) subjects:

   **a.** If the population being studied comprises normal adults whose oldest sib has Hurler syndrome. (Ignore new mutation.)

   **b.** In the population at large, given that Hurler syndrome has a frequency of 1 in 100,000 and assuming Hardy-Weinberg equilibrium (i.e., the number of Hurler genes at the particular locus in a person follows a binomial distribution).

**11.3.** On the basis of the data in Table 1.1 determine the value of birth weight in discriminating between inbred and outbred children from this population.

**11.4.** Examine the merit of the rate of disease "per man-year of exposure."

# PART III
# CATEGORICAL DATA

# 12
# THE BINOMIAL VARIATE

In the chapters that follow, we shall discuss the statistical aspects of categorical data, beginning with the simplest possible nontrivial type, the binomial. The underlying theory of the binomial process and the distribution of outcomes have elsewhere been sufficiently discussed for our purpose (7). However, the individual assumptions call for special emphasis since they are often overlooked. The appropriate section of Chapter 1 should be carefully reviewed before proceeding further.

## THE FUNDAMENTAL PHENOMENON

The notion of a categorization in the modern sense is that members of a population fall unambiguously into one of at least two intrinsically distinct classes that may be effectively used for purposes of inference or predication. Such a grouping may indeed exist, although it is extraordinarily difficult to find illustrative examples that are beyond all objection. Often investigators devise categorizations in the teeth of the evidence, because it is convenient for their purposes. The result is often inelegant, inefficient, or even misleading. We may discuss the ideas involved with several examples.

**Example 12.1. Members of a randomized trial.** In a strict experiment, members of a class are allocated into two groups randomly: say those who are to be vaccinated and those who are not. This is a binary classification. It is not altogether natural: in reality the dose of vaccine might take on a great number of values, and the response might vary with the dose, the quality of the preparation, the health of the patient and so on. But artificially we have circumvented these defects. Of course we should not naively suppose the distribution is binomial; but at least we are comfortable talking about the classes.

**Example 12.2. Sex.** Natural categorizations are much less easy to produce. The best one I can find in medicine is sex. Macroscopically, as it were, the distinction, apart from a small proportion of doubtful cases, is clear enough for practical purposes. But attempts at a rigorous delineation prove difficult and conflicting. There are many criteria: anatomy, psychology, endocrinology, the buccal smear, the karyotype, the H-Y antigen. Unfortunately, from time to time, they may conflict; moreover all of them show gradations such that we could without violence represent all subjects on one unbroken or "almost-continuous" scale. For instance, the $Y$ chromosome or fragments of it of any size may be translocated to an autosome; and if in the Klinefelter ($XXY$) syndrome the $Y$ chromosome were to be translocated, it might be unclear whether the subject were male or female. For all we know, this phenomenon is common and will be disclosed with increasing frequency as we devise more refined methods of karyotypic analysis. Yet the clearest evidence we have of natural grouping by sex is chromosomal.

**Example 12.3. Maturity.** We are accustomed to classify patients into the pediatric and adult groups, a categorization that has firm administrative support. Nor is it my concern to deny its utility in many cases (although I have considerable reservations where chronic disease is concerned). One could argue convincingly that while all body processes change with age, the massive changes in adolescence provide an indisputable watershed between childhood and maturity. Nevertheless, it is a flat-topped watershed with no natural critical point. A completely natural criterion by which we could distinguish a child from an adult probably does not exist. On the whole, investigators would be wiser to try wherever possible to treat age as a continuous regressor variable rather than to make it into a classification. If indeed a relationship of interest is a continuous function and we arbitrarily dichotomize it, we almost certainly lose power (see Chapter 15); and since the members of any particular class are not homogeneous with regard to the regressor variable, it is difficult to say, off hand, what the impact on inference may be. Moreover, one is obliged to avoid a common logical circularity. If I divide a population into two groups about some arbitrary age, say sixteen, and show that gout occurs only in the older group, I cannot infer what the age of onset of gout is. Whether it is twenty, forty, or sixty, there will be a difference between the groups. To be sure investigators do not often make this mistake; but superficial readers of their publications commonly do. "Brown has shown that gout does not occur before sixteen" becomes "Brown has shown that gout occurs after sixteen," and hence by easy stages "Brown has shown that the age of onset in gout is sixteen or over."

**Example 12.4. Hypertension.** Physicians are still so accustomed to thinking of hypertension as a disease that they may find inquiry into alternative conceptualizations disturbing. It cannot be disputed that certain dire consequences—heart failure, strokes, kidney failure, retinal damage—occur more commonly in those with high blood pressure than those without. In principle, the counterargument that there are plenty of exceptions both ways is easily dismissed: even where there is no doubt that grouping occurs, for instance hemophilia, occasional anomalous findings occur. Those who believe that "hypertension" has a status similar to that of hemophilia have three major levels of defense. First, it is obvious that there are separate groups, those with, and those without, the disease and that it is only the incompetent and the inexperienced who cannot tell the difference. This argument—like the emperor's clothes—has nothing to offer to those who are already skeptical. In particular, the geneticist trying to find the frequency of the disease must have much more explicit criteria. Then, second, they may argue that the population figures on blood pressure show evidence of groups that correspond to those with, and those without, the disease. Such groupings have certainly been nothing like universally verified, and. many hold that they can be explained away as factitious. Then, third, it is argued that the grouping exists but our present means of detecting it are too crude to yield more than hints of it. While this third statement may be true, it furnishes evidence of belief rather than warrant. But all disputants would agree that if grouping exists the misclassification rate is high—just as it would be for phenylketonuria if we had no access to chemical evidence (94).

**Example 12.5. Nonspecific mental retardation.** When all the more or less well-defined types of mental retardation are excluded (e.g., genetic, infectious, traumatic), there remains a group of patients who are mentally retarded. The justification for defining this group is preeminently their medical and legal management, issues of utility. Scientific inference is not easily or safely based on it. Statements that mental retardation is present in a larger proportion of one racial group than another raise immense problems which I have discussed elsewhere (95). But even if we can accept both the use of intelligence tests as a reference and their execution as being beyond reproach, we still leave unanswered the question of why we do not treat the result of the test as a measurement and why we should want to draw a dividing line between those with, and those without, "mental retardation." The nub of the problem is, why should we convert an uninterrupted distribution into a binomial one? Statistical theory shows that gratuitous "binomialization" of a continuous measurement commonly leads to loss of precision and power. Suppose we wish to decide whether or not two samples of Gaussian variables, otherwise identical, are from populations with the same mean. The usual criterion would be the $t$ test. But suppose we "binomialize" the data according to whether results fall above $(+)$ or below $(-)$ the sample median of the pooled data and compare the proportions of positive results in the two groups by the binomial test. For large samples the power would be approximately $2/\pi$ (or about 64%) of that of the $t$ test. In practical terms this means that to use a binomial test where a $t$ test is appropriate is equivalent to discarding one-third of the data: one would have done just as well with appropriate analysis of a smaller sample, at less trouble to the investigator and expense to the granting agency.

These issues reduce to one fundamental scientific point. If there is *compelling* evidence that grouping exists, then group. But grouping is often done for the sake of grouping; or because for some separate nonscientific reason (e.g., a decision or policy) a grouping is to be done in any case; or, worst of all, because the analysis is easier. (By using chi-square tests one will escape having to learn how to do multiple regression.) In all such cases binomialization is to be deplored. However, this anathema does not mean that binomialization may not be employed as a nonparametric method in data in such small quantities and with such a distribution that Gaussian theory cannot safely be applied. (See Chapters 15 and 17.)

## LOGICAL PLASTICITY OF THE BINOMIAL VARIATE

In the previous chapters in this book we have dealt exclusively with measurements that have distributions. It must not be supposed that statistical inference can be applied to distributions only. We have seen that a distribution relates to a random variable that takes a particular set of values with associated probabilities or probability densities. However, it is possible for the outcome of an experiment to take a categorical form with a specified set of probabilities even where no natural *value* can be associated with any particular outcome. One can test hypotheses about the probabilities, or estimate them by appropriate statistical methods. To be sure, such outcomes can be made into distributions, by ascribing arbitrary values to the particular classes. In the binomial distribution the two possible outcomes may be arbitrarily scored as "successes" or "failures" with the values 1 and 0 respectively. But this system of scoring is a computational construct that merely counts the number of results there are in each class.

## POINT ESTIMATION

Unlike estimation for the Gaussian variate, then, the concern is not with estimating the scores, but the number of successes. A 0–1 scoring system is so commonly used that the distinction between these two quantities (score and number of successes) is rarely made. But if two gamblers were tossing nickels, the scores in cents would be 5 for each success and $-5$ for each failure. Moreover the scores would change if we converted the winnings into *yen* or *lire*. But obviously the total number of successes would be unchanged. Indeed the central issue is not so much the *number* of successes as the *proportion*. Different sample sizes will yield different distributions; but, as shown elsewhere (7), the quantity being estimated is independent of sample size.

### The standard estimator

The method of moments estimate is most easily found by setting the sample number of successes ($x$) equal to the expected and solving:

$$x = E(\hat{X}) = n\hat{p}$$

$$\hat{p} = \frac{x}{n}$$

We have already shown in a particular case (p. 65) that the expectation of the corresponding estimate of the variance of $\hat{P}$, that is of $\hat{P}(1 - \hat{P})/n$, equals the lower bound and this proves to be true for all sizes of sample and all values of $p$ except where the estimate is zero or unity. Thus the method-of-moments estimator is fully efficient. It is unbiased:

$$E\left(\frac{X}{n}\right) = \frac{1}{n}E(X) = \frac{1}{n}pn = p$$

The variance goes to zero as $n$ increases and these two properties ensure consistency. Finally, provided $p$ is neither zero nor unity, $\hat{p}$ is asymptotically Gaussian. Hence the method-of-moments estimator of $p$ has all the properties of maximum likelihood estimation. In fact the two estimators are, in this instance, identical.

**Bayesian estimation of the binomial parameter**

In Chapter 5 we consider the phenomenon of regression toward the mean, which we may regard as a rational modification of the sample estimate in the light of what is known about the behavior of the population at large. It is, in fact, a Bayesian estimator; and the same argument can be applied to the binomial. Indeed, if it could be done satisfactorily, we would be in a position to make true (subjective) probability statements about the parameter in place of the more ambiguous confidence statements. The problem is that we can rarely make any precise statements about the prior probability; and in this state, statisticians tend to fall into two major groups. The "know-nothing" statisticians adopt the attitude that if one has no exact knowledge, the data (which are sound) should not be contaminated with second-rate information. The Bayesians would argue that information which, however rough, is definite, cannot responsibly be ignored. We may not know in any detail what the distribution of weights in adult African elephants may be; but we would be certain that a measurement of 300 tons is impossible. I cite two instances of Bayesian modification of a binomial estimator, both genetic. The first is empiric, unarticulated, and largely negative; the second is quasi-theoretical, explicit, and positive.

**Example 12.6. The estimator of mutation rate.** Vast experience at a rather coarse-grained level and, more recently, better biochemical understanding of mutagenesis, suggest that the probability of a particular gene undergoing mutation in any one generation is about 1 in 100,000. There is a variation of perhaps an order of magnitude on either side of that figure, and there are ready explanations why the magnitude should vary so; but more extreme values than that would set the geneticist thinking hard about cause and interpretation. Without any explicit calculations, he would tend to suspect that 1 in 10,000 is an overestimate and his first line of attack would probably be to examine the source of the data more closely than usual.

**Example 12.7. Genetic linkage analysis.** At each autosomal genetic locus, a prospective parent has two genes. To any particular offspring one, and only one, gene from each locus is normally transmitted and the probability for each is 50%. Genetic loci not carried on the same pair of chromosomes are unlinked, i.e., the choice of the genes at each locus transmitted to any particular child is independently made. If they are on the same chromosome, the probability that, if one is transmitted the other one will not be, is the recombination coefficient, a probability usually denoted by $\theta$. For independent transmission, $\theta = 1/2$; but for linked loci $\theta$ may lie anywhere between zero and $1/2$. Since there are many chromosomes and they vary in size, it is roughly estimated from chromosome lengths that the prior probability of linkage is about 1/18.5. Geneticists, such as Renwick (96), have argued that this information cannot be ignored in linkage analysis. Further, certain not implausible assumptions lead to a prior probability density for $\theta$, given that there is linkage (22, 97). The latter argument has certainly been contested by the "know-nothing" geneticists; and largely because the Bayesian estimator is cumbersome, it has rather fallen into disuse. But as knowledge of the human chromosome improves, we may well see revival of more refined Bayesian estimators.


## CONFIDENCE INTERVALS ON A BINOMIAL PARAMETER: "EXACT" TREATMENT

In the discussion about confidence intervals on the estimate of the mean of a Gaussian distribution in Chapter 6, I suggested an alternative way of viewing the interval as that set of values within the acceptance regions of which the sample value lies. This representation has a certain appeal; but a logically strict definition is that given in Chapter 3; and where discrete distributions are concerned, the two formulations are not equivalent. The problem is somewhat difficult to grasp and I shall first revert to the idea for the Gaussian case. If the true mean is $\mu$, and the standard deviation of the mean is $\sigma_M$, then we know that the probability of a (future) sample mean being within $1.96\sigma_M$ of the mean is 95%. That is,

$$P[\mu - 1.96\sigma_M \leq M \leq \mu + 1.96\sigma_M] = 0.95 \tag{12.1}$$

Now the equalities are still true if we subtract the same quantity from each term. Thus, subtracting $(M + \mu)$ from each,

$$P[-M - 1.96\sigma_M \leq -\mu \leq -M + 1.96\sigma_M] = 0.95 \tag{12.2}$$

and since inequalities are reversed by changing the sign* we have (writing 12.2 in reverse order and changing the signs)

$$P[M - 1.96\sigma_M \leq \mu \leq M + 1.96\sigma_M] = 0.95 \tag{12.3}$$

as stated in Chapter 6.

The equation (12.3) is *not* a probabilistic statement about $\mu$ but about some *future* sample value $M$ being such as to satisfy the inequalities.

The difficulty about applying this argument to a discrete distribution is that, in general, statements of the form (12.1) cannot be made because the probabilities in the tails can assume only a limited number of values, given that the parameter has a particular (though unknown) value.

**Example 12.8.** In a binomial variate let $n = 7$ and $p = 0.5$. Then a priori, the number of successes, $X$, has the following distribution

| $x$ | $P(x)$ | $F(x)$ |
|---|---|---|
| 0 | 0.007,812,5 | 0.007,812,5 |
| 1 | 0.054,687,5 | 0.062,500,0 |
| 2 | 0.164,062,5 | 0.226,562,5 |
| 3 | 0.273,437,5 | 0.500,000,0 |
| 4 | 0.273,437,5 | 0.773,437,5 |
| 5 | 0.164,062,5 | 0.937,500,0 |
| 6 | 0.054,687,5 | 0.992,187,5 |
| 7 | 0.007,812,5 | 1.000,000,0 |

Because of the small size of the sample we shall be content with 90% confidence limits. The statement corresponding to 12.1 is now

$$P[0 < X < 7] > 0.90 \tag{12.4}$$

which is an inequality, not a precise probability statement.

Although the sample space is discrete ($X$, the number of successes can, in fact, assume only the values 0, 1, 2, 3, 4, 5, 6, or 7), $p$ may assume any value whatsoever between zero and one. So in our problem we may choose as an upper limit on $p$, that value for which the probability of the observed number ($x$) of successes or less is exactly 5%; and as a lower limit that for which the probability of $x$ or more is exactly 5%.

Figure 12.1 may be helpful. The horizontal scale represents the value of the parameter. The curves represent the outcomes, i.e., the number of successes $x$, as indicated by the number to the *left* of each curve. The vertical scale represents the probability of $x$ successes or less. Thus the height of the curve marked "2" is the probability, $F(2)$, of 0, 1, or 2 successes. Since the alternative ("3 successes or more") has associated with it the complementary probability, it is evident that if we view the diagram upside down we can read

---

*For instance, "$3 < 5$" implies "$-3 > -5$."

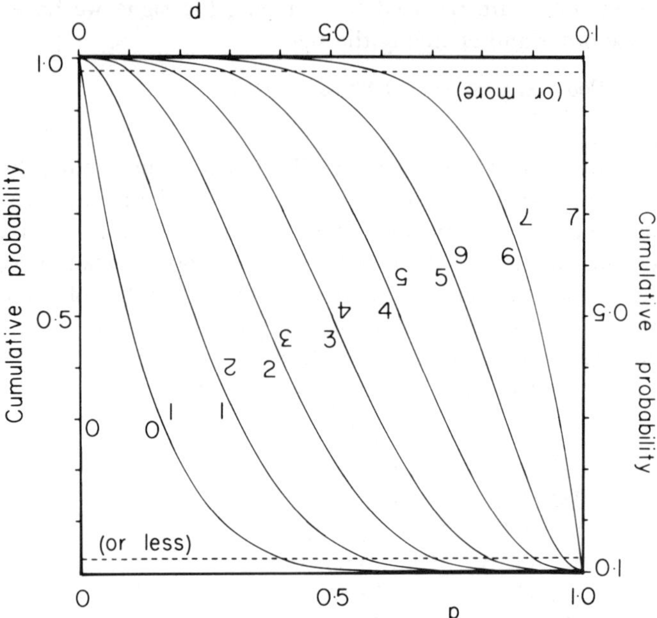



**Figure 12.1.** The binomial distribution of order 7. On the horizontal axis is the probability of success. The curves represent the cumulative probability of the number of successes shown by the number to the *left* of each curve. Thus the curve to the immediate right of "3" is the probability of 3 successes or less. As $p$ increases, this probability gets smaller. The complementary probability ("4 successes or more") represented by the inverted number to the *right* of the same curve can be read off the top scale by inverting the diagram. The dotted lines represent the critical levels for rejection regions with a size of 2½% in that tail.

off exactly the same curve the probability of 3 successes or more, the probability in the upper tail. The curve is therefore labeled with the inverted figure 3 on the *right* side (or the *left* side if the diagram is viewed upside down). The 2.5% lines for each tail size are shown as interrupted lines.

The appropriate solutions for all possible values of $x$ for $n = 7$ are shown in Table 12.1. Given an equal division of error, we are assured that no matter what the value of $p$, so long as we choose the corresponding confidence interval, the prior probability of including the true value is at least 90%. But as we shall see, the limits are conservative. Inspection shows that if the true value of $p$ is (say) 0.2, the 90% confidence limits, if $x$ is 0, 1, 2, or 3, will include 0.2. The corresponding probabilities that it will assume these values can be calculated to be respectively, $0.2097\ldots + 0.3670\ldots + 0.2752\ldots + 0.1146\ldots = 0.966\ldots$. The results are laid out in more detail in Table 12.2. It will be evident that the actual probability is substantially greater than the 0.9 that we aimed for. Ultimately this discrepancy is due to

**Table 12.1. "90%" confidence intervals on a binomial parameter based on sample of size 7**

| Sample number | Confidence limits | |
| of successes | Lower | Upper |
|---|---|---|
| 0 | 0.000,000 | 0.348,164 |
| 1 | 0.007,301 | 0.520,703 |
| 2 | 0.053,375 | 0.658,739 |
| 3 | 0.128,756 | 0.774,678 |
| 4 | 0.225,322 | 0.871,244 |
| 5 | 0.341,261 | 0.946,625 |
| 6 | 0.479,297 | 0.992,699 |
| 7 | 0.651,836 | 1.000,000 |

NOTE: As nearly as possible the errors in the tails are all equal to 5%.

the incongruity of trying to estimate a quantity from a continuous parameter space by observations in a discrete sample space. It will be evident that the process of computing probabilities in the tails is tedious (especially for large samples) and we readily turn to continuous approximations (see below).

## HYPOTHESIS TESTING: EXACT TREATMENT

Some theory may exist about how the numbers in an enumerative experiment ought to behave—for example that two events occur independently or that they should be distributed according to some model, or some such. In general, we do not expect that the actual numbers found in the categories will exactly equal those predicted, because there are sampling effects; and sometimes the numbers predicted by the theory may not be integers and therefore

**Table 12.2. Outcomes in which the 90% confidence limits include the parameter, in binomial processes of order 7**

| Value of parameter | Outcomes for which the "90%" confidence limits include the parameter | Combined probability of these outcomes |
|---|---|---|
| 0.05 | 0, 1 | 0.955,620 |
| 0.1 | 0, 1, 2 | 0.974,308 |
| 0.2 | 0, 1, 2, 3 | 0.966,656 |
| 0.3 | 0, 1, 2, 3, 4 | 0.971,204 |
| 0.4 | 1, 2, 3, 4, 5 | 0.953,165 |
| 0.5 | 1, 2, 3, 4, 5, 6 | 0.984,375 |

*cannot* be in agreement with observed values. The point at issue, then, is to decide how much discrepancy may be allowed between the observed and predicted values before the theory is discarded as untenable. This problem is basically the same as in other tests of hypotheses, but the logistical details are considerably different.

**Example 12.9.** Suppose that an investigator has set up the theory that a form of thyrotoxicosis, thought to be an autosomal trait, is equally common in men and women. Population figures show that, in the age group at risk, 50% of all subjects are males and 50% are females. Then according to the theory, in a sample of 20 subjects with this form of thyrotoxicosis he would expect 10 to be male. However, in his actual sample he finds that 2 are male and 18 are women. Is the hypothesis tenable?

Let us use a two-tailed test at the 5% level with equal errors in the tails. Because the binomial parameter is $1/2$, the distribution is symmetrical and we need tabulate the cumulative probabilities for one tail only. Convention requires that the size of the rejection region should be (say) 5% *or less*;* and since we cannot attain this value exactly, our test has to be somewhat conservative.

| $x$ | Probability | Cumulative probability |
|---|---|---|
| 0 | .000,001 | .000,001 |
| 1 | .000,019 | .000,020 |
| 2 | .000,181 | .000,201 |
| 3 | .001,087 | .001,288 |
| 4 | .004,621 | .005,909 |
| 5 | .014,786 | .020,695 |
| 6 | .036,964 | .057,659 |

We are thus obliged to make the rejection region "5 or less" at one tail and "15 or more" at the other. We are led to reject the hypothesis.

With a suitable calculator—even a pocket calculator—this problem represents some ten minutes' work. However, if we were dealing with a large sample, this method would be tedious and we might cast around for something simpler. Two classical solutions are the Gaussian and the Poisson approximations. The latter will be deferred to Chapter 14.


## HYPOTHESIS TESTING: THE GAUSSIAN APPROXIMATION

### The one-tailed test

**Example 12.10.** Consider the binomial distribution $B(10, 1/2)$ that is, of order 10 and with a parameter of $1/2$. Then the mean and variance of the proportion of successes are

---

*However, there is no particular reason why we should not choose a more convenient value such as the 0.041,39 level, which would correspond to the outcomes (0–5) and (15–20). The only point that matters is that the choice should not depend on the data.

$$E\left(\frac{X}{n}\right) = \frac{10 \times \frac{1}{2}}{10} = 0.5$$

$$\mathrm{var}\left(\frac{X}{n}\right) = \frac{10 \times (\frac{1}{2})(1 - \frac{1}{2})}{10 \times 10} = 0.025$$

respectively.

The individual terms, $P(x)$ and the corresponding exact (binomial) cumulative distribution function, $F_B(x)$, are shown in Table 12.3 and $F_B(x)$ is plotted in Figure 12.2. Note that because $X$ is discrete, as it passes from (say) 3.0 to 3.999 the function $F_B(x)$ does not change; but at $x = 4$ there is an instantaneous rise of $0.205,078\ldots$, and so for each integer value from zero to ten. There is a possible ambiguity as to what $F_B(x)$ might be. We define it, in accordance with convention, to be the probability that the random variable assumes the value $x$ or less.

Also shown in Figure 12.2 as a continuous line is the Gaussian distribution function $F_G(x)$ having the same mean and variance. Thus for $x = 4$ we compute the standardized normal variate

$$z(4) = \frac{\frac{4}{10} - 0.5}{\sqrt{0.025}} = -0.632,45$$

and from Gaussian tables $F_G(4) = 0.263,544$, which is duly recorded in the fourth column of Table 12.3. There is reasonable agreement between $F_B(x)$ and $F_G(x)$, even for so small a value as $n = 10$; and as $n$ becomes larger, suitably scaled, the steps become shorter and there will eventually be perfect agreement.

Note, however, that in any particular experiment, the probability associated with $x$ or less will always be greater than the corresponding $F_G(x)$; and while the difference for large $n$ may be trivial, in cases such as the present

**Table 12.3. The Gaussian approximation to the binomial distribution $B$ (10, 1/2)**

| | | | Gaussian approximation | |
| --- | --- | --- | --- | --- |
| $x$ | $P(x)$ | $F_B(x)$ | $F_G(x)$ | $F_G(x + 1/2)$ |
| 0 | 0.000,977 | 0.000,977 | 0.000,783 | 0.002,213 |
| 1 | 0.009,766 | 0.010,742 | 0.005,706 | 0.013,428 |
| 2 | 0.043,945 | 0.054,688 | 0.028,890 | 0.056,923 |
| 3 | 0.117,188 | 0.171,875 | 0.102,952 | 0.171,391 |
| 4 | 0.205,078 | 0.376,953 | 0.263,544 | 0.375,915 |
| 5 | 0.246,094 | 0.623,047 | 0.500,000 | 0.624,085 |
| 6 | 0.205,078 | 0.828,125 | 0.736,455 | 0.828,609 |
| 7 | 0.117,188 | 0.945,312 | 0.897,048 | 0.943,077 |
| 8 | 0.043,945 | 0.989,258 | 0.971,110 | 0.986,572 |
| 9 | 0.009,766 | 0.999,023 | 0.994,294 | 0.997,787 |
| 10 | 0.000,977 | 1.000,000 | 0.999,217 | 0.999,383 |

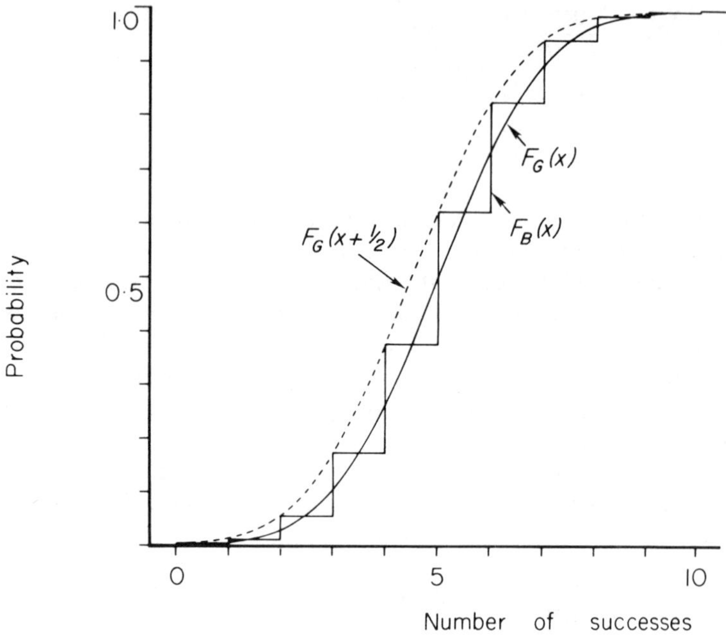


**Figure 12.2.** The (cumulative) distribution function for a binomial variate, $F_B(x)$ of order 10 and with parameter 0.5 is shown. By convention, "$F(x)$" is the probability of $x$ or less. Hence, an ideal approximation should go through the "points" of the steps. The Gaussian approximation, $F_G(x)$ having the same mean and variance (the continuous curve) misses these points. If, however, it is shifted half a unit to the left (interrupted line, $F_G(x + \frac{1}{2})$: this is known as *Yates's correction for continuity*) it is an excellent approximation.

where $n$ is small it may be troublesome, as witness the discrepancy between the third and fourth columns of Table 12.3.

So we shall try to find a more accurate approximation for use, especially when $n$ is small. First note that because of the stepped nature of the true distribution function the steps characteristically straddle the approximation given by the Gaussian curve; and since it is always the larger value at the step that is applicable to the "probability of $x$ or less," the approximation is characteristically an underestimate. Then, secondly, the approximation more or less bisects both the horizontal and the vertical parts of each step. These facts suggests that we might try shifting the approximating curve some way to the left, to make sure that the approximation is adequate where it is required to be accurate, that is, at the sample values that are actually encountered (integer values of $x$). Symmetry suggests that half a step might be an appropriate shift, so that before computing the approximation we add one half to the value of $x$. The results are shown numerically in the last column of

Table 12.3 and as the interrupted line in Figure 12.2. Considering the small size of the sample, the fit is satisfactory. This adjustment is known as Yates's correction (for continuity).

### The two-tailed test

Extension of Example 12.10 to the two-tailed test calls for a little care. If the allocations of the error to the two tails are to be equal, then we pick some appropriate critical value which we shall denote by $x_c$ rejecting the null hypothesis if the sample value $x$ is—

equal to, or less than, $x_c$;

equal to, or greater than, $(n - x_c)$.

Note here that we have a double dose of the problem we encountered above. If for instance we decide that the appropriate critical value for the lower tail is 2, by symmetry we will reject the hypothesis if $x = 0, 1, 2, 8, 9,$ or 10. The size of this test is found with the Gaussian approximation by computing the area in the lower tail of the curve using $x = 2.5$ and the upper tail (which from symmetry would have exactly corresponding probabilities) by finding the area with $x = 7.5$. In effect we would find the values for each tail by shifting the sample value half a unit nearer the mean; that is, finding the area corresponding to the standard Gaussian deviate.

$$z = \frac{|x - \mu| - \frac{1}{2}}{\sqrt{0.025}}$$

Note that this approximation will prove absurd if the sample value is 5 (or, more generally, the expected value) since then one is approximating the combined probabilities of "$x = 5$ or less" and "$x = 5$ or more." The probability of $x = 5$ is being added twice and the whole probability is greater than unity. However, if the sample value equals the mean exactly, no test is needed.

In Example 12.10 we have capitalized on the symmetry of binomial distributions of which the parameter is $\frac{1}{2}$. For other values, symmetry about the mean is assured if both $np$ and $nq$ are large: a common rule of thumb in practical statistics is that both quantities should be at least 5. Asymptotically, the contributions from the two tails with limits equidistant from the mean will be equal.

## HYPOTHESIS TESTING: THE CHI-SQUARE APPROXIMATION

The square of a standard normal deviate is a chi-square variate with one degree of freedom. Thus one may compute exactly the same statistic as in the previous section and square it. Yates's correction may be used when appropriate. That is, given

$$\frac{|X - \mu| - \frac{1}{2}}{\sqrt{npq}} \sim N(0, 1)$$

with a critical region at the 5% level of less than $-1.96$, or greater than 1.96, then

$$\left[\frac{|X - \mu| - \frac{1}{2}}{\sqrt{npq}}\right]^2 \sim \chi_1^2 \tag{12.5}$$

with a critical region of $(\pm 1.96)^2 = 3.84$ or greater.

The statistic in (12.5) is usually calculated somewhat differently. As is easily verified

$$\frac{[|X - \mu| - \frac{1}{2}]^2}{npq} = \frac{[|X - \mu| - \frac{1}{2}]^2}{np} + \frac{[|X - \mu| - \frac{1}{2}]^2}{nq} \tag{12.6}$$

Now the expected number of successes is $np = \mu$, and of failures is $nq = n - \mu$. Furthermore since the sum of the observed numbers of failures and successes must also equal $n$ and

$$\mu - X = -[(n - \mu) - (n - X)]$$

then $|X - \mu|$ may equally well be viewed as either the absolute value of the deficit of successes or of the excess of failures. Thus we may write (12.5) as

$$\frac{[|O_s - E_s| - \frac{1}{2}]^2}{E_s} + \frac{[|O_f - E_f| - \frac{1}{2}]^2}{E_f} \tag{12.7}$$

where $O$ and $E$ denote observed and expected numbers respectively and the subscripts $s$ and $f$, successes and failures. Nothing has been gained by this maneuver, but this general form is commonly used for multinomial generalizations. (See Chapter 13.)

**Example 12.11.** In a sample of 25 subjects aged 30–39 with one parent heterozygous for the autosomal dominant disorder familial polyposis coli (12) the number affected was 8. Test the hypothesis that the true proportion is $\frac{1}{2}$, as predicted by theoretical genetics.

We shall test the hypothesis in several ways.

**1. The exact binomial test** gives the following tail probabilities:

| x | P(x) | Cumulative probability<br>P(x or less) = F(x) |
|---|---|---|
| 0 | 0.000,000,029,8 | 0.000,000,029,8 |
| 1 | 0.000,000,745,0 | 0.000,000,774,8 |
| 2 | 0.000,008,940,6 | 0.000,009,715,4 |
| 3 | 0.000,068,545,3 | 0.000,078,260,7 |
| 4 | 0.000,376,999,3 | 0.000,455,260,1 |
| 5 | 0.001,583,397,3 | 0.002,038,657,5 |
| 6 | 0.005,277,991,3 | 0.007,316,648,8 |
| 7 | 0.014,325,976,4 | 0.021,642,625,3 |
| 8 | 0.032,233,446,9 | 0.053,876,072,2 |

If we have at a maximum $2\frac{1}{2}\%$ error in each tail, the acceptance region is 8 to $(25 - 8) = 8$ to 17 and we will accept the null hypothesis.

**2. The Gaussian approximation.** The mean is $^{25}/_2 = 12.5$ and the variance of $X$ is $npq = 6.25$, giving a standard deviation of 2.5. Then

$$z = \frac{8 - 12.5}{2.5} = -1.8$$

which being inside the 95% acceptance region (1.96 standard deviations on either side of the mean) leads us to accept. Note that the significance level according to this approximation is $2 \times 0.03593 = 0.07186$ compared with the exact figure $0.053,87\ldots \times 2 = 0.107,75$.

**3. The Gaussian approximation with Yates's correction.** The test statistic here is

$$\frac{8 + 0.5 - 12.5}{2.5} = -1.6 \tag{12.8}$$

Again we accept. The significance level here is $2 \times 0.054,799\ldots = 0.109,59\ldots$ which tallies well with the exact figure.

**4. The chi-square approximation with Yates's correction.** The test statistic is simply the square of the result in (12.8), i.e., 2.56. Consulting standard tables yields the same significance level (which now, as a result of squaring, represents a one-tailed test).

**5. The chi-square test with Yates's correction by comparing the observed and expected.** In each of the two groups the expected number is 12.5 and the observed numbers are 8 (affected) and 17 (unaffected). The chi-square statistic is thus

$$\frac{[|8 - 12.5| - 0.5]^2}{12.5} + \frac{[|17 - 12.5| - 0.5]^2}{12.5}$$

$$\frac{4^2}{12.5} + \frac{4^2}{12.5} = 2.56$$

as before.


## THE GAUSSIAN APPROXIMATION TO CONFIDENCE LIMITS

The problems in computing confidence limits for the binomial parameter are only too evident. It is logically confusing and arithmetically tedious; and at the end of it all, the answer is not exact but errs on the side of conservatism. An approximation would clearly be welcome, and we will now explore how the Gaussian distribution may be used. Let us suppose that the sample is large enough to invoke asymptotic normality, i.e., that both $np$ and $n(1 - p)$ are, say, five or over. Then as a preliminary we recognize two special features. First, the binomial variance that we shall use in the approximation depends on the parameter $p$ which we have to find. Second, just as in using the Gaussian approximation for testing hypotheses we bring the sample value half a unit closer to the mean in order to correct for continuity, here we must move the sample value half a unit *further* from the conjectured mean so that after we shall have applied Yates's correction for continuity we would use as the Gaussian deviate the value that the sample result has taken. This should be verified in Figure 12.1. Then if we set the true value of the parameter at $p_u$ at which we would just reject the hypothesis in the lower tail, we find

$$\frac{x - np_u - \frac{1}{2}}{\sqrt{np_u(1 - p_u)}} = -1.96 \tag{12.9}$$

(The right hand side is the critical value of a two-tailed test at the 5% level. Any other suitable value might be substituted.) Equation 12.9 is then squared and solved for $p$, making sure that the correct root is chosen. Likewise the lower confidence limit $p_b$ is the appropriate root of the equation

$$\frac{x - np_b + \frac{1}{2}}{\sqrt{np_b(1 - p_b)}} = 1.96 \tag{12.10}$$

**Example 12.12.** Consider again the problem in Example 12.11. For comparison, in what follows direct calculation for the binomial distribution gives $p_b = 0.120,716,6\ldots$ and $p_u = 0.493,876\ldots$

Using equation (12.9) we get for the higher value

$$\frac{7.5 - 25p_u}{\sqrt{25p_u(1 - p_u)}} = -1.96 \tag{12.11}$$

Squaring,

$$(7.5 - 25p_u)^2 = 96.04p_u(1 - p_u)$$

$$56.25 - 375p_u + 625p_u^2 = 96.04p_u - 96.04p_u^2$$

Collecting terms and solving,

$$p_u = \frac{471.04 \pm \sqrt{471.04^2 - 4(721.04)(56.25)}}{2 \times 721.04}$$

$$= 0.157,284 \quad \text{or} \quad 0.495,993$$

Obviously it is the larger value we require since the upper confidence limit must exceed the sample estimate. (The ambiguity arises because in squaring [12.11] to solve for $p$ we have obscured the sign on the right hand side.)

Note two points here. First, that the true value of $np_u$ is $0.493,876\ldots \times 25 = 12.347$ which is comfortingly greater than 5 and would allow us to use the Gaussian approximation with assurance. Second, the estimate so obtained in fact differs little (about $+0.4\%$) from the "exact" binomial estimate.

The approximate lower limit, $p_b$, is the appropriate root of

$$\frac{8.5 - 25p_b}{\sqrt{25p_b(1 - p_b)}} = 1.96$$

$$721.04p_b^2 - 521.04p_b + 72.25 = 0$$

Whence $p = 0.535,505$ or $0.187,118$. (Obviously, it is the latter root we require.)

This estimate is not so satisfactory and I think we must ascribe it to the small value of $np = 0.120,716,6 \times 25 = 3.017,915$. This illustration should serve as a warning about the glib use of large sample properties.


## COMBINING BINOMIAL VARIABLES

Sometimes we have available multiple sources of information about binomial variables and there is reason to believe that they may jointly tell us more than any component separately. Some of the best examples are genetic. The issue may arise in either estimation or hypothesis testing.

Yet once again I urge the reader to review the assumptions underlying the binomial distribution. I propose to confine discussion to independent sets of data. Broadly speaking, there are two ways of dealing with such data.

1. Pool the results within categories and treat them as if they are a sample from the same population. The principal advantage of this method is simplicity and ready intelligibility. Readers like clean statements such as "the mortality rate in pneumonia is 2%" or "Polyposis coli is an autosomal dominant disorder." However, to warrant this method and to apply the result we must satisfy the condition that the binomial parameter is the same in all groups pooled. Those, statisticians or physicians, who like to get below the surface of things, would mistrust any such simple statement about pneumonia, which clearly has a higher mortality among some groups of patients (the old, the debilitated, those with structural lung defects) than others. Likewise, the statement about dominance is one of behavior, including segregation ratios. But polyposis coli is age-dependent, so the probability of being affected changes with age. In Example 12.11 we sinned mildly by taking subjects of a range of ages rather than one particular age;* but the non-binomial variation, which in fact yields a Lexis distribution (7), distorts results little in this case, where the range of ages within classes is so small that it may be ignored. Indeed except where the data have been deliberately "binomialized" by random assignment to one of two groups (as in Example 12.1), we are never completely assured that the same parameter applies to all cases and hence that binomial theory is fully justified. In practice, we are less

---

*There is no assurance that even this restriction would meet all criticisms: for instance it has been suggested that there are two populations of patients with familial polyposis coli, one having a lower mean age of onset than the other.

fastidious than strict mathematics would require, and with warrant. But that we can allow some leeway is no endorsement of carelessness. The more variation in the parameter, the greater the distortion. The analyst should be on the lookout for this defect and try to keep it to a minimum. On the other hand, there is some risk of becoming so obsessed with heterogeneity as to abandon all attempts at scientific coherence and to substitute for sound generalization an infinite particularity. The right spirit is tentative inquiry and analysis, which involves a perpetual openmindedness: neither blind assurance on the one hand, nor neurotic skepticism on the other.

2. If it is known, or at least suspected, that the data sources differ systematically but are reasonably homogeneous within, one may analyze results separately and pool the statistics.


### Test of a genetic hypothesis

Consider the hypothesis that A and B are harmless allelic genes with A dominant to B. Homozygous BB are easily recognized by phenotype. All offspring of phenotype A with one parent of phenotype B must be heterozygous. Then the hypothesis predicts that from the mating AB $\times$ AB, one-quarter will be of type B, and from the mating AB $\times$ BB, one-half will be of type B. Analyses of both types provide tests of the hypothesis; but since the segregation ratios differ, we cannot pool the entire data. Nevertheless, we may compute a chi-square test for both types, each having one degree of freedom. Then under the null hypothesis the sum of these values, being independent, should follow a chi-square value with the sum of their degrees of freedom. (This is a general property of independent chi-square statistics, see 7.) Unfortunately it is hard to find human data of this kind since, although readily identified, such matings are not usually distinguished. My illustration must be fictitious.

**Example 12.13.** From the mating AB $\times$ BB, 40 progeny result, of whom 24 are of type A; and from the AB $\times$ AB matings, 56 progeny result of whom 49 are of type A. Test the Mendelian hypothesis.

Note that we have not said how many families are involved, and if we are *sure* that in each the same locus and the same genes are involved there is no need to. From the data the following results are obtained:

|  | Phenotype of progeny | |
|  | A | B |
|---|---|---|
| First type | | |
| Expected | 20 | 20 |
| Observed | 24 | 16 |
| Second type | | |
| Expected | 42 | 14 |
| Observed | 49 | 7 |

Then the chi-square statistics (using Yates's correction) are—
For the first

$$\frac{[|20 - 24| - \frac{1}{2}]^2}{40 \times \frac{1}{2} \times \frac{1}{2}} = 1.225$$

For the second

$$\frac{[|42 - 49| - \frac{1}{2}]^2}{56 \times \frac{1}{4} \times \frac{3}{4}} = 4.024$$

The sum of the chi-square values, 5.249, is read against values for the chi-square distribution with two degrees of freedom, and the hypothesis is not rejected at the 5% level. (Note, however, testing the second statistic separately as a $\chi^2$ with one degree of freedom would lead to rejection of that part of the data. But this argument is not altogether legitimate because the decision to pay particular attention to this component is made after the data have been inspected.)

## Estimation

**Example 12.14.** Suppose that either because of the foregoing data (in both of which the phenotype B is underrepresented) or for some other reason, the geneticist suspects that the phenotype B is less fit than A and that in consequence only some proportion of them live long enough to be counted. As a result, the segregation ratios are not $1:1$ and $3:1$ respectively but $1:p$ and $3:p$ in the two types of mating. How is $p$ to be estimated? In the first mating, the segregation ratios are converted into binomial form, preserving the same ratios but adding to 1.

$$\frac{1}{1 + p} : \frac{p}{1 + p}$$

and likewise in the second

$$\frac{3}{3 + p} : \frac{p}{3 + p}$$

Then the likelihood function for all the data is the product of the likelihood of two binomials

$$\binom{40}{24}\left(\frac{1}{1 + p}\right)^{24}\left(\frac{p}{1 + p}\right)^{16}\binom{56}{49}\left(\frac{3}{3 + p}\right)^{49}\left(\frac{p}{3 + p}\right)^{7} \tag{12.12}$$

Subsuming the terms $\binom{40}{24}$ and $\binom{56}{49}$ and $3^{49}$ under a common factor $K$, which is of no interest, we may write L, the (natural) logarithm of (12.12), as

$$L = -40 \log(1 + p) + 23 \log p - 56 \log(3 + p) + \log K$$

which we proceed to maximize with respect to $p$ in Table 12.4.

**Table 12.4. Likelihood of a set of heterogeneous genetic data**

| $p$ | $\log_e(L/K)$ |
|---|---|
| 0.2 | $-109.446$ |
| 0.4 | $-103.065$ |
| 0.6 | $-102.281$ |
| 0.8 | $-103.404$ |
| 0.54 | $-102.234,674,6$ |
| 0.55 | $-102.229,514,0$ |
| 0.56 | $-102.229,848,8$ |
| 0.553 | $-102.229,048,9$ |
| 0.554 | $-102.229,002,6$ |
| 0.555 | $-102.229,010,2$ |

Applying the equations (3.2) and (3.3) from Chapter 3

$$\hat{p} = 0.554,718$$

and

$$\text{var}(\hat{p}) = 0.018,539$$

giving an estimated standard deviation of 0.136.

## PROBLEMS

**12.1.** Suppose that in the population of all circulating pennies, the probability of heads varies from coin to coin according to the beta probability density function.

$$f(p) = Kp^{20}(1 - p)^{20}$$

where $K$ is an (unimportant) normalizing constant. A coin, picked at random, is tossed 10 times coming up heads ("success") 4 times.

    **a.** Compute the prior density function for suitable values of $p$ as a multiple of $K$.

    **b.** Compute the likelihood of $p$ in the sample.

    **c.** Hence, find the joint density and the most probable value of $p$ for the particular coin.

**12.2.** Prove the identity (12.6) in the text.

**12.3.** Consider the hypothesis that $X \sim B(n, p)$ against the alternative hypothesis that $X \sim B(n, p + h)$ where $h > 0$. Prove that the rejection region (constructed in accordance with the Neyman-Pearson lemma) should be in the right-hand tail.

# 13

# THE MULTINOMIAL DISTRIBUTION: CONTINGENCY TABLES

In Chapter 12, we recognized that the score assigned to a Bernoulli process (0 for a failure, 1 for a success) need have no meaning other than as a counting device. The absence of any fixed meaning is even more evident once we consider three or more categories. Thus, arranging continents in alphabetical order (to avoid diplomatic friction), we might attach the arbitrary score of 0 to Africans, 1 to Americans, 2 to Asians, 3 to Australians, and 4 to Europeans. But now the total score for a sample of persons taken at random tells us very little about the composition of the sample. Of course, a total score of 0 can only mean that all the subjects are Africans; and a score of four times the sample size can only mean that all the subjects are Europeans. But obviously, for most intermediate values, the same total would be produced by samples of quite different compositions. It is difficult to see in what sense, other than mere Jingoism, one could say that one European is equivalent to two Asians or four Americans.

If the arbitrary number of the category has no useful meaning, then it is difficult to see in what nontrivial sense its total score can be said to be a distribution. Nevertheless, we can restore useful properties to the process by formulating it in the somewhat different manner that we shall now present. We should first note that there are *some* instances in which the scores assigned to the classes do have a meaning other than arbitrary; for instance, there are multiple categories of family sizes and the mean and variance of the scores have interpretable meanings. There may even be a structural mathematical relationship, both among the scores and the assigned probabilities, as in the Poisson distribution (Chapter 14).

Suppose that instead of representing the outcome for $n$ Bernoulli trials as a single number (the successes), we represent both the number of successes $x_1$ and the number of failures $x_2$. Now, of course, their sum must equal $n$; and in consequence, they must be perfectly dependent quantities. To know the one in any experiment is to know the other exactly. So, we can think of the distribution as being bivariate, but because of this perfect dependence, degenerately so. If one were to plot the outcome in a plane, with the number of successes as one coordinate and the number of failures as the other, the points representing all possible outcomes would fall on a straight line; that is, although displayed in two dimensions, the outcome would in fact occupy only one.

If there are three categories, we speak of a trinomial variate. For instance, at the ABO blood group locus, a gene is either of type A, or B, or O. There, the outcome for a sample of size $n$ may be represented by a set of three figures $(X_1, X_2, X_3)$. There is no longer perfect dependence; but this trivariate distribution is still degenerate since, given any two of the results for a sample of specified size, the third can be found by subtraction. Plotted in a three-dimensional space, all points would fall on one plane (75). And, in general, if we have $k$ categories, the outcome of this so-called multinomial distribution will be presented by a set of $k$ quantities $(X_1, X_2, \ldots, X_k)$ which is degenerate by one degree; that is, given $(k - 1)$ of them, the last may be found by subtraction.


## ESTIMATION

We distinguish two types of multinomial distribution from the statistical standpoint.

### Unstructured estimation

There may be no logical relationship among the categories other than that the sum of their probabilities must be unity. This is sometimes called (in the context of the multinominal distribution) "nonparametric estimation."

**Example 13.1.** Granted that any gene at the ABO blood group locus must be of type A, B, or O, there is no other constraint on what the individual probabilities should be. We are obliged to estimate, without further assistance, two of the three parameters. It transpires that the problem is, in principle, easy.* Given independence of sample data, the maximum likelihood estimate for the probability of belonging to type A is simply the proportion

$$\hat{p}_A = \frac{\text{Number of genes of type A}}{\text{Total number of genes in sample}} = \frac{n_A}{n}$$

and likewise for $\hat{p}_B$ and $\hat{p}_O$. This estimate is unbiased, efficient, consistent, and asymptotically Gaussian. Moreover, we can readily state the variance of each of them separately by the ingenious device of lumping together the other categories and hence making it a binomial problem, e.g., the outcomes are "A" and "not A."

*I say "in principle." The difficulty is in counting the number of genes. In practice we must make do with phenotype, and the problem becomes somewhat more complicated (see below).

$$\text{var}(\hat{p}_A) = \frac{p_A(1 - p_A)}{n}$$

which we estimate in practice by substituting for $p_A$, the sample value of $\hat{p}_A$. We may use the ideas from Chapter 12 to test hypotheses or set confidence limits on $p_A$. Furthermore, we can treat $p_B$ and $p_O$ in exactly the same fashion. However, it will be evident that these estimates are not independent. In general (7), the true covariance of the estimates of $p_i$ and $p_j$ in a multinominal is

$$\text{cov}(\hat{p}_i, \hat{p}_j) = -\frac{p_i p_j}{n}$$

and the maximum likelihood estimate of this quantity is found simply by inserting sample estimates for true values of $p_i$ and $p_j$.

**Structured estimation**

The probabilities associated with the various cells of the multinomial table may be functions of some underlying parameters which may be the real quantities of interest. Thus, the estimation may no longer be altogether straightforward.

**Example 13.2** One of the most famous theorems in genetics is the Hardy-Weinberg principle which states that, under certain specified conditions, the probability of a particular genotype at a genetic locus is equal to that of two corresponding genes being assembled at random from the gene pool. If, for instance, there are two types of genes, A with probability $p$ and B with probability $q$, then under the specified conditions the probabilities associated with the three possible genotypes are:

| Genotype | Probability |
|---|---|
| AA | $p^2$ |
| AB | $2pq = 2p(1 - p)$ |
| BB | $q^2 = (1 - p)^2$ |

Here, there is a clear structural relationship among the cells. If there are only two alleles so that $p + q = 1$, instead of having to estimate two parameters (one less than the number of cells) we have to estimate only one, $p$.

**Example 13.3** Omoto and associates (98) in a study of Philippine Negritos found the following frequencies for the genotypes at the locus for glyoxylase (GLO):

| GLO genotype | Number of cases |
|---|---|
| 1-1 | 8 |
| 1-2 | 47 |
| 2-2 | 74 |
| Total | 129 |

These may readily be reduced to binominal form since the individual genes can be counted. GLO-1 genes are present on both chromosomes in eight subjects and on one chromosome in 47. By this argument:

$$8 \times 2 + 47 = 63 \text{ genes are of type 1}$$
$$74 \times 2 + 47 = 195 \text{ genes are of type 2}$$

The appropriate estimate of the probability of the type 1 gene is, thus;

$$\frac{63}{63 + 193} = \frac{63}{2 \times 129} = 0.244\ldots$$

Let us now consider a case where the issues cannot be so readily resolved.

**Example 13.4** Zanardi and associates (99) give the following frequencies for the ABO group locus in 239 subjects from Migliarino in Ferrara, Italy: 119 of type A; 21 of type B; 87 of type O; 12 of type AB. We assume Hardy-Weinberg equilibrium. The traits A and B are dominant to O, that is, they will obscure it; but they are (mutually) codominant; that is, they do not obscure each other. Thus, if, in accordance with custom, we denote the probability of A by $p$, of B by $q$, and of O by $r$, with $p + q + r = 1$, we have the following probability structure:

| Genotype | Probability | | Phenotype | Probability | Observed proportion |
|---|---|---|---|---|---|
| AA | $p^2$ | $\left.\right\}$ | A | $p^2 + 2pr$ | 0.497,91 |
| AO | $2pr$ | | | | |
| BB | $q^2$ | $\left.\right\}$ | B | $q^2 + 2qr$ | 0.087,87 |
| BO | $2qr$ | | | | |
| OO | $r^2$ | | O | $r^2$ | 0.364,02 |
| AB | $2pq$ | | AB | $2pq$ | 0.050,21 |

We may estimate the several frequencies by various methods.

**Method of moments (known as Berstein's method).** Set the observed proportions equal to the probabilities and solve.

$$\hat{p}^2 + 2\hat{p}\hat{r} = 0.497,91$$

$$\hat{r}^2 = 0.364,02$$

Adding,

$$\hat{p}^2 + 2\hat{p}\hat{r} + \hat{r}^2 = (\hat{p} + \hat{r})^2 = 0.861,9\ldots$$

whence taking square roots,

$$\hat{p} + \hat{r} = 0.928,40$$

We substract this from unity:

$$1 - \hat{p} - \hat{r} = \hat{q} = 1 - 0.928,40 = 0.071,60.$$

Likewise,

$$\hat{p} = 1 - \sqrt{(\hat{q} + \hat{r})^2} = 1 - (0.087,87 + 0.364,02) = 0.327,78$$

We may estimate $r$ directly;

$$\hat{r} = \sqrt{0.364,02} = 0.603,34$$

Note that these estimates, though easily attained, are somewhat contradictory since their sum is 1.002,71, not 1. There exists a procedure for reducing this joint bias of estimate.

**Maximum likelihood.** The joint likelihood is (ignoring, as usual, the combinatorial coefficient, which is simply a constant multiplier):

$$(p^2 + 2pr)^{119}(q^2 + 2qr)^{21}(r^2)^{87}(2pq)^{12}$$

it may be simplified somewhat, substituting $(1 - p - q)$ for $r$:

$$p^{131}\, q^{33}\, (1 - p - q)^{174}(2 - 2q - p)^{119}(2 - q - 2p)^{21}$$

which we may maximize by trial and error. This is a tedious process and it is somewhat simplified by taking as starting points the method-of-moments estimates given above. Eventually, we attain the values for the logarithm of the likelihood shown as natural logarithms in Table 13.1. The least negative number (and hence, the largest likelihood) is given by $\hat{p} = 0.327,3$ and $\hat{q} = 0.071,5$; whence (by subtraction) $\hat{r} = 0.601,2$. These calculations could be further pursued to obtain more exact solutions.

**Minimum chi-square estimation.** This is the first time we have used this method, mentioned, but not expounded, in Chapter 3. We first set up the appropriate chi-square test of goodness of fit, which is analogous to the alternative form given in Chapter 12, that is,

$$\chi^2 = \sum_i \frac{(O_i - E_i)^2}{E_i} \tag{13.1}$$

where the $O$ and $E$ are the observed and expected numbers of subjects. Here it is

$$\frac{(239p^2 + 478pr - 119)^2}{239p^2 + 478pr} + \frac{(239q^2 + 478qr - 21)^2}{239q^2 + 478qr}$$

$$+ \frac{(239r^2 - 87)^2}{239r^2} + \frac{(478pq - 12)^2}{478pq}$$

**Table 13.1. Natural logarithms of the likelihood for various trial values of $p$ and $q$ in Example 13.4**

| | $p$ | | |
|---|---|---|---|
| $q$ | 0.327,4 | 0.327,3 | 0.327,2 |
|---|---|---|---|
| 0.0716 | −257.917,950 | −257.917,941 | −257.917,950 |
| 0.0715 | −257.917,913 | −257.917,910 | −257.917,925 |
| 0.0714 | −257.917,947 | −257.917,950 | −257.917,971 |

Minimizing this expression, even after substituting $(1 - p - q)$ for $r$ is more complicated than ever. However, it is, like maximum likelihood, an efficient estimator that is also consistent and asymptotically Gaussian and has the advantage that it furnishes a test of the genetic hypothesis that a Hardy-Weinberg equilibrium exists. With a prolonged search, we obtained the estimates

$$\hat{p} = 0.327,3$$
$$\hat{q} = 0.071,6$$
$$\chi^2 = 0.094,354$$

**Modified minimum chi-square estimation.** Those who have carried through the foregoing calculations* will have recognized that a large part of the difficulty lies in the fact that the unknown quantities appear in both the numerator and the denominator. An ingenious modification is to insert in the denominators not the expected, but the observed, values. Remarkably enough, this much simplified estimator also is efficient, consistent, and asymptotically Gaussian. (The only peril is if the observed number in any of the groups should be zero, which would, of course, make the chi-square statistic infinite. However, most investigators would refuse to make estimates under those circumstances and either combine groups or [if possible] eliminate some parameters.)

It will be appropriate to summarize the estimates obtained by these various methods (Table 13.2). There is reasonable agreement among them; the most discrepant is method of moments. For the three efficient estimators, the only differences lie in the fourth decimal place. Thus, we need have little compunction about using an estimate of the variance based on the likelihood curve.

### Interval estimation

Under ordinary circumstances, maximum likelihood, minimum chi-square, and modified minimum chi-square estimation are all efficient and, for large samples at least, it is customary to use the lower bound for interval estimation (cf. Chapter 3).

## TEST OF HYPOTHESIS

In order to interpret the foregoing results and to apply them to testing hypotheses, we must give some thought to two fundamental ideas.

### Degrees of freedom

We have previously (see Chapter 7) thought of the number of degrees of freedom as the number of dimensions mutually at right angles in which a

---

*The difficulties are even more evident to those who attempt to find the minimum chi-square by the more penetrating methods of the calculus.

**Table 13.2. Estimates of the parameters for the data of Example 13.4**

| Method | $\hat{p}$ | $\hat{q}$ | $\hat{r}$ | Qualifying comment |
|---|---|---|---|---|
| Moments | 0.327,8 | 0.071,6 | 0.603,3 | Sum of estimates $= 1.002,7$ |
| Maximum likelihood | 0.327,3 | 0.071,5 | 0.601,2 | Natural logarithm of maximum likelihood is $-$ 257.917,9 |
| Minimum chi-square | 0.327,3 | 0.071,6 | 0.601,1 | $\chi^2 = 0.094,35$ |
| Modified minimum chi-square | 0.327,3 | 0.071,3 | 0.601,4 | $\chi^2 = 0.091,23$ |

point may move. Another way of looking at it is as the number of pieces of information one must have to locate the point inside a system of coordinate axes with a specified origin. Given some point of reference ("the origin") one measurement locates a point on a line; two define a point in a plane; three, in a space. Thus, if we take samples (of unspecified size) of individuals and class them into three groups, the result of any such experiment could be represented by one point in a (discrete) three-dimensional space, the coordinates of which correspond to the numbers in each category. Any point (with integral-valued coordinates) in the space may be occupied by the sample point, which therefore has three degrees of freedom.

### Constraints

However, suppose that we fix in advance at (say) $n$ the number of individuals to be included in the sample. Then the point representing the sample must lie on a particular plane (Fig. 13.1). If it did not, the sum of the numbers in the three classes would not add up to the predetermined total. Thus by incorporating this restriction on size (which is an example of a constraint) we limit the movement of the sample point to the two dimensions of the plane; that is, it has now only two degrees of freedom. This result illustrates

**Principle 1.** Every independent constraint on a sample value reduces the number of its degrees of freedom by one.

Suppose a geneticist is testing a genetic hypothesis about the three genotypes AA, AB, and BB at a particular locus—for example that Hardy-Weinberg equilibrium exists. In Figure 13.1 this condition is fulfilled by every point on the "Hardy-Weinberg" line. The geneticist collects his sample of predetermined size, so that the point must lie on the plane. We may envision two possibilities:

1. The null hypothesis involves a joint conjecture that Hardy-Weinberg equilibrium exists, and that the frequency of the A gene is (let us say) 0.4. In the diagram are shown a set of lines on the plane that connect all points for which the gene frequency would be the same—they are labeled accordingly. There is only one point in the plane at which both the conjectures in the null

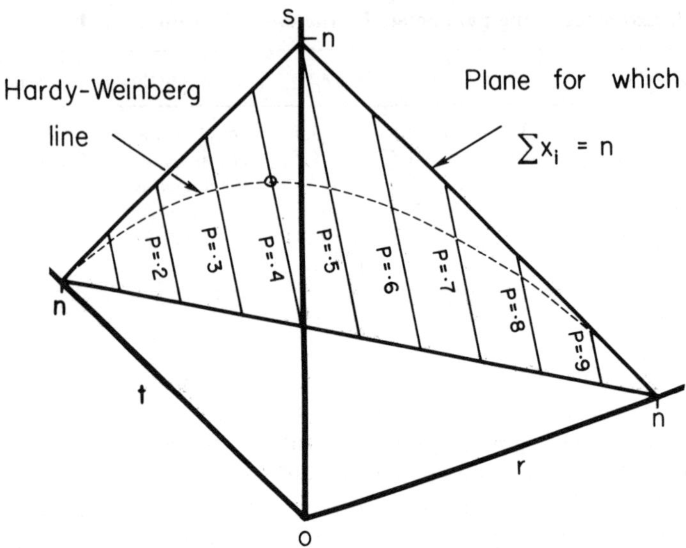


**Figure 13.1.** Degrees of freedom in a trinomial sample space. There are three classes of outcomes, so any sample result may be plotted as a point with positive integral coordinates $(r, s, t)$ in three-dimensional space. The sample point has three degrees of freedom. If the sample size is fixed at $n$, then the result is limited to a plane intercepting the three axes at a distance $n$ units from the origin, and shown here as a triangle. The sample point has now two degrees of freedom. If we suppose that a Mendelian trait with two alleles is concerned, the constraint that the frequency of the one is some fixed quantity, $p$, confines the sample space to a single line, e.g., $p = 0.4$. If we impose the condition that a Hardy-Weinberg equilibrium must exist (random selection of spouse) the expected outcome is constrained to fall on the dotted line. With either of these constraints, the sample point has one degree of freedom only. If *both* constraints are imposed, there will be no degrees of freedom and the structure of the population will be uniquely determined.

hypothesis are fulfilled: where the line "$p = .4$" intersects the "Hardy-Weinberg" line. Thus the sample value might differ from it either because the estimate of the gene frequency is wrong or because the ratios are wrong. The sample point, then, might either be on the wrong "$p$" line or not on the Hardy-Weinberg line, or both. Thus it can vary in two different directions and there are two degrees of freedom for error.

2. Now suppose that the geneticist uses the sample to estimate the value of $p$—the null hypothesis leaves this quantity unspecified. Clearly the sample point must be on the right "$p$" line because, in effect, this $p$ line has resulted from the estimate; and the only degree of freedom for error (that is, for discrepancy between the data and the null hypothesis) is whether the data point lies on the Hardy-Weinberg line. We infer

**Principle 2.** For every independent parameter estimated from the data we sacrifice one degree of freedom.

We may note two points about Principle 2:

1. The loss is attached to *independent* parameters. In the last example $p$ was estimated from the data and so was $q$; but $\hat{p}$ and $\hat{q}$ are not independent, and indeed to know the one is to know the other exactly. This point is illustrated geometrically in Figure 13.2. The joint parameter space may be represented as a quandrant of a plane, $p$ on one axis and $q$ on the other. However, it will not, as might first appear, be two-dimensional but will consist simply of all the points on the straight line that represents the graph relating $q$ to $p$. A line is, of course, one-dimensional. The constraint on the parameter space (that $p$ and $q$ must add to 1) transfers the degree of freedom used up in estimating the second parameter back into the error space. In the same way, in the ABO blood group example where a trinominal is concerned, any two of the parameters determine the third; so the parameter space is two-dimensional. Then we would lose only two degrees of freedom for the parameters estimated, even though there are actually three parameters involved.

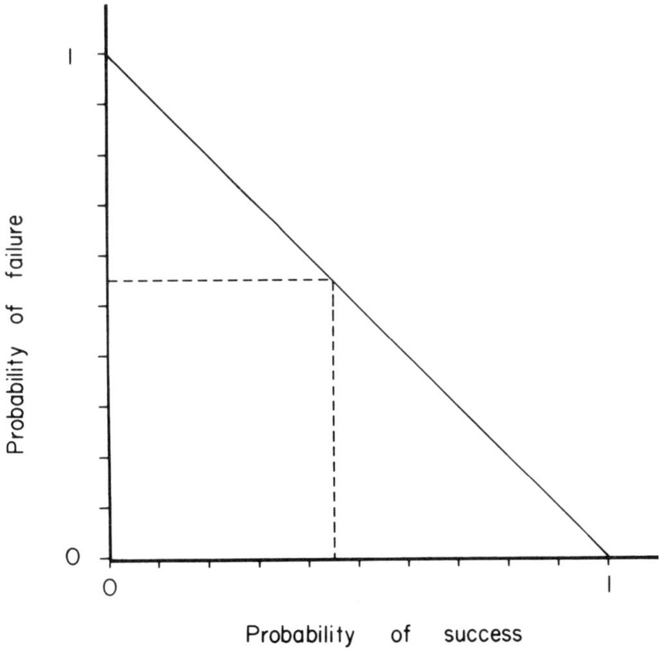


**Figure 13.2.** Completely dependent parameters. The probability of success and that of failure may be displayed in two-dimensional space; but their sum is unity (no other class of outcome being admitted). For any point the sum of the coordinates (the interrupted lines) is unity. This means that all possible outcomes fall on a straight line, which has only one degree of freedom.

2. The loss applies only to parameters estimated from the data on which the test of hypothesis is to be done. If the parameter is to be estimated from another source, the hypothesis may still be rejected for two kinds of discrepancy (which $p$ line it falls on and whether it is on the Hardy-Weinberg line); thus there are two degrees of freedom for error. Estimating a parameter from another source will make the distribution of the statistic different and altogether more complicated.

Let us at this stage revert to the topic of generalizing the chi-square test to deal with data that fall into three or more classes. The chi-square statistics can be computed in the form (13.1) above. However, it is usually simpler to use the following form where $n$ represents the total size of the sample:

$$\chi^2 = \sum \frac{O_i^2}{E_i} - n \tag{13.2}$$

It can be shown (though the proof is somewhat difficult) that for a reasonably sized sample the resulting statistic is distributed approximately as a chi-square with the appropriate number of degrees of freedom. "Appropriate number of degrees of freedom" means calculated in accordance with principles 1 and 2 given above: that is, one degree of freedom for each category minus one degree of freedom for every independent constraint and one for every parameter independently estimated from the data upon which the test is to be performed. In concrete terms, by a constraint in this sense is meant the requirement that some total or subtotal for the observed and expected values must be equal. Thus, if the observed total has to equal the expected, there is one constraint involved. If the sum of the observed equals the sum of the expected for men, and the two sums must also be equal for women, then two constraints are involved and two degrees of freedom are lost. (Note in the latter example that there is also the constraint that the observed and expected totals must be equal for men and women combined; but this constraint is certainly not independent of the other two constraints and in fact can be deduced from them. So no further degree of freedom is lost on its account.)

## TESTS OF INTERNAL RELATIONSHIPS: CONTINGENCY TESTS

In the preceding examples we have considered categorical tests of goodness of fit of the data to values predicted by prescribed ratios, or to values deduced from a model and from estimates of parameters derived from the data. It remains to consider the test of an internal and, as it were, private, relationship among the categories. The chief of these is independence of two or more characteristics.

**Example 13.5** Suppose that a sample of subjects from a defined population can be divided into two groups in two different ways: for example, married or not married; and graduate from a university or not. We set up an appropriate table in which algebraic terms are used for the numbers (Table 13.3). The null hypothesis is that these two traits are stochastically independent.

Probabilistically, as we have seen (7):

1. The criterion of independence is whether each joint probability is equal to the product of the marginals.

2. The expected number in each cell is the size of the sample ($n$) times the joint probability for that cell.

Again because of sampling error we cannot expect the observed number in each cell to be exactly equal to the expected, especially since in most cases the latter will not be an integer.

The solution to the problem is obvious enough. First we estimate the proportion of graduates, $p$, by the estimator

$$\hat{p} = \frac{e}{n}$$

This step is justified because, under the null hypothesis the proportions are the same whether the subjects are married or not; so, in the interests of a larger sample and hence a better estimate, we use the marginal total. Likewise we estimate the proportion, $m$, of those married, by the estimator

$$\hat{m} = \frac{g}{n}$$

The estimated joint probabilities for each cell are then computed from the product of the estimated marginal probabilities. Thus the married-graduate class has the estimated joint probability

$$\hat{p}\hat{m} = \left(\frac{e}{n}\right)\left(\frac{g}{n}\right) = \frac{eg}{n^2}$$

**Table 13.3. The contingency table**

| (1) Data | Married | Not married | Total |
|---|---|---|---|
| Graduate | $a$ | $c$ | $e$ |
| Nongraduate | $b$ | $d$ | $f$ |
| Total | $q$ | $h$ | $n$ |

| (2) Estimates of probabilities | Married | Not married | Total |
|---|---|---|---|
| Graduate | $\dfrac{eg}{n^2}$ | $\dfrac{eh}{n^2}$ | $\dfrac{e}{n}$ |
| Nongraduate | $\dfrac{fg}{n^2}$ | $\dfrac{fh}{n^2}$ | $\dfrac{f}{n}$ |
| Total | $\dfrac{g}{n}$ | $\dfrac{h}{n}$ | $1$ |

The expected number is this quantity multiplied by the sample size, i.e.,

$$\frac{eg}{n^2} \times n = \frac{eg}{n}$$

Thus, we may express the expected number in terms of the original data as "the product of the two marginal totals, divided by the size of the sample total."

The chi-square test is then performed in the usual way. For small sizes of sample it is customary to put in a Yates correction for continuity in $2 \times 2$ contingency tables, using

$$\chi^2 = \sum_i \frac{(|O_i - E_i| - \frac{1}{2})^2}{E_i}$$

The number of degrees of freedom is

| | |
|---|---:|
| The number of categories: the number of subclasses | 4 |
| Minus the number of constraints: | $-1$ |
| Minus the number of parameters independently estimated from the data: $p$, $m$ | $-2$ |
| | |
| Total | 1 |

We may readily generalize this method to tests for independence in a two-way classification, one of the factors being a multinominal with $r$ kinds of outcome ($r$ an abbreviation for "rows") the other with $c$ kinds ($c$ for "columns"). The possible marginal outcomes, being multinomial, call for $(r - 1)$ and $(c - 1)$ independent parameters to be estimated respectively, and there is the usual constraint that the totals, observed and expected, must be equal. There are altogether $r \times c$ classes, each with a degree of freedom; thus the number of degrees of freedom for the test statistic is

$$rc - (r - 1) - (c - 1) - 1 = (r - 1)(c - 1)$$

**Example 13.6** Consider the data of Keys and associates (100) about the relation between the country of origin and the probability of having sustained a myocardial infarction. (I have explored this problem probabilistically elsewhere [7]). We might ask whether the geographical origin of the person and having or not having coronary disease are independent. In Table 13.4 are displayed the numbers observed, and expected on the basis of the marginal totals. The chi-square statistic is evaluated at

$$(3 - 1)(2 - 1) = 2 \text{ degrees of freedom}$$

and leads to emphatic rejection of the null hypothesis of independence at any reasonable size of test.

A word of caution. The reader must be careful in interpreting this result.

**Table 13.4. Analysis of the data of Keys et al. (100)**

| | Northern European men | | Southern European men | | United States railway men | | Total expected & observed |
|---|---|---|---|---|---|---|---|
| | Observed | Expected | Observed | Expected | Observed | Expected | |
| With CHD | 206 | 135.215 | 214 | 361.404 | 212 | 135.381 | 632 |
| Without CHD | 2,233 | 2,303.785 | 6,305 | 6,157.596 | 2,230 | 2,306.619 | 10,768 |
| Total | 2,439 | 2,439.000 | 6,519 | 6,519.000 | 2,442 | 2,442.000 | 11,400 |

$$\chi_2^2 = \frac{206^2}{135.215} + \cdots + \frac{2230^2}{2306.619} - 11{,}400$$

$$= 148.788$$

$$p = 4.9 \times 10^{-33}$$

1. The result is a *test* of independence. It is not a *measure* of the degree of nonindependence. (The expected value of the $\chi^2$ statistic will depend on sample size, whereas the degree of relationship can scarcely depend on sample size. Also the $\chi^2$ value is, of course, based on the assumption of independence. Once we assume there is some degree of nonindependence and attempt to measure it, the logical basis for computing the chi-square test has been implicitly abandoned and the statistic ceases to have its original meaning.) The procedure falls within the domain of hypothesis testing, not estimation. The common index of the degree of nonindependence is the risk ratio. It is discussed in the textbooks of epidemiology. I have my own reservations about it (7).

2. The result is a formal analysis: it is a mathematical treatment of data that prescinds entirely from the scientific merits of the sample. The nonindependence of the categories may be due to geographical differences; or to the peculiar way in which the samples were constructed; or to different diagnostic criteria; or to fifty other factors.* A formal test does not solve the problem of *why* the factors are not independent: that is a scientific problem, not a mathematical one.

## PROBLEMS OF SMALL SAMPLES

The following principles are usually observed by practicing statisticians where sample sizes are not very big and anything larger than a $2 \times 2$ contingency table is involved. They were laid down by Cochran in a classic paper (101).

1. Arrange—by grouping cells if necessary—that the expected number in each cell is at least 1 and, for secure inference, at least 5; and that

2. Not more than 20% of the expected values are less than 5.

3. Do *not* use continuity corrections, which introduce more bias than they remove.

4. Do not make life-and-death decisions from statistics near the critical rejection value. (This is, indeed, an excellent rule in any case!)

5. Ignore sample size in determining the degrees of freedom (to which it is irrelevant except insofar as it may be necessary to pool categories and hence reduce the number of cells.)

## PROBLEMS

**13.1.** Prove formula (13.2) and verify it for the data on the ABO blood group, using the maximum likelihood estimates.

**13.2.** Interpret the chi-square value given by problem 13.1—

---

*I imply no criticism of the actual study: I am arguing in principle.

    **a.** statistically;

    **b.** logically.

**13.3.** Verify the calculations in Table 4.2.

**13.4.** Suppose that in the data for Example 13.3, GLO type 1 were dominant so that the genotypes 1-1 and 1-2 were indistinguishable.

    **a.** How would you have estimated the frequencies of the genes?

    **b.** Construct a chi-square test under these circumstances and interpret the result.

**13.5.** It is stated in the text that there is a negative correlation between the sample numbers falling into any pair of cells in a multinomial distribution. Consider a $\chi^2$ test done in Table 13.2. Two degrees of freedom were sacrificed because two of the three parameters ($p$, $q$, and $r$) must be independently estimated. But (as we have noted) the estimates are not independent. Resolve this contradiction.

# 14

# THE POISSON DISTRIBUTION

The Poisson distribution is, like the binomial and multinomial distributions, not only discrete but enumerative: it deals with things that can be counted and where the actual counts are interpretable. It is encountered, and its use warranted, in three ways which are formally distinct but notionally closely related.

First, it arises whenever events occurring randomly, but at constant hazard, are counted over a fixed period of time. For instance, cosmic rays are generated throughout the known universe, and because of their hardness they are little affected by local conditions (the ionosphere, the cloud layer, etc.). They also do not cluster in time; so the number of cosmic rays to which a person is exposed during the course of a year, although a random variable, has a fixed mean value.

Second, it occurs as an extreme ("limiting") case of several distributions. Thus if the binomial distribution is of very high order ($n$) but the probability of success ($p$) is very small, the Poisson distribution is a good approximation to, and much easier to compute than, the exact binomial (7). If the binomial mean, $np = \lambda$, is fixed and $n$ becomes indefinitely large the Poisson distribution results. This fact has wide application in nuclear medicine. Again consider the Lexis distribution (7), which resembles the binomial in all respects except that the probability of success varies nonrandomly from trial to trial. Where the variation is not too wide, say an order of magnitude or so, the limiting form as $n$ becomes large will again follow this distribution. (For an informal proof, see the appendix to this chapter.) This property is valuable in the theory of genetic load. Yet again, in the negative binomial distribution (7), which is that of the number of trials that must be carried out to attain a predetermined number of successes, the number of failures will, under suitable circumstances, follow a Poisson distribution.

Thirdly, the scientist may justify the use of the Poisson distribution because, although there is no prior warrant, empirically it is found to fit a set of data. Thus, the number of house calls that a British general practitioner is called upon to make at night fits this distribution. It has been found (with unconscious humor) that the number of fish caught by an angler does likewise.

There is some risk of lax argument where the empiric fit is concerned. There is a tendency to argue, because certain assumptions lead to a Poisson distribution, that if a variate has this distribution, then the assumptions must be true. There are two main weaknesses in this argument. For one thing we have seen that this distribution may result from several different starting points, and the fine differences are lost in the process. For another, the scientist is obliged to be concerned with power: Is the sample large enough that departures from the Poisson distribution would be readily detected? There may be other distributions that fit the data as well, or better.

## COMPUTATION

The Poisson variate has a single parameter, $\lambda$, which is equal to both its mean and its variance. We shall denote such a distribution by $\mathcal{P}(\lambda)$. The probability of $x$ successes is

$$P(x, \lambda) = \frac{e^{-\lambda} \lambda^x}{x!}$$

It is most easily calculated recursively, that is, by first finding $e^{-\lambda}$ and then computing each subsequent term from the present one by the relationship

$$P(x + 1) = \frac{\lambda}{x + 1} P(x)$$

Details are given elsewhere (7). Tables of the terms of various Poisson distributions are available, e.g., Pearson and Hartley (102); but of course they are for selected values of $\lambda$ only. It is convenient to be able to compute one's own values as needed; and with most modern desk calculators, it is little trouble to do so.

Note also that the cumulative terms can be found (exactly) by using chi-square tables with the chi-square value equal to $2\lambda$ and the degrees of freedom equal to (2 plus twice the number of terms to be summed). Pearson and Hartley is a good source.

**Example 14.1.** Suppose that we wish to sum the terms of a Poisson with $\lambda = 3$. We enter the table with $\chi^2 = 2\lambda = 6$ and read off the significance levels as follows

| Degrees of freedom | $x$ | $F(x)$ |
| --- | --- | --- |
| 2 | 0 | 0.049,79 |
| 4 | 1 | 0.199,15 |
| 6 | 2 | 0.423,19 |
| 8 | 3 | 0.647,23 |
| 10 | 4 | 0.815,26 |

etc.

## ESTIMATION

### Point estimation

Ordinarily, estimation is very simple. Both the method-of-moments and the maximum likelihood estimate of the only parameter is simply the sample mean:

$$\hat{\lambda} = \frac{\sum\limits_{i=1}^{n} x_i}{n}$$

where the $x_i$ are the numbers of successes on the various trials and $n$ is the total number of trials. (In many cases, $n$ may be 1.)

**Example 14.2.** Consider the data of Lewitus and Neumann (103) on the number of patients with myocardial infarction admitted to Israeli hospitals on 1,642 particular days (Table 14.1). The mean is found by multiplying the number of days on which there were no such patients by "0," those with one admission by "1" and so on, as shown in the third column. The sum of such products divided by 1,642 is the mean number of admissions of this type per day. Then a priori, the variance of the estimate is

$$\text{var}\left[\frac{\Sigma X}{n}\right] = \frac{1}{n^2}\text{var}(\Sigma X) = \frac{1}{n^2}(n\lambda) = \frac{\lambda}{n} \tag{14.1}$$

and this is true whatever the distribution may be. For the Poisson distribution (14.1) proves to be exactly equal to the lower bound; so the estimator is absolutely efficient.

What is the appropriate estimate of the variance? We can of course use the usual method-of-moments estimator, but there are reasons for not doing so. In the first place it involves unnecessary effort: given that the variate follows a Poisson distribution we know the variance equals the mean, and since our estimator of the mean is absolutely efficient, it cannot be improved on. In the second place it turns out that the method-of-moments estimator of the variance has a larger variance. Of course, the former argument falls through if we are uncertain as to whether the variate follows the Poisson distribution. Indeed, we may compare the method-of-moments estimates of the mean and of the variance as a test of conformity to the distribution (see below).

### Interval estimation

The variance of the Poisson variate, as we have seen, is most efficiently estimated by the sample mean. Moreover as the mean increases, we use the Gaussian approximation with increasing assurance.

**Example 14.3.** The number of radioactive counts on a mass of material in one minute is 900. Find approximate 95% confidence limits on the mean.

Here the observed value itself, 900, is the best estimate of the mean, and

**Table 14.1. Distribution of number of heart attacks admitted per day to Israeli hospitals**

| Number of attacks (x) | Number of days (n) | xn | $x^2n$ |
|---|---|---|---|
| 0 | 1,246 | 0 | 0 |
| 1 | 336 | 336 | 336 |
| 2 | 54 | 108 | 216 |
| 3 | 6 | 18 | 54 |
| 4 or more | — | — | — |
| Total | 1,642 | 462 | 606 |

$$m = \frac{462}{1642} = 0.281,364\ldots$$

$$s^2 = \frac{606 - 462m}{1642 - 1} = 0.290,072\ldots$$

$$\frac{s^2}{m} = 1.030,951,985\ldots$$

Source: Data of Lewitus and Neumann (103).

its square root, 30, the estimate of the standard deviation. Then we treat the measurement as coming from a Gaussian distribution: $N(900, 900)$

$$P(900 - 1.96 \times 30 \leq \lambda \leq 900 + 1.96 \times 30) = 0.95$$

whence the confidence limits are approximately 841.2 and 958.8. Note that since the variance equals the mean, the observed value is in fact

$$\frac{900 - 841.2}{\sqrt{841.2}} = 2.0273$$

standard deviations above the lower limit if the latter is taken as the exact mean; and

$$\frac{958.8 - 900}{\sqrt{958.8}} = 1.8989$$

below the upper. These values are not quite so accurate as we would like,* and we are led to a second approximation. We want a quantity $p$ such that

$$\left| \frac{p - 900}{\sqrt{p}} \right| = 1.96$$

*For a Gaussian variate, the probability more than 1.8989 standard deviations above the mean is $0.0287\ldots$, whereas that more than 2.0273 standard deviations below the mean is $0.0213\ldots$. Thus the combined confidence level would be the complement of their sum or $0.9489\ldots$. The asymmetry is largely self-correcting with two-tailed confidence limits.

Squaring both sides and cross-multiplying, we get the quadratic equation

$$810,000 - 1,800p + p^2 = 3.841,6p$$

Whence

$$p = 843.089 \quad \text{and} \quad 960.752$$

which (as can be verified directly) satisfies the equation and are hence the new approximate upper and lower limits. For comparison we may quote the "exact" confidence limits (in the sense used in Chapter 13), which are $p_L = 842.153,8$ and $p_U = 960.772,9$; that is

$$\sum_{0}^{899} \frac{e^{-p_L}(p_L)^i}{i!} = 0.975$$

$$\sum_{0}^{900} \frac{e^{-p_U}(p_U)^i}{i!} = 0.025$$

These quantities are vastly more tedious to calculate.

### The problem of background noise

It is shown elsewhere (7) that the sum ("convolution") of two Poisson variates is itself a Poisson variate with parameter equal to the sum of the parameters. Thus if a stable radioactive source gives, on average, 10 emissions per minute, the actual number will follow a Poisson distribution $\mathcal{P}(10)$. Likewise for one with a mean rate of 15 per minute. So long as we maintain independence, any combination will itself remain Poisson. Thus their combined counts in one minute will follow $\mathcal{P}(10 + 15) = \mathcal{P}(25)$. The total number of emissions in *three minutes counting of the first mass and four minutes of the second will follow* $\mathcal{P}(90)$; and so on. We recall (7) also that if two variates $X$ and $Y$ are uncorrelated, the variance of their sum *or difference* is the sum of the variances.

Commonly in studies involving radionuclides it is not possible even by the most stringent methods—shielding, coincidence counting, etc.—to eliminate all extraneous sources of activity. This irreducible minimum is called the "background." The problems of analysis encountered are of three general types.

First, the background may be deterministic. For instance, it may represent "sixty-cycle" interference from an alternating current. Then the observed count (over a suitable interval) will not follow a Poisson distribution whereas $(X - 60)$ will. Note that if this variate is $\mathcal{P}(\lambda)$ then

$$E(X) = \lambda + 60$$
$$E(X - 60) = \lambda$$
$$\text{Var}(X) = \lambda + \text{var}(60) = \lambda$$
$$\text{Var}(X - 60) = \lambda$$

Obviously

$$\frac{\text{var}(X)}{E(X)} < 1$$

so that, in principle at least, one could readily not only identify that non-Poisson noise is being introduced, but even by computing

$$E(X) - \text{var}(X)$$

one could estimate the value for the background noise.

Second, it may be that the noise is of Poisson type, but that it is well known for the conditions of the experiment, with a mean of (say) 5 per minute. Then from each observation lasting one minute, 5 is subtracted. The result is what we may call a "shifted" Poisson distribution, which has the correct mean, $\lambda$, but a variance of $(\lambda + 5)$. There is a finite probability of a negative result since the observed number of counts may be less than five. If this probability is not very small, it is doubtful that the method should be used.

The third, and most perilous, method is to count both the preparation and a comparable empty container, for a specified period, and take the difference between the two readings, which we will suppose have means of $\lambda$ and $\mu$ respectively. The resulting distribution is somewhat complicated. The only simple exact statement that I can make about it is that all the odd cumulants are $(\lambda - \mu)$ and all the even ones are $(\lambda + \mu)$. Thus the mean and variance are

$$E(X) = \lambda - \mu$$

$$\text{var}(X) = \lambda + \mu \tag{14.2}$$

The third cumulant is $(\lambda - \mu)$; so the index of skewness (see Chapter 1) is

$$\frac{\lambda - \mu}{(\lambda + \mu)^{3/2}} \tag{14.3}$$

which clearly goes to 0 as $(\lambda + \mu)$ becomes large. Likewise for kurtosis

$$\frac{\lambda + \mu}{(\lambda + \mu)^2} \tag{14.4}$$

and so for higher cumulants.

The saving grace for the last two distributions is that if $\lambda$ is large, all the higher cumulants become negligibly small and we may treat the difference as approximately a Gaussian variate with known mean and variance. Note that this is true, regardless of whether $\mu$ is small or not.

Where $\lambda$ is not large we are obliged to be more precise. Of course, the cumulants given in (14.2) to (14.4) are still correct, but the actual distribution is somewhat complicated to work out. It is certainly not Poisson, and the Gaussian approximation is not accurate. Consider the typical term $x$. It is, of course, possible that $x$ should be negative, but neither count can be negative. The difference will be $x$ if, for example, the specimen gives the count $x$ and the background 0, or the former $(x + 1)$ and the latter 1 and so forth. Each pair of outcomes being independent, their joint probability may be found by multiplication (Table 14.2). The probability of $x$ is then the sum of the probabilities of all possible pairs of values differing by $x$. The expression in the bottom of Table 14.2 is not readily evaluated. However, it is easy to see that beyond any particular value of $j$ the ratios of successive terms will be less than

$$\frac{\lambda\mu}{(j + 1)(x + j + 1)} < \frac{\lambda\mu}{j^2}$$

so that using the well-known geometric series, we are sure that the sum of the $(j + 1)^{\text{st}}$ term and higher cannot exceed

$$\frac{1}{1 - \dfrac{\lambda\mu}{j^2}}$$

times the $j^{\text{th}}$ term and certainly cannot be less than 0. Hence we may always find within acceptable limits of accuracy the probability of $x$.

**Example 14.4.** In Table 14.3 is shown the distribution where $\lambda = 5$ and $\mu = 3$ for all terms with probabilities exceeding $0.5 \times 10^{-6}$. This treatment,

**Table 14.2. The probability that the difference between two Poisson processes, with means $\lambda$ and $\mu$, assume the value $x$**

| First Poisson process $\mathscr{P}(\lambda)$ | | Second Poisson process $\mathscr{P}(\mu)$ | | |
|---|---|---|---|---|
| Count | Probability | Count | Probability | Joint probability |
| $x$ | $\dfrac{e^{-\lambda}\lambda^x}{x!}$ | 0 | $e^{-\mu}$ | $\dfrac{e^{-(\lambda+\mu)}\lambda^x}{x!}$ |
| $x + 1$ | $\dfrac{e^{-\lambda}\lambda^{x+1}}{(x + 1)!}$ | 1 | $\mu e^{-\mu}$ | $\dfrac{e^{-(\lambda+\mu)}\lambda^{x+1}\mu}{(x + 1)!}$ |
| $x + 2$ | $\dfrac{e^{-\lambda}\lambda^{x+2}}{(x + 2)!}$ | 2 | $\dfrac{\mu^2 e^{-\mu}}{2!}$ | $\dfrac{e^{-(\lambda+\mu)}\lambda^{x+2}\mu^2}{(x + 2)!2!}$ |

etc.

The marginal probability of $x$ is thus

$$P(x) = e^{-(\lambda+\mu)}\lambda^x \sum_{i=0}^{\infty} \frac{(\lambda\mu)^i}{(x + i)!i!}$$

**Table 14.3. The probability distribution of the difference, $X$, between two Poisson variates with parameters 5 and 3 respectively**

| $x$ | $P(x)$ | $x$ | $P(x)$ |
|---|---|---|---|
| $-12$ | 0.000,001 | 1 | 0.136,270 |
| $-11$ | 0.000,005 | 2 | 0.143,130 |
| $-10$ | 0.000,020 | 3 | 0.131,697 |
| $-9$ | 0.000,074 | 4 | 0.106,853 |
| $-8$ | 0.000,257 | 5 | 0.077,025 |
| $-7$ | 0.000,810 | 6 | 0.049,714 |
| $-6$ | 0.002,319 | 7 | 0.028,947 |
| $-5$ | 0.005,989 | 8 | 0.015,314 |
| $-4$ | 0.013,848 | 9 | 0.007,409 |
| $-3$ | 0.028,447 | 10 | 0.003,297 |
| $-2$ | 0.051,527 | 11 | 0.001,357 |
| $-1$ | 0.081,762 | 12 | 0.000,519 |
| 0 | 0.113,132 | 13 | 0.000,185 |
|  |  | 14 | 0.000,062 |
|  |  | 15 | 0.000,020 |
|  |  | 16 | 0.000,006 |
|  |  | 17 | 0.000,002 |

although as exact as need be (especially since the cumulants are known and do not have to be computed from the distribution) is tedious; no doubt the experimentalist will wish to avoid it by making $\lambda$ large, by counting for a sufficiently long time if need be. But, for technical reasons, it is not always possible to do so and it is then comforting to have an exact analysis to fall back on.

The results have been presented rather fully since they give some important insights into this distribution. It is somewhat positively skewed, there being 15 terms, attaining my arbitrary minimum, to the right of the mean (which is 2), compared with 14 terms to the left. But the skewness is conspicuously less than for a simple Poisson with a mean of 2. Moreover, the general configuration is more nearly Gaussian. We shall show later how well the Gaussian approximation behaves. The mean is, of course, $5 - 3 = 2$ and the variance $5 + 3 = 8$, which provide useful checks on the accuracy of the kind of figures given in Table 14.3.

The maximum likelihood estimates are straightforward. If the combined count is denoted by $y$ and the background count by $x$ the estimates of $\lambda$ and $\mu$ are $y$ and $x$ respectively, and these are also the estimated variances. The mean and the variance of the difference are estimated by $(y - x)$ and $(y + x)$ respectively. Exact confidence limits (in the sense of Chapter 12) are tedious to compute. In most real situations the number of counts will be, or can be made, large enough to justify the Gaussian approximation (see below).

## APPROXIMATION TO THE CUMULATIVE DISTRIBUTION FUNCTION

We have seen that the probability of any particular value, $x$ or less, of a Poisson variate can be found exactly by the chi-square distribution. We also know that as the parameter becomes large we may use the Gaussian distribution as an approximation. The arguments from Chapter 12 suggest that the Gaussian approximation might be improved by a continuity correction.

**Example 14.5.** In Figure 14.1 are shown three cumulative distribution functions for Poisson variates with means 5, 10, and 25 respectively. The first is chosen because (as we noted in Chapter 12) the common wisdom says that if the normal approximation is to be used for the binomial the mean should be at least 5 and the Poisson is an extreme case of the binomial. The Gaussian approximation with the same mean and variance is shown by the lower interrupted line in each case, and that with Yates's correction by the upper interrupted line. It will be recalled that the object is to have the ideal approximation pass through the point of each step. On the whole the fit is rather better with the Yates correction, especially in the middle range of the distribution; but the discrepancies are such that one should not use the Gaussian approximation too glibly, even when the parameter is as large as 25.

However, in the "background" problem of the third type (see above) the behavior for small mean values is more nearly exact, notably for small means. In Figure 14.2 are shown results for three cases: the distribution in the previous example where the means are 5 and 3 respectively; that of the difference between two Poisson processes with the same parameter (which is naturally symmetrical about zero); and the case where the combined count is 50 and the background 5.


## HYPOTHESIS TESTING

### Test of conformity to the Poisson model

Two tests are in common use: the usual goodness of fit test for categorical data using the chi-square approximation, and a comparison between the sample mean and variance.

**Example 14.6.** Consider further the data for Example 14.2. Using the maximum likelihood estimator for $\lambda$, i.e., $0.281,364,19\ldots$, we get the expected values shown in Table 14.4.

There are four nonempty categories, hence four degrees of freedom from which we subtract 1 (because the totals are equal) and 1 for the parameter estimated. Finally, we note that the expected number in each class is satisfactorily large. The chi-square statistic with two degrees of freedom is acceptably small, the significance level (as for all $\chi^2$ variates with two degrees of freedom, which are exponential) is

$$e^{-1.222/2} = 0.54$$

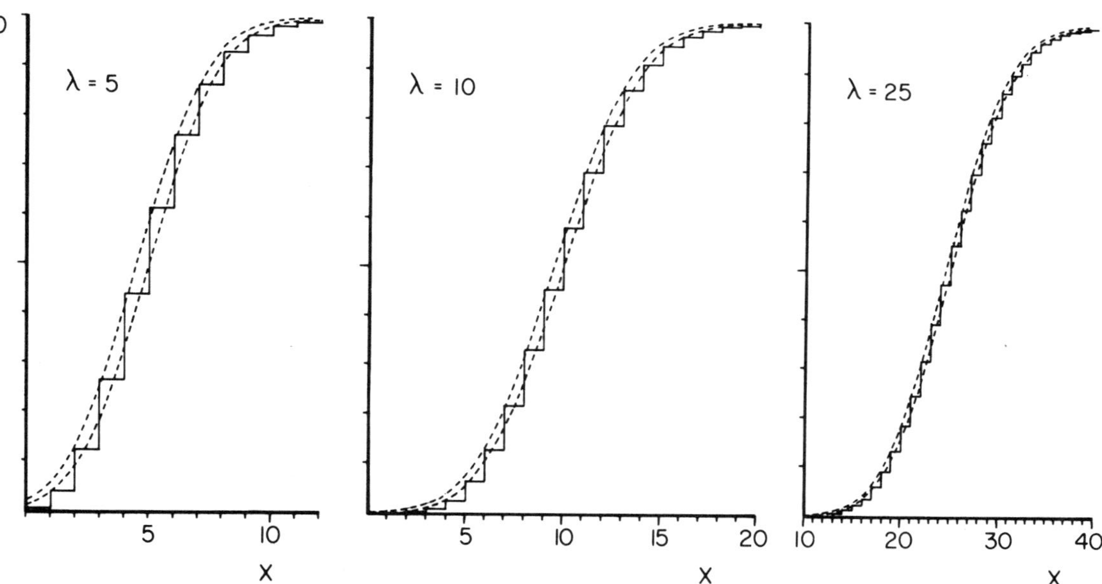


**Figure 14.1.** Distribution functions for three Poisson variates with the parameters shown. To each have been fitted Gaussian approximations with the same mean and variance (interrupted lines). In each case, the curve to the *left* (including Yates's correction for continuity) is a closer fit than the simple Gaussian curve to the *right*.

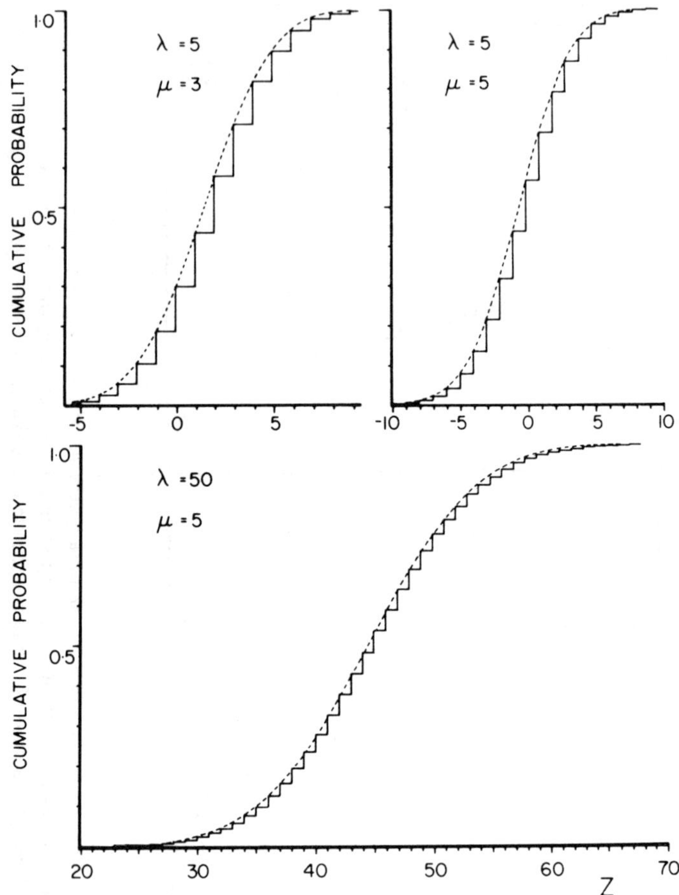



**Figure 14.2.** Three cumulative distribution functions for the difference between two Poisson variates with the parameters shown. In each case a cumulative Gaussian curve with the same variance and a mean increased by ½ a unit (Yates's correction) is fitted.

**Example 14.7.** The second test is said to be more powerful, especially where the tabulation of observed values is short tailed, as in the Lewitus-Neumann data of Example 14.2. All that is required is to compute the sample mean and the sum of squares from the mean. Under the null hypothesis (i.e., that it fits a Poisson model) the ratio of the sample sum of squares to the sample mean follows a chi-square distribution with 1,641 degrees of freedom. A chi-square distribution with so many degrees of freedom may be regarded as Gaussian with mean 1,641 and variance twice as great. In the present problem we have

$$m = 0.281,364$$

$$\Sigma(x - m)^2 = 476.009,744,2$$

**Table 14.4. Goodness of fit of the data of Lewitus and Neumann to the Poisson distribution**

| Number of attacks ($x$) | Number of days | | $\dfrac{O_i^2}{E_i}$ |
| --- | --- | --- | --- |
| | Observed ($O_i$) | Expected ($E_i$) | |
| 0 | 1,246 | 1,239.305,1 | 1,252.731,1 |
| 1 | 336 | 348.696,1 | 323.766,2 |
| 2 | 54 | 49.055,3 | 59.443,1 |
| 3 or more | 6 | 4.943,5 | 7.282,2 |
| Total | 1,642 | 1,642.000,0 | 1,643.222,6 |
| | | | 1,642 |
| | | | $\chi^2 = 1.222,6$ |

The ratio is $1,691.792,208$. Under the null hypothesis

$$\frac{1,691.792,208 - 1641}{\sqrt{2 \times 1641}} \sim N(0, 1)$$

The left hand side is $0.886,6$ which is acceptably close to zero ($p = 0.375,3$). We thus have no reason to reject the hypothesis that the distribution is Poisson.

**Tests of hypotheses about the parameter**

Granted that $X$ is known to follow a Poisson distribution, how do we test whether its parameter is equal to some specified quantity?

**Example 14.8.** Test the hypothesis that the Lewitus-Neumann data come from a Poisson distribution with parameter 0.25.

If the hypothesis is true, then a sample of 1,642 subjects from this distribution should follow a Poisson distribution with mean and variance $(0.25 \times 1,642) = 410.5$. Such a large size allows us to use the Gaussian approximation with confidence. The observed number of patients is 462. Thus under the null hypothesis

$$z = \frac{462 - 410.5}{\sqrt{410.5}}$$

The right-hand side $2.541,8\ldots$ corresponds to $p = 0.011$, two tailed. We might equally well have used the sample mean $0.281,364$, recognizing that its variance is $0.281,364/1,642$. However, since the aptness of the Gaussian approximation depends on the parameter for the sum, we examine the sum.

Testing a hypothesis where the parameter is small involves exact computations.

**Example 14.9.** Suppose the National Safety Council were to set a goal of not more than three fatal traffic accidents in Baltimore next Thanksgiving day. There are five such accidents. Has the goal been attained? At a trivial level the answer is no, since five is greater than three. But we may reasonably agree that the council's primary aim is to decrease risks and since accidents are (by definition) random we cannot infer certainly that risks have not decreased because the outcome is five, any more than we could infer the converse if the outcome were one. A test of a statistical hypothesis is the issue. Moreover, it should be one tailed, for we may suppose the council would be dissatisfied only if the mean were greater than three, not less.

May we plausibly use the Poisson distribution? The justifications would be

1. The population that drives in Baltimore is very large.

2. The number who will be involved in accidents will be relatively small.

3. The risks for each person will not be equal, but they will not be grossly discrepant and all will be small.

4. Accidents are random.

5. While the number driving is not constant, it probably varies little on Thanksgiving Day itself.

6. The probability of anybody being in two separate accidents on that day is negligible.

The cumulative distribution function for a Poisson with parameter three is shown in Table 14.5. As usual with discrete distributions, we cannot find rejection regions with exactly prescribed sizes. The rejection region for the test of size 0.05 would be 7 accidents or more, and of size 0.01 would be 9 or more. We would not reject the null hypothesis at either level.

### Comparison of the parameters of two Poisson variates

Suppose that $X \sim \mathcal{P}(\lambda)$ and $Y \sim \mathcal{P}(\mu)$. We have a set of $n$ values from the first distribution yielding a sample mean $m_X$; and $r$ from the second giving a sample mean $m_Y$. The problem is to test the null hypothesis that $\lambda$ and $\mu$ are equal.

**Table 14.5. The Poisson distribution for $\lambda = 3$**

| $x$ | $F(x)$ |
|---|---|
| 0 | 0.049,8 |
| 1 | 0.199,1 |
| 2 | 0.423,2 |
| 3 | 0.647,2 |
| 4 | 0.815,3 |
| 5 | 0.916,1 |
| 6 | 0.966,5 |
| 7 | 0.988,1 |
| 8 | 0.996,2 |
| 9 | 0.998,9 |

We may consider three cases (Fig. 14.2).

1. The sample sizes and parameters are all large. Here we have recourse to the Gaussian approximation, using the following facts

$$\frac{M_X - M_Y}{\sqrt{\dfrac{M_X}{n} + \dfrac{M_Y}{r}}} \sim N[(\lambda - \mu),\, 1]$$

Under the null hypothesis this reduces to

$$\frac{M_X - M_Y}{\sqrt{\lambda\left(\dfrac{1}{n} + \dfrac{1}{r}\right)}} \sim N(0,\, 1)$$

Of course, even if the null hypothesis is true we do not know $\lambda$; but if $n$ and $r$ are large, we have, at least approximately,

$$\frac{M_X - M_Y}{\sqrt{\dfrac{rM_X + nM_Y}{nr}}} \sim N(0,\, 1)$$

2. $\lambda$ and $\mu$ are large but $n$ and $r$ are small. It was shown in Chapter 4 that if $\lambda$ and $\mu$ are (say) 15 or greater the variances of $\sqrt{X}$ and $\sqrt{Y}$ are very near $\frac{1}{4}$. Thus

$$\mathrm{Var}(\sqrt{X} - \sqrt{Y}) = 0.5$$

Moreover, while $E(\sqrt{X})$ is not $\lambda$, if $\lambda$ and $\mu$ are equal

$$E(\sqrt{X}) = E(\sqrt{Y})$$

Also the square-root transformation makes the Poisson variate, if anything, more nearly Gaussian than the natural value. Thus under the null hypothesis the test statistic, $D$, the difference between the means of the square roots

$$D = M_{\sqrt{X}} - M_{\sqrt{Y}}$$

has mean zero and variance

$$\frac{0.25}{n} + \frac{0.25}{r}$$

and

$$\frac{D}{0.5\sqrt{\dfrac{1}{n} + \dfrac{1}{r}}} \sim N(0,\, 1)$$

Note that the variance here is not a function of either parameter and would be unchanged if they differ, provided they are large. (It is arguable that this would be a preferable test to that given under [1] above even where $n$ and $r$ are large.)

3. If $n$ and $r$ are small and $\lambda$ and $\mu$ small we might wish to use a more complicated transformation on $X$ and $Y$ such as that of Freeman and Tukey (q.v.). If precision is imperative we may use the exact distribution for the difference of two Poisson variates, which, however, depends on the value of the parameters.

### Hypotheses involving three or more sets of data

The issue here is whether the distributions do or do not have the same parameter. By an extension of the argument above, we may take the appropriate transform of all the data, which stabilizes the variance and then proceed to do a regular analysis of variance. However, it seems reasonable to make a slight modification, capitalizing on the fact that the variance of the transformed values is known and need not be estimated from the data. This implies that the denominator of the $F$ test is based on the equivalent of an infinite number of degrees of freedom. Such an $F$ test is equivalent to a $\chi^2$ statistic with degrees of freedom equal to those in the numerator. Also the tests for any set of orthogonal contrasts (q.v.) will be truly independent.

## INVERSE POISSON SAMPLING

The investigator often finds it convenient with Poisson processes not to count the number of events in a standard time, but to time how long it takes to reach (say) $n$ counts. The time is obviously a continuous variable. It follows the gamma distribution (7) which has the probability density function

$$f(t) = \frac{a^n}{(n-1)!} e^{-at} t^{n-1} \tag{14.5}$$

Once the $n^{\text{th}}$ event occurs, the process stops and there is, as it were, no random interval following it. I offer this an an intuitive explanation for the obvious similarity of equation (14.5) to the probability of exactly $n$ hits in a prescribed (nonrandon) interval $t$. In fact if we substitute $\lambda$ for $at$ and $x$ for $n$, the density may be written

$$\frac{ae^{-\lambda}\lambda^{x-1}}{(x-1)!}$$

which, apart from the factor $a$, is precisely the probability of exactly $(x-1)$ hits within the interval.

The problem of maximum likelihood estimation where $n = 1$ is discussed in Chapter 3. Since $n$ is known we have only one parameter to estimate, $a$, and the MLE proves to be

$$\hat{a} = \frac{n}{t}$$

or, more generally,

$$\hat{a} = \frac{\Sigma n}{\Sigma t}$$

The estimated variance of the estimate is

$$\frac{\Sigma n}{(\Sigma t)^2}$$

## APPENDIX

### Proof that the limiting distribution of a Lexis variate with small probabilities is Poisson

The probability-generating function (PGF) of a Bernoulli trial with a probability of success of $p$ and $1 - p = q$ is $P_X(t)$, where

$$P_X(t) = (q + pt) = 1 - p + pt$$

$$= 1 + p(t - 1)$$

The Lexis variate (7) is the convolution of such trials where not all the $p$ are equal, but vary nonrandomly. We shall write the corresponding variate as $X*$. Then, since the PGF of a convolution is the product of the PGF's and the logarithm of the PGF the sum of their logarithms, we have

$$\log[P_{X*}(t)] = \sum_i \log[1 + p_i(t - 1)]$$

which we may expand in a standard logarithmic series, which, being absolutely convergent may be summed term by term

$$\log P_{X*}(t) = \Sigma p_i(t - 1) - \sum \frac{p_i^2(t - 1)^2}{2} + \cdots \tag{1}$$

Now if $n$ is indefinitely large and the $p$'s indefinitely small in such a way that the first term in (1) is finite and fixed, and all the other terms are negligible, in the limit for large $n$ we have

$$\log P_{X*}(t) = \Sigma p_i(t - 1)$$

$$P_{X*}(t) = e^{\Sigma p_i(t-1)}$$

which, being the PGF of a Poisson variate (7) indicates that the convolution is a Poisson variate with parameter $\Sigma p_i$. If the $p_i$ are identical, the sum is $np$ as in the more familiar problem of the binomial variable. If any one or more $p$ remain large we could not ignore higher terms in (1) and the proof would fail. In fact, the value of $p$ must be inversely related to $n$; and we may best achieve this by setting fixed ratios among the $p$'s.

(A more rigorous and compact proof is possible using cumulant-generating functions. However, this method takes us beyond material covered in reference 7.)

## PROBLEMS

**14.1.** Find 95% confidence limits on the parameter for the data in Table 14.1.

**14.2.** In an analysis of the relationship of breast cancer to the use of diethylstilbesterol (DES) Brain et al. (104) cite the following mortalities from breast cancer: 4 of 668 controls, and 12 of 693 exposed to DES.

    **a.** Compute point estimates and 95% confidence intervals on the rates.

    **b.** Test whether the rates differ at a size of 0.05.

**14.3.** Crandall and associates (105) report the following frequencies of new chromosomal abnormalities in 1,822 amniocenteses done because of maternal age.

| Mothers' ages | Number of mothers | Chromosomal Abnormalities Found | | |
|---|---|---|---|---|
| | | Trisomy 21 | Others | Total |
| 34–36 | 876 | 6 | 4 | 10 |
| 37–38 | 501 | 3 | 3 | 6 |
| 39–40 | 309 | 4 | 3 | 7 |
| 41–42 | 136 | 2 | 4 | 6 |

Analyze these results for evidence of effect from maternal age.

# PART IV
# METHODS NOT INVOLVING EXPLICIT DISTRIBUTIONS

# 15

# NONPARAMETRIC METHODS

The elaboration of small-sample statistical methods for Gaussian variables has helped to take statistics from its public place of origin ("the study of affairs of state") to many areas, notably, compact, but intensive, studies in laboratories. There is a strange attitude among the seminumerate that elaborate statistical analyses of small samples are out of place; they forget that small-sample methods were devised expressly for small samples and that where massive studies are concerned, the impact of random fluctuation may be minor and demand little sophistication in analysis.

The main problems about small samples are that they may furnish little information about not only the parameters but also the mathematical form of the distribution of the variate; and unless data on the latter are available from other sources to provide assurance about the distribution, it may be hazardous to use methods that are sensitive to departures from particular assumptions. The latter topic will be discussed in Chapter 17. Thus when doing pioneer work in some unfamiliar field, the investigator may wish to use methods that make little demand on empirical knowledge and involve only very weak (and hence plausible) assumptions. Such methods are commonly called nonparametric or distribution-free. Neither term is entirely beyond criticism, but I have no better one to suggest.

Nonparametric methods span the field of statistics in general: characterization and description of data; estimation of descriptors; tests of hypotheses; and the making of decisions. It is an active field of research and has quite an extensive bibliography. In the space I can afford to allot the subject, I can only deal with the field superficially, aiming to illustrate, rather than to expound, its possible scope. The more orthodox methods are admirably discussed in no-nonsense practical detail by Siegel (106), who also furnishes excellent tables. Tukey (107) has many interesting new ideas that have so far not gained wide use.

Inevitably the question of meaning will arise in at least some instances where nonparametric methods are used: not so much the meaning of the analysis, as the meaning of the data. Ranking methods, for example, depend on the general notion that the evidence, if not actually quantitative, can at

329

least be put in some order that has meaning. There are plenty of instances in which it is not at all clear what it means to say that the order has meaning. This is fundamentally a philosophical problem. It is often difficult to imagine any criteria by which the assertion could be tested and disproved if false. In such instances we may question whether the assertion has value. To quote an old maxim, what is asserted without evidence may be denied without evidence. Where hospital beds are required in an emergency, does it mean anything to list the present patients in ascending order of their need to be in hospital? Can the geneticist prescribe which human genetic stock is most valuable and should be encouraged to reproduce? Based as it is on an arbitrary weighing of arbitrary component tests, does the ranking of the results in intelligence tests have any meaning? (And they, whatever their meaning, are usually supposed to qualify as parametric tests.) Are the assessments by physicians so culturally and professionally conditioned that multiple opinions have no corroborative value? The fundamental difficulty in such cases (not encountered in authentic discrimination: see Chapter 11) is that the imprecision is not in the test, or even perhaps in the measurement, but in the conceptualization: there exists no known referent against which the measurement or ranking may be validated. It is not my intent to degenerate into second-rate philosophy. But I hope that the reader will not abdicate the responsibility of inquiring into the meaning of his problem in favor of some index, perhaps hastily constructed with little insight, or some facile judgment of the imperceptive, or worse.

## DESCRIPTORS

In discussing the multinomial process (Chapter 13), we recognized one type in which the random variable is purely enumerative: the number of trials falling into each class in a sample of specified size. Each individual future trial has specified probabilities of falling into each class. One can easily imagine generalizations of this process where the results of individual trials are not independent or where the probabilities associated with each outcome vary from trial to trial.

Here, however, we shall confine ourselves to orthodox distributions, i.e., those for which every outcome has both probability (or probability density) and a value of its own. However, we shall assume nothing about the form of the distribution, except that it does have some form. We may attempt to characterize the distribution by some of the following terms.

**Nonparametric measures of location**

**Definition 15.1.** *If the probability that a random variable $X$ assumes the value $x$ or less may be written $k/100$, then $x$ is the $k^{th}$ percentile of $X$.*

**Definition 15.2.** *The median of a random variable $X$ is its 50th percen-*

*tile.* The outcome of any future single realization of the process is as likely to be above, as below, the median.

Note that the median or any other percentile is not necessarily a unique value. In a binomial distribution of order 3 with parameter $1/2$, any number between 1 and 2 would qualify as the median; even for a continuous distribution the median may be ambiguous. For instance a Mendelian trait may in many instances segregate into two equally probable types: the results in both types of progeny may show considerable variation, and yet be so widely separated that the median may be thought of as any point in the "daylight" between the two component curves of the distribution.

**Definition 15.3.** *The lower quartile is the 25th percentile.*

**Definition 15.4.** *The upper quartile is the 75th percentile.*

**Definition 15.5.** *The mode is that value in the sample space associated with the maximum probability or probability density.* For example the (empirical) mode for the Lewitus-Neumann study (Example 14.2) is 0; the mode of the Gaussian distribution is the mean and the median as well. There may be multiple modes; and in the continuous rectangular distribution all possible values of $X$ have the same probability density and hence are modes. Nevertheless, a great many distributions have a single well-defined mode.

**Note.** Nonstatisticians commonly talk of the mode as if it is a peak in a curve representing the distribution. But properly it is not the peak itself but the point on the baseline ("the value of $X$") at which it occurs. This is not a mere pedantic difference. For instance, what we have defined as the maximum likelihood estimator of a parameter is akin to a mode (although the likelihood function is, of course, not a distribution). We know from theory, for large samples at least, that the distribution of the "mode" (i.e., *the location, on the abscissa,* of the maximum) is approximately Gaussian; but that does not tell us anything about *the shape* of the likelihood curve in the neighborhood of the maximum.

*The Sample Mean* will of course be understood in the usual fashion as the sum of the values divided by the size of the sample. It has broad utility but we must not attach any mystical significance to it. For the Gaussian distribution, the sample mean has tremendous importance; for the rectangular distribution (as we saw in Problem 3.1) it does not. For some distributions—such as the $t$ distribution with one degree of freedom—a sample value of the mean (computed in accordance with the above prescription) usually exists; but it is a meaningless statistic since the parent distribution does not have any mean (so there does not seem to be the slightest point in trying to estimate it). To show why the mean for this distribution does not exist would require some rather formidable calculus. Instead, I shall give another, simpler, example that will be readily understood.

**Example 15.1.** We do two experiments. In the first, we toss a fair coin twice. The combined number of heads is a binomial of order two with three possible values, 0, 1, and 2, with the probabilities shown in the left marginal distribution of Table 15.1. The second experiment is precisely similar but it is done with another fair coin and independently of the first. The distribution of the number of heads is in the other marginal of Table 15.1. Because of independence, we may find the joint probability of any pair of outcomes by simple multiplication of the marginal probabilities. There are nine outcomes. The problem is: What is the distribution of the random variable $X$, defined as the ratio of the second score to the first? The answer is shown in the lower half of Table 15.1. Clearly we can assign no mean to $X$ because where the denominator is zero, two of the outcomes have a ratio of infinity* and one is indeterminate. Likewise we cannot identify the percentiles of $X$ which would depend on where we assign the indeterminate result. By a fortunate accident, the fiftieth percentile is 1 no matter where we put the indeterminate value, so the median is 1. But this problem would not be so readily resolved in the general case. Likewise a clear mode exists, unity; but if we were to pool the last two categories, there would be a close rival.

Of course the foregoing example is artificial, but not for that reason

**Table 15.1. A discrete distribution with no mean**

|  |  | Second experiment |  |  |  |
| --- | --- | --- | --- | --- | --- |
| First experiment |  | Score | 0 | 1 | 2 |
|  |  | Probability | 1/4 | 1/2 | 1/4 |
| Score | Probability |  |  |  |  |
| 0 | 1/4 | $x$ | — | $\infty$ | $\infty$ |
|  |  | probability | 1/16 | 2/16 | 1/16 |
| 1 | 1/2 | $x$ | 0 | 1 | 2 |
|  |  | probability | 2/16 | 1/4 | 2/16 |
| 2 | 1/4 | $x$ | 0 | 0.5 | 1 |
|  |  | probability | 1/16 | 2/16 | 1/16 |

|  Distribution of $X$ |  |  |
| --- | --- | --- |
| $x$ | $f(x)$ | $x \cdot f(x)$ |
| 0 | 3/16 | 0 |
| 0.5 | 2/16 | 1/16 |
| 1 | 5/16 | 5/16 |
| 2 | 2/16 | 4/16 |
| $\infty$ | 3/16 | $\infty$ |
| Indeterminate | 1/16 | Indeterminate |
| Total | 1 | Indeterminate (mean) |

*A distribution such as this in which there is a probability mass at infinity is said to be dishonest. Distributions such as the Poisson, Gaussian, lognormal, and gamma, which have at least one infinite limit, are not dishonest because at infinity the density is zero; but this saving property must never be taken for granted and clearly does not apply in the ratio of two discrete variates of which the denominator may assume the value zero. The atomic bomb explodes because it reflects a distribution that for practical purposes may be viewed as dishonest. I am sometimes afraid that the National Debt will do likewise.

unreal. The cardinal problem is that both the denominator and the numerator may assume the value zero. And there are plenty of practical problems where this may occur: for instance where an index depends on the ratio of two radioactive counts. With a reasonably sized mean for the denominator it is unlikely that it will take the value zero. In practice, I am sure most investigators would discard the results where it did. Logically, this means that they are using a truncated Poisson distribution and this fact should be reflected in *all* their analyses and not merely in the case where the outcome is zero. The adjustment may in fact be negligibly small; but it is not always possible to get reassuringly high mean counts in an experiment.

## Nonparametric measures of dispersion

The familiar parametric measure of dispersion is the standard deviation. In the very strictest sense, this quantity is a parameter only for those distributions (notably the Gaussian and the Poisson) in which this quantity is equal to an actual constant included in the form of the distribution. But even among other types of data one distinguishes between those that assume numerical values (however bizarre their behavior) for which one can at least formally compute the standard deviation, and those, such as rankings pure and simple, in which one could merely compute the standard deviation of the ranks (which, having as such nothing to do with the data and being determined exclusively by the sample size, would be meaningless). It will be obvious that if the mean of the distribution does not exist (as in Example 15.1) neither does the standard deviation; and there are common distributions—such as the $t$ distribution with two degrees of freedom—that have a mean but no variance. Still, if the variance does exist and is finite, so will the standard deviation and the standard deviation of the sample mean; and these properties will be put to good use in Chapter 17. Moreover, the standard deviation does have certain distribution-free properties, such as the relationship to the sample range and the Chebychev inequality (see Chapter 17). But these do not compare in precision with the properties ascribable to the standard deviation computed for Gaussian data.

Perhaps for this reason, many investigators like to substitute statements about direct probabilities. The probability that the sample value will lie between the 5th and 95th percentiles is (by definition) 90%. Two common measures of dispersion are in use.

**Definition 15.6.** *The sample range is the set of two quantities taken jointly, the smallest and the largest values encountered: thus (46–113).* Defined in these terms the sample range depends on the size of the sample. The term *range* is usually taken also to include all the admissible values between these extremes. (I say "admissible" because to say that the range of family sizes is 2–11 does not include 2.5.) Obviously, the range leads to no definite positive conclusions if the domain is infinite, as it is for the Gaussian. There are instances (notably the rectangular distribution) where the range is finite, and it is then useful to know it. But in biology and medicine the sample range is given too much emphasis at the expense of more informative measures.

**Example 15.2.** The hematocrit (i.e., that proportion of whole blood consisting of red cells) cannot be less than 0% or greater than 100%. Doubtless both extremes are lethal. It would be reasonable to discuss the range of hematocrit readings compatible with survival.

**Definition 15.7.** *The interquartile range is the interval between the 25th and 75th percentiles.* It circumvents the problems of the range unless the distribution is grossly dishonest.


## ESTIMATION

Where the objects being studied have no associated measurement, merely rank (see below), it would be meaningless to pursue estimation; for the properties of the ranks depend only on the size of the sample and not on the data. But where we have measurements that we have temporarily coded as ranks for purposes of analysis, then we can make at least approximate statements about certain key characteristics which can be converted back to the original measurements.

### The median

By definition, the median of a continuous variate is the value such that any sample value is as likely to exceed it as to be exceeded by it. Thus we can always code a distribution with a median as a binomial, values below the median being failures, those above, successes. There is some gain of simplicity and some loss of information. I offer the following heuristic argument for the nonparametric estimator of the median. If the sample size is even, $2n$, the most probable split is that half of the values will be above and half below the median; so the most likely estimate of the median is that it lies between the two most central values in the sample. (Note that because we have ordered the values there is a monotonic relationship between them and their ranks. Hence, the MLE of the rank [which is trivial] is also the MLE of the median.) If the sample size is odd $(2n + 1)$, the two most likely outcomes are that there should be $n$ below and $(n + 1)$ above, or conversely. (They are equally likely.) This argues that it is immaterial whether we regard the median as just above, or just below, the central value. Hence, the central value is as useful as any for the sample median. However, these arguments are scarcely rigorous, and theoretical analysis suggests the problem is not well defined: the rational estimate cannot be distribution-free. Confidence limits on the median may be set by the old familiar binomial argument.

**Example 15.3.** Suppose we have a sample of size 10 and wish to set symmetrical confidence limits about the median. We have the following probability statements about the possible pairs of values which embrace the median, where the numbers in parentheses are ranks

$$P(5) = 252/1{,}024 = 0.246{,}1$$
$$P(4\text{-}6) = 672/1{,}024 = 0.656{,}2$$
$$P(3\text{-}7) = 912/1{,}024 = 0.890{,}6$$
$$P(2\text{-}8) = 1{,}002/1{,}024 = 0.978{,}5$$
$$P(1\text{-}9) = 1{,}022/1{,}024 = 0.998{,}0$$
$$P(0\text{-}10) = 1{,}024/1{,}024 = 1.000{,}0$$

To set "95% (actually 97.85%) confidence limits" on the median we say that it is equal to, or greater than, the third ranking value and equal to, or less than, the eighth. Of course we are not bound to make the limits symmetrical; we might wish to put all the uncertainty into one tail.

### Quartiles

These quantities may be estimated by the same kind of rough and ready argument. If the sample size is $(4n + 3)$ where $n$ is an integer then the probabilities of the split ($n$ below the lower quartile) and the split ($n + 1$) below the lower quartile are equally likely and we might take the average as the estimate of the lower quartile. Contrariwise, for the upper quartile we would take the mean of the $(3n + 2)^{\text{th}}$ and the $(3n + 3)^{\text{th}}$ values. If the sample size is not of this type we should take as the quartile whatever rank, $n$ maximizes the binomial term. We may summarize these somewhat complicated instructions in Table 15.2. Confidence limits on the quartiles may be obtained by exactly the same kind of arguments as for the median.

In practical terms I would suggest the following consolations. When the sample size is large, adjacent terms are likely to have the same or similar values; so refinements of estimates for the quartiles are of little importance. When the samples are small, it is usually no great problem to lay out the appropriate terms in the corresponding binomial distribution. The probabil-

**Table 15.2. Rules for estimating the median and quartiles of a distribution nonparametrically**

| Sample size | Lower quartile | Median | Upper quartile |
|---|---|---|---|
| $4n$ | $\dfrac{R(n) + R(n + 1)}{2}$ | $\dfrac{R(2n) + R(2n + 1)}{2}$ | $\dfrac{R(3n) + R(3n + 1)}{2}$ |
| $4n + 1$ | $\dfrac{R(n) + R(n + 1)}{2}$ | $R(2n + 1)$ | $\dfrac{R(3n + 1) + R(3n + 2)}{2}$ |
| $4n + 2$ | $\dfrac{R(n) + R(n + 1)}{2}$ | $\dfrac{R(2n + 1) + R(2n + 2)}{2}$ | $\dfrac{R(3n + 2) + R(3n + 3)}{2}$ |
| $4n + 3$ | $R(n + 1)$ | $R(2n + 1)$ | $R(3n + 3)$ |

$n$ is an integer.
$R(i)$ denotes that value that is ranked $i^{\text{th}}$ from the bottom

ities are available in standard tables such as those of the National Bureau of Standards (108).

## The mode

In explaining the difference between a sample space and a parameter space (Chapter 3) we pointed out that the one may be continuous and the other discrete, even where we are dealing with the same distribution. The parameter of a binomial distribution of order 10 may assume any of an infinite set of values between 0 and 1; the outcome can assume 11 possible values only. But one combination to which the statistician has paid surprisingly little attention is that in which both the sample estimate and the parameter space are necessarily discrete.

**Example 15.4.** Consider the question of how many intestinal polyps there may be in a patient. We will suppose for simplicity that there is no ambiguity as to what is, and what is not, a polyp. The information consists of observations on X-rays, on colonoscopy, and on the surgical specimen after colectomy. We shall make two plausible assumptions.

1. Whatever the answer to the question is, it is some perfectly definite whole number.

2. Observers and methods of observation are fallible, and not all counts agree perfectly.

The second of these assumptions seems to imply that we need some kind of statistical criterion to get a suitable estimator of the true number with known properties.

Now in general the sample mean for all the counts on the patient will not do, because most often it will not be an integer, as (we are agreed) the answer most certainly must be. If we knew what the distribution of counting errors is, we could use parametric estimation, e.g., MLE, using a discrete parameter space. In fact, what most scientists do is to make multiple (at least two) counts until some concordant counts result. Then they use the mode of the sample counts. It would be interesting to explore what the statistical properties of this estimator are.*

Careful statisticians act likewise when they do a set of calculations more than once (as a check on the accuracy of the arithmetic). As far as I can discern, all of statistical theory is based on the tenuous assumption that statisticians never make mistakes in their calculations or if they do, that there is in principle no problem in recognizing mistakes or resolving discrepancies. The possibility that the same erroneous answer may be reached twice, and all doubts *spuriously* allayed, is not considered. Even assuming independence, there is a possibility of this repeated mistake happening; but I am quite sure in my own experience that I may repeatedly make the same mis-

---

*We may note a few highly abstract papers dealing with the topic, e.g., that of Venter (109).

take—for instance misreading my own handwriting. This possibility of error is all the more disturbing when we realize that in practice statisticians have *selective* checks on mistakes—that they can identify particular results as absurd, e.g., negative estimates of variance, a standard deviation which is greater than the distance of the farthest point from the mean, a correlation coefficient greater than unity. Such results will lead them to go back over their calculations until the error (which must exist) is discovered. But they may never do this for incorrect results that are not impossible or unlikely. Thus there is an uneven editing of mistakes. So far as I know, the impact which this may have on statistical inference has been little explored. However, there is one point on which I am quite sure: that when a reputable statistician, after doing the calculations several times, finds discrepant results, he does not average, but uses the modal value.

## RANK

We have mentioned rank casually, without comments or qualifications. It is worthwhile to discuss it a little more fully. When used without comment, it usually means ascending order of quantities (the smallest being the first rank) and descending orders of quality (Château Haut-Brion, Miss America, and J. S. Bach being of the first rank in their classes). For any particular degree of precision, the more elements there are in the sample, the more likely it is that two of them will have identical values. Since this result violates the assumptions of many nonparametric tests, there are usually three courses open to us.

First and best, closer scrutiny may show that the results are not identical but appear so because of rounding. Baseball averages, ordinarily expressed to three places may be expressed to four to break a tie; ages that are equal when expressed as "age at last birthday" may be distinguishable when expressed in days or even (in the case of twins) minutes.

Second, otherwise identical results may be distinguished by a randomizing device, such as the tossing of a coin. This solution, although probabilistically sound, offends the biostatistical conscience for two reasons. It is resolving a problem of analysis by an ancillary result that has no scientific meaning whatsoever; and it is uncomfortably idiosyncratic—another statistician reanalyzing the data and incorporating another toss of a coin may reach a different result. A computer would arrive at no result at all, except with a preprogrammed "randomizing" device (which is in fact deterministic). It might be argued that this second solution is no more capricious than the first, that it is a matter of chance which twin was born first. However, which twin was indeed born first is part of the data, not of the analysis; and it is a bold scientist who would dismiss the birth order of twins as having no scientific significance.

Third, some nonparametric procedures have special provision made for dealing with ties.

*Notation:* Whereas we ordinarily use $x$, $y$, $z$, and sometimes $w$, to denote natural measurements, we shall use $u$ and $v$ to denote ranks.

### Nonparametric correlation

Consider pairs of random variables, $(U, V)$ which can be ordered each in its own (marginal) sample space. In a sample of size $n$ we arrange the $u$'s in ascending order. Then if there is a perfect correspondence between the variables, the corresponding $v$'s should be in ascending order also; or, if there is a perfect inverse relationship, they should be in descending order. If the relationship is imperfect, we will find some "inversions," that is, $v_5$ instead of being after $v_4$ may be before it (one inversion) or perhaps before $v_3$ (two inversions); or the order may be $v_5$, $v_4$, *and* $v_3$ (three inversions). Then we may use the number of inversions for our statistic. It may range from 0 (perfect correspondence) to $n(n - 1)/2$ (perfect negative, or inverse, correspondence). It is customary, however, to give the statistic something of the form of the usual parametric coefficient of correlation (see Chapter 9); and this is achieved by the form

$$T = 1 - \frac{4Q}{n(n - 1)}$$

where $Q$ is the number of inversions (a random variable) and T is referred to as Kendall's nonparametric rank coefficient. When $Q = 0$ (perfect correspondence) $\tau = 1$ and when $Q = n(n - 1)/2$ the coefficient becomes $-1$. So while the precise implication of $\tau$ may not be altogether clear (for example, to those accustomed to observing and evaluating regular correlation coefficients), the sample value may be used as a kind of estimate of a measure of correspondence. Large-sample theory shows that in the limit under the null hypothesis of independence, $Q$ follows a Gaussian distribution with mean

$$E(Q) = \frac{n(n - 1)}{4}$$

and variance

$$\text{var}(Q) = \frac{2(2n + 5)}{9n(n - 1)}$$

Thus we may use Gaussian theory to interpret it. As we shall see, the conformity to the Gaussian distribution is helped somewhat by using a correction for continuity. No provision is made for large-sample theory if the null hypothesis is rejected (e.g., if we wish to set confidence limits on $\tau$ and wish to predict what sample values are likely to result in future samples).

## NONPARAMETRIC TESTS OF HYPOTHESES

In Chapter 2, I carefully stressed that testing hypotheses by the Neyman-Pearson principle was directed to surmised values of parameter(s) for a well-defined distribution, against a clear alternative or class of alternatives of the same mathematical form but with different parameters. One was not called upon to compare the hypothesis that the distribution of a random variable is exponential, against the alternate hypothesis that it is Gaussian. The nearest approach to that is the goodness-of-fit test such as those used in Chapters 4 and 14 to test conformity to the Gaussian and Poisson distributions. In neither case can Neyman-Pearson theory be used, as the reader will clearly see by attempting to do so.

Nonparametric tests of hypotheses differ in at least three major ways from the parametric type.

1. Since true parameters (as distinct from moments) do not exist, the test cannot be about parameters. A fortiori it cannot be about values of parameters.

2. Since no sufficiently precise statement about the alternative hypothesis may be made, we cannot assess power, and hence cannot select a most powerful rejection region. The usual practice is then to take as the rejection region those outcomes with the lowest probabilities (usually in the tails of the distribution under the null hypothesis). But this solution does not, in general, have a rational defense. Also, we must insist more firmly than ever, that the rejection region must be selected independently of the data, which means in practice before the data have been gathered. If the choice was not made beforehand, the rejection region should be equally divided between the tails. The issue of power is commonly dealt with by the following artificial criterion: If we were to apply the test to data from a Gaussian distribution, how would its power compare with that of the admissible (parametric) test? But, of course, if we knew that the population were Gaussian, we would use the admissible parametric test.

3. The greater part of nonparametric tests deal with relationships between things: whether the data agree with such-and-such a distribution; whether two variates have the same distribution; whether two medians are the same; whether two variates are independent (which is not the same as determining whether the correlation coefficient is zero), and so forth. Thus, apart from the freedom which these tests have from assumptions about the underlying distribution, they are more adventurous (though less well characterized) than parametric theory.

### Goodness-of-fit tests

This class of tests has probably been sufficiently described. I am in some doubt how far they may be viewed as parametric. Most of the chi-square methods in Chapters 4, 13, and 14 are parametric in the sense that they are about conformity to distributions of specified form; also the test criterion, the chi-square statistic, is a specified distribution. But in some senses, at least, they are nonparametric. The fact that one is using a chi-square test has nothing to do with the form of the parent distribution, or even whether the expected values are determined from any distribution whatsoever. Again, the test is based on multinomial data and the multinomial distribution has parameters.* But the appeal to these parameters is often only implicit; and in contingency tables they need not be invoked at all: they are "estimated" (under the null hypothesis) from the marginal data. Finally, if we reject the hypothesis, there is nothing in the logical structure to help us decide whether the mathematical form of the model is incorrect or the estimates of the parameters are faulty. It is perhaps best to see them as having an intermediate position between parametric and nonparametric tests. Note, also, that such tests are of goodness of fit, not badness of fit; and hence, power is more important than size; so the lack of a good criterion of power is even more telling than it would be in the practice of Neyman-Pearson theory.

### The binomial test

I have earlier deplored the tendency of some investigators to force continuously distributed data gratuitously into categories. However, categorizing may be useful as a rough and ready test in many situations, and it makes for a safe procedure if the measurements are specious or even totally unavailable. For instance, suppose we inquire whether a singing voice is made more beautiful by taking cough syrup before a performance. There may be no explicit criterion of beauty, and if somebody were to propose one it might be widely rejected. Nevertheless, judges may have little difficulty in agreeing whether the voice is better or worse, even though the performance is not quantifiable.

Even for a clearly metrical character, we may wish to use the test because we are unsure about certain assumptions.

**Example 15.5.** We wish to test the hypothesis that first-born and second-born members of a pair of twins are equally heavy, against the hypothesis that the first is the heavier. It would be easy to imagine a line of inquiry which would prompt the test, for instance that the heavier twin would be more likely to engage early. Now, because of the peculiarity of twinning, we might be reluctant to invoke common experience. Twins are commonly premature; they may have a higher rate of birth defects; they may be more subject to defective nutrition; some twins are identical and therefore more alike genetically than others; there may be different degrees of similarity between twins of the same sex and of different sexes; some forms of twinning run in families, others do not (which may have implications for fitness); and so on. Thus not only would we wish to avoid assumptions about the form of

---

*It is arguable that in this context they are purely ad hoc parameters (q.v.).

the distribution of weight; we would not even wish to suppose that the distributions are the same for all classes of twins.

So we argue as follows. If weight is irrelevant, both directly and indirectly, to birth order of twins, it is equally probable that the lighter, and that the heavier, twin be born first. The soundness of this statement does not depend in any way on the variance, or any other moments for that matter. Thus if we term the former event a failure and the latter a success, the number of successes, $X$, in $n$ independent pairs of twins should, under this null hypothesis, follow the binomial distribution $B(n, \frac{1}{2})$.

To be sure, there may be some difficulty in stating a coherent alternative hypothesis; it may be that, in some circumstances, the heavier twin is usually born first, and in some the lighter. A purist may object that a global answer to so promiscuous a question is uninterpretable. But that way lies madness. For "homogeneity" is a tentative scientific construct, not an inalienable metaphysical truth. If we always insisted on homogeneous groups before we attempted any general proposition we would make little scholarly progress. Indeed, it is not impossible that all things are incurably particular and, in the words of W. C. Fields, "No generalization is worth a damn, including this one."

If we admit the broad plausibility of the arguments, the test is a straightforward application of the binomial model as discussed in Chapter 12. Note that it is quite immaterial whether the means, variances, or any other moments differ among twins. We have automatically converted them all to the same form by binomializing them. Tied results (which for so finely graded a characteristic as weight must be rare) are evidently uninformative and are discarded.

As to the power of the test, if the differences are in fact Gaussian and have the same variances, the theory of large samples shows that the power of the binomial test is in the ratio $2:\pi$, or about 64%. To use the binomial test on 100 sets of paired data known to be Gaussian and homoscedastic, instead of the $t$ test which would be appropriate, is equivalent to not collecting the last 36 sets of data. When the test is seen in this light, the investigator may be persuaded that measurements are to be preferred to categories wherever the more powerful $t$ method is appropriate.

### The Mann-Whitney test

To the assumptions made for the binomial test (that the results can be compared for magnitude, and that over the appropriate sample space the results are independent) let us add the assumption, under the null hypothesis, that two samples of results have the same distribution. This assumption naturally implies that all moments that exist are identical. Needless to say, the assumptions are a good deal stronger than those for the previous procedure, and the test must be used with more circumspection. But it is not required that the form of the distribution be known, which is fortunate if the data are few and experience scanty. The method would be entirely appropriate for a randomized trial. This test, also known as the signed-rank test, has been explored in various disguises by diverse investigators.

Suppose that we have $n_1$ observations in the one sample and $n_2$ in the other. They are pooled and ranked in ascending order. The test statistic, $U$, is found by counting the number of values of one sample that are greater than each particular value of the other, and adding these numbers. It will be convenient to display the two kinds of results in order but on different lines (or columns). For instance, suppose we are comparing some characteristic in four men ($M$) and three women ($W$) and the ordering proves to be:

$$m_1 \qquad m_2 \qquad m_3 \quad m_4$$
$$\quad w_1 \qquad w_2 \qquad\qquad w_3$$

To the right of

$m_1$ are 3 women
$m_2$ are 2 women
$m_3$ is  1 woman
$m_4$ is  1 woman
Total  7

We might alternatively have computed the complementary statistic (three men to the right of $w_1$, two to the right of $w_2$ and none to the right of $w_3$, total five). Note that the sum of these two complementary statistics $7 + 5 = 12$ is always the product of the sample sizes ($3 \times 4$). Naturally, then, the $U$ statistic cannot be less than zero or greater than $n_1 n_2$. The distribution of outcomes is symmetrical, so that it is immaterial whether $U$ or $(n_1 n_2 - U)$ is computed. The hypothesis of identity implies that the ordering is random and is least plausible when $U$ is either very low or very high. Because of symmetry, it is clear that under the null hypothesis

$$E(U) = \frac{n_1 n_2}{2}$$

and it can be shown that

$$\mathrm{var}(U) = \frac{n_1 n_2 (n_1 + n_2 + 1)}{12}$$

The probability distribution, or even the critical values, for the various sizes of sample, call for a separate table for each combination of $n_1$ and $n_2$; and the number of entries increases rapidly. Detailed tables up to $n_1 = n_2 = 10$, and critical values at common levels of significance up to $n_1 = n_2 = 20$ are given by Owen (110). Further tabulation up to samples of size 40 is given by Milton (111). There seems no point in reproducing them here. Many statisticians are content to use the Gaussian approximation if the sizes of both samples are ten or greater. The asymptotic power of the test for Gaussian data is $3/\pi$ or about 95% of that for the $t$ test.

**Example 15.6.** In a study of the effect of repeated injections of low doses of heparin on atherogenesis in rabbits (112), one of the outcome variables was the percentage of aorta involved by atheroma. There was prior reason to believe that such low doses might enhance atherogenesis. No transformation was forthcoming that would reduce the data to a form suitable for analysis by Gaussian theory. The results obtained are analyzed by the Mann-Whitney test in Table 15.3.

### Correlated samples: the Wilcoxon test

In the typical before-and-after experiment such as we have discussed above under the binomial test, the Mann-Whitney test is not applicable since the two samples are correlated. On the other hand we may wish to squeeze out of the $d$ paired differences more information than what their signs are. The Wilcoxon test (which logically is closely related to the Mann-Whitney test) is commonly used. Any pairs for which the difference is zero are discarded. The paired differences are then ranked in ascending order of absolute magnitude (that is, regardless of sign). The sum $T_p$ of the ranks of the positive values is used as the test statistic; or equivalently, the sum of the ranks of the negative differences, $T_n$. The sum of $T_p$ and $T_n$ is $d(d + 1)/2$, and from symmetry, under the null hypothesis the mean is

$$E(T_p) = E(T_n) = \frac{d(d + 1)}{4}$$

and the variance proves to be

$$\mathrm{Var}(T_p) = \mathrm{var}(T_n) = \frac{d(d + 1)(2d + 1)}{24}$$

Critical values are tabled by Owen (110) and Siegel (106). Gaussian approximations are used for large sizes of samples. As for the Mann-Whitney test, the asymptotic power for Gaussian data is $3/\pi$ or 95% of that for the $t$ test.

### Kendall's rank correlation

The statistic previously discussed may be interpreted under the null hypothesis of no correlation as follows. After ordering on the $x$'s there are $n!$ possible permutations of the corresponding $y$ values of the $y_1, y_2, \ldots, y_n$, which *ex hypothesi* are all equally likely. The number of inversions of order in the $y$'s may be anything from 0 to $n(n - 1)/2$; and the number of ways in which $q$ inversions may occur can (with some tedium) be built up recursively. Thus, for a sample of size 2 we have $2! = 2$ arrangements, one having no inversion and the other having one. The distribution may thus be written

| $q$ | $n_2(q)$ | $\dfrac{n_2(q)}{2!}$ |
|---|---|---|
| 0 | 1 | 0.5 |
| 1 | 1 | 0.5 |

The distribution for $n = 3$ is found by computing the cumulative sums for the terms for $n = 2$ but nowhere including more than the last three $(n)$ terms. The sum is divided by $3! = 6$ to give the corresponding probability. Thus

| $q$ | $n_2(q)$ | $n_3(q)$ | $\dfrac{n_3(q)}{3!}$ |
|---|---|---|---|
| 0 | 1 | 1 | 1/6 |
| 1 | 1 | $1 + 1 = 2$ | 2/6 |
| 2 | 0 | $1 + 1 + 0 = 2$ | 2/6 |
| 3 | 0 | $1 + 0 + 0 = 1$ | 1/6 |
| 4 | 0 | $0 + 0 + 0 = 0$ | 0/6 |

For $n = 4$ we have $n! = 24$ arrangements and the numbers are built up from the cumulative numbers from $n = 3$, but nowhere exceeding more than the last four $(n)$ terms.

| $q$ | $n_3(q)$ | $n_4(q)$ | $\dfrac{n_4(q)}{4!}$ |
|---|---|---|---|
| 0 | 1 | 1 | 1/24 |
| 1 | 2 | $1 + 2 = 3$ | 3/24 |
| 2 | 2 | $1 + 2 + 2 = 5$ | 5/24 |
| 3 | 1 | $1 + 2 + 2 + 1 = 6$ | 6/24 |
| 4 | 0 | $2 + 2 + 1 + 0 = 5$ | 5/24 |
| 5 | 0 | $2 + 1 + 0 + 0 = 3$ | 3/24 |
| 6 | 0 | $1 + 0 + 0 + 0 = 1$ | 1/24 |
| 7 | 0 | $0 + 0 + 0 + 0 = 0$ | 0/24 |

And so on. In Table 15.4 are shown probabilities for individual values of $q$ and cumulative probabilities for all quantities exceeding $0.5 \times 10^{-6}$ for $n = 2$ to $n = 20$. Clearly the distribution is symmetrical; so there is no need to tabulate more than one term beyond the fiftieth percentile.

The tedium (and the larger numbers) involved in computing exact values by this recursive method leads to the use of the normal approximations, which is reasonably good for samples of moderate size (Table 15.5). As so often the case in these approximations, the agreement is best in the center of the distribution; and while the *absolute* errors in the tails are small, their relative sizes are large. A Yates correction for continuity improves the overall accuracy.

**Table 15.3. The impact of repeated low doses of heparin on the extent of atheroma formation in rabbits**

| Control animal | Treated animal | Number of treated greater than controls | Number of controls greater than treated |
|---|---|---|---|
| 0.0 | | 10 | |
| 0.0 | | 10 | |
| 0.4 | | 10 | |
| 7.4 | | 10 | |
| 10.0 | | 10 | |
| 10.6 | | 10 | |
| 14.6 | | 10 | |
| 15.4 | | 10 | |
| | 48.0 | | 2 |
| | 52.0 | | 2 |
| | 55.3 | | 2 |
| | 56.3 | | 2 |
| 56.9 | | 6 | |
| 57.1 | | 6 | |
| | 57.3 | | |
| | 58.5 | | |
| | 61.4 | | |
| | 63.2 | | |
| | 75.3 | | |
| | 83.0 | | |
| Totals | | 92 | 8 |

SOURCE: Data of Rowsell et al. (112).
NOTE: The figures shown are percentages of the total area of the arterial wall that are involved.

$u = 92$ or $8$

Significance level:

One-tailed $p < 0.001$

Gaussian approximation:

$$E(U) = \frac{10 \times 10}{2} = 50$$

$$\mathrm{var}(U) = \frac{10 \times 10 \times 21}{12} = 175$$

$$z = \frac{92 - 50}{\sqrt{175}} = 3.1749$$

Significance level:

One-tailed $p < 0.001$

## Table 15.4. Kendall's Tau statistic

| $n$ | $q$ | $f(q)$ | $F(q)$ | $n$ | $q$ | $f(q)$ | $F(q)$ | $n$ | $q$ | $f(q)$ | $F(q)$ |
|---|---|---|---|---|---|---|---|---|---|---|---|
| 3 | 0 | 0.166,667 | 0.166,667 | | 2 | 0.000,096 | 0.000,121 | | 15 | 0.009,819 | 0.030,085+ |
| | 1 | 0.333,333 | 0.500,000 | | 3 | 0.000,306 | 0.000,427 | | 16 | 0.013,196 | 0.043,281 |
| | | | | | 4 | 0.000,785+ | 0.001,213 | | 17 | 0.017,205− | 0.060,485 |
| 4 | 0 | 0.041,667 | 0.041,667 | | 5 | 0.001,731 | 0.002,943 | | 18 | 0.021,801 | 0.082,287 |
| | 1 | 0.125,000 | 0.166,667 | | 6 | 0.003,390 | 0.006,333 | | 19 | 0.026,887 | 0.109,173 |
| | 2 | 0.208,333 | 0.375,000 | | 7 | 0.006,038 | 0.012,370 | | 20 | 0.032,310 | 0.141,483 |
| | 3 | 0.250,000 | 0.625,000 | | 8 | 0.009,937 | 0.022,308 | | 21 | 0.037,872 | 0.179,356 |
| | | | | | 9 | 0.015,281 | 0.037,588 | | 22 | 0.043,335+ | 0.222,691 |
| 5 | 0 | 0.008,333 | 0.008,333 | | 10 | 0.022,131 | 0.059,719 | | 23 | 0.048,439 | 0.271,130 |
| | 1 | 0.033,333 | 0.041,667 | | 11 | 0.030,371 | 0.090,090 | | 24 | 0.052,918 | 0.324,048 |
| | 2 | 0.075,000 | 0.116,667 | | 12 | 0.039,669 | 0.129,759 | | 25 | 0.056,527 | 0.380,575+ |
| | 3 | 0.125,000 | 0.241,667 | | 13 | 0.049,485− | 0.179,244 | | 26 | 0.059,060 | 0.439,635− |
| | 4 | 0.166,667 | 0.408,333 | | 14 | 0.059,110 | 0.238,354 | | 27 | 0.060,365+ | 0.500,000 |
| | 5 | 0.183,333 | 0.591,667 | | 15 | 0.067,747 | 0.306,101 | | | | |
| | | | | | 16 | 0.074,606 | 0.380,707 | 12 | 3 | 0.000,001 | 0.000,001 |
| 6 | 0 | 0.001,389 | 0.001,389 | | 17 | 0.079,021 | 0.459,728 | | 4 | 0.000,002 | 0.000,003 |
| | 1 | 0.006,944 | 0.008,333 | | 18 | 0.080,545− | 0.540,272 | | 5 | 0.000,006 | 0.000,008 |
| | 2 | 0.019,444 | 0.027,778 | | | | | | 6 | 0.000,014 | 0.000,022 |
| | 3 | 0.040,278 | 0.068,056 | 10 | 1 | 0.000,002 | 0.000,003 | | 7 | 0.000,032 | 0.000,054 |
| | 4 | 0.068,056 | 0.136,111 | | 2 | 0.000,012 | 0.000,015 | | 8 | 0.000,066 | 0.000,120 |
| | 5 | 0.098,611 | 0.234,722 | | 3 | 0.000,043 | 0.000,058 | | 9 | 0.000,129 | 0.000,249 |
| | 6 | 0.125,000 | 0.359,722 | | 4 | 0.000,121 | 0.000,179 | | 10 | 0.000,238 | 0.000,487 |
| | 7 | 0.140,278 | 0.500,000 | | 5 | 0.000,294 | 0.000,473 | | 11 | 0.000,414 | 0.000,902 |
| | | | | | 6 | 0.000,633 | 0.001,106 | | 12 | 0.000,689 | 0.001,591 |
| 7 | 0 | 0.000,198 | 0.000,198 | | 7 | 0.001,237 | 0.002,343 | | 13 | 0.001,099 | 0.002,690 |
| | 1 | 0.001,190 | 0.001,389 | | 8 | 0.002,231 | 0.004,574 | | 14 | 0.001,698 | 0.004,379 |
| | 2 | 0.003,968 | 0.005,357 | | 9 | 0.003,759 | 0.008,333 | | 15 | 0.002,507 | 0.006,885+ |
| | 3 | 0.009,722 | 0.015,079 | | 10 | 0.005,972 | 0.014,305− | | 16 | 0.003,605− | 0.010,490 |
| | 4 | 0.019,444 | 0.034,524 | | 11 | 0.009,007 | 0.023,311 | | 17 | 0.005,035− | 0.015,525+ |
| | 5 | 0.033,532 | 0.068,056 | | 12 | 0.012,964 | 0.036,275+ | | 18 | 0.006,843 | 0.022,368 |
| | 6 | 0.051,389 | 0.119,444 | | 13 | 0.017,882 | 0.054,157 | | 19 | 0.009,066 | 0.031,435− |
| | 7 | 0.071,230 | 0.190,675 | | 14 | 0.023,714 | 0.077,871 | | 20 | 0.011,724 | 0.043,159 |
| | 8 | 0.090,278 | 0.280,952 | | 15 | 0.030,316 | 0.108,187 | | 21 | 0.014,817 | 0.057,975+ |
| | 9 | 0.105,357 | 0.386,310 | | 16 | 0.037,437 | 0.145,624 | | 22 | 0.018,320 | 0.076,295+ |
| | 10 | 0.113,690 | 0.500,000 | | 17 | 0.044,736 | 0.190,360 | | 23 | 0.022,180 | 0.098,475− |
| | | | | | 18 | 0.051,796 | 0.242,156 | | 24 | 0.026,315− | 0.124,790 |
| 8 | 0 | 0.000,025− | 0.000,025 | | 19 | 0.058,170 | 0.300,327 | | 25 | 0.030,615+ | 0.155,405 |
| | 1 | 0.000,174 | 0.000,198 | | 20 | 0.063,418 | 0.363,745− | | 26 | 0.034,947 | 0.190,352 |
| | 2 | 0.000,670 | 0.000,868 | | 21 | 0.067,156 | 0.430,900 | | 27 | 0.039,160 | 0.229,512 |
| | 3 | 0.001,885− | 0.002,753 | | 22 | 0.069,100 | 0.500,000 | | 28 | 0.043,090 | 0.272,602 |
| | 4 | 0.004,315+ | 0.007,068 | | | | | | 29 | 0.046,578 | 0.319,181 |
| | 5 | 0.008,507 | 0.015,575 | 11 | 2 | 0.000,001 | 0.000,002 | | 30 | 0.049,472 | 0.368,653 |
| | 6 | 0.014,931 | 0.030,506 | | 3 | 0.000,005+ | 0.000,007 | | 31 | 0.051,641 | 0.420,294 |
| | 7 | 0.023,834 | 0.054,340 | | 4 | 0.000,016 | 0.000,023 | | 32 | 0.052,985+ | 0.473,280 |
| | 8 | 0.035,094 | 0.089,435 | | 5 | 0.000,043 | 0.000,066 | | 33 | 0.053,441 | 0.526,720 |
| | 9 | 0.048,115+ | 0.137,550 | | 6 | 0.000,101 | 0.000,167 | | | | |
| | 10 | 0.061,830 | 0.199,380 | | 7 | 0.000,213 | 0.000,380 | 13 | 5 | 0.000,001 | 0.000,001 |
| | 11 | 0.074,826 | 0.274,206 | | 8 | 0.000,416 | 0.000,796 | | 6 | 0.000,002 | 0.000,003 |
| | 12 | 0.085,565+ | 0.359,772 | | 9 | 0.000,758 | 0.001,553 | | 7 | 0.000,004 | 0.000,007 |
| | 13 | 0.092,659 | 0.452,431 | | 10 | 0.001,300 | 0.002,854 | | 8 | 0.000,009 | 0.000,016 |
| | 14 | 0.095,139 | 0.547,569 | | 11 | 0.002,119 | 0.004,973 | | 9 | 0.000,019 | 0.000,035 |
| | | | | | 12 | 0.003,297 | 0.008,270 | | 10 | 0.000,037 | 0.000,073 |
| 9 | 0 | 0.000,003 | 0.000,003 | | 13 | 0.004,922 | 0.013,192 | | 11 | 0.000,069 | 0.000,142 |
| | 1 | 0.000,022 | 0.000,025 | | 14 | 0.007,074 | 0.020,266 | | 12 | 0.000,122 | 0.000,264 |

| n | q | f(q) | F(q) |
|---|---|---|---|
| | 13 | 0.000,207 | 0.000,471 |
| | 14 | 0.000,337 | 0.000,808 |
| | 15 | 0.000,530 | 0.001,338 |
| | 16 | 0.000,807 | 0.002,145− |
| | 17 | 0.001,194 | 0.003,339 |
| | 18 | 0.001,720 | 0.005,059 |
| | 19 | 0.002,416 | 0.007,475+ |
| | 20 | 0.003,316 | 0.010,791 |
| | 21 | 0.004,450+ | 0.015,241 |
| | 22 | 0.005,850− | 0.021,091 |
| | 23 | 0.007,538 | 0.028,628 |
| | 24 | 0.009,530 | 0.038,158 |
| | 25 | 0.011,832 | 0.049,990 |
| | 26 | 0.014,436 | 0.064,426 |
| | 27 | 0.017,318 | 0.081,744 |
| | 28 | 0.020,440 | 0.102,183 |
| | 29 | 0.023,745+ | 0.125,929 |
| | 30 | 0.027,164 | 0.153,092 |
| | 31 | 0.030,610 | 0.183,702 |
| | 32 | 0.033,988 | 0.217,690 |
| | 33 | 0.037,197 | 0.254,887 |
| | 34 | 0.040,133 | 0.295,020 |
| | 35 | 0.042,696 | 0.337,717 |
| | 36 | 0.044,796 | 0.382,512 |
| | 37 | 0.046,354 | 0.428,867 |
| | 38 | 0.047,314 | 0.476,181 |
| | 39 | 0.047,638 | 0.523,819 |
| 14 | 7 | 0.000,000 | 0.000,001 |
| | 8 | 0.000,001 | 0.000,002 |
| | 9 | 0.000,003 | 0.000,004 |
| | 10 | 0.000,005 | 0.000,010 |
| | 11 | 0.000,010 | 0.000,020 |
| | 12 | 0.000,019 | 0.000,039 |
| | 13 | 0.000,034 | 0.000,072 |
| | 14 | 0.000,058 | 0.000,130 |
| | 15 | 0.000,096 | 0.000,226 |
| | 16 | 0.000,153 | 0.000,379 |
| | 17 | 0.000,238 | 0.000,617 |
| | 18 | 0.000,361 | 0.000,979 |
| | 19 | 0.000,534 | 0.001,512 |
| | 20 | 0.000,771 | 0.002,283 |
| | 21 | 0.001,088 | 0.003,371 |
| | 22 | 0.001,505+ | 0.004,877 |
| | 23 | 0.002,042 | 0.006,919 |
| | 24 | 0.002,720 | 0.009,639 |
| | 25 | 0.003,561 | 0.013,200 |
| | 26 | 0.004,583 | 0.017,783 |
| | 27 | 0.005,805+ | 0.023,588 |
| | 28 | 0.007,241 | 0.030,829 |
| | 29 | 0.008,899 | 0.039,728 |
| | 30 | 0.010,782 | 0.050,510 |
| | 31 | 0.012,883 | 0.063,394 |
| | 32 | 0.015,188 | 0.078,582 |

| n | q | f(q) | F(q) |
|---|---|---|---|
| | 33 | 0.017,672 | 0.096,254 |
| | 34 | 0.020,302 | 0.116,556 |
| | 35 | 0.023,034 | 0.139,590 |
| | 36 | 0.025,816 | 0.165,406 |
| | 37 | 0.028,588 | 0.193,994 |
| | 38 | 0.031,287 | 0.225,282 |
| | 39 | 0.033,845− | 0.259,126 |
| | 40 | 0.036,193 | 0.295,320 |
| | 41 | 0.038,267 | 0.333,587 |
| | 42 | 0.040,007 | 0.373,594 |
| | 43 | 0.041,361 | 0.414,955− |
| | 44 | 0.042,287 | 0.457,242 |
| | 45 | 0.042,758 | 0.500,000 |
| 15 | 10 | 0.000,001 | 0.000,001 |
| | 11 | 0.000,001 | 0.000,002 |
| | 12 | 0.000,003 | 0.000,005 |
| | 13 | 0.000,005− | 0.000,010 |
| | 14 | 0.000,009 | 0.000,019 |
| | 15 | 0.000,015+ | 0.000,034 |
| | 16 | 0.000,025+ | 0.000,059 |
| | 17 | 0.000,041 | 0.000,100 |
| | 18 | 0.000,065+ | 0.000,165 |
| | 19 | 0.000,101 | 0.000,266 |
| | 20 | 0.000,152 | 0.000,418 |
| | 21 | 0.000,225− | 0.000,643 |
| | 22 | 0.000,325+ | 0.000,968 |
| | 23 | 0.000,461 | 0.001,429 |
| | 24 | 0.000,642 | 0.002,071 |
| | 25 | 0.000,879 | 0.002,951 |
| | 26 | 0.001,184 | 0.004,135 |
| | 27 | 0.001,570 | 0.005,705 |
| | 28 | 0.002,050+ | 0.007,755 |
| | 29 | 0.002,640 | 0.010,395 |
| | 30 | 0.003,352 | 0.013,748 |
| | 31 | 0.004,201 | 0.017,949 |
| | 32 | 0.005,198 | 0.023,146 |
| | 33 | 0.006,352 | 0.029,498 |
| | 34 | 0.007,670 | 0.037,167 |
| | 35 | 0.009,154 | 0.046,321 |
| | 36 | 0.010,802 | 0.057,124 |
| | 37 | 0.012,608 | 0.069,731 |
| | 38 | 0.014,558 | 0.084,289 |
| | 39 | 0.016,632 | 0.100,921 |
| | 40 | 0.018,808 | 0.119,729 |
| | 41 | 0.021,054 | 0.140,783 |
| | 42 | 0.023,334 | 0.164,117 |
| | 43 | 0.025,608 | 0.189,725 |
| | 44 | 0.027,834 | 0.217,559 |
| | 45 | 0.029,966 | 0.247,525 |
| | 46 | 0.031,958 | 0.279,483 |
| | 47 | 0.033,764 | 0.313,247 |
| | 48 | 0.035,343 | 0.348,591 |
| | 49 | 0.036,657 | 0.385,248 |

| n | q | f(q) | F(q) |
|---|---|---|---|
| | 50 | 0.037,673 | 0.422,921 |
| | 51 | 0.038,365− | 0.461,285 |
| | 52 | 0.038,715− | 0.500,000 |
| 16 | 13 | 0.000,001 | 0.000,001 |
| | 14 | 0.000,001 | 0.000,002 |
| | 15 | 0.000,002 | 0.000,004 |
| | 16 | 0.000,004 | 0.000,008 |
| | 17 | 0.000,006 | 0.000,014 |
| | 18 | 0.000,010 | 0.000,025 |
| | 19 | 0.000,017 | 0.000,041 |
| | 20 | 0.000,026 | 0.000,067 |
| | 21 | 0.000,040 | 0.000,108 |
| | 22 | 0.000,060 | 0.000,168 |
| | 23 | 0.000,089 | 0.000,257 |
| | 24 | 0.000,129 | 0.000,387 |
| | 25 | 0.000,184 | 0.000,571 |
| | 26 | 0.000,258 | 0.000,830 |
| | 27 | 0.000,356 | 0.001,186 |
| | 28 | 0.000,484 | 0.001,670 |
| | 29 | 0.000,649 | 0.002,320 |
| | 30 | 0.000,858 | 0.003,178 |
| | 31 | 0.001,120 | 0.004,297 |
| | 32 | 0.001,443 | 0.005,740 |
| | 33 | 0.001,837 | 0.007,578 |
| | 34 | 0.002,313 | 0.009,890 |
| | 35 | 0.002,878 | 0.012,769 |
| | 36 | 0.003,544 | 0.016,313 |
| | 37 | 0.004,318 | 0.020,631 |
| | 38 | 0.005,208 | 0.025,838 |
| | 39 | 0.006,218 | 0.032,057 |
| | 40 | 0.007,354 | 0.039,410 |
| | 41 | 0.008,615 | 0.048,025 |
| | 42 | 0.009,999 | 0.058,024 |
| | 43 | 0.011,501 | 0.069,525 |
| | 44 | 0.013,113 | 0.082,638 |
| | 45 | 0.014,821 | 0.097,458 |
| | 46 | 0.016,608 | 0.114,067 |
| | 47 | 0.018,456 | 0.132,523 |
| | 48 | 0.020,340 | 0.152,863 |
| | 49 | 0.022,234 | 0.175,098 |
| | 50 | 0.024,110 | 0.199,207 |
| | 51 | 0.025,935 | 0.225,142 |
| | 52 | 0.027,680 | 0.252,822 |
| | 53 | 0.029,311 | 0.282,134 |
| | 54 | 0.030,799 | 0.312,933 |
| | 55 | 0.032,114 | 0.345,048 |
| | 56 | 0.033,230 | 0.378,278 |
| | 57 | 0.034,123 | 0.412,401 |
| | 58 | 0.034,775 | 0.447,176 |
| | 59 | 0.035,172 | 0.482,347 |
| | 60 | 0.035,305 | 0.517,653 |

**Table 15.4.** *(continued)*

| $n$ | $q$ | $f(q)$ | $F(q)$ | $n$ | $q$ | $f(q)$ | $F(q)$ | $n$ | $q$ | $f(q)$ | $F(q)$ |
|---|---|---|---|---|---|---|---|---|---|---|---|
| 17 | 15 | 0.000,000 | 0.000,001 | 18 | 19 | 0.000,000 | 0.000,001 | | 73 | 0.028,794 | 0.411,412 |
| | 16 | 0.000,001 | 0.000,001 | | 20 | 0.000,001 | 0.000,001 | | 74 | 0.029,264 | 0.440,676 |
| | 17 | 0.000,001 | 0.000,002 | | 21 | 0.000,001 | 0.000,002 | | 75 | 0.029,582 | 0.470,258 |
| | 18 | 0.000,001 | 0.000,003 | | 22 | 0.000,001 | 0.000,004 | | 76 | 0.029,742 | 0.500,000 |
| | 19 | 0.000,002 | 0.000,006 | | 23 | 0.000,002 | 0.000,006 | | | | |
| | 20 | 0.000,004 | 0.000,010 | | 24 | 0.000,004 | 0.000,009 | 19 | 23 | 0.000,000 | 0.000,001 |
| | 21 | 0.000,006 | 0.000,016 | | 25 | 0.000,005 | 0.000,015 | | 24 | 0.000,000 | 0.000,001 |
| | 22 | 0.000,010 | 0.000,026 | | 26 | 0.000,008 | 0.000,023 | | 25 | 0.000,001 | 0.000,002 |
| | 23 | 0.000,015 | 0.000,041 | | 27 | 0.000,012 | 0.000,035 | | 26 | 0.000,001 | 0.000,003 |
| | 24 | 0.000,023 | 0.000,064 | | 28 | 0.000,017 | 0.000,052 | | 27 | 0.000,002 | 0.000,005 |
| | 25 | 0.000,034 | 0.000,097 | | 29 | 0.000,025 | 0.000,077 | | 28 | 0.000,003 | 0.000,008 |
| | 26 | 0.000,049 | 0.000,146 | | 30 | 0.000,035 | 0.000,113 | | 29 | 0.000,004 | 0.000,012 |
| | 27 | 0.000,070 | 0.000,216 | | 31 | 0.000,049 | 0.000,162 | | 30 | 0.000,006 | 0.000,018 |
| | 28 | 0.000,098 | 0.000,314 | | 32 | 0.000,068 | 0.000,230 | | 31 | 0.000,009 | 0.000,026 |
| | 29 | 0.000,136 | 0.000,451 | | 33 | 0.000,093 | 0.000,323 | | 32 | 0.000,012 | 0.000,039 |
| | 30 | 0.000,187 | 0.000,638 | | 34 | 0.000,125 | 0.000,449 | | 33 | 0.000,017 | 0.000,056 |
| | 31 | 0.000,253 | 0.000,890 | | 35 | 0.000,167 | 0.000,615 | | 34 | 0.000,024 | 0.000,079 |
| | 32 | 0.000,337 | 0.001,228 | | 36 | 0.000,220 | 0.000,835 | | 35 | 0.000,032 | 0.000,112 |
| | 33 | 0.000,445 | 0.001,673 | | 37 | 0.000,287 | 0.001,122 | | 36 | 0.000,044 | 0.000,155 |
| | 34 | 0.000,581 | 0.002,254 | | 38 | 0.000,371 | 0.001,493 | | 37 | 0.000,059 | 0.000,215 |
| | 35 | 0.000,750 | 0.003,003 | | 39 | 0.000,475 | 0.001,967 | | 38 | 0.000,079 | 0.000,293 |
| | 36 | 0.000,957 | 0.003,961 | | 40 | 0.000,602 | 0.002,570 | | 39 | 0.000,103 | 0.000,397 |
| | 37 | 0.001,210 | 0.005,170 | | 41 | 0.000,757 | 0.003,326 | | 40 | 0.000,135 | 0.000,532 |
| | 38 | 0.001,514 | 0.006,684 | | 42 | 0.000,943 | 0.004,270 | | 41 | 0.000,175 | 0.000,707 |
| | 39 | 0.001,876 | 0.008,560 | | 43 | 0.001,166 | 0.005,436 | | 42 | 0.000,224 | 0.000,931 |
| | 40 | 0.002,303 | 0.010,863 | | 44 | 0.001,429 | 0.006,865 | | 43 | 0.000,286 | 0.001,217 |
| | 41 | 0.002,802 | 0.013,665 | | 45 | 0.001,739 | 0.008,604 | | 44 | 0.000,361 | 0.001,577 |
| | 42 | 0.003,380 | 0.017,044 | | 46 | 0.002,098 | 0.010,702 | | 45 | 0.000,452 | 0.002,029 |
| | 43 | 0.004,041 | 0.021,085 | | 47 | 0.002,513 | 0.013,216 | | 46 | 0.000,561 | 0.002,590 |
| | 44 | 0.004,791 | 0.025,877 | | 48 | 0.002,989 | 0.016,204 | | 47 | 0.000,693 | 0.003,283 |
| | 45 | 0.005,635 | 0.031,511 | | 49 | 0.003,528 | 0.019,732 | | 48 | 0.000,849 | 0.004,132 |
| | 46 | 0.006,573 | 0.038,085 | | 50 | 0.004,136 | 0.023,868 | | 49 | 0.001,033 | 0.005,164 |
| | 47 | 0.007,609 | 0.045,693 | | 51 | 0.004,814 | 0.028,682 | | 50 | 0.001,248 | 0.006,412 |
| | 48 | 0.008,739 | 0.054,432 | | 52 | 0.005,566 | 0.034,248 | | 51 | 0.001,497 | 0.007,909 |
| | 49 | 0.009,962 | 0.064,394 | | 53 | 0.006,393 | 0.040,642 | | 52 | 0.001,786 | 0.009,695 |
| | 50 | 0.011,272 | 0.075,667 | | 54 | 0.007,296 | 0.047,937 | | 53 | 0.002,115 | 0.011,810 |
| | 51 | 0.012,662 | 0.088,329 | | 55 | 0.008,272 | 0.056,209 | | 54 | 0.002,491 | 0.014,301 |
| | 52 | 0.014,121 | 0.102,450 | | 56 | 0.009,319 | 0.065,528 | | 55 | 0.002,914 | 0.017,215 |
| | 53 | 0.015,637 | 0.118,086 | | 57 | 0.010,434 | 0.075,961 | | 56 | 0.003,390 | 0.020,605 |
| | 54 | 0.017,194 | 0.135,280 | | 58 | 0.011,610 | 0.087,571 | | 57 | 0.003,919 | 0.024,525 |
| | 55 | 0.018,777 | 0.154,057 | | 59 | 0.012,841 | 0.100,412 | | 58 | 0.004,505 | 0.029,030 |
| | 56 | 0.020,366 | 0.174,423 | | 60 | 0.014,118 | 0.114,530 | | 59 | 0.005,150 | 0.034,180 |
| | 57 | 0.021,941 | 0.196,364 | | 61 | 0.015,430 | 0.129,960 | | 60 | 0.005,853 | 0.040,033 |
| | 58 | 0.023,479 | 0.219,843 | | 62 | 0.016,765 | 0.146,725 | | 61 | 0.006,615 | 0.046,648 |
| | 59 | 0.024,960 | 0.244,803 | | 63 | 0.018,111 | 0.164,837 | | 62 | 0.007,436 | 0.054,084 |
| | 60 | 0.026,360 | 0.271,164 | | 64 | 0.019,453 | 0.184,290 | | 63 | 0.008,314 | 0.062,398 |
| | 61 | 0.027,658 | 0.298,822 | | 65 | 0.020,777 | 0.205,067 | | 64 | 0.009,247 | 0.071,645 |
| | 62 | 0.028,832 | 0.327,654 | | 66 | 0.022,065 | 0.227,132 | | 65 | 0.010,230 | 0.081,875 |
| | 63 | 0.029,862 | 0.357,516 | | 67 | 0.023,302 | 0.250,434 | | 66 | 0.011,259 | 0.093,134 |
| | 64 | 0.030,731 | 0.388,247 | | 68 | 0.024,472 | 0.274,906 | | 67 | 0.012,328 | 0.105,461 |
| | 65 | 0.031,424 | 0.419,671 | | 69 | 0.025,560 | 0.300,466 | | 68 | 0.013,430 | 0.118,892 |
| | 66 | 0.031,928 | 0.451,598 | | 70 | 0.026,549 | 0.327,014 | | 69 | 0.014,558 | 0.133,449 |
| | 67 | 0.032,234 | 0.483,832 | | 71 | 0.027,426 | 0.354,440 | | 70 | 0.015,702 | 0.149,151 |
| | 68 | 0.032,336 | 0.516,168 | | 72 | 0.028,178 | 0.382,618 | | 71 | 0.016,852 | 0.166,003 |

**Table 15.4.** *(continued)*

| $n$ | $q$ | $f(q)$ | $F(q)$ | $n$ | $q$ | $f(q)$ | $F(q)$ | $n$ | $q$ | $f(q)$ | $F(q)$ |
|---|---|---|---|---|---|---|---|---|---|---|---|
| | 72 | 0.017,999 | 0.184,002 | | 40 | 0.000,027 | 0.000,098 | | 68 | 0.005,738 | 0.042,751 |
| | 73 | 0.019,130 | 0.203,132 | | 41 | 0.000,035 | 0.000,133 | | 69 | 0.006,414 | 0.049,165 |
| | 74 | 0.020,235 | 0.223,367 | | 42 | 0.000,047 | 0.000,179 | | 70 | 0.007,137 | 0.056,302 |
| | 75 | 0.021,302 | 0.244,669 | | 43 | 0.000,061 | 0.000,240 | | 71 | 0.007,905 | 0.064,207 |
| | 76 | 0.022,318 | 0.266,987 | | 44 | 0.000,079 | 0.000,319 | | 72 | 0.008,715 | 0.072,922 |
| | 77 | 0.023,272 | 0.290,259 | | 45 | 0.000,101 | 0.000,420 | | 73 | 0.009,566 | 0.082,488 |
| | 78 | 0.024,153 | 0.314,412 | | 46 | 0.000,129 | 0.000,550 | | 74 | 0.010,453 | 0.092,942 |
| | 79 | 0.024,950 | 0.339,363 | | 47 | 0.000,164 | 0.000,714 | | 75 | 0.011,373 | 0.104,314 |
| | 80 | 0.025,654 | 0.365,016 | | 48 | 0.000,206 | 0.000,920 | | 76 | 0.012,319 | 0.116,633 |
| | 81 | 0.026,254 | 0.391,271 | | 49 | 0.000,258 | 0.001,177 | | 77 | 0.013,287 | 0.129,920 |
| | 82 | 0.026,745 | 0.418,015 | | 50 | 0.000,320 | 0.001,497 | | 78 | 0.014,269 | 0.144,189 |
| | 83 | 0.027,118 | 0.445,134 | | 51 | 0.000,394 | 0.001,891 | | 79 | 0.015,259 | 0.159,448 |
| | 84 | 0.027,370 | 0.472,503 | | 52 | 0.000,483 | 0.002,374 | | 80 | 0.016,249 | 0.175,697 |
| | 85 | 0.027,497 | 0.500,000 | | 53 | 0.000,588 | 0.002,962 | | 81 | 0.017,231 | 0.192,929 |
| | | | | | 54 | 0.000,711 | 0.003,673 | | 82 | 0.018,197 | 0.211,125 |
| 20 | 27 | 0.000,000 | 0.000,001 | | 55 | 0.000,855 | 0.004,528 | | 83 | 0.019,137 | 0.230,262 |
| | 28 | 0.000,000 | 0.000,001 | | 56 | 0.001,022 | 0.005,551 | | 84 | 0.020,043 | 0.250,305 |
| | 29 | 0.000,001 | 0.000,002 | | 57 | 0.001,216 | 0.006,766 | | 85 | 0.020,906 | 0.271,211 |
| | 30 | 0.000,001 | 0.000,003 | | 58 | 0.001,437 | 0.008,203 | | 86 | 0.021,718 | 0.292,929 |
| | 31 | 0.000,002 | 0.000,004 | | 59 | 0.001,689 | 0.009,892 | | 87 | 0.022,470 | 0.315,399 |
| | 32 | 0.000,002 | 0.000,006 | | 60 | 0.001,975 | 0.011,867 | | 88 | 0.023,155 | 0.338,554 |
| | 33 | 0.000,003 | 0.000,009 | | 61 | 0.002,297 | 0.014,164 | | 89 | 0.023,764 | 0.362,318 |
| | 34 | 0.000,004 | 0.000,013 | | 62 | 0.002,658 | 0.016,822 | | 90 | 0.024,292 | 0.386,610 |
| | 35 | 0.000,006 | 0.000,018 | | 63 | 0.003,059 | 0.019,881 | | 91 | 0.024,732 | 0.411,341 |
| | 36 | 0.000,008 | 0.000,026 | | 64 | 0.003,503 | 0.023,384 | | 92 | 0.025,079 | 0.436,421 |
| | 37 | 0.000,011 | 0.000,037 | | 65 | 0.003,992 | 0.027,377 | | 93 | 0.025,330 | 0.461,751 |
| | 38 | 0.000,015 | 0.000,051 | | 66 | 0.004,527 | 0.031,904 | | 94 | 0.025,482 | 0.487,233 |
| | 39 | 0.000,020 | 0.000,071 | | 67 | 0.005,109 | 0.037,013 | | 95 | 0.025,533 | 0.512,767 |

## COMMENT

The foregoing discussion by no means exhausts the nonparametric tests that are in even common use. My aim has been illustrative rather than exhaustive. I warmly recommend the lucid text by Siegel (106) with its excellent set of tables. Numerous examples are provided from the author's own field of psychometrics.

Let us leave the reader with a general liberating thought. In general, problems involving finite samples from discrete distributions (or distributions rendered discrete, e.g., by ranking or categorizing) of which the probabilistic structure under some null hypothesis is specified can, with sufficient labor, be formalized and appropriate tests or estimation procedures devised. Modern computers make such solutions much more readily available than they were in the past; but solutions by pencil and paper may be found by those with the right combination of desperation and industry. Never lose sight of this cardinal principle. However, those who use this approach in every case are apt to find themselves solving the same fundamental problem over and over again. They may be able to avoid this repetition by identifying

**Table 15.5. The Gaussian approximation (with and without Yates's correction) to the distribution function of Q in Kendall's $\tau$ statistic for samples of size 20**

| q | $\tau$ | F(q) | $F_G(q)$ | $F_G(g + 1/2)$ |
|---|---|---|---|---|
| 27 | 0.716 | 0.000,001 | 0.000,005 | 0.000,006 |
| 40 | 0.579 | 0.000,098 | 0.000,179 | 0.000,203 |
| 46 | 0.516 | 0.000,550 | 0.000,738 | 0.000,823 |
| 49 | 0.484 | 0.001,177 | 0.001,419 | 0.001,576 |
| 56 | 0.411 | 0.005,551 | 0.005,693 | 0.006,241 |
| 59 | 0.379 | 0.009,892 | 0.009,746 | 0.010,624 |
| 69 | 0.274 | 0.049,165 | 0.045,765 | 0.048,997 |
| 75 | 0.211 | 0.104,314 | 0.097,183 | 0.102,877 |
| 81 | 0.147 | 0.192,929 | 0.181,823 | 0.190,516 |

an instance as a particular case of a general class which one of the standard nonparametric procedures solves in generality, with appropriate preexisting tables and large-sample approximations. The standard tests may thus save much labor. But it is wise to see them in this perspective. The tests were made for the statistician, not the statistician for the tests.


## PROBLEMS

**15.1.** Consider the following measurements of the mean survival of blood platelets (in days) in seven subjects when they were on a diet rich in vegetable fat and when they were on a diet low in fat (3).

| Subject | Vegetable fat diet | Low fat diet |
|---|---|---|
| 1 | 4.44 | 4.50 |
| 2 | 3.99 | 4.45 |
| 3 | 4.80 | 5.64 |
| 4 | 2.98 | 3.88 |
| 5 | 4.82 | 5.36 |
| 6 | 3.55 | 5.71 |
| 7 | 3.54 | 3.92 |

Test the hypothesis that diet has no effect—
    **a.** parametrically (assuming a Gaussian distribution);
    **b.** by the sign test.
Comment on the results.

**15.2.** Calculate the Kendall $\tau$ statistic for the three possible pairs of variables in Table 9.9.

**15.3.** A geneticist believes that a trait is under the control of three loci at each of which there is a high-level (H) and low-level (L) gene. The loci behave independently and the phenotypic effects are additive. They are as follows:

| Locus | Gene | Probability of gene | Effect |
|-------|------|---------------------|--------|
| 1 | H | .1 | 3 |
|   | L | .9 | 1 |
| 2 | H | .5 | 1 |
|   | L | .5 | 0 |
| 3 | H | .02 | 5 |
|   | L | .98 | 1 |

**a.** Write down all possible combinations of genes with associated probabilities and phenotypic values.

**b.** Arrange them as a distribution and find the mean and variance.

**c.** Construct tests of size .02 and .10 that a particular observation comes from this distribution.

**15.4.** The atomic weight $(a)$ of an element is the product of the valency $(v)$—which is necessarily an integer—and the equivalent weight $(w)$. In many cases it is easy to determine $w$ chemically, but $v$ is more elusive. However the law of Dulong and Petit (LDP) states that for a metal the product of the specific heat $(s)$ and $a$ is approximately 6.4. For copper, $s$ is found in an experiment to be 0.092 with a standard deviation for this estimate of 0.001. The estimate of $w$ from the average of 200 measurements is 31.773 and has an estimated standard deviation of 0.02. Estimate the atomic weight of copper and furnish approximate 95% confidence limits. (Assume that errors of measurement are independent and approximately Gaussian.)

# 16

# SURVIVORSHIP

We have seen that the distribution of the sample values may be represented in two ways.

First one may display the concentration of cases at particular values of the variate, denoted by $f(x)$. For a discrete variable, such as family size, we may write simple proportions, e.g., the probability of three children. For a continuous variate such as height the concentration of cases may be represented by probability density that the height lies within some specified neighborhood of (say) 70 inches. (The probability is, of course, related to the width of the neighborhood.)

The second way is to express the proportion that assumes the value *x or less*. For both discrete and continuous variables it is written $F(x)$ and is known as the (cumulative) distribution function. (See, e.g., Table 4.1.)

Constructing such curves empirically is a matter of obtaining an adequate and sound sample and arranging the results in ascending order. Then we may group them and draw a histogram to give a sample approximation to $f(x)$ that has the advantage of familiarity of form. To be explicit, the probability of falling in the class interval is represented in the histogram by the area of the cell. If the scale of baseline varies (as in Example 4.3) the height will vary reciprocally; but if it does not vary but is expressed on a unitary scale, the height of the cell in the histogram will represent the mean probability density over the interval represented by the cell. Or better, we may construct a cumulative curve as we have done repeatedly, and especially in Chapter 4, which gives an approximation to $F(x)$. If we wish to compare either construction with some particular model, we may superimpose the best-fitting curve, taking due care over the choice of scale.

The cumulative representation has four major advantages over the histogram. First, since there is no need to make decisions about grouping we do not have to be concerned about loss of information. A comparison of Figures 1.3 and 4.3b will show how much detail is lost by grouping. Second, grouping, even of well-behaved data, causes some distortion, notably in the sample estimate of the variance, which may or may not be readily correctable (see below). Third, any grouping is arbitrary; and careful (but unlawful) choice among groupings may be used to bolster up some preconceived idea—e.g., that the distribution is bimodal, or leptokurtic. (This special pleading is a violation of a sound, although not absolute, principle of statistical inference that the formal conclusions from an analysis should not be idiosyncratic: that statisticians—or computers, for that matter—using the same methods should all reach the same results in any particular case.) Fourth, we may examine the effect of transformation of the "horizontal" scale (the variate) without having to be concerned about vertical height. The latter problem, illustrated in the two parts of Figure 4.5, stems from the fact that, in a histogram, a probability is represented, not by a height, but by an area, the product of two dimensions; and since, during transformation, probability must be conserved, one cannot change the horizontal scale without changing the vertical. But in the cumulative distribution, probability is represented in one direction only—vertical height—and it requires no adjustment for a transformation of the variate. (The probability density is represented by the slope of the curve $F(x)$ or the depth of the steps.)

## SHEPPARD'S CORRECTION

This is as convenient a place as any to discuss the impact of grouping on estimation. Let us suppose that the distribution from which the sample has been taken

1. Is continuous. This will usually be warranted, or not, from the nature of the measurement.

2. Has a finite range. In practice we may be assured of this property provided we avoid ratios and certain transformations. (See Chapter 15 for some exceptions.)

3. Is smooth, not jagged. I can state this property here in vague terms only. Roughly, it means that the slope of the probability density does not change abruptly over any short distance, including at the ends of the distribution. Because of sampling error, digit preference, and other defects of measurement, we cannot often test in practice whether the underlying distribution is smooth in this sense.

We now divide the distribution into segments of equal width, $h$. All the values within each interval are treated as if they were concentrated at its midpoint. For these discrete points, the mean and variance are computed. Then the sample mean is unbiased; but the sample variance systematically overestimates the variance by one-twelfth of the square of the class interval.

**Example 16.1.** Consider the distribution

$$f(x) = kx^p(1 - x)^q \qquad 0 \le x \le 1$$

It belongs to a class known as the beta distribution: $k$ is a suitable normalizing constant. I use this distribution because it has a finite range, it may be symmetrical, and, although it bears some superficial resemblance to the Guassian, as the reader can easily verify by plotting it out, the area under it between 0 and $x$ denoted by $F(x)$ may be given by a relatively simple formula.

If $p = q = 3$ then this formula is

$$F(x) = 35x^4 - 84x^5 + 70x^6 - 20x^7 \tag{16.1}$$

The area between any two points, $x_2$ and $x_1$ is simply

$$F(x_2) - F(x_1)$$

The exact mean of such a distribution is given by

$$E(X) = \frac{p + 1}{p + q + 2}$$

the variance by

$$\text{var}(X) = \frac{(p + 1)(q + 1)}{(p + q + 3)(p + q + 2)^2}$$

Here the mean is

$$\frac{4}{4 + 4} = 0.5$$

and the variance

$$\frac{4 \times 4}{9 \times 8^2} = 0.02777\ldots$$

In Table 16.1 are given calculations when the range is split into 10 equal strips and the probabilities, computed from formula (16.1), assigned to the midpoints of the intervals. The "grouped estimate" of the variance of $X$ is calculated. After Sheppard's correction the grouped estimate of the variance is close to the true variance. (It is about 0.004% too high). The estimate of the mean is unbiased. These being the properties of the *exact* distribution, we expect them to be exhibited by adequate samples from it.

In contrast, let us consider a distribution which violates the assumption of smoothness. We might examine such distributions as the weight of thrombus (Table 4.8); but as an illustration of precise effects we shall appeal again to a finite continuous distribution, also a beta variate with $p = 4$, $q = 0$.

**Example 16.2.** The distribution

$$f(x) = 5x^4 \qquad 0 \le x \le 1$$

is far from smooth. At $x = 1$, the density abruptly changes from 1 to 0. It may be taken as a type of the $J$-shaped distribution. We shall again group at intervals of 0.1. The calculations are shown in Table 16.2. They may be compared with the following exact results

$$E(X) = \frac{4 + 1}{4 + 0 + 2} = 0.833\ldots$$

$$\text{var}(X) = \frac{(4 + 1)(0 + 1)}{(4 + 0 + 3)(4 + 0 + 2)^2} = 0.019,841,269,8\ldots$$

The grouped estimate of the mean is negatively biased by about 0.5%

$$\frac{0.829,175}{0.833,\ldots} = 0.995,01\ldots$$

The variance is underestimated by about 7%. Sheppard's correction has, in fact, done more harm than good: without the correction the bias in the estimate of the variance is only $-2.79\%$.

While for clear demonstration I have used well-defined distributions, we should realize that Sheppard's correction is applicable to a broad class of distributions, and that beyond checking that the three underlying assumptions are fulfilled we are not obliged to inquire further into the form of the distribution. Indeed in the world of empirical experience we are not able to do so.

## ACCURACY IN COMPUTATION

Suppose that we are given a measurement of blood cholesterol as 248.3 mg/100 ml. For purposes of computation, should we use this figure or round it to 248 or even to 250? This is a question on which many hold doctrinaire views for which they offer no justification. In fact it is a surprisingly elusive issue which I have considered in detail elsewhere (3). I would propose two simple principles, which I think can be justified and which do not conflict with informed practice.

1. If the datum is a direct measurement, then retain as many digits such that the penultimate figure is known to be correct. The main issue here is what the resolving power of the measurement is. Competent observers can usually interpolate to about one order of magnitude beyond the calibrations of the scale. Values beyond that are probably observer noise and will merely further inflate the variance by an amount which (under certain reasonable circumstances) turns out, remarkably enough, to equal Sheppard's correction. (See reference 3 for proof.)

2. If the datum is the resultant of any nontrivial manipulation of measurements (e.g., transformations, ratios, etc.), retain as many digits as possible in the *intervening* calculations and leave whatever may appear to be a reasonable rounding to the last step. Rounding always introduces errors and in a chain of calculations, they may propagate alarmingly. Those skeptical might review the variance-stabilizing transformation on the Poisson distribu-

**Table 16.1. Sheppard's correction applied to the distribution** $f(x) = 140x^3(1 - x)^3$

| Range of $x$ | Midpoint $(x_m)$ | Probability | $x_m f(x)$ | $x_m^2 f(x)$ |
|---|---|---|---|---|
| 0.0 − 0.1 | 0.05 | 0.002,728 | 0.000,136,4 | 0.000,006,82 |
| 0.1 − 0.2 | 0.15 | 0.030,616 | 0.004,592,4 | 0.000,688,86 |
| 0.2 − 0.3 | 0.25 | 0.092,692 | 0.023,173,0 | 0.005,793,25 |
| 0.3 − 0.4 | 0.35 | 0.163,756 | 0.057,314,6 | 0.020,060,11 |
| 0.4 − 0.5 | 0.45 | 0.210,208 | 0.094,593,6 | 0.042,567,12 |
| 0.5 − 0.6 | 0.55 | 0.210,208 | 0.115,614,4 | 0.063,587,92 |
| 0.6 − 0.7 | 0.65 | 0.163,756 | 0.106,414,4 | 0.069,186,91 |
| 0.7 − 0.8 | 0.75 | 0.092,692 | 0.069,519,0 | 0.052,139,25 |
| 0.8 − 0.9 | 0.85 | 0.030,616 | 0.026,023,6 | 0.022,120,06 |
| 0.9 − 1.0 | 0.95 | 0.002,728 | 0.002,591,6 | 0.002,462,02 |
| | | | 0.500,000,0 | 0.27,861,232 |

$$\text{Sheppard's correction} = (0.1)^2/12 = 0.000,833,33$$
$$\Sigma x_m^2 f(x) = 0.27,861,232$$
$$[\Sigma x_m f(x)]^2 = 0.25$$
$$\text{Grouped estimate of var } (X) = 0.028,612,32$$
$$\text{Sheppard's correction} = 0.000,833,33$$
$$\text{Corrected grouped estimate} = 0.027,778,99$$

tion (Chapter 4). There, rounding that represents (where means are concerned) only a small percentage may have a devastating effect on the estimates of variances. The theory of the variance-stabilizing transformation is based on the assumption that the digits to the right of the decimal point are accurately retained regardless of how large those to the left may be.

## POINT PREVALENCE

The function $F(x)$ which we have used above is the probability that the random variable assumes the value $x$ or less. Suppose that $x$ represents the age at which the first of the permanent canine teeth erupts. Then to avoid certain distortions discussed later, we suppose that there is no mortality rate from any cause before this event. It is obvious that the expected proportion of people aged twelve in whom this tooth has erupted will be $F(12)$. This quantity is termed the age-specific prevalence of the trait. In principle or in fact, the study is best done at a particular point in time. One then speaks of point prevalence. However, certain types of data are studied over time and that proportion of subjects who exhibit the condition at some stage is the "term prevalence," e.g., lifetime prevalence. It is, of course, no longer age-specific. What the empiricist considers in assessing point prevalence is the proportion in those people who are available to be studied.

In fact, there is always mortality from many causes, including (competing) diseases; let us denote the cumulative probability of mortality by $G(x)$ so that the probability of surviving is the complement $[1 - G(x)]$. Now if mortality is in no way related to the canines, the joint probability of surviving to age $x$ and having erupted canine teeth is simply the product

$$[1 - G(x)] F(x) \tag{16.2}$$

The conditional probability that the tooth has erupted, given survival to this age, is simply (16.2) divided by $[1 - G(x)]$. It gives the prevalence, which here reduces to $F(x)$. If delayed or premature eruption of canines is related to mortality, that is, we cannot assume independence, (16.2) is no longer correct and the age-specific prevalence is no longer an unbiased estimate of $F(x)$. In particular this is true where the trait itself causes, or is at least followed by, mortality.

**Example 16.3.** An illustration which has been of special interest in my researches is familial polyposis of the colon (113). The polyposis, which is age-dependent, leads eventually to malignant change and then the patient either undergoes colectomy or dies. In either case, they are no longer available for the study as patients with polyposis. Let us take as our population those with one parent heterozygous, that is, people who a priori are at 50% risk of polyposis; and we make the reasonable assumption that the longest-living offspring will be one of those who did *not* inherit the gene. Then the prevalence of polyposis is 0 at birth; it is also 0 just before the last sampled person dies. The highest prevalence at any age (which in our study is about 40%) is determined by the cumulative probability of developing polyps and of also being removed from the study by cancer, which we shall denote by

**Table 16.2. Sheppard's correction applied to the distribution $y = 5x^4$**

| Range of $x$ | Midpoint $(x_m)$ | Probability $f(x_m)$ | $x_m f(x)$ | $x_m^2 f(x)$ |
|---|---|---|---|---|
| 0.0 − 0.1 | .05 | .000,01 | .000,000,5 | .000,000,025 |
| 0.1 − 0.2 | .15 | .000,31 | .000,046,5 | .000,006,975 |
| 0.2 − 0.3 | .25 | .002,11 | .000,527,5 | .000,131,875 |
| 0.3 − 0.4 | .35 | .007,81 | .002,733,5 | .000,956,725 |
| 0.4 − 0.5 | .45 | .021,01 | .009,454,5 | .004,254,525 |
| 0.5 − 0.6 | .55 | .046,51 | .025,580,5 | .014,069,275 |
| 0.6 − 0.7 | .65 | .090,31 | .058,701,5 | .038,155,975 |
| 0.7 − 0.8 | .75 | .159,61 | .119,707,5 | .089,780,625 |
| 0.8 − 0.9 | .85 | .262,81 | .223,388,5 | .189,880,225 |
| 0.9 − 1.0 | .95 | .409,51 | .389,034,5 | .369,582,775 |

$$\mu' = .829,175 \qquad \mu_2' = .706,819$$

$$\text{Sheppard's correction} = (0.1)^2/12 = 0.000,833$$
$$\text{Approximate } \mu_2' = 0.706,819$$
$$\text{Approximate}(\mu)^2 = 0.687,531,118,0\ldots$$
$$\text{Approximate variance} = 0.019,287,819$$
$$\text{Sheppard's correction} = 0.000,833,333\ldots$$
$$\text{Corrected ``grouped'' variance} = 0.018,454,486$$

$R(x)$. Since polyposis is a necessary intermediate step we are assured $R(x) \leq F(x)$ for all values of $x$. Since half of the progeny inherit the gene, the probability of being affected *and available for study* at any age $x$ is

$$[F(x) - R(x)]/2$$

The probability of being available is (ignoring other mortality)

$$1 - R(x)/2$$

Hence, if both the numerator and the denominator are multiplied by 2 the age-specific prevalence is

$$\frac{F(x) - R(x)}{2 - R(x)}$$

which is not, in general, equal to $F(x)/2$, the cumulative probability of having the disease. Of course, if $R(x) = 0$ it assumes the simpler form $F(x)/2$. The important general principle that is illustrated by this example is that one must not confuse "cross-sectional" data (i.e., point prevalence) with longitudinal on a whole population. This confusion, which is common, would in Example 16.3 lead to the incorrect conclusion that familial polyposis coli regresses in later life.


## INCIDENCE

We now take up briefly the subject of incidence. The word is persistently and inexplicably confused with prevalence. It may help, as a mnemonic device, to remember that inciden*ce* is concerned with inciden*ts,* with things happening; and since the notion of happening involves time, incidence is a function of time, which requires to be specified.

**Example 16.4.** What proportion of men now aged 45 have myocardial scars? The question involves a state and the answer is a prevalence. What proportion of men now aged 45 have developed myocardial infarctions in the last interval of time (e.g., during the last year)? The answer involves incidents of myocardial infarction and the answer is an (annual) incidence.

Where the event is in discrete time, the answer is a simple matter of probabilities.

**Example 16.5.** In 1978, what proportion of men aged 45 had hangovers on New Year's Day? Since this joyous event occurs only once a year, we can associate a true probability with it.

**Example 16.6.** The rate of car accidents in the last year is expressed in continuous time, since accidents may occur at any instant whatsoever. Then the quantity we are concerned with is the probability density: the longer the interval, the more likely an event; in no time, the probability of an accident is zero; and if we suppose a uniform risk of accidents, over any short interval the ratio of probability to time may be regarded as a constant.

**Example 16.7.** If (as the data of Lewitus and Neumann [103] discussed in Chapter 14 suggest) the incidence of coronary disease in men of 45 is not seasonal, an incidence of 1.04 cases per thousand per year and 0.02 cases per thousand per week are virtually equivalent.


## HAZARD

The probabilistic equivalent of instantaneous age-specific incidence of *new* disease, the age of onset of which has the density function $f(x)$, would be the instantaneous rate per unit time at which the condition is occurring divided by the proportion of the population who have not hitherto had the disease, $[1 - F(x)]$. This quantity

$$h(x) = \frac{f(x)}{1 - F(x)}$$

is known as the hazard function. The exponential distribution has the remarkable property that the hazard function is constant, i.e., does not depend on $x$. It is in this sense the only completely "memoryless" system there is. If the age of onset of a disease, or even death itself, is truly exponential, then the past history is irrelevant to the prognosis. Many heedless claims have been made that a process is exponential—for instance, that blood platelets have an exponential survival. This could only mean that they never wear out—that the only cause of death is a single external insult at constant hazard and operating indiscriminately on platelets regardless of how long they have already been in circulation. I have serious doubts that any biological system behaves in this way, and am sure that few investigators would admit this claim—even those who implicitly do so by using the exponential model.

Where many diseases and many types of data are concerned, $F(x)$ never attains much size.* In this case $[1 - F(x)]$ scarcely differs from unity at any stage and we may use $h(x)$ and $f(x)$ more or less interchangeably. But logically the two should not be confused. Ashley (114), for instance, is content to equate the prevalence at age $x$ with $F(x)$ where common bowel cancer is concerned, which in normal human life span rarely attains a value of 0.02 or higher. But in familial polyposis coli where $F(x)$ eventually attains 100% and age of death varies widely the distortion would be considerable. In consequence, cross-sectional and longitudinal data give very different patterns (113).

The hazard functions are algebraically rather inconvenient to work with for most distributions in common use. One reason why they have been used is that actual data available are often in this form. We commonly have no ade-

---

*This means that for practical purposes the distribution of $X$ may be regarded as dishonest (see Chapter 15).

quate longitudinal data, but merely studies, over a brief period of time, on subjects at various stages of their disease. Thus the denominator usually consists, not of the original population at risk, but the current population at risk. The hazard function merits three further comments here.

First, where statistically *independent* risks are involved—whatsoever their mathematical form may be, or even whether they are of diverse forms—the hazard of the risks taken jointly is the sum of the hazards. If, in particular, all the risks taken separately are exponential, then their hazard functions are constant and their joint hazard is the sum of these constants, which is, itself, a constant and thus must represent an exponential process.

Second, if the hazard function is explicitly known as a function of $x$, then $F(x)$ can in principle be found, and hence $f(x)$.

Third, hazard functions can be used only for continuous distributions. For discrete distributions the hazard function would assume the value infinity at discrete points and zero elsewhere.

More specifically gamma (multiple-hit) processes tend to become nearly linear over a large part of the domain, when they are displayed in a log-log plot (Fig. 16.1). It seems to me highly suggestive that oncologists have so often found similar results in mortality statistics. The terminal hook that is so characteristic may not be encountered unless the hazard is looked at

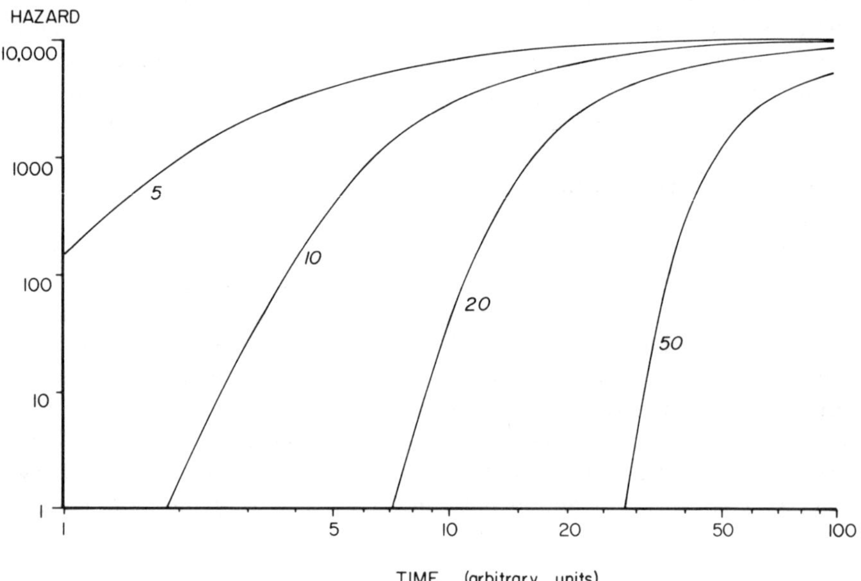



**Figure 16.1.** The hazard functions of gamma distributions of various orders are plotted against time (in arbitrary units). Both scales are logarithmic. The curves are close to linear throughout most of the course, but with a perceptible terminal hook. The plot of age-specific incidence against age on such a double logarithmic plot follows a closely similar pattern in empirical data.

carefully throughout life, but it has been found empirically, for instance by Knudson (115).

## AMBIGUOUS USAGE

I have mentioned the deplorably common confusion of the terms *prevalence* and *incidence.* For those who still have not a clear idea of the difference, I recommend the nontechnical term *frequency* which is mostly what they mean anyway. But epidemiologists speak of (for example) "life-time incidence" of some disease. Since their information (unlike the statistician's conceptualization of it) comprises data collected over time, one can in some degree understand their usage; however, one should realize that life incidence (or any other "term" incidence) is not a hazard rate but a probability. I would be much happier if they had another word for it. But in particular, what do we call the frequency of birth defects (say anencephaly)—prevalence or incidence? In vital statistics we encounter either term used indifferently. It is undoubtedly a point prevalence in the sense that it is the frequency of a trait at a precise instant in ontogeny. It may be viewed as a cumulative incidence of whatever it is that causes anencephaly, provided we assume that there is no differential loss of such fetuses and that once anencephaly appears, it remains.

For my part I would avoid all terms except *point prevalence* and *age-specific incidence,* and I would recommend readers do the same. If any other notion is to be used, I would state explicitly what has been studied, and designate it in algebraic symbols rather than words.

## SURVIVORSHIP CURVES

When a distribution has time as its random variable, we commonly deal with the fate of the process by the complement of the (cumulative) distribution function. This so-called survivorship function is a statement not of the probability of dying (or falling sick, or whatever may be involved) but the probability of remaining intact until at least time $x$. Since every element either does or does not reach the endstate, it follows that the survivorship function, $G(x)$, for a continuous variate has the relationship

$$F(x) + G(x) = 1$$

$$G(x) = 1 - F(x)$$

where $F(x)$ has the usual statistical meaning. The probabilistic aspects of $G(x)$ have been laid out in some detail elsewhere (7). It suffices to say here that such a curve must (in the nature of the process) start out with the value 1 (that is, survival cannot be less than zero) and, barring immortality, must eventually end at zero. Also, granted that the endstate is irreversible biologically (as death is) or conceptually (as having sustained one's first attack of pneumonia is), as $x$ increases $G(x)$ may become smaller, or may stay unchanged, but it can nowhere increase. This monotonicity, and the fact that survival must be positive, account for the remarkable properties of such curves and are the theoretical basis for the methods by which such curves are constructed empirically (see below).

## Cohorts

It is well at this stage to form a clear idea of what we are about. The ideal curve $G(x)$ is a predictive statement about the fate of *one* person, animal, or thing (as the case may be), yet to be picked at random. Hard-nosed empiricists will deny that such a statement has any verifiable meaning since even the most detailed study of this individual will not allow us to construct such a curve. The actual *data* on the patient consist of an event which occurs at one particular point in time. In practice, we estimate $G(x)$ by a study of the collective behavior of a group of individuals. But that approach by no means solves all our problems. For the idealized $G(x)$ may be different for each individual; and when we consider how many sources of variation in fitness there are, actual idiosyncrasy of individual patterns of survival is highly plausible. This heterogeneity leads us to two difficulties.

If all the $G(x)$ functions from which we are making observations are different, it is far from clear what we are estimating from pooled data. What can we make of observations on one apple, one banana, one Drosophila, one elephant, one postexistentialist philosopher, and so on, however long the list?

Also, even if we know very accurately what this composite $G(x)$ is, it will be of no value in prognosis for any particular future patient who presumably has his own unique $G(x)$. This kind of reasoning rapidly leads to a total scientific nihilism, something which, because of its superadded probabilistic nature, is even more disruptive than nominalism.

In reality, the situation is not nearly so intractable as that argument sounds. In the first place, uniqueness and nominalism are not interchangeable ideas (1); this false supposition stems from a confusion between cardinality and dimensionality (3). It is quite possible that $G(x)$ might be expressible as a function of (say) six universal parameters and six particular values of variables, and still be unique for each individual. In the second place, we may perfectly well be able to devise a reference population from which the subjects may be regarded as a random sample. In practice, we have a choice between making our population of reference very general and being content with very general conclusions; or aiming for highly precise statements about narrowly defined groups. Obviously there may be an infinite gradation from one extreme to the other. A somewhat narrowed group that we define for study over time is commonly (if somewhat facetiously)*

---

*A cohort was originally an enclosed garden and later one-tenth part of a Roman legion.

referred to as a cohort. It is a relative term. It might be applied, for instance, to a group born on the same day, a single graduating class in college, or all physicians currently resident in hospitals. We do not in practice demand absolute homogeneity in a cohort and perhaps could not define what we mean by the term *absolute homogeneity*. Usually, members of a cohort have one major feature in common; but there is no particular requirement that they should be contemporaneous. We may talk about a cohort of male patients aged 50 that have undergone pneumonectomy for carcinoma whom we have encountered in the last ten years. We suppose the group is sufficiently homogeneous to allow some useful generalization about the fate of such subjects. In doing so, we recognize that operative techniques are changing over time, diagnoses are being made earlier, etc. Thus some compromise is involved.

## PATTERNS OF SURVIVORSHIP DATA

Obviously, if we knew any of the above-mentioned three functions—$f(x)$, $F(x)$, or $G(x)$—exactly, either from theory or because we had an infinite number of data free of error, in a homogeneous population there would be no *statistical* problems whatsoever. In practice, we rarely have any theoretical warrant for our conclusions and are driven to data which are limited, or imperfect, or both.

The question of the sources of our uncertainties bears profoundly on the method of the analysis. There are typically three sorts of data.

### Synchronic (cross-sectional) data

The pattern of survival may be inferred from data representative of the surviving population at large. We may find, for instance, that the proportion of people aged 70–71 is one-half what it is at ages 0–1, and conclude that half of the population dies between birth and 70. More generally we may conclude that the age distribution of the population, suitably scaled, is exactly equal to $G(x)$. The soundness of these inferences depends on three main assumptions.

1. That the pattern of survival, $G(x)$, has remained unchanged over all the years spanned by the subjects in our sample. This means that the behavior of a cohort over time is the same as the behavior of the population over age.

2. That a state of equilibrium exists. This assumption has two components: that the birth and death rates are constant and that they are equal to one another.

3. That the information on the age of individuals at 70 and at birth are equally accurate, and the sample size is large enough that we may ignore sampling error at all ages.

The violations of these assumptions where populations of people are concerned are so numerous and widespread as to make "longitudinal" inferences from "cross-sectional" data rarely, if ever, sound. The first assumption has been profoundly disturbed in recent years, particularly by the containment of infectious disease, and by antibiotics. The second is demonstrably false where the population is growing, and since both birth and death rates are seasonal. The third is known by demographic experience to be false even in sophisticated societies with elaborate official records. In primitive societies, the data are so much more inaccurate as to lead at times to grossly mistaken estimates of life span. It is a pity that such flaws exist. Cross-sectional data have two enormous advantages. The first is that one does not have to wait a lifetime to order to study the complete survivorship. The other is that only one observation is included on each subject: the complications of time-series analysis (see Chapter 8) are largely avoided.

### Identified diachronic (longitudinal) data

Where human populations are concerned, each person has legal, usually conscious, identity. Then we may do longitudinal studies either prospectively (as in clinical trials) or retrospectively. Unfortunately, the meaning of the term *retrospective* is ambiguous. It is sometimes used to mean tracing back the surviving population to its origins.* I mean it here in the other sense of reconstructing a prospective study after the event. For instance, we have done such a study (117–119) on the progeny of a group of 2,320 non-agenarian probands collected some fifty years ago by Pearl and Pearl (120). With sufficient patience and enterprise in exploiting the channels available, it is surprising how complete such data can be made. There are, of course, several highly specialized populations—the French Canadian population of Quebec, the Northern Italian villages, the Latter-Day Saints of Utah, the Amish—who keep genealogical records quite accurate enough for study.

The main problems here are serial correlations in the data. For instance, suppose according to the function $G(x)$ which we are trying to estimate, half the population should be dead at 70 and three-quarters at 80. Now for any finite sample we cannot expect that the true proportions will be exactly realized; there will be a greater or lesser sampling deviation from these proportions. It is possible that, just by chance alone, only 20% survive to age 70. Then the proportion in the sample surviving to 80 cannot be greater than 20% and will in consequence be less than the 25% expected at that age. Thus, the low estimate of the first proportion imposes a low estimate of the second. The two sample proportions are necessarily correlated. If the proportion living to 70 is excessive, say 100%, the implications are less certain. But it seems that the percentages surviving at the two ages will be in general correlated. However with sound data we may be assured that the estimates of survival, whatever their defects, are at least exactly known. In a particular therapeutic cohort of 100 subjects, at age $x$, it may be known that 57 subjects

*This type has been designated a *trohoc* by Feinstein (4, 116). This term is the word *cohort* spelt backwards.

are still alive. Then the estimate of $G(x)$ is *exactly* 0.57. This may tell us nothing about the properties of this estimate, but at least the figure 0.57 is not in doubt.

## Anonymous diachronic data

With certain populations it is not possible to preserve identities as it is in man or in distinguishable laboratory animals. Two common instances are: wild animals that are trapped, marked in some way but not individually, released, and recaptured (randomly, it is hoped); and circulating cells such as erythrocytes or platelets. Here we have, superimposed on those imperfections of the data encountered in previous methods, the further problem of uncertainty in the estimate. When we do longitudinal studies of the effect of a treatment, the 100 patients are—at least ideally—always kept in sight and they are always the same cohort. But in the study of the survival of red cells, it is a different sample of cells that is taken each time; and this is necessarily so whenever (as here) each process of studying frequencies is destructive: the sampled cells that are counted are not put back into the circulation. Of course, there are two other fundamental sources of difference. The number of red cells labeled in the first place is so enormous that (in the absence of collective catastrophe or contagion) we may ignore binomial variance altogether and treat the behavior of the true activity of the sample as deterministic. But also we are not counting labeled cells individually as we were the 57 survivors in the therapeutic series; and the measurement we do on the sample of red cells may have its own type of imprecisions (errors of measuring or weighing, variation in handling them, and so on). The important distinction between these two causes of error is well illustrated by the decay of a radioactive mass. The number of atoms available for decay is so large that we ignore variance in *decay* and for all practical purposes suppose that *exactly* half the atoms will have decayed after one half-life. But when we time a specimen over one minute and get 10,000 counts, Poisson error alone will attach a standard deviation of 100 to that value. Blind duplicate determinations of radioactivity on aliquots from the same sample of platelets labeled in vivo show that a technical error of 5% is not uncommon. Such an error has no equivalent where human mortality is involved: there is no error in measuring whether a person is dead or not (provided he has been kept under supervision). However, there may be some such error where a "softer" endpoint is concerned, e.g., where there may be distant metastasis of a cancer.

## Segmental diachronic studies

We may add a fourth type of study that is beginning to acquire some popularity and has long been known to vital statistics. Granted that, without elaborate and implausible assumptions, one cannot make longitudinal inferences from cross-sectional data alone, and granted that it takes at least seventy years to do a prospective seventy-year follow up, the method proposes that we may at least assume steady forces of mortality, and can thus study diachronic patterns in fragments. For example, if limited to a study lasting five years, we may take four groups aged, at the start of the study, zero, five, ten, and fifteen years old, and watch how they behave as they grow to be five, ten, fifteen, and twenty years old, respectively. Then we can reconstruct behavior over the first twenty years of life by supposing the pattern in the youngest cohort in the next five years will be the same as that of the second youngest cohort in the last five years, and so on. This argument may be criticized on the grounds that since patterns of mortality are constantly changing we cannot believe the main assumption of the argument. But this criticism invites the rejoinder, "Why, then, do we want to study the whole life pattern?" If we are to argue that the predictive value of the whole study is out of date as soon as it is completed, what conceivable practical value can there be in doing it? If we are using diachronic data as an index of current *trends* then it may well be argued that in the interests of homogeneity we should be studying current trends—at all ages— over as short a period as possible. It would not mean much to assess the consequences, over seventy years, of a policy of management which is constantly changing. The argument could be elaborated at great length. I trust the reader will not accept, uncritically, prefabricated opinions which he may imagine me to hold.

## ESTIMATION OF SURVIVORSHIP CURVES

The problems of estimation are classified according to the type of data available, the components of uncertainty, and the pattern of death. An exhaustive study of this topic would involve statistical methods that are not only elaborate but technically far beyond the scope of this book. They would include, for instance, complicated mathematical modeling, time-series analysis, nonlinear estimation, iterative weighted least squares methods, and so forth. Much of this subject is the territory of the demographic theorist; much of the modeling is the concern of the theoretical biologist or the geneticist. I know of no textbook that treats the subject comprehensively, although certain authorities (121, 122) deal admirably with human survivorship. We shall have to be content here with a brief outline and some discussion of the standard nonparametric method.

### Cell survivorship

There is an extensive literature on the survivorship of cells of the body. For the most part the cells studied are those with an active turnover: red cells, platelets, intestinal epithelium, and skin epithelium. We are dealing here with enormous numbers of units, so that for all practical purposes the errors are those of measurement and the system may in other respects be treated as deterministic. For obvious reasons the cells cannot be individuated and we study them by collective labeling. It is sometimes possible to label a group of cells that are close enough in age to be treated as a cohort. But, in hematological studies, most often we have to be content with labeling an entire population, regardless of age. Interpreting such curves involves some theoretical explanations that I have laid out elsewhere (7) together with some convenient nonparametric properties which enable us readily to estimate the mean and variance of the length of life. However, these properties can be invoked only when technical error is small, data points are numerous, and the cell has a relatively long minimum survival. They are fulfilled in high quality data on the survival of red cells provided that equilibrium exists. However, we have documented in detail (123) why they are not appropriate for blood platelets. Blind duplicate determinations among, and even within, observers are embarrassingly large (124).

For these reasons it has always seemed to us best to use a mathematical model with a structure faithful to the known facts of physiology of the cell. Early investigators proposed both exponential and Gaussian models of survival with no physiological rationale. These models describe some instances adequately, but there are many intermediate forms. However, it occurred to us (123) that the gamma model provides a nice unifying theory since it encompasses the two early proposed models as special cases, as well as many intermediate forms; and the episodic nature of depletion of cells as they age seems to call for a multiple-hit process of some type. Unfortunately, I do not know of any such model which both is biologically plausible and lends itself to elementary statistical analysis. However, the reader should be aware that such methods do exist and that they (or something of comparable sophistication) are indispensable to sound analysis. Computer programs are available for the actual calculations, which are extremely tedious (76).

## INDIVIDUATED SURVIVORSHIP

When each individual member of the population can be identified, a different set of problems arises. Sufficiently detailed records will allow us to construct directly a table of actual individual lengths of life, which may then be treated like realizations of any other random variable. The problems in practice are that we must wait to complete the analysis until all members of the cohort are dead, although certain characteristics such as the median and the interquartile ranges may be estimated somewhat earlier; and that since the variable is being measured over time, there is a constant risk that some members will be lost to follow-up. Indeed, where a human population is concerned, attrition may be a formidable problem, partly because the use (and misuse) of human liberty results in migration, change of name, falsification of birthdate and other identifying information; and partly because the subjects of study may outlive the observer. In practice these problems give rise to the peculiarity of the nonparametric methods for estimation of survival (see below). Nevertheless, it is still possible to explore the pattern of survival from fragmented data, using rational mathematical models.

It is shown readily enough (7) that if deaths occur at $n$ discrete points in time, $A_1, A_2, \ldots, A_n$, the survivorship curve can be constructed if either

1. The probability of survival to each of the $n$ points is known, $P(A_i)$, or

2. The probability of surviving to each point, conditional on survival to the previous point, is known, $P(A_i|A_{i-1})$ for all values of $i$ from 1 to $n$, and if $P(A_0)$ is unity. In the second case, we have in general

$$P(A_j) = P(A_0) \prod_{i=1}^{j} P(A_i|A_{i-1}) \tag{16.3}$$

We do not assume that the $n$ points are equally spaced.

It is also shown that the mean survival time is simply the area under the survivorship curve, which may, as appropriate, be found graphically or by calculation.

**Example 16.8.** In Figure 16.2 is shown the set of data on survivorship of women with malignant renal hypertension (21). The survival of each is represented by a segment of line. A graphical display consists simply of arranging

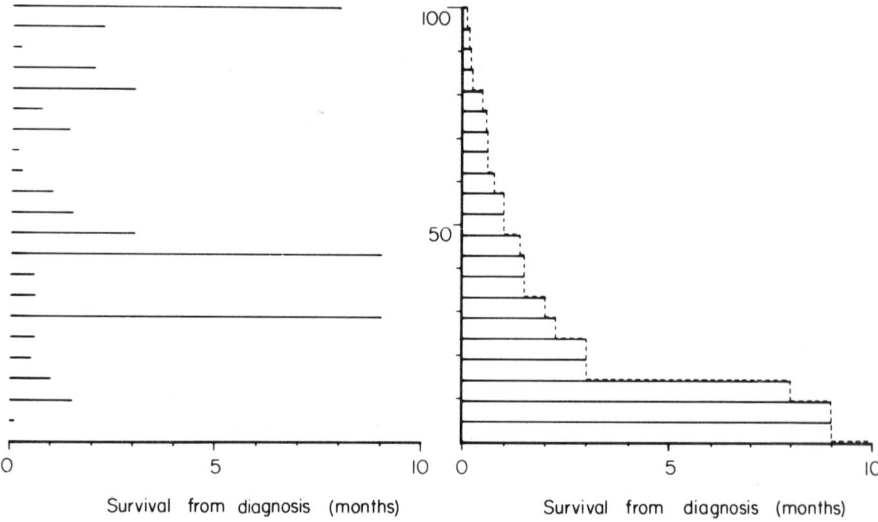


**Figure 16.2.** Survivorship in months from diagnosis in 21 women with malignant renal hypertension. The individual readings are plotted as segments of line on the *left*. On the *right* they are arranged from above down in increasing order of survival. The dotted line represents the sample estimate of $1 - F(x)$, the vertical axis being scaled so that the initial value is 1, or 100%. (Note that if the actual length of survival were unknown in some or all cases, no such simple relationship would exist between the lines and the estimated survivorship.)

them in descending order of length. A little thought will make it clear that the step function shown dotted is the complement of the sample distribution function $F(x)$.

**Point estimation**

In practice, we do not, of course, have the exact values of either $P(A_0)$ or $P(A_i|A_{i-1})$ and we must be content with sample estimates. These may be obtained in various ways. I shall be content here to use the standard non-parametric method. We distinguish two cases, the first being in fact a special case of the second, singled out merely because the calculations are much simpler.

**Method 1.** If the data are complete and the actual survival of each unit in the sample is known, the results are plotted individually as in Figure 16.2 *right*.

**Method 2.** More commonly, the results are incomplete and we have to obtain makeshift estimates of the various conditional probabilities (see below). The missing data are of two broad kinds.

a. The duration of the natural survival may be such that the investigator must abandon his study while some units are still alive. Then the survivorship curve cannot be completed, because there may be a subset of the population which has a very long survival and, for all the investigator knows, the distribution may be dishonest (see Chapter 15). Of course, certain descriptors may be found even from incomplete data. For instance, provided that half or more have died, one can get at least a point estimate of the sample median, and analogously one may estimate other percentiles. In practice, demographers commonly assume some simple form for the tail, e.g., that it is exponential; and if the tail is small enough, in practice the estimate of the terminal is much the same for any reasonable shape of curve.

b. The more common gap, especially for human data, is that the intermediate data are incomplete, characteristically for one of three reasons:

i. The patient is known to be alive after a limited period of time. Typically in a therapeutic study, recruiting of patients is staggered over time. Each time the investigator is giving a lecture or renewing his grant, he reanalyzes his data. But no matter how well a treatment is going, no patient recruited 6 months ago can possibly have survived more than 6 months from the start of treatment: he is "living for ever as fast as he can."

ii. The patient may have disappeared from observation at some stage and his further fate unknown. A common cause is that the patient has moved to another town, or become disaffected.

iii. The patient may have died from some other cause that the investigator is prepared to regard as incidental, e.g., a person on treatment for cancer may have been killed in a train crash. Treating the patient as "lost to follow-up" at this stage is equivalent to "survivorship adjusted for competing causes of death."

In all three instances, the argument is the same: that we want to give full credit for the survival that the patient has achieved while recognizing that we cannot extrapolate beyond the observation as to how the patient would have behaved if not lost. Note that according to the latter adjustment, a treatment that prevented death from the disease of interest would lead us to estimate the patient's survival as infinite. It must thus be used with discernment if absurdity is to be avoided.

Thereafter the calculations proceed on the principle that of those who start an interval, any person who because of factors $b$ i, ii, or iii, is "withdrawn alive" counts as "a fraction (usually half) of a person at risk" and all others (including those who die) count as one person at risk for that period. The estimated mortality rate is the number who die divided by the number of persons or partial persons at risk. The survival rate for the interval is the complement of the mortality rate.

Actuaries laying out the calculations use what always seems to me a dismayingly complex notation with pre- and post-subscripts. I propose to avoid notation altogether, relying on the table headings to identify terms and the common sense of the foregoing statements to keep the reader informed.

**Example 16.9.** Suppose that we are estimating survival during the seventh year of treatment for lymphoma. There are 200 subjects still alive who were recruited at least six years ago and 180 at least seven years ago. Thus, 20 subjects have not had the opportunity to complete seven years of survival. We may designate them as subjects with insufficient follow-up. In the year of study, five are lost sight of, three die of unrelated disease, and ten die of the lymphoma. Then we have

| | | |
|---|---:|---|
| Subjects beginning the period | | 200 |
| Subjects withdrawn alive: | | 28 |
|    Insufficient follow-up | 20 | |
|    Lost sight of | 5 | |
|    Unrelated deaths | 3 | |
| Subjects at risk throughout | | 172 |
| Subjects at half risk (28) | | 14 |
| Total subject-risks | | 186 |
| Deaths | | 10 |
| Estimated conditional mortality | | $10/186 = 0.05376$ |
| Estimated conditional survival rate | | $1 - 0.05376 = 0.94624$ |

There are two implicit assumptions in this treatment that may "stick in the throat."

First, and the more easily confronted, is that the cases withdrawn alive are adequately allowed for by making them half a unit at risk. This step would be justified formally by supposing something about the patterns of both attrition and mortality. If 80% of the sample are withdrawn alive on the second day of the year and the heaviest mortality in the seventh year is in the last month, the estimate of the mortality may be too low. The distortion might be serious if there was a sudden massive recruitment at one stage in a rapidly expanding program. One might deal with this anomaly by more or less elaborate methods; but to my mind, the simplest and most robust is simply to keep the intervals short. It is little more effort (especially on a computer) to make computations at monthly, rather than yearly, intervals: it merely supposes somewhat more detailed data. (I have known of one surgeon who studied follow-up by the ingenious device of asking his patients to send him a Christmas card each year. One would require more detailed information than that for our purposes.)

The second set of assumptions are a good deal more troublesome. They center on representativeness and are twofold. First, is the quality of the patients being recruited changing over time? I have discussed and illustrated this point elsewhere (3). It is an interesting exercise (Problem 16.2) to think up reasons why this might be so. The remedy for it would be a more or less complicated stratification. A much greater problem arises over the assumption of the randomness of attrition. Those lost to follow-up may include disproportionately many of those who have died unbeknownst to the investigator; or who have been cured; or those who have crippling symptoms or side effects; or those who have none at all and fancy themselves cured. Since these patterns of causes are unclear and evidently conflicting, many investigators argue that they "cancel each other out" and may be ignored. But (as I have pointed out [7]) *our ignorance* does not make the loss random. The best remedy is prophylaxis: do not lose touch with patients. The few stray sheep may require disproportionate effort and (with sound biblical precedent) are a disproportionate cause for joy should they be found again after they have been lost. Where investigators are going to do a randomized therapeutic trial, it is a common and wise practice to keep the patients on a placebo for a suitable period before randomizing, so that the faint-hearted will drop out *before* they may distort the balance of the experiment by doing so. Time will deal with the problem of withdrawal due to insufficient follow-up.

**Example 16.10.** In Table 16.3 are laid out some data on survival in 103 patients who received kidney transplants and who were HB Ag negative at the time of the operation. They were kindly furnished to me by my colleague, Dr. W. D. Hillis. Note that it is necessary to calculate survival rate only in an interval in which a death or withdrawal occurs. The losses (*b* i–iii above) do not affect the point estimate of the conditional survival, but, by changing the denominator, influence interval estimation.

### Computing the standard deviation of the estimate

Interval estimation for survivorship is complicated by two features. First, binomial variance depends on sample size, which is decreasing as time passes. Secondly, each estimated proportion in Column 6 carries its own

**Table 16.3. Survivorship in recipients of kidney transplants who were HB Ag-negative at the outset**

| (1) | (2) | (3) | (4) | (5) | (6) | (7) | (8) | (9) | (10) |
|---|---|---|---|---|---|---|---|---|---|
| Month | Number at start | Withdrawn alive | Mean number | Number dying | Fraction dying | Fraction surviving | Cumulative probability of survival | $h = \dfrac{\text{Col 6}}{\text{Col 4} - \text{Col 5}}$ | H |
| 1 | 149 | 33 | 132.5 | 2 | .01509 | .98491 | .98491 | .00012 | .00012 |
| 2 | 114 | 7 | 110.5 | 4 | .03620 | .96380 | .94925 | .00034 | .00046 |
| 3 | 103 | 3 | 101.5 | 2 | .01970 | .98030 | .93055 | .00020 | .00065 |
| 4 | 98 | 3 | 96.5 | 1 | .01036 | .98964 | .92091 | .00011 | .00076 |
| 5 | 94 | 4 | 92.0 | 3 | .03261 | .96739 | .89088 | .00037 | .00113 |
| 6 | 87 | 6 | 84.0 | | | | | | |
| 7 | 81 | 4 | 79.0 | | | | | | |
| 8 | 77 | 2 | 76.0 | 1 | .01316 | .98684 | .87915 | .00018 | .00130 |
| 9 | 74 | 2 | 73.0 | | | | | | |
| 10 | 72 | 5 | 69.5 | | | | | | |
| 11 | 67 | 3 | 65.5 | 1 | .01527 | .98473 | .86573 | .00024 | .00154 |
| 13 | 63 | 5 | 60.5 | | | | | | |
| 14 | 58 | 0 | 58.0 | 1 | .01724 | .98276 | .85081 | .00030 | .00184 |
| 16 | 57 | 2 | 56.0 | | | | | | |
| 17 | 55 | 1 | 54.5 | | | | | | |
| 18 | 54 | 3 | 52.5 | | | | | | |
| 20 | 51 | 1 | 50.5 | | | | | | |
| 21 | 50 | 2 | 49.0 | | | | | | |
| 22 | 48 | 4 | 46.0 | | | | | | |
| 24 | 44 | 1 | 43.5 | | | | | | |

| | | | | | | | | | |
|---|---|---|---|---|---|---|---|---|---|
| 25 | 43 | 1 | 42.5 | | | | | | |
| 26 | 42 | 3 | 40.5 | 1 | .02469 | .97531 | .82980 | .00063 | .00247 |
| 27 | 38 | 1 | 37.5 | | | | | | |
| 28 | 37 | 3 | 35.5 | | | | | | |
| 30 | 34 | 1 | 33.5 | | | | | | |
| 33 | 33 | 1 | 32.5 | | | | | | |
| 34 | 32 | 4 | 30.0 | | | | | | |
| 35 | 28 | 1 | 27.5 | | | | | | |
| 36 | 27 | 3 | 25.5 | | | | | | |
| 37 | 24 | 3 | 22.5 | | | | | | |
| 38 | 21 | 2 | 20.0 | | | | | | |
| 39 | 19 | 1 | 18.5 | | | | | | |
| 40 | 18 | 0 | 18.0 | 1 | .05556 | .94444 | .78370 | .00327 | .00574 |
| 42 | 17 | 1 | 16.5 | | | | | | |
| 45 | 16 | 2 | 15.0 | | | | | | |
| 46 | 14 | 3 | 12.5 | | | | | | |
| 48 | 11 | 2 | 10.0 | | | | | | |
| 49 | 9 | 1 | 8.5 | | | | | | |
| 51 | 8 | 1 | 7.5 | | | | | | |
| 54 | 7 | 1 | 6.5 | | | | | | |
| 55 | 6 | 1 | 5.5 | | | | | | |
| 57 | 5 | 1 | 4.5 | | | | | | |
| 58 | 4 | 1 | 3.5 | | | | | | |
| 61 | 3 | 1 | 2.5 | | | | | | |
| 62 | 2 | 1 | 1.5 | | | | | | |
| 65 | 1 | 1 | 0.5 | | | | | | |

SOURCE: Data of W. D. Hillis

component of variance, and when they are all multiplied together, other things being equal, one expects the amount of variation to increase. The data being binomial, there are naturally finite limits on the size of the variance.

The calculations are simplified by finding a quantity (denoted by $h$ in Table 16.3) which is for each interval

$$h = \frac{\text{Estimated probability of dying}}{\text{Mean number at risk minus the number of deaths}}$$

This quantity is accumulated *(H)* in the last column. The standard error $S_t$ at any time $t$ is then approximately the proportion of survivors at time $t$ multiplied by the square root of $H$

$$S_t = p_t\sqrt{H} \tag{16.4}$$

Formula (16.4) is only approximate, and justification for it is not easy.


## PROBLEMS

**16.1.** Grouping the data of Table 4.1 by months, estimate the mean and variance, using Sheppard's correction. Compare the results with the estimates from the ungrouped data. Comment.

**16.2.** What are some likely reasons why the quality of patients recruited into a study may change over time?

**16.3.** Suppose in familial polyposis coli the age at which polyps appear, $X$, has an approximately Gaussian distribution with mean 15 and standard deviation of 4. The further interval, $Y$, before cancer appears has an approximately Gaussian distribution with mean 12 and a standard deviation of 3 years. Suppose $X$ and $Y$ to be independently distributed.

    **a.** Plot the cumulative distributions for the onsets of polyposis and cancer respectively between ages 0 and 50, assuming that the probability of inheriting the gene is 50%.

    **b.** Plot the hazard of cancer over the same range.

    **c.** Assuming that once cancer appears the patient is removed from the study, plot the prevalence of polyposis as a function of age. (Ignore deaths from other causes.)

**16.4.** Consider a distribution bounded by two straight lines with density 0 at 20 and 40 and density 0.1 at 30. Estimate the variance by grouping at intervals of **(a)** 5 units, **(b)** 4 units, **(c)** 1 unit. Discuss.

**16.5.** Perform the same calculations as in Problem 16.4 where the two halves of the distribution are interchanged, i.e., density is 0 at 30 and 0.1 at 20 and 40. Discuss.

**16.6.** Recalculate the data of Table 16.3 regrouping at intervals of a year instead of at one month. Discuss the results.

**16.7.** Consider the following argument. The faster a car travels through an intersection, the less time it takes. Therefore the probability that a southbound car and a westbound car traveling on independent schedules will occupy the intersection at the same time, and hence crash, is inversely related to their velocities. Hence the faster one drives through an intersection the safer.

# 17

# MAKESHIFT

In this chapter I shall attempt to discuss the most difficult aspect of the statistician's art, difficult because it addresses methods of dealing with the unexpected, the indeterminate, and the uncodifiable. There are a few sound general principles, a few tricks, and a great many rough-and-ready methods of dealing with disorderly data. It is not a topic such as makes a great reputation among mathematicians; but, for all that, it may do much to maintain good will between the statistician and the scientist. The reader might wonder why I should choose to lower the tone of this book, such as it is, by entering the field. My reply would be twofold.

The first point is the more urgent because the more deplorable. Gresham's law of economics says that bad money drives out good; and it applies by analogy to practical statistics. Even where perfectly sound methods of analysis are available, we have cause to be concerned about the heathen who will use inappropriate methods, or will make up his own, paying no attention to their properties. As an instance, with which I have some familiarity, the older literature in genetic analysis is full of makeshift tests and estimators. Sometimes their properties turn out to be remarkably good; and many of the others have fallen into disuse. My point is that in the happy-go-lucky days of 1910, such an approach could at least be indulged; but with the logically powerful machinery for evaluating statistical procedures that now exists—much of it developed by the founding fathers of genetic theory themselves—in wide areas we have scarcely any defense for the amateur approach now.*

I take it that any conscientious analyst, before getting into these make-shift methods, will have explored three major questions and met with no success from any of them.

1. Does some well-established procedure already exist? (Even accom-

---

*I say this with the reservation that there are topics of major interest to the biologist that the statistician largely ignores and that the amateur must handle as best he can. I have mentioned several in this book: the statistics of modes, of treatment to effect, of regressions with discontinuities, among others. If I refer to them often enough, I may goad the professional into doing something about them.

plished statisticians can hardly expect to know the answer in all instances and will consult their erudite colleagues. Those of us brought up in Medicine and the admirable methods for tracing pertinent references find statistical literature, on the whole, poorly equipped in this respect. Fortunately the subject is more circumscribed.)

2. Can the data be manipulated formally (principally by transformation; see Chapter 4) in such a way as to make them conform to the assumptions of standard models?

3. Can the problem be resolved adequately and with reasonable statistical properties by nonparametric methods? (See Chapter 15.) In general, hypothesis testing is much better catered for in this respect than estimation.

Granted that it may not be possible to produce a perfect solution to some statistical problem, there exist degrees of imperfection. A statistician will know what he is letting himself in for when invoking the central limit theorems (see below), and when it is better to abandon the endeavor and use nonparametric methods. The inexperienced amateur is likely to swing wildly from the extreme of insouciance about all defects to that of a catatonic paralysis because there is some minor departure from the underlying assumptions.

The second reason for discussing makeshift methods is that excessive fastidiousness is likely to delay indefinitely the first step in what is inevitably an iterative relationship. Difficulties between the statistician and the scientist arise in the same way as the analogous difficulties in our genetics clinic. In both cases the client—

1. wants an unrealistically simple and clean answer and
2. wants it here and now.

Only too often the clinician cannot give such an answer because he needs more data; or because he needs to see the patient periodically over some time; or because he wants to search the literature or do some elaborate calculations. But also he needs to see the patient more than once because the patient has to be educated and because the barriers to understanding are by no means all intellectual. Now if I am consulted by a colleague who "only wants about five minutes" of my time, to deal with some statistical trifle or other, it may well be that the issue is an open-and-shut one which can be handled promptly. But commonly the problem is not of that type; and it is only from a considerable give-and-take between the other scientist and me that there is enough information even to ask the right questions, let alone answer them. I could quote at length a complicated problem I became involved in about standardizing thromboplastin for prothrombin time studies. In order to test out the assumptions underlying the analysis which would solve the problem, we required extensive ad hoc studies about the components of variance, the distribution of results, homoscedasticity, and so forth, which Dr. Denson carried out expressly. (Some of these data are used in Example 7.7.) The notion of statistical consultation as a one-shot affair, either some hopeful guess about adequacy of sample size before the experiment, or a coat of varnish on an experiment already (and often irrevocably) performed, does not conduce to fruitful collaboration. Indeed the history of science is against such shallow compartmentalization: most of our profound insights come from a regular alternation between experiment, analysis, and observation. The most distinguished scientists of the past are those who have been both their own theorists and their own experimentalists. After all, if the biostatistician is asked, about some prospective study which has never been done before, how large a sample is required, and how it should be constructed, the odds are that he has little to offer but guesswork founded on analogies, hunches, or perhaps nothing whatsoever. If these guesses have any standing at all, it is in the domain of the scientist. Answers that the biostatistician gives them—and answers must be given by somebody or else we never shall have any data—are tenuous and tentative; and he would not relish the thought of having such highly tentative opinions regarded as still binding twenty experiments later on. Of course, I argue as a scientist; but I imagine that a useful and sympathetic biostatistician will have a similar general attitude.

I am moved to inject a comment on what seems to me a fundamental blind spot, if not in the way that statisticians operate, at least in the way they expound their art. Anybody who has ever done experimental research recognizes two components that are not merely notionally separate but operationally distinct as well. They are: making discoveries; and formalizing them. The former is a delicate art for which the only qualification is genius. The traps are mainly logical and most good discoverers (they are scarce) know of them by training or intuition. At this stage the statistician can give little technical help. The responsible scientist will not publish preliminary results with no further evidence. The second step is a formal exploration of the preliminary findings, when the scientist will already know something of what he expects to find and will wish either to test a hypothesis or to estimate some parameters. Consultation with the statistician at this stage is prudent, likely to be much better informed, better polarized, and profitable. In practice, the scientist is most vulnerable during the transition between the two stages.


## PROPHYLAXIS

A great many of the difficulties of analysis are due to lack of constructive foresight. I would stress the word *constructive*. There is some risk that planning becomes mere attitudinizing and that its only effect is to demand detail where there are inadequate grounds for obtaining it, while the really germane and difficult problems are brushed aside as of minor importance. I could write at gruesome length about agonizing over the construction of samples, the appropriate size of test, the power, the experimental method, and what not, that a certain type of investigator will lavish on a study of diabetes in American Negroes; and give no thought to the two fundamental problems (both of which as a geneticist I suspect are not only unanswerable but even meaningless): What is meant by *diabetes*? and what is meant by *Negro*? The latter term might just be salvageable if the investigator made up his mind whether he was concerned with a sociological type or a genetic constitution; and if (in the latter event) he made some conscientious effort to estimate what proportions of the genomes in the individual subjects were derived from African ancestry and from European ancestry, respectively. As to *diabetes*, despite much energetic data gathering I have yet to see any evidence that blood sugar levels, glucose tolerance tests, or any other pertinent characteristics naturally segregate subjects into two classes other than those artificially constructed for administrative convenience. The issue is, perhaps, to find what the distribution of the blood glucose level, fasting or in response to a challenge, or blood insulin secretion, or glucagon levels, or what not, may be. This question is more readily, more accurately, and more efficiently approached, analyzed, and interpreted, by a random sample from the population than by a two-stage screening procedure involving a *fallible* method of exploring an *arbitrarily* truncated distribution.* I would not condemn universally all arbitrary truncations. They are appropriate where—as in blood cholesterol levels—sound reason exists for believing that there is a subsidiary population (familial hypercholesterolemia, 125) concentrated in the upper range of values, and the object is to enrich the yield of it from the sample.

It has been my chastening, but nonetheless exasperating, experience on several occasions to see prolonged planning of the logistical, rather than the scientific, aspects of a study. On one occasion a committee met regularly for three years and, in the event, completely overlooked certain fundamental scientific points. One obvious preventive measure is, instead of large experiments, to do many small ones, each designed to answer intermediate questions only, but in such a way as to provide the essential information for the next step in the planning. This method is, I regret to say, an undeveloped art. On the whole, large-scale therapeutic studies by reputable investigators are well carried out. By the time a definitive study is ready to be done, the issues are, or should be, relatively clear; and competent design should then be easily available. But in the heuristic, trail-blazing stage of research, at which the investigator thinks "he is on to something interesting," rapid and agile iteration is essential. In my experience, research of this kind calls for diffidence on the part of the investigator, if he is not to be trapped into premature claims. On the whole, it is much more likely to lead to good questions than to good answers.


## PRECISION

There are three aspects of precision to consider: of facts, of analysis, and of ideas. But precision is time-consuming, and maximal efficiency demands an even-handed attention to all three.

*Of course if we have a special interest in some *part* of the distribution we may sample otherwise. But investigators arbitrate so at their peril.

**Example 17.1.** Suppose, for instance, we are investigating the claim that a high blood level of fructose favors atherosclerosis. We can take each of the aspects of the study and explore it at great length. As experimental: What chemical or other method should be used? Should time of day and season be taken into consideration? Should dietary intake of fructose be standardized? Are age, sex, and race germane? What other extraneous and important factors affect levels? And so forth. Or analytical: How is the blood level distributed? Is the variance stable? Or is it systematically related to the mean? How much intrafamilial correlation is there? How much serial correlation? What are the components of variance in random samples? And much more of the same. Or conceptual: What do we mean by atherosclerosis? What is a sound criterion to use as an index of it? What causal relationship is surmised? Do we suppose there are feedback mechanisms? On what aspect of the cell does the fructose act? Is dietary fructose equivalent to blood fructose level? Etc. To answer any one of these sets of questions completely would involve much work. A sensible investigator will require some intermediate assurance on all three aspects that the idea is worth pursuing at all. But the reassurance can only come from makeshift (because incomplete) analysis of flawed data.

It is not uncommon for the statistician to be presented with a complete set of data about which he could ask literally thousands of possibly useful questions. This situation commonly means that the confrontation has come too late. The statistician must then be guided in his explorations by what the scientist thinks important. But, equally, the scientist should have been guided by the statistician as to which data should be collected in the first place. Some few investigators instinctively collect the appropriate data; I suppose this is no accident but happens because they have really thought the questions out scientifically. I believe that sensible approaches to sound scientific inquiry will always lead to sound designs and unambiguous analyses. But some investigators are forever getting distracted from the main issue, adding on control groups without any clear idea of what they are controlling for or why; or asking more questions than they have degrees of freedom; or getting trapped into very complicated designs in the interests of some totally misplaced ideal of efficiency.


## PROVENANCE

It is a good maxim "Never get far from your data." In the excesses of my youth I have occasionally ignored it and usually to my cost. What does it mean? No reputable analyst should ever analyze data that he does not know and that he does not have every reason to believe were collected soundly and by reputable methods.

That this warning is but a short one does not make it the less important. If this were a text on broad scientific inference I would devote several chapters to it. But brevity is the soul of wit and sometimes may lend emphasis, lacking in pages of low-density platitude.


## INSPECTION

A step further yet from the data is not even to inspect them but to proceed with computation forthwith. The acme of this attitude is number-crunching by prefabricated computer programs that are designed for speed rather than discernment, and that attract their own kind of technician. For my part, I favor old-fashioned intellectual intervention between the collection and the computation.

It is no easy matter to inspect data perceptively. Simple graphical methods—histograms, scatter diagrams, and other such devices, which have been

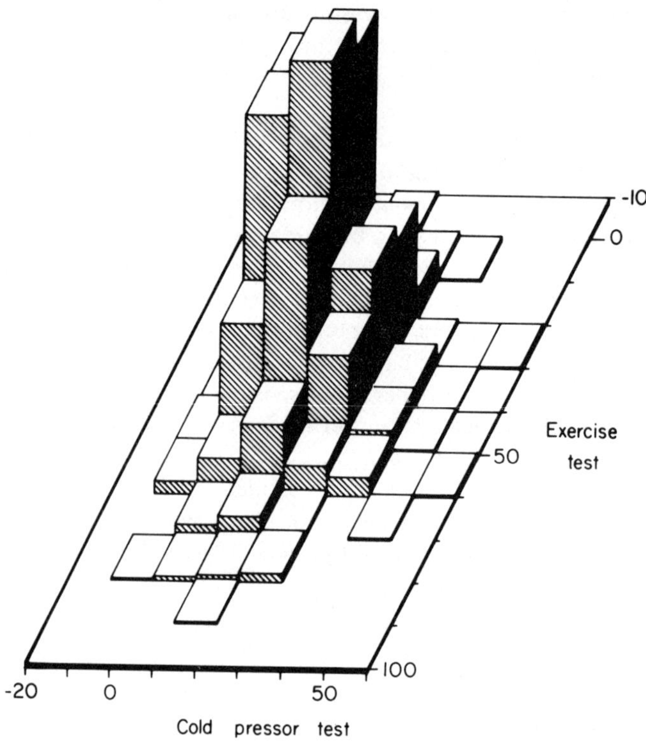


**Figure 17.1.** A bivariate histogram. Two proposed tests of "hypertensive tendency" were done in over a thousand medical students (19). One test was the response of the blood pressure to exercise, the other the response to putting an arm in ice water ("cold pressor test"). One of the issues was the correlation between (or the degree of redundancy in) these measurements. The other was whether the joint distribution displayed evidence of clustering not apparent in the marginal distributions of either.

amply illustrated elsewhere in this book—are often valuable. They tell us, for instance, what the distribution may plausibly be; they may give good evidence of clustering, which may have profound scientific, logical, and analytical implications. It is also useful to find the "range," the median, the mode, and the interquartile range. Comparing them with each other and with sample moments may throw light on structure not immediately obvious from the data, particularly if massive. They may suggest remedial transformation; or they may provide warrant for some powerful nonparametric test, or even a parametric test somewhat off the beaten track. Tukey (107) has devised a quite elaborate system of general descriptors which I can merely refer to here.

It would be false to suggest that the main inspiration for inspection of the data lies in the data themselves. This policy in the past has led to many ornate wild goose chases, which may prove expensive. I have yet to see any *formal* procedures for analyzing idea-free data that have led to a discovery of any importance. Conversely, we all have experience of phenomena with quite clear patterns of behavior that can be formally demonstrated only with great difficulty or perhaps not at all, e.g., esthetic qualities.

It is not uncommon to find that an adequate display of the data is tedious, because there are so many units in the sample. Although making a histogram of a thousand measurements on height is not too troublesome, an accurate scatter diagram of a thousand pairs of readings on height and weight is. One may salve one's conscience by displaying a *properly* constructed sample of the thousand points. There are computer programs and programmable desk calculators that can carry out this task. Also, a two-way table allows a simple sorting at the expense of some rounding. From it a two-way histogram may be constructed (Fig. 17.1).

## ASSUMPTIONS

It is necessary, before applying any test, to verify certain explicit assumptions. I have been at some pains to make these assumptions clear in each method discussed in the previous chapters, and I hope that my readers will pay at least as much attention to this issue as I have done. There is no such thing as a statistical method which is free of assumptions. The assumptions may not be very demanding; but that is all the more reason for treating them gently. It will be worthwhile to devote some discussion to them.

### Independence

This assumption is almost universally demanded. For instance, one of the nearest there are to a perfect nonparametric test, the sign test (see Chapter 15), requires that the data be mutually independent. Now, strictly, independence is a probabilistic, not a statistical, quality. For instance, it is readily shown that if the number of elements in the reference population is prime, no two traits can be independent except in the trivial case that one (or both) has the probability zero or one.* In the circumstances, it is quite useless to test statistically for independence between any two attributes whatsoever in any sample from it. Yet I suspect most investigators would not be deterred from doing such a test; and I do not think that this is mere obtuseness on their part. But in making sense of their attitude, we shall be led to a somewhat different view of independence. What the investigator does is to appeal to the superpopulation and to notional, as distinct from formal, independence.

**The superpopulation.** For the most part, the investigator is interested in universals, and particulars are to be seen as means to that end.

**Example 17.2.** Suppose we ask the question whether the average age of students graduating in medicine from Johns Hopkins in 1978 is greater than that for Harvard. What test should we use? (The reader might, with advantage, think seriously about this question before reading further.)

The ages are on record; they may be averaged; and a direct comparison will answer the question. Since the populations are finite and supposing there are no missing data, there is no sampling error; the population means are not estimated, but known. Taken on its face value, the problem is statistically trivial. But those who feel that they have been tricked by this question will have more to say about the matter. The scientist (they will argue) is not interested in gossip. Even the historian would be interested in such a comparison only if it has a wider meaning: for instance, that it tells us something about the *kind* of students graduating from the two institutions. This standpoint is equivalent to the notion that somewhere, "out there," in the admission committee's minds, perhaps, there is a population of which the students can be considered a random sample. Even events that are known to be unique are deemed of scholarly interest only if they admit of some generalization that transcends the sample. I have published a detailed discussion on this problem (70).

**Notional independence.** Elsewhere (7) I have given a broad justification for considering that the formal criterion of independence of two events (that their joint probability be equal to the product of their marginal probabilities) and the notional criterion (that the outcomes are quite unrelated to each other) are in accord. Indeed, the investigator, as distinct from the statistician, is overwhelmingly concerned with the latter meaning. The fact that the superpopulation is both infinite and prime and hence that formal independence is impossible for any conceivable pair of characteristics will leave the scientist unmoved. Whether or not A can have caused B or conversely, or they are both effects of some common cause, can have no notional relationship to whether the size of the reference set is prime or not.† This discordance

---

*This interesting fact was pointed out to me by Dr. S. H. Walker.
†Suppose, for instance, that the number of particles, quanta, etc., in the universe is prime. Then formal independence is empirically meaningless, except in contrived subspaces, such as a deck of cards. But, furthermore, since one can show from first principles (1) that we can never know the exact number of such units in the universe, we can never know whether or not formal independence has any empirical meaning.

does not lead to the abdication of rationality; it is merely yet one more instance in which the formal mathematical, and the notional, treatments of two aspects of a topic are shown to be close but not identical. It means that the rational investigator will appeal to, but not canonize, the mathematical index of independence. It is only the naive mind, or the mind dedicated to tautologies, which expects that, having made a commitment to statistics, it will never encounter any problems other than formal. As it is, I dare say that a clever enough mathematician could resolve the paradox by showing that as prime reference sets become infinite, there exist proper, non-null, subsets that are, in the limit, independent. My main point is that the formal and the notional meanings of independence, although in general reconcilable, are not in fact the same or necessarily mutually consistent.

And it is fortunate that that is so. The statistician is forever appealing to the mutual independence of his data. But to require a formal proof of independence before statistical analysis, would make statistics unwieldy and even (in my view of the world) impossible. One can know a priori from the probabilistic structure of the sample space whether events are independent—for instance, what the suit and denomination of a playing card picked from a regular deck of playing cards are. Furthermore, such a statement can be made about a single card yet to be picked. But we have agreed (in Example 17.2) on the idea that science is not concerned with the particular except insofar as it represents the general; and of course in real life, the whole unsolved scientific issue is precisely what the probabilistic structure is. Our difficulties in elucidating it cannot be resolved by looking at a single sample value. Indeed, we could not demonstrate independence with certainty from any finite subsample from the reference set. Even with an infinite subsample we would be making uncomfortably demanding assumptions about the homogeneity of the population (see below). In the face of these difficulties, the investigator is driven to appealing to notional independence—that is, to invoking independence of his data, not from stochastic evidence, but from the relationship between things in their scientific context. Although this appeal is commonly made, it calls for cultivation of judgment.

**Example 17.3.** The output of sperm in man is enormously large. Some are X-bearing and some Y-bearing. The proportions of them and their relative chances of fertilizing an ovum are not germane to our problem. It is usually supposed that (over the sample space of the husband's sperm) the types of sperm that fertilize ova in successive pregnancies are independent. There is a good deal of supporting evidence; but this supposition is made as much on the basis of broad biological knowledge as on formal analysis. The pertinent facts are that the sperm are so numerous and the turnover so rapid that the proportions should not be perturbed by any one sperm being used up in fertilization; and that no known feedback mechanisms exist that might change the probabilities as a result of the sex of the previous offspring. However, we are obliged to be careful about the definition of the sample space. There are genetic disorders (e.g., incontinentia pigmenti or focal dermal hypoplasia) in which affected male fetuses are spontaneously aborted. Thus the liveborn sex ratio is not approximately $1:1$ but $1:2$. Now the sexes of successive offspring of a carrier mother are independent *over the sample space of her children*, but they are not independent if we define the sample space as that of all progeny of all mothers in the population at large.

**Example 17.4.** Dante's *Divina Commedia* is in three sections. All three of them end in the same word *stelle* ("stars"). Were the choices of these three words independent events or are they related? I have deliberately picked what is clearly not a scientific problem. Word counts in this poem (which I do not propose to do) would doubtless show that despite the many references to stars, the joint probability of the same word in these three strategic positions is extremely small. But I think a literary scholar would be appalled at the idea of analyzing the point statistically. The relatedness of these ideas is artistically clear: Dante, "the most symmetrical of poets," does not produce such patterns by chance. One would adopt a very different attitude to such a parallel in one of the novels of Zane Grey. The scientist will be uncomfortable about this kind of argument. Yet there is little doubt in my mind that creative scientists use arguments which are no more explicit or rigorous: not as a formal proof of a proposition, but as a guide to plausible conjecture and to the most promising line of inquiry.

What can we do about analysis where there is nonindependence among at least some of the data? I have miscellaneous remedies to propose.

**Example 17.5.** A careful definition of the sample space may eliminate the problem, as illustrated in Example 17.3.

**Example 17.6.** There may be a *systematic* nonindependence which can be handled by a corresponding analysis. The most familiar example is the paired comparison (see Example 5.3). Problems of correlation are always of this type. Sometimes time-series problems may be circumvented by treating readings on each patient as realizations from a separate sample space.

**Example 17.7.** If there is certain or good or (in the case of testing a hypothesis) even conjectural evidence of nonindependence, one may *randomly* discard all but one of each set of related readings. This would be appropriate for a genetic trait where a group of subjects, otherwise mutually unrelated, included a pair of identical twins. The result would have lost a little information, but, if the correlation between twins is high, very little. However, note again that such problems may be dealt with systematically, e.g., suitable analysis of variance, interclass correlation methods, and so on.

**Example 17.8.** Where it is not clear whether two results should be regarded as independent, one may do one's calculations on all the results and also after discarding (as in Example 17.7) the questionably superfluous values. Roughly speaking, we expect the truth to lie somewhere in between.* If the two results are much the same, then our misgivings are allayed. For instance, if under both extremes a hypothesis is rejected, or under both it is ac-

---

*This is a very *rough* statement, which I would have difficulty in formulating, let alone proving.

cepted, our problem is solved. If the results with and without discarding are in conflict, there may not be grounds for making a decision. Publish both analyses in sufficient detail and "let who will, be clever"; but be more careful next time!

**Example 17.9.** Where a clear issue of size or power exists, do whatever is *least* favorable to your contention and argue a fortiori. However, this argument may be abused (3).

**Example 17.10.** For large samples where means are concerned, one may adjust the total variance by adding (or subtracting) covariance. Then under the usual assumptions one may appeal to the central limit theorems (see below).

### Metricity and categoricity

All statistical methods which I know or can conceive, require that (at least arbitrarily) the data are unambiguously categorizable, that is, can be put into exhaustive, and mutually exclusive, groups. (I say "at least arbitrarily" because it may be necessary to divide a continuous domain of values into nonoverlaping parts.) This property seems to be the irreducible minimum requirement to allow any kind of statistical abstraction whatsoever. To be able to categorize we need make no appeal to values, quantities, or even magnitudes, merely quality. Handedness, for instance, allows categorization without any appeal to which state is best, biggest, fastest, or to any other quantity. But it is impossible to make a distribution on the basis of classification alone; we cannot (for instance) compute the average handedness or what its variance may be. Instead, we may construct the distribution out of the *numbers* in the categories, and this is the foundation for much (although not all) of the discussion in Chapters 12 and 13. Handedness is not a random variate; the number of left-handed people in a sample is; and of course we could make statements about the mean, variance, etc., of this number.

Many variates can be more subtly dealt with. We may be able to rank what we cannot metricate. Pain is a notoriously difficult quantity to measure yet it may be clear that $A$ has a greater tolerance for pain than $B$ has. This contention would not necessarily be destroyed simply because of disagreement among assessors, any more than a linear metrical measurement is undermined by discrepancies among observers in the results they obtain for the same quantity. There would at least be agreement that what is being measured in both instances has meaning. But *total* lack of agreement where either type of criterion is concerned would lead to its dismissal. I would be most reluctant to compute means and variances for any purported measurement for pain; but I would have no hesitation about using the ranking methods discussed in Chapter 15. Nevertheless, while we could find medians, percentiles, etc., we would be back to the old problem of finding a meaning for them, and my misgivings would return. Still other quantities, such as height, may be treated literally as measurements. In the latter case, we will certainly capitalize on this literal interpretability, and abandon parametric efforts for nonparametric only when all efforts to regularize the data have failed.

The reader may share much of my misgiving about the tenuousness of the distinction between those quantities with meaning and those without. There is an analogy here with the distinction in Chapter 3 between "real" and ad hoc parameters. Although the issues are philosophical rather than scientific, they cannot be totally brushed aside by the investigator. We might safeguard the meaning of a measurement in two major ways. It may be useful because it corresponds faithfully to some quality that is well understood. Thus physical height or weight or (less certainly) blood pressure would be such a measurement; an index of capacity-to-get-along-with-one's-peers would not. The second, and more important, mechanism is the intelligent and perceptive editing of meaning in the light of experience. The meaning of physical height has been well edited. We know, for instance, that it is a sound means of identification, that it differs in the two sexes, that it is per-turbed in various diseases. But we know also that (except in extreme cases)* it tells us almost nothing about intelligence, musical taste, blood groups, or length of life. Intelligence tests, on the other hand are in a much more parlous state, because there is no reference standard against which they can be validated. It is arguable that analyses of intelligence quotients should be done with nonparametric methods. For although I.Q. has a distribution close to Gaussian, this shape has been produced artificially by weighting of quite diverse tests. Thus, while formally we are justified in using Gaussian theory, I have reservations as to what (for example) comparisons among cases may mean. I can readily grasp and believe the claim that A (with I.Q. 120) is more intelligent than B (with I.Q. 100) and B than C (with I.Q. 80). But I am not sure I understand, let alone believe, that B is as much less intel-ligent than A as he is more intelligent than C. But if the issue were height or body temperature, I would have no such reservations. I believe the monoto-nicity of I.Q. tests but not the literal values; and these are the circumstances in which I would use ranking tests.

Should we metricate or not? This question is not readily dismissed in a few lines. The investigator should be guided at least as much by the state of the science as by statistics. (But *please* do not confuse "the state of the science" with the fashions of the scientists!) Where the trait of interest is categorical—for instance where one is considering inheritance at a single, clearly established genetic locus—variation within types may, and should, be regarded as noise and ignored, just as static in the broadcast of a concert should be regarded as noise. But taking the responsibility for making this decision is no trivial one and is not a matter for the statistician, as such, to make.

On the other hand, if the characteristic truly is metrical, while arbitrary

---

*For instance, the height might be such as to show that the subject is not human. This theme has been developed elsewhere (7).

classification may be condoned as a crude device, it is to be deplored as an *elective* procedure. The commonest causes for the latter solecism are laziness and ignorance. The third commonest is slavishly imitating one's elders and betters. Some characteristics may have both categorical and metrical components and should be analyzed accordingly. The common example is the genetic trait which exhibits the effects of one main genetic locus (categorical) and miscellaneous effects (from other loci and the environment, Lyonization, allelic exclusion, errors of measurement, etc.) which tend to blur the basic categorization. It calls for a metrical analysis.

### Homogeneity

What has just been said about meaning carries over by analogy to the idea of homogeneity. Those investigators who imagine they are carving imperishable truths on marble, can easily fall into the trap of editing their data out of existence. In principle, we assume that the reference population is "all of one type." But, in fact, this is a polite fiction. It works well enough provided we do not try to make the inferences "more refined" than the data on which they are based. Conversely, the usefulness of the sample depends on the population to which one wishes to generalize the results. If the data are a random sample from "all boys," then any conclusions reached may be aptly applied to "all boys" and samples from them. They would not generally apply to any particular subgroup. For instance in Table 1.1 we constructed a histogram from, and subsequently based several statistical analyses on, a sample of babies, some from consanguineous matings and some from non-consanguineous. In the event, this diversity seems to be not too much of a defect. A purist might object to such pooling. But once we start to chop the population up, where do we end? Smoking habits bear on fetal size; so does dietary intake; so does duration of the pregnancy; so does birth order; so, no doubt, do many other factors known and unknown. It is easy to talk oneself into not doing any analyses whatsoever. Ultimately, I suppose, every unit is unique and we give up the effort at generalization, and thus abandon the scientific endeavor altogether.

The remedy for this nihilism is to be content with *provisional* conclusions at each step. There is no objection to comparing the effect of supplementary vitamins on a mixture of sheep and goats provided that the treated and control groups are randomly selected from the same reference set, and if the conclusions are confined to a population comprising a mixture of sheep and goats in that proportion. I do not *recommend* this method and we might argue that the conclusion may have limited value; but the logic (as distinct from the perceptiveness of the science) is unobjectionable. In any case, it would prompt the next question: whether sheep and goats respond differently; or whether there are differences among diverse types of sheep; and so on as the evidence leads.

For example, in the early history of a diagnostic test we are content to look at its usefulness in telling the sick from the well. It may prove useless. It may prove infallible. Commonly it is useful but imperfect; and methods of refining its use are to eliminate extraneous sources of variation ("standardizing the procedure") and to consider results in separate classes which bear on the result (sex, age, race, occupation, etc.). Refinement of an empiric relationship is a natural quest of science. So long as we never mistake the present state of knowledge for "the last word" on a topic, we will not go too far astray. The purist will not be happy with this attitude; but the purists are often undertakers (of a fact or a science) rather than their progenitors.

However, for my part I am apprehensive about limited experiments that had not been stratified in the first place, being artificially converted into stratified experiments after the event. There is some danger that the "questions" explored reflect no sound scientific conjecture, merely a meaningless *post hoc* comparison of mutual extremes *because* they are extreme, a state which had not been surmised beforehand.

There are several common methods for dealing with data *known* to be heterogeneous.

*Single or multiple adjustments* may be applied to the data. Pooled estimates of variance, for instance, are found by computing the residual sum of squares for each group from its own mean. Thus it does not matter whether or not the data are heterogeneous as to means, provided we know which belong to the same group. The method of covariance analysis is a slightly more elaborate procedure. Where we are concerned to test hypotheses we may use matrix algebra and the method of conditional error, which has been extensively discussed in Chapter 9.

*Analysis of variance* may be carried out in such a way as to acknowledge the existence of heterogeneity. For example, consider the randomized block design discussed in Chapter 7. In the interests of enhancing power we group on one (commonly a "nuisance") variable while testing on another. The only feature about which I have misgiving is that this method involves us in interaction: it may be a necessary evil to provide a test of significance but (as I have remarked at length) it is in some danger of being reified and creating figmented scientific problems. One method of dealing with this issue is sufficiently important to merit a separate section.


## COMBINING LEVELS OF SIGNIFICANCE

The problem of interaction can be skirted, in some cases at least, when the issue is one of testing hypotheses. The general method is based on a simple argument (7). It is that if the null hypothesis is true, then a priori the distribution of the significance level for any continuous statistic is uniform over the interval zero to one. The (sample) significance level, $P$ (which, in this context, being a random variable must be written as an upper case letter) is as likely to lie between 0.10 and 0.11 as it is between 0.98 and 0.99. The same is true for any other pair of intervals of equal width between 0 and 1. Thus the form of the distribution is a simple rectangle; and under the null hypothesis this is exactly so for every sample statistic with a continuous distribution. It will also be approximately so for discretely distributed sample statistics based on samples of reasonable size. Thus we might have done several separate, but related, experiments each giving a significance level, no one of which is decisive; and the issue is how to combine them to give an overall test, when the data themselves are so heterogeneous that they cannot be pooled. We can of course *add* the sample significance levels and compare the result with the distribution of the sum (or convolution) of identical rectangular variates. However, there is a simpler and more exact treatment. It is shown with the aid of a little calculus that if $P$ has a rectangular distribution

$$-2 \log_e P \sim \chi_2^2$$

and since the convolution of $\chi^2$ variates is a chi-square variate with degrees of freedom equal to the sum of their degrees of freedom (7), if there are $k$ such $P$ values, under the null hypothesis

$$-2 \sum_{i=1}^{k} \log_e P_i \sim \chi_{2k}^2$$

**Example 17.11.** Consider again the relationship between age and survivorship in malignant hypertension (Example 10.3). Various methods of testing for an overall relationship exist. None of the correlation coefficients for the four subgroups is significantly different from 0 at (say) the 5% level. In fact we have the values in Table 17.1. The total is distributed (under the null hypothesis that $\rho = 0$) as a chi-square variate with 8 degrees of freedom.

**Table 17.1. Correlation between age and logarithm of survival from diagnosis in four categories of sex and diagnosis**

| Group | | Correlation coefficient | $t$ | $p^*$ | $-2\log_e p$ | $-2\log_e(1 - p)$ |
|---|---|---|---|---|---|---|
| Male essential | 19 | $-0.2910$ | $-1.326$ | .900 | 0.211 | 4.599 |
| Male renal | 12 | $+0.2043$ | 0.723 | .242 | 2.839 | 0.554 |
| Female essential | 11 | $-0.1420$ | $-0.476$ | .678 | 0.777 | 2.268 |
| Female renal | 18 | $-1.1951$ | $-0.844$ | .795 | 0.459 | 3.171 |
| Total | | | | | 4.286 | 10.592 |
| Combined significance level | | | | | 0.830 | 0.226 |

SOURCE: Data of Kincaid-Smith et al. (21).
*The probability of the sample value or higher under the hypothesis $\rho = 0$.

The critical value for a test of size 5% is 15.507. We are thus led to accept the hypothesis.

Let us consider a little further the meaning of this test. We do well to recall the confusion exposed in Chapter 2 between a statistical hypothesis about the value of a parameter and a scientific hypothesis about relationship. A statistician as such is not equipped to cope with a question like "Is prognosis in malignant hypertension related to age?" It has to be cast in some such form as "Is the correlation coefficient between $X$ and $Y$ zero?": for him the evidence has to be not only quantitative but homogeneous. We may use the method we are discussing for data which cannot be pooled, where differences among samples are a matter of scaling or calibration of measurement. We cannot for statistical purposes pool data on blood levels of urea in man, creatinine in dogs and nonprotein nitrogen in rats, even if we have perfectly good data and significance levels. But insofar as all three measurements reflect renal functions, as a scientist I may quite legitimately inquire from such miscellaneous evidence whether phenacetin damages the kidney. Evidence by several different criteria in several species may perfectly well be *notionally* pooled to try to answer that question. In fact scientists do this frequently, for example in writing review papers.

The formal problem is, how one is to construct a most powerful rejection region; and the main misgiving is whether one is to arrive at a homogeneous conclusion about such a heterogeneous question. For my part, I cannot afford the luxury of much fastidiousness. I do not believe that any actual scientific question which I am ever likely to ask will address a homogeneous question or that any evidence I can obtain will (except for scale) be homogeneous either. Yet to be discouraged by these facts would paralyze most of biological inquiry. Part of the problem in the present context is that what we really need in many problems is a test of *independence*, not of *correlation*. If two variables are correlated they are not independent, but the converse does not hold.

**Example 17.12.** Consider the notional question "Is the level of heparin in blood in any way related to the mean length of survival of blood platelets in dogs?" Detailed studies (126) show that the relationship is a complicated one: at low, and extremely high, dosages, heparin shortens platelet survival (although the mechanisms are quite different); at intermediate levels it lengthens it. It would scarcely be surprising if several groups of investigators exploring different levels of drug came up with highly significant, but mutually contradictory, results. The more telling problem arises where several investigators obtain mutually contradictory results none of which quite attains a convincing level of significance. It is all very well for the purist to throw his hands up in despair at this evidence. Would he not be prepared to say that, *whatever* the mechanism may be, there is *some* relationship between these two quantities? The inference is, of course, not to be seen as a conclusion, but as an incentive to explore the question more deeply.

The policy I would base on these ideas is that tests of the combined hypothesis should be two-tailed.

**Example 17.13.** In a collaborative study on the diagnostic value of the Salzman test in von Willebrand's disease (127), the upper-tail significance levels for the correlation coefficients of the platelet adhesive index and rate of flow of blood into the syringe were obtained (Table 17.2). Those dedicated to homogeneity will not like the fact that some of the sample correlation coefficients are negative, i.e., $p > 0.5$; and will insist that the only alternative to the null hypothesis would be that the correlation coefficient is positive for all or negative for all and not positive for some and negative for others. Thus, the significance levels should be based on one-tailed tests. I have stated my scientific objections to this restriction. Let me now state some statistical objections.

1. If, instead of doing a $t_{n-2}$ test on the correlation coefficient, I do an $F(1, n - 2)$ test on $r^2$, a perfectly legitimate operation, the effect of the sign of the $t$ test will be obliterated. An analogous effect results whenever we substitute a chi-square for a binomial test.

2. If we compute one-tailed significance levels, some of the values will be less than $\frac{1}{2}$ and some greater. According to which tail we put the rejection region in, under the null hypothesis the significance for the $i^{th}$ value will be either $p_i$ or its complement $(1 - p_i)$. For instance, suppose in a particular result the $p$ value given $\rho = 0$ against the alternate $\rho > 0$ is 0.9. Then the $p$ value against the alternate $\rho < 0$ is $1 - 0.9 = 0.1$. The reader can readily verify this statement by drawing a suitable diagram.

In the present case both

$$-2\Sigma \log_e p_i = 30.45$$

and

$$-2\Sigma \log_e (1 - p_i) = 60.33$$

lead to rejection of the null hypothesis. I confess I have problems in making sense of this conclusion on the supposition of homogeneity. To be sure we can say that one wild result contributes the largest part of the former $\chi^2$ value and that there is something peculiar about this sample, which should be treated separately. But as a scientist I get no inspiration from this arbitrary decree. I had much rather suppose the relationship between the two variables is complicated than the otherwise unjustified supposition that the populations sampled at the several participating centers are diverse.

3. Doubtless it will be objected that the sample correlation coefficients are heterogeneous. But there is no formal requirement that they should be homogeneous. And what would we do about it in analysis, anyway? We cannot legitimately decide which tail to put the alternative value into on the basis of the data. On the other hand there seems to be overwhelming evidence that there is *some* relationship, perhaps a very complicated one.

**Table 17.2. Significance levels for the linear regression coefficient of Salzman's platelet adhesive index on rate of flow obtained in twelve research centers**

| Center | $p^*$ | $-2\log_e p$ | $(1 - p)$ | $-2\log_e(1 - p)$ |
|---|---|---|---|---|
| 1 | 0.00003124 | 20.7476 | 1.0000 | .0000 |
| 2 | 0.9623 | 0.0769 | 0.0377 | 6.5562 |
| 3 | 0.3437 | 2.1360 | 0.6563 | 0.8423 |
| 4 | 0.9957 | 0.0086 | 0.0043 | 10.8983 |
| 5 | 0.9541 | 0.0940 | 0.0459 | 6.1626 |
| 6 | 0.9995 | 0.0010 | 0.0005 | 15.2018 |
| 7 | 0.9530 | 0.0963 | 0.0470 | 6.1152 |
| 8 | 0.9753 | 0.0500 | 0.0247 | 7.4019 |
| 9 | 0.1146 | 4.3326 | 0.8854 | 0.2434 |
| 10 | 0.8519 | 0.3206 | 0.1481 | 3.8197 |
| 11 | 0.4448 | 1.6203 | 0.5552 | 1.1769 |
| 12 | 0.6160 | 0.9690 | 0.3840 | 1.9142 |
| Sum | | 30.45 | | 60.33 |
| Combined significance level | | 0.0333 | | 0.000,056 |

*Probability, under the null hypothesis, of the sample value or higher.

## GAUSSIAN APPROXIMATION

In fragmentary fashion, we have encountered the use of the Gaussian distribution as an approximation to many distributions: the binomial, the Poisson, Kendall's $\tau$, the Mann-Whitney $U$, the chi-square, the gamma, etc. The time is come to make a coherent statement of a general property which covers a wide variety of cases, including these particular examples, and many more. The property is stated more formally elsewhere (7) but proof calls for some considerable knowledge of theory. It is proved in many standard texts in mathematical statistics. The statements apply to the distribution of the sum and product of independent random variables.

### The sum of random variables

**Theorem 17.1.** *The sum of any number of identically distributed Gaussian variates is itself Gaussian.*

Since dividing that sum by $n$ (the number of values in the sample) merely involves a difference in scale, the mean of these values is itself Gaussian. The Gaussian is the only variate for which this is *exactly* true.

**Theorem 17.2.** *The mean of* n *independent, identically distributed values from* any *distribution (provided only that the variance of each is finite) gets closer and closer to Gaussian as* n *increases.* Because of independence, the mean and the variance are simple functions of the mean and variance of the individual values. This general proposition constitutes the so-called classical central limit theorem.

By way of illustration, there are shown in Figure 17.2 the histograms of the sample means of sets of 10,000 samples of various sizes for three distributions of very different forms, all far from Gaussian. The samples are generated artificially by the method of Monte Carlo simulation, using a random number generater. It will be evident that for small samples the distributions are very irregular. But with even moderate sizes of sample the distributions tend to become symmetrical and to be heaped up near the mean, giving a pattern which approaches the Gaussian. This is a very remarkable property, which is of the greatest value to the practical biostatistician.

Now remember that we are in the field of approximations and are making only rather weak assumptions which are readily met. It is no easy matter to make ironclad generalizations. By and large, the properties of the central limit theorem are most readily invoked—

1. if the sample is large;

2. if the acceptance region or the confidence limits (as the case may be) are not too large (they will be more accurate at 95% than at 99%);

3. if the parent distribution is close to normal. In particular, marked skewness seems to give a good deal of trouble.

A source of concern may be doubt as to whether the units being sampled even come from the same parent distribution. Where they are not, there are more or less complicated theorems which warrant using Gaussian theory, but they are tedious to state and in practice it is very difficult to verify the assumptions. Certain cases can be given blanket warrant.

**Example 17.14.** The convolution of independent, nonidentical Poisson variates is always a Poisson variate with parameter equal to the sum of the individual parameters. As this sum becomes large, the distribution of the sum becomes more and more nearly Gaussian. Any linear combination of Gaussian variates is Gaussian; hence, the mean is also Gaussian.

**Example 17.15.** The convolution of independent, binomial variates with parameters which differ but are fixed (i.e., they vary, but not randomly) follows not a binomial but a Lexis distribution (7). However, the Lexis distribution of order $n$ becomes ever more nearly Gaussian with increasing $n$. For instance, if a genetic trait is controlled by many equally important loci, the effects of which are additive, the trait will tend to have a Gaussian distribution. (But note that additivity here is a very strong assumption.)

### The product of random variables

We have encountered this problem earlier and it can be dealt with promptly. Let us suppose that a number of independent, identically distributed, random variables contribute multiplicatively to some measurement. Suppose that each may assume positive but finite values only and that the variance of the logarithm of each is finite. Then clearly:

1. Since they are all from the same distribution, their logarithms are all from the same distribution.

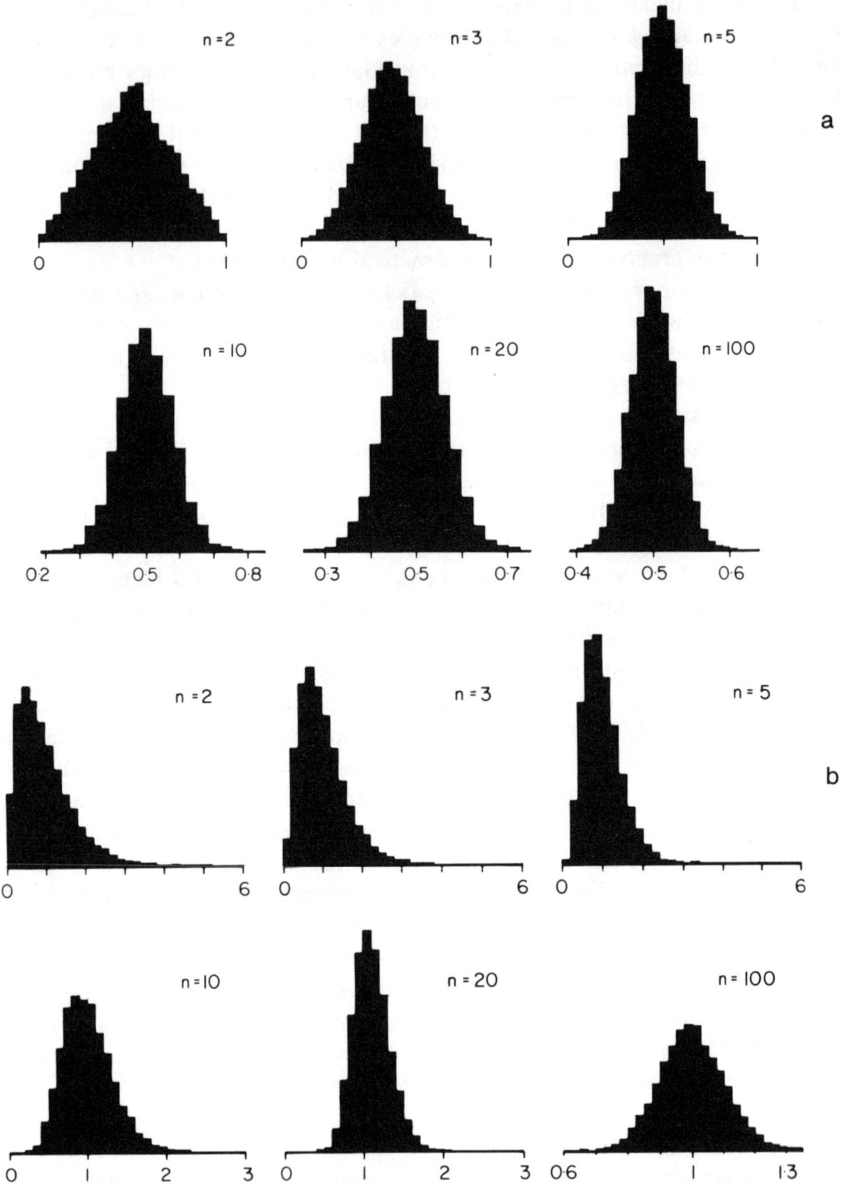

**Figure 17.2.** Illustration of the central limit theorem by Monte Carlo simulation. These three sets of histograms each display 10,000 simulations, each simulation being the sample mean of $n$ independent identically distributed variates. (*a*) For the continuous rectangular distribution over the interval [0,1]. (*b*) For the exponential variate with unit mean. (*c*) for the $U$-shaped distribution

$$f(x) = 5.946x^4 \qquad -\sqrt[4]{0.5} \le x \le \sqrt[4]{0.5}$$

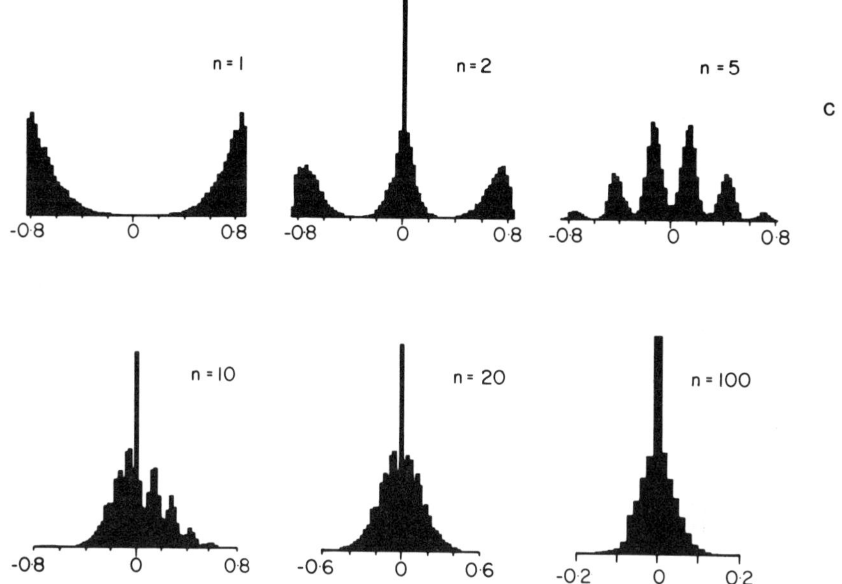



**Figure 17.2.** *(continued)*

2. The convolution of their logarithms will thus, according to the central limit theorem approach the Gaussian distribution.

3. The sum of their logarithms is the logarithm of their product.

4. Hence the logarithm of their product will tend to follow a Gaussian distribution.

5. That is (by definition), their product will tend to be lognormal (7).

This argument warrants the bold use of logarithms as a normalizing and variance-stabilizing transform for biological variables where the law of mass action, enzyme cascades, and autocatalytic processes are commonplace. Needless to say, at least approximate appropriateness of the distribution must be empirically verified in the individual case.

### Other patterns

There are other, and more complicated, ways in which the components of a system may be put together. There are at least three instances that come to mind in which the components may be made additive by taking reciprocals of some type of measurement:

1. The strengths of lenses measured in focal lengths. (Expressed as diopters they are additive.)

2. Redundant metabolic pathways with activities expressed as biological half-lifes. (Expressed as rates of action they are additive.)

3. Electric resistances in parallel. (Expressed as conductances they are additive.)

In all such cases the reciprocal of the combined effect, $Y$ is the sum of the reciprocals of the $X_i$.

$$\frac{1}{Y} = \Sigma \frac{1}{X_i}$$

Thus, granted that the $X$ are random variables, their reciprocals will also be random variables and, under the usual conditions, the reciprocal of $Y$ will approach a Gaussian distribution, or $Y$ itself will follow a "harmonic-normal" distribution, an interesting distribution which has, by and large, been ignored. Indeed it is largely circumvented by doing all the manipulations on the transformed scale. However, it deserves mention as a warrant for using the reciprocal transformation to normalize data. At least once I have used this transformation empirically: in the "plasma thromboplastin time" (128), a crude measure of coagulation. It is noteworthy that many coagulationists use the reciprocal of the prothrombin time as a measure to control dosage of dicumarol. (In fact, because of the monotonic relationship, it is quite indifferent which measure is used.)

This subject of nonadditive variates could be elaborated extensively. Genetic linkage analysis, for instance, has been the subject of extensive theoretical development to show how recombination fractions should be compounded (129). The whole idea of a linkage map is to devise a monotonic transformation that will make the quantities additive and hence pave the way for the operation of the central limit theorem. There is an added problem, which is too complex to go into here, that estimates for large values of the recombination fraction have pathological statistical properties (130).

It will be evident that on an appropriate transform we may find scope for the central limit theorem in a great many disguises. I shall be interested in hearing of other biological examples from readers.


## PATHOLOGIES OF THE MEAN

We are so accustomed to use the sample mean and variance that it is salutary to realize that they, too, have their shortcomings. I may mention briefly three interesting pathologies of the mean.

It may not exist at all. For example under certain plausible assumptions the distribution of hits about a target follows a Cauchy distribution (7), which is much the same as a $t$ distribution with one degree of freedom. Since, like all $t$ distributions, the latter is symmetrical, it obviously has a center of symmetry; but some rather complicated theory shows this is not its mean: that in fact there is no mean. Hence, there is not the slightest point in trying to estimate it.

Second, as we have seen in Chapter 15, the mean may be infinite. The reciprocal of a Poisson variate has a positively infinite mean, the logarithm of a Poisson variate a negatively infinite mean. Again there is no point in estimating it. In both this and the previous instance we must attempt some more profitable task, such as estimating the median.

Third, the sample mean may *not* be the best estimate of the true mean. We have seen in problem 3.1 that for a rectangular distribution the best estimate (the MLE) is not the mean of *all* the sample values but the mean of the largest and smallest sample values.

## PATHOLOGIES OF VARIANCE

In a fragmentary way throughout this text we have encountered the seeds of strange behavior in the variance.

The variance may not exist at all. Obviously since the variance is the mean square about the mean, if the mean does not exist, the variance cannot even be defined.

Even if the mean exists, the variance may be infinite. For example the $t$ distribution with two degrees of freedom has a mean (0) but the variance is infinite. This distribution has the curious property that the standard deviation of the sample mean ("the standard error") does not become smaller as the size of sample increases. We are no nearer the true value of the mean after averaging a million values than we are after one value. Of course we know much more about the distribution. But this pathology should drive home to readers the perils of unthinking routine in analyzing data. Put not your trust in means and variances. And it is a strong reason for *inspecting* one's data carefully and not merely handing it over to a computer.*

### Heterogeneity of variance

In Chapter 5 we have seen that the comparison of sample means from any two mutually independent Gaussian distributions may be readily made by converting the difference in the sample statistics into a standard Gaussian variate. Where the variances are unknown but may be assumed to be equal, we use the $t$ distribution with a pooled estimate of the common variance. When the variances are unknown and we have reason to believe them unequal—for example by comparing them by means of an $F$ test—then there is no simple solution. We must deal with the problem piecemeal.

**Example 17.16.** If the sample sizes are equal and there are any reasonable grounds for pairing, then pair and do a one-sample $t$ test on the paired differences, on the hypothesis that their mean is 0. In some cases, of course,

---

*One of my clinical teachers used to say, "The methods of physical examination are inspection, palpation, percussion and auscultation; and the greatest of these is inspection."

we *must* pair, for instance if they are readings on the same patient under two circumstances or "normal" and "abnormal" sibs are involved in some study. The grounds for pairing may be accidental; however, it is of the first importance that the pairing be done *independently of the measurement*. For instance, it would not be legitimate to rank both samples and match the ranks in the two groups. But we might match them on some covariable (e.g., age or sex).

**Example 17.17.** If both samples are large, the sample estimates of the variances will be close to the true variances and we may proceed as if they were exact and using formula 5.3.

**Example 17.18.** It may be possible to stabilize the variance by suitable transformation (see Chapter 4). This may involve slight distortion of a Gaussian variate.

**Example 17.19.** If the sample sizes are equal but there are no grounds for matching, then match *randomly* and analyze the paired differences as before.

**Example 17.20.** If the size of one sample is a multiple, $k$, of the other, one may match randomly (or, of course, systematically if this is possible) the individual cases of the smaller group and the mean of $k$ values from the larger group. The usual one-sample $t$ is then done on the paired differences.

**Example 17.21.** If the two sample sizes are not quite equal or not quite a multiple one of the other, one may omit unmatched values *randomly*.

The last three procedures are all quite sound, but have the disadvantage of being idiosyncratic. A second random allocation might produce a quite different result. In addition, the last two procedures are somewhat inefficient.

**Example 17.22.** If we cannot circumvent the problem and do not wish to do random matching—either because we are afraid of abuses, or of being charged with them, or because the data are too precious to discard any—we may use one of several formal exact tests described and tabled in advanced textbooks. But there are good approximate methods. Of these, the Aspin-Welch method (131) will ordinarily be quite adequate for our purposes. It is in effect something like a two-sample $t$ test, but the variances are handled separately and the equivalent number of degrees of freedom is calculated approximately.

Let us suppose that we have two samples. The one is of the Gaussian variate $X$, and is of size $n$. The sample estimate of the mean is $m_X$ and that of the variance is $s_X^2$. The other is a sample of size $r$ from the Gaussian variate $Y$ with a sample mean $m_Y$ and sample variance $s_Y^2$. Then the estimated variances of the sample means will be respectively

$$\frac{s_X^2}{n} = W_X$$

$$\frac{s_Y^2}{r} = W_Y$$

Then we read the test statistic*

$$\frac{m_X - m_Y}{\sqrt{w_X + w_Y}}$$

against a $t$ distribution with degrees of freedom, $f$, given by the formula

$$\frac{1}{f} = \left(\frac{w_X}{w_X + w_Y}\right)^2\left(\frac{1}{n - 1}\right) + \left(\frac{w_Y}{w_X + w_Y}\right)^2\left(\frac{1}{r - 1}\right)$$

## SOME MEMORABLE VALUES

The statistician properly equipped has extensive tables of test statistics. For quick approximations (for example on field trips) it is worthwhile to have at one's finger tips certain values.

First, the Gaussian distribution. One may, with reasonable effort, remember the key values in Table 17.3.

Second, for the chi-square distribution, the mean is the degrees of freedom and the variance is twice as great. Even for very few degrees of freedom the Gaussian approximation works quite remarkably well. (Table 17.4.)

Third, the $F$ distribution for any reasonable number of degrees of freedom for the denominator can be reasonably approximated by the chi-square.

Fourth, the $t$ distribution rapidly converges on the Gaussian as the number of degrees of freedom increases. (Here I may interject a comment on misplaced fastidiousness. One commonly sees investigators, even statisticians,

**Table 17.3. Some approximate values for the standard Gaussian distribution**

| $z$ | $P(Z > z)$ | Approximate probability |
|---|---|---|
| $2/3$ | 0.2525 | 1/4 |
| 1 | 0.1586 | 1/6 |
| $1\frac{1}{3}$ | 0.0913 | 1/11 |
| $1\frac{2}{3}$ | 0.0478 | 1/21 |
| 2 | 0.0227 | 1/44 |
| $2\frac{1}{3}$ | 0.0098 | 1/100 |
| $2\frac{2}{3}$ | 0.0038 | 1/260 |
| 3 | 0.0013 | 1/740 |

*This quantity is commonly called the Behrens-Fisher statistic. Mickey and Brown (132) showed formally that the appropriate significance level for any particular value of it must lie between the significance level for a $t$ statistic with degrees of freedom given by $(n + r - 2)$ and whichever of $(n - 1)$ and $(r - 1)$ is the smaller.

**Table 17.4. The Gaussian approximation to the $\chi^2$ distribution**

| Degrees of freedom | $\alpha$ (one-tailed) | | | | | |
|---|---|---|---|---|---|---|
| | 0.05 | | | 0.01 | | |
| | Exact | Approximation | Ratio | Exact | Approximation | Ratio |
| $(\nu)$ | | $(\nu + 1.645\sqrt{2\nu})$ | | | $(\nu + 2.326\sqrt{2\nu})$ | |
| 1 | 3.841 | 3.326 | 1.155 | 6.635 | 4.289 | 1.547 |
| 2 | 5.991 | 5.290 | 1.133 | 9.210 | 6.652 | 1.385 |
| 3 | 7.815 | 7.029 | 1.112 | 11.345 | 8.698 | 1.304 |
| 4 | 9.488 | 8.653 | 1.097 | 12.277 | 10.579 | 1.255 |
| 5 | 11.070 | 10.202 | 1.085 | 15.086 | 12.355 | 1.221 |
| 10 | 18.307 | 17.357 | 1.055 | 23.209 | 20.402 | 1.138 |
| 20 | 31.410 | 30.404 | 1.033 | 37.566 | 34.711 | 1.082 |

who will appeal to the central limit theorem to analyze data which they have every reason to know are sampled from a non-Gaussian distribution. Having made this approximation they go through an absurd parade of wondering about how many degrees of freedom they should use in the resulting "$t$" statistic. This is to strain at a gnat and swallow a camel. As one trenchant theologian put it, "If you sin, sin boldly.")

### The Chebychev inequality

For any distribution the probability that any particular sample value is more than $k$ (true) standard deviations from the (true) mean is not greater than $1/k^2$. Thus the probability of a value being more than three standard deviations from the mean is less than $\frac{1}{9}$. This statement is a weak one compared with what one could say if the distribution were known to be Gaussian. Nevertheless, although conservative, it is still valuable in the general case; and it is certainly a rule which is easy to remember.

## SIMPLE CHECKS

Finally, we may note a few simple checks which may be applied to calculations and protect one from at least the more crass and obvious errors. (I shall assume that the sample values are not all equal.)

1. The sample mean is always less than the maximum reading and greater than the minimum.

2. The sample standard deviation is always greater than the absolute value of the smallest deviation of a sample value from the sample mean and always less than that of the furthest.

3. For practical purposes, the standard deviation is always less than half the range of the sample values. (A precise statement of the relationship is given in 7.)

4. The sample variance is always positive.

5. The absolute value of the sample correlation coefficient never exceeds one.

## PROBLEMS

**17.1.** An investigator (whose identity, even whose topic, I shall tactfully conceal) selected for further analysis the top 10% and the bottom 10% of a sample of several hundred values he had collected. These analyses included a formal test of whether their means differed significantly. Comment.

**17.2.** In analyzing the conformity to some Mendelian pattern of the occurrence of some trait in numbers within sibships, one encounters several sets of twins. How would you treat them statistically if they are (**a**) monozygous (identical); (**b**) dizygous (nonidentical); (**c**) of doubtful zygosity?

**17.3.** Compute the point estimate on the mean of the data of Table 4.1.

  **a.** Using the method of moments. Compute the 95% confidence limits using the central limit theorem. Also compute the method-of-moments estimates of the parameters.

  **b.** Comment on the results.

**17.4.** Compare the means in Table 4.9 using—

  **a.** the logarthmic transformation;

  **b.** the Aspin-Welch statistic;

  **c.** random pairing.

**17.5.** Prove the statement in the text that no two categorical attributes defined on a reference population with a prime number of elements can be independent unless either one or both are certain or impossible.

**17.6.** In a metabolic pathway comprising 100 steps, the rate at which any one step occurs is a random variable which is determined independently for each at birth. Suppose that the distribution is the same at each of the steps, each assuming the arbitrary value 0.1 with probability 0.001 and 1 with probability 0.999. At equilibrium, what is the distribution of the value for the whole system? (Assume that the system is strictly serial, that there are no alternative pathways, and that the system always works at the maximum rate possible.)

# COMMENTS
# ON THE PROBLEMS

## CHAPTER 1

**1.1.** To some extent, almost all of these types are ambiguous.

**a.** Given that it is paid in cash, the distribution of the hospital bill is discrete: it could not be more finely divided than one cent. One could imagine problems if it could be paid in kind!

**b.** If the result is expressed as the *number* of nodes (even partially) involved, the distribution would be discrete. However, if some such measures as the distance the metastases had traveled, or the weight of malignant tissue, were used, it would be continuous.

**c.** Intelligence is presumably continuously distributed. However the measurements of it are characteristically made on a scale of whole units. One may have an assigned intelligence quotient of 88 or 89 but not of an intervening quantity. It is therefore discrete.

**d.** Discrete.

**e.** Traditional treatment of recombination is as a continuous variable. Now that the nature of the genetic material is better understood, we are better able to see it as a finely divided but discrete process, the indivisible unit being the cistron. Even before the elucidation of DNA and RNA, we must presume that the geneticist believed the process discrete at the atomic level. However, the coarser discreteness at the level of the gene makes continuous approximations more precarious.

**f.** Characteristically the lesions are discrete. However, when the condition is very severe, the lesions may be confluent, and their extent can no longer be adequately described by enumeration.

**g.** Glossing over the fact that the ultimate problem is expressed in the ratios of molecules (which ensures that it is discrete) we might suppose the threshold continuously distributed. However, in the usual method of testing, certain standard solutions of fixed strength are used and the results of the testing will be discretely distributed.

**1.2. a.** There are seven possible outcomes. There may be no successes, with probability $(2/3)^6 = 64/729$. There are six ways in which there may be one success, which thus has the probability $6(2/3)^5(1/3) = 192/729$. There are $\binom{6}{2}$ ways of getting two successes so the probability is $(6)(5)/2$ times $(2/3)^4(1/3)^2 = 240/729$. And so forth. The other probabilities are $160/729$, $60/729$, $12/729$, and $1/729$ respectively.

**b.** The probability density is partly the invariant fraction $1/(5)\sqrt{(2\pi)} = 0.079788\ldots$, and partly a variable component, i.e., the natural antilogarithm of $[-(x - 10)^2/(2)(5)^2]$. We may calculate it for selected values. However since it is symmetrical about 10, we may give the following results:

| $x$ | $\exp[-(x-10)^2/50]$ | $f(x)$ |
|---|---|---|
| $-5$ or 25 | $0.011,108,9\ldots$ | $0.000,886\ldots$ |
| 0 or 20 | $0.135,335,2\ldots$ | $0.010,798\ldots$ |
| 5 or 15 | $0.606,530,6\ldots$ | $0.048,394\ldots$ |
| 10 | $1.0$ | $0.079,788\ldots$ |

**c.** There are $(2)^5 = 32$ possible outcomes, the maximum winnings being $15. The various outcomes may be written in full according to the number of coins won. We may capitalize on the obvious symmetry of the individual winnings about 3, which means that we need only reckon up the possibilities giving winnings of $7 or less. The coins won in the various possible outcomes will be shown as figures, those lost as blanks.

| Winnings | Combinations | | | Probability |
|---|---|---|---|---|
| 0 | ----- | | | 1/32 |
| 1 | 1---- | | | 1/32 |
| 2 | -2--- | | | 1/32 |
| 3 | 12--- | --3-- | | 2/32 |
| 4 | 1-3-- | ---4- | | 2/32 |
| 5 | 1--4- | -23-- | ----5 | 3/32 |
| 6 | 1---5 | -2-4- | 123-- | 3/32 |
| 7 | 12-4- | -2--5 | --34- | 3/32 |

**Comment.** This is an example of a discrete distribution which is the convolution (i.e., the "sum") of Bernoulli trials with differing scores. It is therefore not binomial, since it violates assumption 3. It is unimodal, and already has something the shape of a Gaussian distribution. Its form and that of the convolution of two such distributions are shown in Figure A1.

**1.3.** Beware of such statements! In the first place one should be very careful about making any inferences from patients seen in a clinic, who are apt to be atypical instances of the disease. (The more severely affected, for instance, will be overrepresented.) But also, random patients may be "picked" but not "picked up." Formally, they are selected in accordance with a sampling procedure. More than likely, the data are not representative even of the patients coming to the clinic. Most people making such statements would have to admit, when pressed, that they have not the foggiest idea what they do mean by "at random."

**1.4.** I trust that the reader will have had a pleasant time speculating. In fact, these 50 reading were taken at random from a book of random numbers, each digit being equally likely. There should thus be no clustering. An infinite sample, suitably scaled, would give a uniform discrete distribution. This exercise is a lesson in the vagaries of small samples. For the behavior of histograms for small samples, see (3).

## CHAPTER 2

**2.1. a.** The ratio of the densities is

$$\frac{\dfrac{1}{10\sqrt{2\pi}}\,\exp[-x^2/200]}{\dfrac{1}{20\sqrt{2\pi}}\,\exp[-(x-15)^2/800]}$$

which with a little manipulation reduces to

$$2 \exp[-3(x + 15)(x - 5)/800]$$

This function can be readily shown to be symmetrical about $x = -5$, attaining its maximum at that point.

Thus

| $x$ | Density ratio |
|---|---|
| $-60$ or $50$ | 0.000,034,463,7 |
| $-40$ or $30$ | 0.029,434,058,6 |
| $-20$ or $10$ | 1.251,568,019 |
| $-6$ or $-4$ | 2.899,090,829 |
| $-5$ | 2.909,982,829 |

The Neyman-Pearson lemma says we must pick the lowest density ratios, which means that the rejection region must be chosen in two parts, one in each tail of the distribution under the null hypothesis.

**b.** The possible pairs of values defining the acceptance region must be symmetrically placed about $-5$ because, as we have noted, the density ratio is symmetrical about this point. With the help of Table 2.1 and remembering that the standard deviation under the null hypothesis is 10, we get the following results.

| Lower limit | Upper limit | Contribution to the size | | |
|---|---|---|---|---|
| | | Lower tail | Upper tail | Total |
| $-30$ | 20 | 0.0013 | 0.0228 | 0.0241 |
| $-28$ | 18 | 0.0026 | 0.0359 | 0.0385 |
| $-27$ | 17 | 0.0035 | 0.0446 | 0.0481 |
| $-26.45$ | 16.45 | 0.0041 | 0.0500 | 0.0541 |

Interpolation suggests about $-26.81$ and $16.81$ as the appropriate limits, and these figures can be confirmed from more detailed tables of the Gaussian probability function. These quantities are respectively

$$\frac{-26.81 - 15}{20} = -2.0905 \quad \text{and} \quad \frac{16.81 - 15}{20} = 0.0905$$

standard deviations from the mean under the alternative hypothesis. The probability below the former figure is 0.0183 and that above the latter is 0.4639. These two sectors lie in the rejection region, and the combined power is thus 0.4822. If the alternative hypothesis is true, the chance of rejecting the null hypothesis is thus rather less than 1/2.

**2.2.** This is a recurrent vexed problem. Scientifically it appears that there is a significant effect on antibody formation and hence that the response to dosage is at least biphasic: that there are at least two mechanisms involved, one of which tends to dominate the picture with smaller doses and conversely for higher doses. One could quote many biological examples of the same kind of nonlinear effect. It can even be argued that since the effect was the opposite of what was expected the outcome is secure from the charge of observer bias. Most scientists would, I think, consider this appraisal reasonable.

But there are logical problems. If the use of a formal test of hypothesis in such a case is justified by appeal to Neyman-Pearson theory, which prescribes what rejection region should be used, we must abide by the rules which, among other matters require that we specify an alternative hypothesis *beforehand*, i.e., we do not make the choice of the alternative hypothesis depend on the data. Now here no explicit alternative was chosen in advance: the most that we can say is that the investigators "expected" the outcome to be a reduction, not an increase, in antibody formation. May we regard this as an implicit decision that the rejection region must lie in the lower tail of the curve for control subjects? If so, none of the upper tail is in the rejection region and any value in it, *however large*, must be ascribed to chance. This amounts to saying that it is inconceivable that any dose of serum should *enhance* antibody formation. If this is what is believed, then the data in the upper tail, however extraordinary, should not lead us to change our minds. Naturally, if all the rejection region is put in one tail, the critical sample value that would lead us to reject is less far from the mean under the

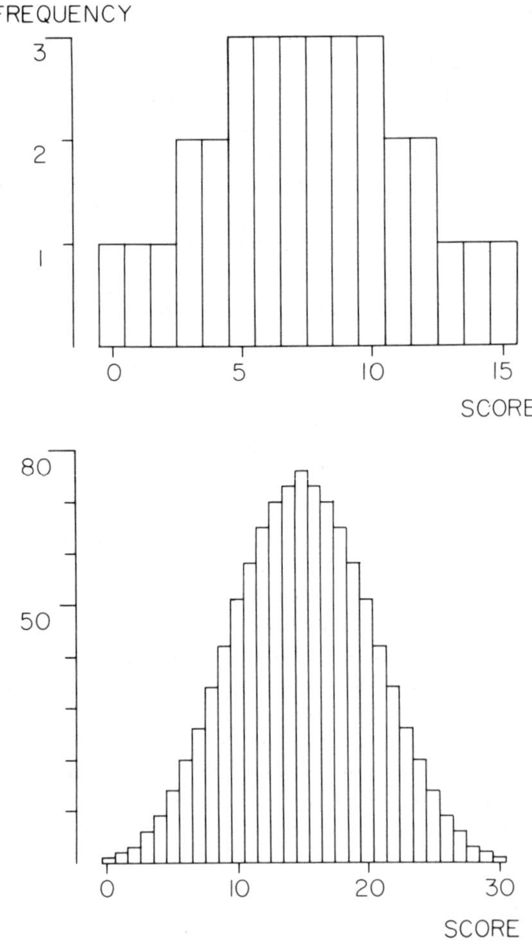


**Figure A.1.** *Upper* diagram—the convolution of five Bernoulli trails with scores 1, 2, 3, 4, and 5. *Lower* diagram— the convolution of two variates with the above distribution. Note its "bell-shaped" character.

null hypothesis. For instance, for a test of size 5% on normally distributed data a value 1.645 standard deviations from the mean would be required for a one-tailed test, whereas 1.9600 standard deviations would be required for a symmetrical two-tailed test of the same total size. The power of the test will be greater in the one-tailed test. Not having made the decision in advance, the investigators must struggle with their consciences in attempting to reconstruct whether they would have used a one-tailed or a two-tailed test before they knew the results.

Every investigator who wishes to use hypothesis testing of this type must at some stage establish a policy about one-tailed and two-tailed tests. If there are *good* prior grounds for the decision, he may have the problem solved for him. He might argue that amputating a patient's leg cannot possibly increase the patient's weight; that blood transfusion cannot make a patient more anemic; that giving somebody money cannot make him poorer. But this kind of absolute certainty is excessively difficult to attain. The patient with the amputation will likely take less exercise and may console himself for his disability by overeating; the blood transfusion may lead to hemolysis of the patient's own cells; giving a man money may induce him to give up his job and live riotously. It may be argued that these are freakish results and that we are interested overwhelmingly in one tail of the curve. But that they are freakish does not mean they are impossible. A reasonable policy might be to assign different probabilities to the alternative hypotheses, putting, say, 4% of the rejection region in one tail and 1% in the other. This seems to be what common prudence dictates; but there is rarely a rational basis for deciding exactly how the rejection region should be divided up to make some kind of provision for surprise.

**2.3.** The ratio of the densities is

$$\frac{\dfrac{1}{10\sqrt{2\pi}}\exp[-x^2/200]}{\dfrac{1}{5\sqrt{2\pi}}\exp[-x^2/50]} = \exp[-x^2/200]/2$$

The lowest density ratios occur for small absolute values of $x$; we therefore pick zone II from Figure 2.1a.

**Comment.** To those to whom it is "obvious" that the rejection region should be in the tail of the curve, this conclusion comes as an unpleasant surprise. To the conscientious analyst nothing is obvious. Be prepared!

**2.4.** Let us identify the two tests as I and II. For definiteness let us pick a particular size of test, say 5%. Consider the sample space, $S$, of the experiment. Let $A$ be the event "rejection by test I" and $B$ "rejection by test II." Then, by definition, under the null hypothesis

$$P(A) = P(B) = 0.05$$

Now in general (7) the probability of $A$ or $B$ or both, i.e., of the union of $A$ and $B$, denoted by $(A + B)$, is

$$P(A + B) = P(A) + P(B) - P(AB)$$

where $P(AB)$ is the probability that $A$ and $B$ are both positive.

Clearly since

$$P(AB) + P(\bar{A}B) = P(B)$$
$$P(AB) \leq P(B)$$

the equality being true only if $B$ is a subset of $A$; i.e., $B$ always implies $A$. Likewise

$$P(AB) \le P(A)$$

Thus in general

$$P(A + B) \ge P(A) = 0.05$$

So the actual size of the conditional test procedure will be greater than 0.05 except where one test leads to rejection only when the other one does. Examples of the latter are uncommon and usually absurd. For instance:

1. If a test is based on the range of sample values, $(x_{max} - x_{min}) = d$, the range of any subsample of the data will never be greater than $d$. The power of a test on a range increases with sample size; hence tests based on any subset would be inadmissible. To reject the hypothesis that the range is not $d$ against the hypothesis that the range is greater than $d$ on the basis of a subsample will always lead to rejection on the whole sample. In this case the size of the test is what it purports to be, 0.05.

2. If a two-tailed test leads to rejection, then a one-tailed test of the same size in whichever tail of the distribution the result lies, will always lead to rejection.

Where no such necessary logical relationship exists and $A$ is not a subset of $B$ or conversely, the size of test will be greater than 5%. One particular instance I have published elsewhere (3) in which for a set of seven paired readings following a Gaussian distribution, the two-tailed $t$ test (see Chapter 5) would not lead to rejection at the 0.02 level, but the sign test (see Chapter 15) would lead to rejection, since all seven changes were of the same sign.

A common abuse arises in the experiment with data that follow a well-known distribution, e.g., the Gaussian. The hypothesis about the mean should be tested on all the readings; but if "outliers" (individually in the rejection region) are arbitrarily discarded and an inadmissible (and blatantly biased) test done on the remaining subsample, the power will be systematically increased. Unwarranted censoring of this type is commonly practiced, doubtless without conscious dishonesty, but none the less illegally. (I have known one investigator who, in an excess of zeal, published only those points that fell within one standard deviation of the mean.)

## CHAPTER 3

**3.1. a.** The method-of-moments estimate ($MME$) of the mean, denoted by $m$, is

$$(4 + 17 + 3 + 0.8)/4 = 24.8/4 = 6.2$$

**b.** The likelihood of any value $x$ is

$1/b \qquad 0 \le x \le b$

$0 \qquad$ elsewhere

The joint likelihood is thus

$(1/b)^4 = 1/b^4 \qquad 0 \le$ all $x \le b$

$0 \qquad$ otherwise

Now the smaller $b$ the greater its reciprocal and the larger the joint likelihood;

but if $b$ becomes less than the largest sample value, the likelihood becomes zero. Thus the *MLE* of $b$ is the largest sample value, 17, and, from symmetry, the *MLE* of the mean is the midpoint between this value and zero, i.e., $(17 + 0)/2 = 8.5$.

**c.** By a similar argument, the *MLE* of $a$ and $b$ are those quantities that minimize the width of the domain and yet contain all the sample points. Thus the *MLE* of the mean is the mean of the largest and smallest values in the sample, i.e.,

$$(0.8 + 17)/2 = 8.9$$

**Comment.** In this case, the *MLE* takes an unexpected form that makes only indirect use (as it were) of all but the extreme values, whereas the *MME* uses them all explicitly. If the likelihood curve is plotted out, it will be evident that it is not "smooth" at the *MLE* and we cannot appeal to the usual formula to find the variance of the estimate. This variance can be found by other means with some difficulty.

**3.2.** The invariance properties of the *MLE* tell us that the estimate is the same whether we use the likelihood curve or its logarithm. On the logarithmic scale, the only effect of the combinatorial coefficient is to shift the curve upwards an amount equivalent to the logarithm of 56 which does not change the location of the maximum. (The highest point above sea level on an island at low tide is still the highest point at high tide.) For those who solve this problem by using calculus, any such constant term will disappear on differentiation.

In the same way, the variance depends on the curvature of the logarithm of the likelihood curve, which is clearly not changed by moving the curve bodily upwards. (The shape of the top of the mountain does not vary with the tides either.) Again the formal approach is to find the value of the second derivative of the logarithm in the neighborhood of the maximum, in which, again, the effects of the combinatorial constant will have been removed by differentiation.

**3.3. a.** Assuming that all the data values are independent and ignoring the combinatorial constant (in accordance with the conclusion from Problem 3.2) we may write the joint likelihood as the product of the individual likelihoods

$$(p^2)^7(2pq)^9(q^2)^9$$

Collecting terms we obtain the likelihood

$$2^9(p^{23})(q^{27})$$

This is essentially the likelihood of a binomial variate and we know that the *MLE* is simply the proportion of genes of the particular type, in this case $23/50 = 0.46$.

**b.** Since all three genotypes are distinguishable, we may simply count the number of genes of the first type: the 7 subjects in the first group all have two genes of this type and all those of the second group one each, twenty-three in all. Since there are 25 persons in the sample and each has two genes we may readily reach the same estimate as before, 0.46. Note furthermore that we may apply the second means of estimation without making any assumptions at all about the *relative* frequencies of the three genotypes in the population. It is therefore a more elastic method, but with no loss of power. (Technically the proportions quoted in the problem are those given by the Hardy-Weinberg principle, which is widely violated in nature. However, implicitly, we have assumed they are independent, which will lead to Hardy-Weinberg equilibrium in the expectations for the next generation.)

**3.4. a.** Since we cannot distinguish the first two types, all we can say about the likelihood of each person of phenotype $A_1$ is that it is the sum of the likelihoods of the two (disjoint) explanations of the phenotype, $A_1A_1$ and $A_1A_2$ with combined probability $(p^2 + 2pq)$. The remaining 9 cases have an unambiguous genotype. Thus the likelihood may be written

$$L = (p^2 + 2pq)^{16}(q^2)^9$$

Now we may maximize this by exploring values of $q$. However, we may view the problem in a somewhat different light. A person taken at random is either a recessive, i.e., of type $A_2A_2$, with probability $q^2$, or not a recessive, with the complementary probability. This is in fact a binomial process in which the parameter takes the somewhat unusual form $q^2$. Its *MLE* may thus be readily found by inspection as the sample proportion of successes, i.e., $9/25 = 0.36$. The estimated variance of $\hat{q}^2$ will thus assume the usual binomial form

$$\frac{(1 - q^2)q^2}{n}$$

estimated here as

$$\frac{0.64 \times 0.36}{25} = 0.009{,}216$$

**b.** Clearly $q^2$ must be nonnegative since it is a probability. Thus over the range of interest it is a monotonic function of $q$ which is also a probability (though of a gene rather than of a genotype). Thus because of the invariance property, the *MLE* of $q$ is simply the square root of the *MLE* of $q^2$, i.e., 0.6.

**c.** We get the following values:

| $q$ | $\log L$ | $q$ | $\log L$ |
|---|---|---|---|
| .2 | $-29.623$ | .599 | $-16.335{,}532{,}940{,}8$ |
| .4 | $-19.283$ | .600 | $-16.335{,}454{,}869{,}8$ |
| .6 | $-16.335$ | .601 | $-16.335{,}533{,}049{,}3$ |
| .8 | $-20.363$ | | |

The maximum is found by formula (3.2) to be approximately

$$0.6000 + \frac{0.000{,}000{,}108{,}5 \times 0.001}{-0{,}000{,}156{,}250{,}5 \times 2} = 0.6000$$

The variance (based on formula 3.3) is approximately $0.006{,}399{,}9\ldots$. Asymptotic theory shows that the appropriate formula is

$$\frac{1 - q^2}{4n} = \frac{1 - 0.36}{100} = 0.0064$$

**3.5.** A sample of size $n$ gives an estimate of the mean with variance $\sigma^2/n$. Thus the relative variances of two such estimates will be in inverse ratios to their sample sizes. Now the rule for minimizing the variance of a linear combination of estimates is to weight inversely as their variances: in this case, to weight inversely as the reciprocal of their sample sizes, or more simply, to weight directly as sample size. The linear combination with the smallest variance is simply the sum of all the values divided by the sum of the sample sizes.

## CHAPTER 4

**4.1.** For clarity, let us use natural logarithms.

$$\text{Antilog}_e[(1/2)(\log x + \log y)] = \sqrt{\exp(\log x + \log y)}$$

$$= \sqrt{xy}.$$

It is easily seen by extending this argument that

$$\text{Antilog}\left[\left(\sum_{i=1}^{n} \log x_i\right)\Big/ n\right] = \sqrt[n]{\Pi x_i}$$

(The above results can be equally well derived where the logarithms are to any base but unity.)

**4.2.** Let the numbers be $x$ and $y$. Their difference, we are told, is not zero. Consider the quantity $(x - y)^2$ which, being a square, is necessarily greater than zero. Thus expanding it

$$x^2 - 2xy + y^2 > 0$$

Let us add $4xy$ to both sides

$$x^2 + 2xy + y^2 > 4xy$$

Factoring

$$(x + y)^2 > 4xy$$

Taking the positive square root of both sides

$$x + y > 2\sqrt{xy}$$

Hence dividing by 2

$$\frac{x + y}{2} > \sqrt{xy}$$

The generalization is that the arithmetic mean of any number of positive quantities not all equal exceeds their geometric mean. (This result is a useful check on the accuracy of one's calculations where the logarithms of the values are being used.)

**4.3. a.** Let the increment be the positive quantity $h$.

$$\text{Log}(x + h) - \log x = \log\left(\frac{x + h}{x}\right) = \log\left(1 + \frac{h}{x}\right)$$

which exceeds 0, being the logarithm of a number greater than unity.

**b.** Let us again increase $x$ by a positive quantity $h$

$$\frac{1}{x + h} - \frac{1}{x} = \frac{x - (x + h)}{x(x + h)} = -\frac{h}{x(x + h)}$$

which is less than zero since $x$ and $h$ are both positive, and thus the numerator is negative and the denominator is positive.

**4.4.** Note first that any nonzero multiple of a Gaussian variate is itself Gaussian. Now we may argue heuristically in the following terms. Coagulant activity, like thrombus-formation, has a lognormal distribution. We may then agree that coagulation time is reciprocally related to coagulant activity, just as the flying time from $A$ to $B$ is, other things being equal, inversely related to the velocity. What is the distribution of the reciprocal of a lognormal variate? If we say $X$ is a lognormal variate we mean $\log X$ is Gaussian. But if so, $-\log X$, being a multiple of a Gaussian variate, is itself Gaussian. But

$$-\log X = \log(1/X)$$

i.e., the reciprocal of $X$ is lognormal.

(This loose biological speculation has been verified empirically.)

**4.5.** Experimenting with this formula, we hit on the device of taking reciprocals of both sides

$$\frac{1}{F} = \frac{x+y}{xy} = \frac{x}{xy} + \frac{y}{xy} = \frac{1}{y} + \frac{1}{x}$$

We have converted the strengths of lenses into a form that makes their effects additive. Furthermore, it is readily extended to any number of lenses, whereas the generalization of the original formula rapidly becomes complicated. Hence the practice of expressing strengths of lenses in the reciprocals of their focal lengths (diopters).

**4.6.** First I compute the sample means, variances, etc., with the following results:

| | Controls | Heterozygotes | Homozygotes |
|---|---|---|---|
| $n$ | 27 | 13 | 12 |
| Sum | 4157 | 1049.4 | 63.6 |
| Mean ($m$) | 153.96 | 80.72 | 5.30 |
| Sum of squares | 658,085. | 91,592.2 | 659.46 |
| (Mean)(sum) | 640,024.04 | 84,710.80 | 337.08 |
| Sum of squares from mean | 18,060.96 | 6,881.40 | 322.38 |
| Variance ($s^2$) | 694.65 | 573.45 | 29.31 |
| Standard deviation ($s$) | 26.36 | 23.95 | 5.41 |
| Variance/mean | 4.512 | 7.104 | 5.530 |

The largest variance is 23.7 times the smallest, an uncomfortably wide spread. Also, if I compare the means and the medians in each case the mean is somewhat the larger. In the first there are 15 values below the mean, 12 above; in the second group 7 and 6; in the third, 8 and 4. None of these is impressive, but collectively they are suggestive. I may if I like, plot Gaussian probability curves or compute sample cumulants (see Chapter 1); but the samples are too small to show very much. Were it not for the variances I would not be unduly worried.

Evidently the variance increases with the mean. The relationship is not quite so clear as we would like it; but it is not very far from being proportionate, considering the sizes of the samples; and we are prompted to try transforming by taking square roots. Logarithmic transformation is interdicted by the zeroes. (No doubt the latter figures really indicate that we are at the limit of resolution of the method; but there is no simple remedy for that.) We get the following results from the square roots of the original readings.

|                          | Controls  | Heterozygotes | Homozygotes |
|--------------------------|-----------|---------------|-------------|
| Sum                      | 333.874,9 | 115.571,3     | 21.806,4    |
| Mean                     | 12.365,7  | 8.890,1       | 1.817,2     |
| Sum of squares from mean | 28.391,9  | 21.959,9      | 23.973,4    |
| Variance                 | 1.092,0   | 1.830,0       | 2.179,4     |
| Standard deviation       | 1.045,0   | 1.352,8       | 1.476,3     |

It seems we have overcorrected a little, but the range of sample variances is a good deal less alarming. The adventurous might like to try out other fractional roots, e.g.,

$$y = x^{0.6}$$

But for practical purposes, I think the square root transform will do well enough.

How well has the transform done in other respects? Because of the monotonicity it will be simplest to transform the new sample means back into the old scale, by squaring:

For normal subjects $(12.365,7)^2 = 152.9$
For heterozygotes $(8.890,1)^2 = 79.03$
For homozygotes $(1.817,2)^2 = 3.30$

The transform affects the relationship of the mean and median in the last case only, which instead of being an 8-4 split is a 4-8 split. This does a little to redress the balance, the same total number, 26, being below the mean as above. It seems we have lost little, and possibly gained something in symmetry, by the transform.

**4.7.** It is not uncommon to have a small sample only, and no background information to which one may appeal as to how the variate may be distributed. The scientist does well to cultivate the art of looking at his data.

In the present instance we note the following points. First, all the values are positive. Second, there is more crowding for the lower values. The values are arranged in ascending order and if one looks at the lowest, 0.1, the 8th, or the sample *lower quartile*, 2.2, the 15th or middlemost or sample median, 4.8, the 22nd or the sample *upper quartile*, 9.3, and the highest value, 32.9, there is a steady increase in the intervals between them. This curve suggests a smoothly positive skewness. I hope that the reader will be persuaded to confirm this impression by plotting out the results on Gaussian probability paper, noting the concave curve of a type similar to that shown in the top diagram of Figure 1.2.

Let us then compute appropriate sample moments, beginning with the natural values, the positive square roots, and the natural logarithms of the data. We get the following results.

|                       | Natural values | Square root transform | | Logarithmic transform | |
|-----------------------|----------------|--------|-------------|--------|-------------|
|                       |                | Values | Retransform | Values | Retransform |
| Mean (*m*)            | 7.097          | 2.352  | 5.531       | 1.351  | 3.863       |
| Standard deviation (*s*) | 7.193       | 1.273  | —           | 1.348  | —           |

We may compute some rough classes.

| Quantity | Natural values | Square root | | Natural logarithms | |
|---|---|---|---|---|---|
| | | Values | Retransform | Values | Retransform |
| $m - 2s$ | $-7.29$ | $-0.20$ | (Not monotonic) | $-1.34$ | 0.26 |
| $m - s$ | $-0.10$ | 1.08 | 1.16 | 0.00 | 1.00 |
| $m$ | 7.10 | 2.35 | 5.53 | 1.35 | 3.86 |
| $m + s$ | 14.29 | 3.63 | 13.14 | 2.70 | 14.87 |
| $m + 2s$ | 21.48 | 4.90 | 24.00 | 4.05 | 57.23 |

We may hope that about ⅔ of the cases will fall within one standard deviation of the mean and 95% within two standard deviations. (These statements will be somewhat spoiled by the fact that we have only estimates of the parameters to work with. But they are only rough statements in any case.) The above five points define six segments of the sample space. The actual numbers falling within these limits are:

| Natural values | Square root transform | Logarithmic transform | Expected under Gaussian model |
|---|---|---|---|
| 0 | 5 | 2 | 0.7 |
| 0 | | 3 | 3.9 |
| 18 | 12 | 6 | 9.9 |
| 6 | 7 | 14 | 9.9 |
| 4 | 4 | 4 | 3.9 |
| 1 | 1 | 0 | 0.7 |
| Total 29 | 29 | 29 | 29 |

The natural values cluster too much in the upper tail, a distortion that both of the transforms overcorrect. The square root transform gives rise to problems of monotonicity, since some of the "estimated" limits are negative. A three-parameter lognormal transform would probably be rather more satisfactory. However, we do well to remember two facts. First, the data are few and I have no background information. In such circumstances elaborate transformations are somewhat pretentious and apt to enjoy a short-lived popularity. One should keep in mind what the object of the maneuverings is. Second, to test hypotheses about means, the simple logarithmic transformation will do well enough, as we shall see in Chapter 17.

**4.8.** Let $y = \sqrt{x} + \sqrt{(x + 1)}$

The problem is to express $x$ as an explicit function of $y$. Squaring both sides, we get

$$y^2 = x + x + 1 + 2\sqrt{x(x + 1)}$$

$$(y^2 - 1) - 2x = 2\sqrt{x(x + 1)}$$

Squaring both sides again, we get

$$(y^2 - 1)^2 - 4x(y^2 - 1) + 4x^2 = 4x^2 + 4x$$

$$(y^2 - 1)^2 = 4xy^2$$

$$x = \frac{(y^2 - 1)^2}{4y^2}$$

If $x = 2$, $y = 1.414\ldots + 1.732\ldots = 3.146{,}264{,}37\ldots$

$$x = \frac{(3.146{,}264{,}37^2 - 1)^2}{4 \times 3.146{,}264{,}37^2} = 2$$

## CHAPTER 5

**5.1.** I do not know the answer; but I suspect that what commonly happens in such cases is that there is no *natural* definition of the disease in question. Accordingly, the campaign managers (who cannot be expected to know anything about science) seek what they suppose to be expert help; and their experts know (what I do not) that the normal is everything between the 2.5 and the 97.5 percentiles. In this way, the organization will never run out of things to do; nor will they ever be overwhelmed by the scale of their task. However little scientific sense this makes, I concede that it has certain practical advantages.

**5.2. a.** An acute abdomen that has been allowed to progress to frank surgical shock may cause a low temperature. Leaving that aside, one would certainly expect that in appendicitis, fever, if anything, will occur. Hence a rejection region in the right-hand tail seems reasonable. Since in this disease high fever is not typical, we cannot expect a high average; and since the sample is small, we are wise not to be too ambitious in the size of the test. Hence the proposal in the problem to do a 5% test also seems reasonable. The standard deviation of the sample mean will be that for single readings divided by the square root of the sample size (we assume independence). Thus the rejection region will comprise sample means greater than

$$37.0 + \left( \frac{0.5}{\sqrt{12}} \right) (1.6449) = 37.237\ldots$$

**b.** If the true mean is 37.5, the rejection region would start at

$$\frac{37.237 - 37.5}{\left( \dfrac{0.5}{\sqrt{12}} \right)} = - \frac{0.2625}{0.1443} = -1.8192$$

That is, 1.8192 standard deviations below the mean. The power would therefore be the probability that under the alternative hypothesis the sample mean should lie above this point, which is 0.9656. Power is worked out for other values likewise.

**c.** The sample mean is 37.48 which lies inside the rejection region. Therefore we reject the hypothesis at the 5% level.

**5.3.** The test (which is one-tailed) will of course be made on the logarithms. The mean value in familial hypercholesterolemia is commonly supposed much higher than normal, so that we might venture a test at the 1% level. Finally, we suppose, since the available data are extensive and since for the "total population" neither the sample size nor the standard deviations of the estimates is given, that the sample for the latter is very large and the errors of estimation negligible. Thus we shall treat the values given for the population as exact, and hence the problem is reduced to a one-sample test with known variance. (The latter assumption is based on the broad, and well-supported, supposition that logarithmic transformation stabilizes the variance.) Then the rejection region would be

$$2.35 + (2.3263) \sqrt{(0.014/36)} = 2.3959.$$

Hence, we just reject the hypothesis at the 1% level.

**b.** The critical point is

$$\frac{2.5 - 2.3959}{0.019,72\ldots} = 5.28$$

standard deviations below the mean and the power is hence, for practical purposes 100%.

**5.4.** The procedure here is clear, the distribution of height being very close to Gaussian, and the sample size large. In principle, we do not know the variance and should use a $t$ statistic; however the sample size, 576, is so large that the departure from Gaussian is, for our purposes, trivial. Having no background information or plausible argument to the contrary, we do a two-tailed test. The standard deviation of the estimate of the mean is $2.4/\sqrt{576} = 0.1$. The acceptance region is thus

$$170 - (1.96)(0.1) \quad \text{to} \quad 170 + (1.96)(0.1)$$

$$= 169.804 \quad \text{to} \quad 170.196$$

The hypothesis is thus rejected.

**Comment.** The reader must not be so enchanted by these maneuverings as to lose his 'satiable curiosity as to what the answer means. It can be argued that heights are measured with sufficient accuracy that we can decide whether the means of two populations differ without any statistical test. The mean for this population of students is what it is, and the 576 should not be seen as a random sample of anything unless (a) the population from which it is drawn can be defined; (b) it is homogeneous; and (c) there is, at least in principle, an actual random sampling procedure involved. Any complete class of medical students that I have ever heard of flagrantly violates all three assumptions. The topic will be discussed further in Chapter 17, and I have given a more detailed analysis of the issues elsewhere (70).

A second point is that wherever big samples are involved, we can afford to take the statistical tests pretty much for granted, but should concern ourselves much more with the biology. The high level of significance attained in this study notwithstanding, there are two major questions to be addressed. Is a mean difference in height of two centimeters of any importance? Are we satisfied that there is not some systematic bias at work—in the observer, in the measuring scale, in the time of day of the measurement—that cannot adequately account for this difference in measurements?


## CHAPTER 6

**6.1.** This is a classical error. It prompts me to utter a universal anathema.
*Never use the patient as his (or her) own control in before-after studies.*
The investigator ignores this dictum at his peril. One might pick out any of a

number of fallacies apt to arise in such a study. For instance, any nonrandom change may be a seasonal variation or due to some other trend (e.g., the effect of increasing age) and have nothing to do with the effect of the treatment whatsoever. It may be that the patient is becoming more accustomed to the clinic and that apprehension about the unfamiliar had increased the blood cholesterol level at the first visit. It may be a drift over time in the baseline for the calibration of the apparatus measuring the level. This type of fallacy, known as confounding, is discussed in detail in another book (3). In such instances it lies in the domain of logic rather than in statistics.

However the intent of this example is to illustrate regression toward the mean. The patients were selected *because* their levels of cholesterol in blood were high. Since there is a modest technical error of measurement, and a large variation from time to time within persons, we might expect the average value for all of them to fall, even if there is *known* to be no trend over time in the mean for the population at large.

One other possibility must be borne in mind. There is good evidence, almost, but not quite, compelling, that at least one disorder, commonly known as familial hyper-cholesterolemia, constitutes a secondary peak in the upper tail of the "normal" curve. If it is sufficiently frequent, we might expect that some of our patients belong to it, and that their levels should rise. If in spite of this trend, their levels fall, then the observed change, far from being explained away, is more remarkable than ever.

**6.2.** This problem is another example of the same phenomenon. Granted that the intelligence in children is on average the same as that in their parents, neverthe-less, there is error of measurement of the I.Q.; and we may suppose that multiple readings in parents, picked in the first place because they are retarded, would on average show a rise, and so should the expected readings in the children. To carry the argument further, one would need more details—e.g., how many times the parents were tested; what the error of measurement for a single test is; how far the difference between parents and children can be accounted for by methods of education (children in recent years being trained to be good at doing intelligence tests), etc.

Note that by the criteria given, we would not expect an unretarded person to have an I.Q. of 100% but the average for the censored population of those who do not fall below 70%. However, in the absence of regression toward the mean, this would raise the average to only about 102% (assuming that the standard deviation for the popula-tion at large is 15%).

**6.3.** Since we have decided from the analysis of the data that the square root of the variate is Gaussian, then the *MLE* of the mean is simply the sample average of the transformed data (12.3657) and the estimated variance of the estimate of the mean is the estimated variance (1.092...) divided by 27 (or 0.04044...). Hence the 95% con-fidence limits will be

$$12.3657 - (2.0551) \sqrt{0.04044\ldots} \; = \; 11.952$$

$$12.3657 + (2.0551) \sqrt{0.04044\ldots} \; = \; 12.779$$

Since all the values are positive, the transformation is monotonic and we merely have to square these confidence limits to get the confidence limits on the square of the mean, i.e., 142.86 and 163.30.

**6.4.** Considering the sizes of the samples (13 and 12 respectively) we may be quite satisfied with the similarity of the estimates of their variances and therefore be pre-pared to compute a pooled estimate of the variance:

$$\frac{21.9599 + 23.9734}{23} = 1.9971$$

The estimated variance of the difference is then

$$1.9971 \left[ \frac{1}{12} + \frac{1}{13} \right] = 0.3200$$

The $t$ value for the 97.5th percentile with 23 degrees of freedom is 2.069. Hence the 95% confidence limits on the difference of the means will be $2.069\sqrt{0.32} = 1.17$ on either side of the point estimate, $8.8901 - 1.8172 = 7.0729$, i.e., 5.90 to 8.24. Since these limits do not include 0, we would not accept the null hypothesis of no difference. The size of the equivalent test (the acceptance region of which would be $-1.17$ to $+1.17$) would be 5%.

**6.5.** It will be recalled that the geometric mean of a variate is the antilogarithm of the mean of the logarithms. If we are satisfied from the answer to Problem 4.7 that to a good approximation the distribution from which the data in Table 4.11 are drawn is lognormal, then the estimates of the moments of the logarithms given in the answer ($m = 1.351,32$, $s = 1.347,89$) are the *MLE*; hence, from monotonicity and the invariance property, the natural antilogarithm of $1.351,32$, or 3.86, is the *MLE* of the geometric mean. The confidence limits on the mean are computed on the logarithms, using normal theory and then transformed back. The sample size is 29, and the 97.5th percentile for $t$ with 28 degrees of freedom is 2.048. Thus the 95% confidence limits on the mean are the natural antilogarithms of

$$1.351,32 - (2.048)(1.347,89)/\sqrt{29} = 1.351,32 - 0.512,6 = 0.838,7$$

$$\text{and} \quad 1.351,32 + 0.512,6 = 1.863,9$$

The natural antilogarithms of these limits are 2.31 and 6.45 which are the 95% confidence limits on the geometric mean.

**6.6.** This instance is the first in which we have addressed the statistics of ratios. In general I do not like ratios for reasons developed in Chapter 11. In this instance, the formal problem is simple (whatever difficulties one may get into in deciding what the *meaning* of the answer may be). The logarithm of any ratio is the logarithm of the numerator minus that of the denominator. Having satisfied ourselves that logarithms stabilize the variance, we use a pooled estimate of this variance

$$\frac{15(0.254,8^2 + 0.273,4)^2}{2 \times 16 - 2} = 0.069,835,3$$

which gives for the standard deviation of the difference

$$\sqrt{(0.069,835,3)\left(\frac{1}{16} + \frac{1}{16}\right)} = 0.093,431$$

The 97.5 percentile for the $t$ statistic with 30 degrees of freedom is 2.042. The difference of the means is $(1.0281 - 0.3423) = 0.6858$. Also, $(0.093,431)(2.042) = 0.1908$.

So we have the following limits:

$$0.6858 - 0.1908 = 0.4950$$

$$0.6858 + 0.1908 = 0.8766.$$

The 95% confidence limits on the ratio are the natural antilogarithms of these quantities, 1.64 and 2.40.

Note the following points:

a. The ratio and its confidence limits are dimensionless.

b. The 95% confidence limits are not symmetrically placed about the point estimate; but the square of the point estimate is the product of the confidence limits. (This is a useful check on the accuracy of the calculations where a lognormal variate is concerned.)

## CHAPTER 7

**7.1.** The analysis is shown in Table A1.

**7.2.** The three contrasts are:

|     | $M$ | $MC$ | $MS$ | $MCS$ | Total |
|-----|-----|------|------|-------|-------|
| (1) | 3   | $-1$ | $-1$ | $-1$  | 0     |
| (2) | 0   | 2    | $-1$ | $-1$  | 0     |
| (3) | 0   | 0    | 1    | $-1$  | 0     |

The dot products are:

|        |   |      |      |   |   |
|--------|---|------|------|---|---|
| (1)(2) | 0 | $-2$ | 1    | 1 | 0 |
| (1)(3) | 0 | 0    | $-1$ | 1 | 0 |
| (2)(3) | 0 | 0    | $-1$ | 1 | 0 |

The matrix may conveniently be made square by adding a line for the contrast to correspond to the mean; and it is to be orthogonal, that is, multiplied by its own transpose it is to yield the identity matrix. The product of the above matrix and its transpose will certainly be diagonal because the dot products (which give the off-diagonal terms) are all zero. It remains to rescale it in such a way that the terms on the principal diagonal are unity. This is assured by dividing throughout each line by the square root of the sum of squares of the above coefficients. Thus:

$$\begin{bmatrix} \dfrac{1}{2} & \dfrac{1}{2} & \dfrac{1}{2} & \dfrac{1}{2} \\[2mm] \dfrac{3}{\sqrt{12}} & \dfrac{-1}{\sqrt{12}} & \dfrac{-1}{\sqrt{12}} & \dfrac{-1}{\sqrt{12}} \\[2mm] 0 & \dfrac{2}{\sqrt{6}} & \dfrac{-1}{\sqrt{6}} & \dfrac{-1}{\sqrt{6}} \\[2mm] 0 & 0 & \dfrac{-1}{\sqrt{2}} & \dfrac{1}{\sqrt{2}} \end{bmatrix} \begin{matrix} \text{(Mean)} \\[2mm] \\[2mm] \\[2mm] \end{matrix}$$

**7.3.** Such a set of data is not at all uncommon. It is the kind of study that gets a committee a bad name. Analysis is not helped by the fact that there is no coherent set of questions that could have given rise to it. Is it a study of the damage done to joints by dead streptococci? If so, why is there no group receiving dead streptococci without

**Table A1. A set of orthogonal contrasts for the data of Table 7.1**

| Scientific question | Alignment of groups | Contrast mean square | $F(1,40)$ | $t_{40}$ | $p$ |
|---|---|---|---|---|---|
| Does dietary cholesterol affect the serum cholesterol level? | $(T_M + T_{MS})$ and $(T_{MC} + T_{MCS})$ | $\dfrac{[(6816 + 5508) - (15,112 + 8368)]^2}{44} = 2,828,553.09$ | 28.699 | 5.357 | $3.771 \times 10^{-6}$ |
| Does dietary sitosterol affect the serum cholesterol level? | $(T_M + T_{MC})$ and $(T_{MS} + T_{MCS})$ | $\dfrac{[(6816 + 15,112) - (5508 + 8368)]^2}{44} = 1,473,516.00$ | 14.951 | 3.867 | $3.965 \times 10^{-4}$ |
| Does any effect of sitosterol depend on the presence of cholesterol in the diet? | $(T_M + T_{MCS})$ and $(T_{MS} + T_{MC})$ | $\dfrac{[(6816 + 8368) - (15,112 + 5508)]^2}{44} = 671,593.09$ | 6.814 | 2.610 | 0.01267 |
| Are there some differences among the groups? | | 4,973,662.18 | | | |
| Error mean square (from Table 7.2) | | 98,559.16 | | | |

any treatment at all? As it is, the best that one could hope to study would be the interaction between streptococci and various forms of treatment: we can make no inferences at all about streptococci. (For instance, the streptococci may have no effect: all the results may be due to treatment.) Is it a study to compare several treatments? Then why is there a control group? Again, since we have no untreated group we have an incomplete factorial experiment on penicillin and butazolidine. If this is the focus of interest, why the cortisone group?

The best sense I can make out of the experiment (and I have little respect for this child of my fancy) is represented in this set of contrasts.

| Contrast | | | | | Scientific question |
|---|---|---|---|---|---|
| $-$ | $SB$ | $SP$ | $SBP$ | $SC$ | |
| 4 | $-1$ | $-1$ | $-1$ | $-1$ | Do injections of dead streptococci have any effect on the hip joint in guinea pigs receiving various treatments? |
| 0 | $-1$ | $-1$ | $-1$ | 3 | Is the effect of cortisone on such animals different from butazolidine and penicillin, separately or together? |
| 0 | $-1$ | $-1$ | 2 | 0 | Is the combined effect of butazolidine and penicillin different from either given alone to such animals? |
| 0 | 1 | $-1$ | 0 | 0 | Is butazolidine alone different from penicillin alone in such animals? |

**7.4.** Inspection shows clear positive skewness in all groups with the exception of the third. Detailed examination (which I shall leave to the reader) suggests that the logarithms are approximately Gaussian. The results given to three decimal places together with the appropriate analysis are shown in Table A2. The choice of contrasts seems quite clear. We have a two-way classification by sex and disease, and there may be an interaction between sex and disease. The "among" sum of squares is almost entirely (92%) accounted for by the heterogeneity in disease.

Note that the sum of the squares for contrasts equals that for the total due to groups. (As a nice point in style of layout, the degrees of freedom and the sums of squares for the contrasts, being subtotals, are displayed out of line with the totals.)

**7.5.** In accordance with the earlier comments, we shall suppose that the data are from lognormal distributions. In Table A3 are shown the transforms to the base 10, with the appropriate analysis. The differences are highly significant. Inspection of the means suggests that there is a fairly general origin of the heterogeneity, that is, it is not due simply to one outlandish set of results. There would be no point in setting up contrasts in the absence of further information as to the possible grouping of subjects on some etiological factor of importance.

**7.6.** The analysis is shown in Table A4. Effects are tested against the mean square for interaction (which is all that is available for comparison). Whether significant results would be obtained if the mean square for technical error were available, I cannot say. Note that, unlike the flow chamber experiments described in the text (and also carried out in the present study), measuring the adhesive index is not a procedure destructive to the pig and, in principle, there is no reason why the test should not be repeated as many times as need be. In this instance, it seems reasonable to do the duplicate determinations on specimens from separate venipunctures. Unfortunately in pigs the ear is almost the only site from which a venipuncture can be done, and this conduces to the tranquillity of neither the pig nor the investigator.

The analysis shown here has a few features of interest. First it is the first illustration I have used in which a set of contrasts has been taken out in a two-way design. It is also noteworthy that although the overall effect of diet is unimpressive, the significance level being between 0.05 and 0.1, there is one contrast significant at a level of less than 0.05.

**Table A2. Analysis of variance of the level of blood urea nitrogen in malignant hypertension**

| | Essential | | Renal | | |
| | Male | Female | Male | Female | Total |
|---|---|---|---|---|---|
| | 1.146 | 1.114 | 0.778 | 1.301 | |
| | 1.204 | 1.114 | 1.613 | 1.301 | |
| | 1.301 | 1.176 | 1.623 | 1.301 | |
| | 1.342 | 1.176 | 1.845 | 1.792 | |
| | 1.362 | 1.301 | 1.940 | 1.806 | |
| | 1.398 | 1.602 | 2.146 | 1.898 | |
| | 1.447 | 1.826 | 2.155 | 1.924 | |
| | 1.519 | 1.851 | 2.155 | 2.000 | |
| | 1.556 | 1.968 | 2.155 | 2.013 | |
| | 1.748 | 1.978 | 2.173 | 2.083 | |
| | 1.833 | 2.000 | 2.190 | 2.230 | |
| | 1.968 | 2.045 | 2.207 | 2.238 | |
| | 2.117 | 2.167 | 2.210 | 2.262 | |
| | 2.190 | 2.276 | 2.332 | 2.398 | |
| | — | — | — | — | — |
| $\Sigma x$ | 22.131 | 23.594 | 27.522 | 26.547 | 99.794 |
| $m$ | 1.5808 | 1.6853 | 1.9659 | 1.8962 | 1.7820 |
| $\Sigma x^2$ | 36.454,457 | 42.111,024 | 56.257,400 | 52.094,533 | 186.917,414 |
| $\Sigma(x - m)^2$ | 1.470,088 | 2.348,393 | 2.153,080 | 1.755,732 | 9.080,942 |
| $\Sigma\Sigma(x - m)^2 =$ | 7.727,293 | | | | |

Etiology and sex is the heading spanning the data columns (Essential over Male/Female, Renal over Male/Female).

**Analysis of variance**

| Source | Degrees of freedom | Sum of squares | Mean square | $F$ | $P$ |
|---|---|---|---|---|---|
| Groups | 3 | 1.353,649 | 0.451,216 | 3.036 | <0.05 |
| Sex | 1 | 0.004,253 | 0.004,253 | 0.029 | |
| Etiology | 1 | 1.243,256 | 1.243,256 | 8.366 | <0.01 |
| Interaction | 1 | 0.106,140 | 0.106,140 | 0.714 | |
| Within (error) | 52 | 7.727,293 | 0.148,602 | | |
| Total | 55 | 9.080,942 | | | |

**Table A3. Analysis of variance of the logarithms of the blood levels of phospholipid among patients**

| | | Subject | | | |
|---|---|---|---|---|---|
| 1 | 2 | 3 | 4 | 5 | 6 |
| 2.540 | 2.310 | 2.307 | 2.215 | 2.267 | 2.377 |
| 2.507 | 2.297 | 2.238 | 2.281 | 2.307 | 2.403 |
| 2.520 | 2.288 | 2.330 | 2.253 | 2.322 | 2.410 |
| 2.540 | 2.312 | 2.301 | 2.199 | 2.318 | 2.431 |
| 2.516 | 2.301 | 2.297 | 2.258 | 2.352 | 2.410 |
| — | — | — | — | — | — |
| 12.623 | 11.508 | 11.573 | 11.206 | 11.566 | 12.031 |
| 2.5246 | 2.3016 | 2.3146 | 2.2412 | 2.3132 | 2.4062 |
| 31.868,905 | 26.487,198 | 26.788,203 | 25.119,360 | 26.758,250 | 28.950,499 |
| 0.000,8792 | 0.000,3852 | 0.001,3372 | 0.004,4728 | 0.003,7788 | 0.001,5068 |

$\Sigma\Sigma x = 70.507$

$\Sigma\Sigma x^2 = 165.972,415$

    Correction term (CT) $= (70.507)^2/30 = 165.707,901,6$

Pooled results:

    Within: $0.000,8792 + \cdots + 0.001,5068 = 0.012,3600$

    Among subjects:

$$\frac{(12.623)^2 + \cdots + (12.031)^2}{5} - \frac{(12.623 + \cdots + 12.031)^2}{30} = 0.252,154$$

Total: $165.972,415 - 165.707,901,6 = 0.264,513,4$

| Analysis of Variance | | | | | |
|---|---|---|---|---|---|
| Source | Degrees of freedom | Sum of squares | Mean square | $F$ | $P$ |
| Among subjects | 5 | 0.252,153,4 | 0.050,430,68 | 97.92 | $\ll 10^{-3}$ |
| Within subjects | 24 | 0.012,360,0 | 0.000,515 | | |
| Total | 29 | 0.264,513,4 | | | |

As to the contrasts, they were chosen to answer the three problems addressed in this study:

1. Does dietary fat influence platelet adhesiveness?

2. Does egg-yolk have any effect on platelet adhesiveness other than is ascribable to its content of neutral fat and cholesterol?

3. Does cholesterol add anything to the effect of neutral fat?

The argument is that eggs have been incriminated in atherogenesis. Some have supposed that it is the cholesterol that matters; others have supposed that a trace phospholipid is to blame. So far as this study goes (and it is too small for firm negative claims) and insofar as platelet adhesiveness is germane to atherogenesis, the effects of fat appear to be nonspecific.

Finally, it might be argued that the effect of age should be treated as a regressor

**Table A4. Analysis of variance of the relationship of platelet adhesiveness to diet in pigs**

| | Totals | | | |
|---|---|---|---|---|
| | (a) By dietary group | | (b) By age | |
| Group | $\Sigma x$ | $\Sigma x^2$ | Age | $\Sigma x$ |
| C | 7.94 | 7.9976 | 10 | 4.73 |
| N | 9.57 | 11.6993 | 11 | 4.38 |
| NC | 8.84 | 10.1046 | 12 | 4.82 |
| Y | 9.65 | 11.7559 | 13 | 4.24 |
| | | | 14 | 5.02 |
| Total | 36.00 | 41.5574 | 15 | 4.40 |
| | | | 16 | 4.50 |
| | | | 17 | 3.91 |
| | | | | 36.00 |

Sum of squares among diets

$$= \frac{7.94^2 + 9.57^2 + 8.84^2 + 9.65^2}{8} - \frac{(36.00)^2}{32} = 0.237,075$$

Sum of squares among ages

$$= \frac{4.73^2 + \cdots + 3.91^2}{4} - \frac{(36.00)^2}{32} = 0.216,450$$

Total sum of squares

$$\frac{41.5574}{1} - \frac{(36)^2}{32}$$

Sum of squares for interaction (by subtraction) 0.603,875

| | Analysis of Variance | | | | |
|---|---|---|---|---|---|
| Source | Degrees of freedom | Sum of squares | Mean square | $F$ | $P$ |
| Diet | 3 | 0.237,075 | 0.079,025 | 2.748 | $<0.1$ |
| 3C$-$N$-$NC$-$Y | 1 | 0.187,267 | 0.187,267 | 6.512 | $<0.05$ |
| 2Y$-$N$-$NC | 1 | 0.016,502 | 0.016,502 | 0.574 | |
| N$-$NC | 1 | 0.033,306 | 0.033,306 | 1.158 | |
| Age | 7 | 0.216,450 | 0.030,921 | 1.075 | |
| Interaction | 21 | 0.603,875 | 0.028,756 | | |
| Total | 31 | 1.057,400 | | | |

variable (see Chapter 8), rather than an unordered categorical one. There seems to be no significant trend with age, so that, in the event, the choice scarcely matters.

**7.7.** Here we may make use of the calculations on the square root transform given in the answer to problem 4.6.

> The total sum is $333.874,9 + 115.5713 + 21.806,4 = 471.252,6$
> The total sum of squares is $4,157 + 1,049.4 + 63.6 = 5,270$
> The residual sum of squares is thus $5,270 - (471.2526)^2/52 = 999.2497$.
> The combined within sum of squares is $28.391,9 + 21.959,9 + 23.9734 = 74.3252$.

This problem differs from the samples worked out in the text in that the numbers of samples in the three groups are unequal. The form of the sum of squares among treatments requires to be somewhat modified:

$$\frac{(333.8749)^2}{27} + \frac{(115.5713)^2}{13} + \frac{(21.8064)^2}{12} - \frac{(471.2526)^2}{52}$$

In fact it is far simpler to work with the total sum of squares and the within sum of squares, and to find the sum of squares for treatments by taking the difference. The rest of the analysis then follows as usual.

Analysis of variance

| Source | Degress of freedom | Sum of squares | Mean square | $F$ | $p$ |
|---|---|---|---|---|---|
| Among | 2 | 924.92 | 462.46 | 304.89 | $<10^{-6}$ |
| Within | 49 | 74.33 | 1.5168 | | |
| Total | 51 | 999.25 | | | |

Note, however, that one is not in a position to set up contrasts that are at once orthogonal and have a readily interpretable meaning. If the sample sizes were equal, one might usefully do two contrasts, one to determine whether there is a linear effect with the increasing number of aberrant genes (first group vs. third) and one for dominance effects (first and third vs. twice the second). In rare disorders such as this, the geneticist must take what he can get and not complain about the unequal sizes of the samples!

It is arguable that this problem might be treated as one in regression, the regressor variable being the number of aberrant genes in the person. (See Chapter 8.)

**7.8.** The analysis is laid out in Table A5. It is perhaps sobering to see how much even highly skilled observers may differ in their estimates based on judgment. I say nothing of a few of the participants that saw fit to express their graphical estimates to two decimal places. In the original paper we quoted a comment from Sir Thomas Lewis about the exquisite nicety with which clinical cardiologists attempt to time heart murmurs. He ends with an ironic comment about the "manifest tendency, which is traditional, for the medical profession to exaggerate the accuracy of its subjective methods of examination." It does well to recognize that this is a vanity of some investigators as well, especially where measurements from fields with low resolution are concerned. Formal statistical analysis, at the least, should furnish some evidence of how bad our estimate may be. But of course analysis will not take care of all our prob-

**Table A5. Analysis of variance of estimates of mean survival time of platelets. Preliminary totals.**

| Data set 1 | | Data set 2 | | Data set 3 | |
|---|---|---|---|---|---|
| Reading 1 | Reading 2 | Reading 1 | Reading 2 | Reading 1 | Reading 2 |
| 110.6 | 109.2 | 110.7 | 109.4 | 109.8 | 107.4 |
| 1,231.46 | 1,199.34 | 1,231.59 | 1,201.26 | 1,208.92 | 1,158.42 |

| Observer | $\Sigma x$ | Observer | $\Sigma x$ |
|---|---|---|---|
| 2 | 68.4 | 8 | 55.4 |
| 3 | 59.9 | 9 | 67.8 |
| 4 | 69.3 | 10 | 68.4 |
| 5 | 67.6 | 12 | 67.3 |
| 6 | 67.7 | 13 | 65.3 |

$\Sigma x$    657.1
$\Sigma x^2$    7,230.99

Correction term (CT) $= (657.1)^2/60 = 7{,}196.340{,}167$

Sums of squares

Among observers: $68.4^2/6 + \cdots + 65.3^2/6 - \mathrm{CT} = 30.334{,}833$

Among data sets: $(110.6 + 109.2)^2/20 + (110.7 + 109.4)^2/20 +$
$(109.8 + 107.4)^2/20 - \mathrm{CT} = 0.254{,}333$

Error: $(11.3 - 11.7)^2/2 + \cdots + (11.0 - 10.8)^2/2 = 2.745$

Analysis of Variance

| Source | Degrees | Sum of squares | Mean square | $F$ | $P$ |
|---|---|---|---|---|---|
| Observers | 9 | 30.334,833 | 3.370,537 | 46.11* | $\ll 0.001$ |
| Data sets | 2 | 0.254,333 | 0.127,166 | 1.74* | |
| Interaction | 18 | 1.315,667 | 0.073,093 | 0.80† | |
| Within (error) | 30 | 2.745,000 | 0.091,500 | | |
| Total | 59 | 34.649,833 | | | |

*Tested against interaction.

†Tested against error.

lems. From inspecting these data, we form a clear impression that some investigators read higher than others; and it might be argued that consistency of performance is as important as accuracy. But this does not help if $A$ wants to use $B$'s values as a control for his own data. "Highly significant" differences may tell us a good deal more about the investigator than about the data!

## CHAPTER 8

**8.1.** Assuming homogeneous variance throughout, the investigator will presumably wish to minimize the standard errors of his estimates. Now $S^2$, the usual estimate of the variance of the residual error from regression, is always unbiased and hence will naturally have the same expected value whatever the value of the regressor variable. Hence, the variance of the regressor slope, that is, $S^2/\Sigma(x - m_x)^2$, is minimized by maximizing the denominator, the deviations of the experimental values of $x$'s from their common mean. For any given domain, this is attained by putting half the values at each extreme (7).

However, this strategy puts a great strain on the assumption of linearity. If the true regression function is (say) quadratic, we have a further parameter, for curvature, to measure; and clearly, however well we estimate the outcome at the ends of the domain, we can say nothing about how curved the line is. Unless there are inescapable reasons for believing that the relationship must be linear, it is wise to have at least some intermediate data points. On the other hand, if all the data were at a single value of the regressor, we could make no inference about the slope.

It is a common fallacy in more naive therapeutics to suppose proportionality of effects; and striking counterexamples come to embarrass us from time to time (e.g., Problem 2.2. See also reference 3.)

**8.2. a.** Let us write $u$ for $\log_e x$, and $V$ for $\log_e Y$, retaining three decimal places. We get the following preliminary results.

| $u$ | $V$ |
|:---:|:---:|
| 5.704 | 5.892 |
| 5.628 | 5.489 |
| 5.521 | 5.425 |
| 5.342 | 5.100 |
| 5.481 | 5.220 |
| 5.697 | 5.521 |

$$\Sigma u = 33.373 \qquad \Sigma v = 32.647$$

$$m_u = 5.5621 \qquad m_V = 5.4412$$

$$\Sigma u^2 = 185.725,575 \qquad \Sigma v^2 = 178.015,251 \qquad \Sigma uv = 181.759,642$$

$$m_u\Sigma u = 185.626,188,2 \qquad m_V\Sigma v = 177.637,768,2 \qquad m_u\Sigma v = 181.588,055,2$$

$$\Sigma(u - m_u)^2 = 0.099,386,8 \quad \Sigma(v - m_V)^2 = 0.377,482,8 \qquad 0.171,586,8$$

The regression slope is $0.171,586,8/0.099,386,8 = 1.726,455$.
The intercept is $5.4412 - 5.5621 \times 1.726,455 = -4.1617$.
The sum of squares from regression is

$$\Sigma d^2 = 0.337,482,8 - \frac{(0.171,586,8)^2}{0.099,386,8} = 0.081,245,976,4$$

giving

$$s^2 = 0.081,245,976,4/(6 - 2) = 0.020,311,49$$

whence $s = 0.1425$. The critical value of $t$ for 4 degrees of freedom and size 0.05 is 2.776.

The regression line and confidence limits may then be set up for selected values of the regressor variable. Note that these values, like the predictors, are expressed in ordinary units, but the intermediate calculations are in logarithms. It is in general wise not to extend the predictions beyond the limits of the original observations, approximately 200–300.

|       |       |       |       | Confidence limits | | | |
|       |       |       |       | On $E(Y)$ | | On $\hat{Y}$ | |
| $x$ | $u$ | $\hat{v}$ | $y$ | | | | |
|-----|-------|--------|-------|--------|--------|--------|--------|
| 200 | 5.298 | 4.9856 | 146.3 | 101.2 | 211.5 | 85.2 | 251.2 |
| 220 | 5.394 | 5.1502 | 172.5 | 132.2 | 225.1 | 107.1 | 277.8 |
| 240 | 5.481 | 5.3004 | 200.4 | 165.5 | 242.7 | 129.1 | 311.0 |
| 260 | 5.561 | 5.4386 | 230.1 | 195.8 | 270.5 | 150.1 | 352.8 |
| 280 | 5.635 | 5.5665 | 261.5 | 217.3 | 314.8 | 168.9 | 404.9 |
| 300 | 5.704 | 5.6857 | 294.6 | 231.7 | 374.6 | 185.4 | 468.0 |

**b.** This problem is intriguing. It will be clear that in general the width of the confidence limits on a future value of $Y$ given $x_f$ will be $2ts\sqrt{1 + g(x_f)}$, where the $x_i$ are the data from a sample of size $n$ and

$$g(x_f) = \frac{(x_f - m_x)^2}{\Sigma(x_i - m_x)^2} + \frac{1}{n}$$

Now other things being equal, the width is smaller if $g(x_f)$ is equal to zero. Furthermore if we do not have to estimate the regression slope, the $t$ statistic has one more degree of freedom and hence for any particular confidence level is somewhat smaller. Hence the critical issue is how much the estimate of $s$ is reduced by including a regression slope. If the latter is in fact zero, on average $S$ will have the same value and the width is increased.

On the other hand, if the regression slope is not zero, we would have no choice but to include it in our confidence limit, however small its contribution. But this argument is to stir up a hornet's nest; for it might be argued that once we are unsure about simple linearity, we should be including an indefinite number of powers (or possibly other functions) of $x$, as far as the number of data we have will allow, each contributing its component and each eroding the number of degrees of freedom for the test statistic. (This problem will be addressed in Chapter 9.) Furthermore, any statement we can make about any particular parameter is purely probabilistic, whereas the decision to include it in our working model and hence in our confidence statement is all-or-none. Hence we are making a critical decision about estimation on the basis of a test of significance that is of arbitrary size.

I know of no statistical escape from this logical difficulty. This problem seems to me to have arisen (and this theme is a recurrent one) because we are trying to use statistics not to *test* or *estimate* the particular parameters of a model, but to *discover* what the *structure* of the model is. In my opinion, the latter is a scientific issue and outside the purview of statistics. I repeat my opinions from Chapters 1 and 2, that the proper coinage of statistics is quantity; and I take no responsibility for what happens when this principle is abandoned.

**8.3. a.** It follows at once from the Gauss-Markov theorem that the least-squares estimator $\sqrt{\hat{\alpha}}$ gives an unbiased estimate of $\sqrt{\alpha}$.

**b.** Let it be granted that for any finite sample $\hat{\alpha}$ has a variance greater than zero. Then by an argument we have used previously

$$\text{var}(X) = E(X^2) - [E(X)]^2$$

whence in general

$$E(X^2) > [E(X)]^2$$

That is, the expectation of the square of a random variable is greater than the square of the expectation of a random variable.

**8.4.** Clearly, the issue here is one of regression, the particular case where the regression function is linear and passes through the origin. As we shall see in Chapter 11, if the counterclaim were substantiated, the test would be no better a discriminator than the simple preliminary reading. The test would be essentially to see whether the best-fitting straight line would pass through the origin or that the origin lies within (say) the 95% confidence limits on the intercept. In general one would be obliged to be careful about variance: it may be proportionate and a transformation may be necessary, or some more complicated procedure such as weighted least squares.

But there is a trap here for the unwary, which has a much wider importance. The investigator, we will suppose, measures the initial pressure $(x)$ and the pressure after immersion $(y)$. He determines the response as $(x - y)$, which he denotes as $z$. Now none of these measurements is free of error. Thus for any given $y$ the more $x$ is overestimated by error, the more $z$ will be underestimated. There will clearly be a negative component of correlation from this source. This would be true even if the response is totally unrelated to the initial value. If, as seems likely, the two are positively correlated, the error in measurement of the initial reading will lead to an underestimate of the regression slope, which may not pass through the origin even when the response to the cold pressor test is proportionate. The simplest appropriate precaution is to estimate the two measurements, the preliminary reading, and the response, independently.


# CHAPTER 9

**9.1.** Let us denote the individual mean value by $X$, the variance by $Y$, and the probability (evolutionary fitness) by $Z$. Then we have the estimates shown in Table A6. In each case "$CT$" denotes the correction term, which when subtracted from the raw sum of squares or products gives the sum of squares or of the products from the sample mean(s).

**9.2.** By this, I hope, the reader will have come to regard calculating the estimates of means and variances by the usual method of moments as a straightforward exercise. Solving the polynomial equation (an idea which, in principle, was well established by 1930) is a good deal more tedious, which may account for the neglect of this obvious measure of fitness. Although in principle (82) for sufficiently bizarre distributions of family size, $Z$ may have only a loose relationship to $Y$ and practically none to $X$, in practice the relationship may be a good deal closer and the geneticist may wish to take a short cut by substituting for $Z$ an empiric regression equation in $X$ and $Y$.

We ordinarily assume either that $X$ and $Y$ are fixed and $Z$ is also fixed with an added random component; or that $X$, $Y$, and $Z$ are jointly multivariate normal, that is, that any linear combination of them is Gaussian. In either case, we would require for each random variable independent errors that, for each individual variable, are homogeneous, and between each possible pair of variables have a homogeneous covariance. If we condition on $X$ and $Y$, we merely have to concern ourselves with the distribution of $Z$. There is no biological reason to suppose that, over the sample space of interest (Amish genealogies), the errors for the individual pedigrees are correlated. But since

**Table A6. Regression of evolutionary fitness on the mean and variance of the size of completed family**

| | | $n = 15$ | |
| --- | --- | --- | --- |
| | Mean $(x)$ | Variance $(y)$ | Evolutionary fitness $(z)$ |
| Sum | 26.319 | 47.139 | 7.361 |
| Mean | 1.7546 | 3.1426 | 0.4907 |
| Sum of squares | 50.266,851 | 171.528,523 | 4.309,409 |
| CT | 46.179,317,4 | 148.139,021,4 | 3.612,288,1 |
| Sum of squares from mean | 4.087,533,6 | 23.389,501,6 | 0.697,120,9 |
| Mean square | 0.291,967 | 1.670,679 | 0.049,794 |
| | $xy$ | $xz$ | $yz$ |
| Sum of products | 91.222,041 | 14.515,347 | 25.913,149 |
| CT | 82.710,089,4 | 12.915,610,6 | 23.132,678,6 |
| Sum of products from means | 8.511,951,6 | 1.599,736,4 | 2.780,470,4 |
| Estimated covariance | 0.607,997 | 0.114,267 | 0.198,605 |

the estimates are based on varying numbers of sibships, the estimate of $Z$ will have one heterogeneous component of error. However, it seems a reasonable supposition that this component is small relative to that of the actual fitness among families (which is, of course, the fundamental concern of evolutionary genetics). Finally, we should at least be able to get a reasonable point estimate of the multiple regression coefficients, even if there is some residual doubt about interval estimation.

With the model

$$Z = a + bx + cy + H$$

$H$ being the random component, we have the usual Gauss-Markov equation

$$\begin{bmatrix} a \\ b \\ c \end{bmatrix} = \begin{bmatrix} 15 & 26.319 & 47.139 \\ 26.319 & 50.266,851 & 91.222,041 \\ 47.139 & 91.222,041 & 171.528,523 \end{bmatrix}^{-1} \begin{bmatrix} 7.361 \\ 14.515,347 \\ 25.913,149 \end{bmatrix}$$

The determinant of the matrix is 347.280,804, and the inverse, found by the method of cofactors is

$$\begin{bmatrix} 0.865,979,173,700 & -0.617,204,877,641 & 0.090,254,939,026 \\ -0.617,204,877,641 & 1.010,256,023,24 & -0.367,654,280,136 \\ 0.090,254,939,026 & -0.367,654,280,136 & 0.176,551,664,514 \end{bmatrix}$$

Premultiplying the vector by this inverse gives

$$
\begin{bmatrix} \hat{a} \\[1em] \hat{b} \\[1em] \hat{c} \end{bmatrix}
=
\begin{bmatrix} -0.245,680,588,480 \\[1em] 0.593,891,490,200 \\[1em] -0.097,253,257,290 \end{bmatrix}
$$

Thus the regression sum of squares is

$$-0.245,68\ldots \times 7.361 + 0.593,89\ldots \times 14.515\ldots - 0.097,25\ldots \times 25.913\ldots$$

$$= 4.291,948\ldots$$

It comprises 99.59% of the total sum of squares for $Z$, 4.309,409. The variance of the residual error is the remainder of the total sum of squares divided by the number of sets of data (15) less 3 for the three parameters that have been estimated:

$$s^2 = (4.309,409 - 4.291,948)/12 = 0.001,455,07\ldots$$

**9.3.** The regression parameter for age may be denoted by $a$, and that for the logarithm of the blood urea nitrogen by $b$, with an intercept $c$. Then the estimates will be given by the usual equation

$$
\begin{bmatrix} \hat{c} \\[1em] \hat{a} \\[1em] \hat{b} \end{bmatrix}
=
\begin{bmatrix} 22 & 1,187 & 35.260 \\[1em] 1,187 & 65,969 & 1,914.962 \\[1em] 35.260 & 1,914.962 & 59.414,004 \end{bmatrix}^{-1}
\begin{bmatrix} -1.942 \\[1em] -144.475 \\[1em] -6.903,016 \end{bmatrix}
$$

The total sum of squares for survival is 9.846,986. The determinant of the matrix is 119,438.924,6. The inverse matrix is

$$
\begin{bmatrix}
2.113,238,789,4 & -0.025,141,407,109 & -0.443,800,429,194 \\[1em]
-0.025,141,407,109 & 0.000,534,503,204,996 & -0.002,306,986,611,96 \\[1em]
-0.443,800,429,194 & -0.002,306,986,611,96 & 0.354,566,152,883
\end{bmatrix}
$$

whence

$$
\begin{bmatrix} \hat{c} \\[1em] \hat{a} \\[1em] \hat{b} \end{bmatrix}
=
\begin{bmatrix} 2.591,956,526,54 \\[1em] -0.012,472,572,442,5 \\[1em] -1.252,413,502,15 \end{bmatrix}
$$

The regression sum of squares is 5.413,825... out of a total of 9.846,986. The residual sum of squares, 4.433,160 divided by three less than the number of sets of data, 22, gives an estimate for the variance of the residual error of 0.233,324.

## CHAPTER 10

**10.1. a.** Let us denote the random variables by $X$ and $Y$. Whenever all the data lie on a straight line, then one may write $Y$ as a linear function of $X$.

$$Y = a + bX$$

(where $a$ and $b$ are not random variables). Now the mean of $Y$ will be $a$ plus $b$ times the mean of $X$. Since the correlation coefficient involves subtracting from each value whatever the mean may be, clearly the value of $a$ does not matter, and we then have to find the correlation coefficient between $X$ and $bX$. That is

$$\frac{E(X - M_X)(bX - M_{bX})}{\sqrt{\mathrm{var}(X)\mathrm{var}(bX)}} = \frac{E(X - M_X)(bX - bM_X)}{\sqrt{\mathrm{var}(X)b^2\mathrm{var}(X)}}$$

$$\frac{bE(X - M_X)^2}{b\,\mathrm{var}(X)} = \frac{b\,\mathrm{var}(X)}{b\,\mathrm{var}(X)} = \frac{b}{b}$$

Clearly if $b = 0$, then $r$ is $0/0$ which is indeterminate.

**b.** And if $b \neq 0$, $r = 1$.

**c.** The first estimate results because $b = 0$ implies that $Y = a$ (i.e., a constant); and it is meaningless to talk about a correlation between anything but two random variables. In the second case, it is clear that with suitable adjustment of origin and scale the problem is to find the correlation coefficient of a random variable with itself, which, as we noted in the text, is unity.

**10.2.** If $r = 0$ (the null hypothesis) $\mathrm{cov}(X, Y) = 0$ and $b = 0$. Now we have noted in Chapter 8 that

$B/S_B$ follows a $t$ distribution with $n - 2$ degrees of freedom.

for clarity write $H$ for $\Sigma(Y - M_Y)^2$, $J$ for $\Sigma(X - M_X)^2$, and $K$ for $\Sigma(X - M_X) \cdot (Y - M_Y)$. Then substituting we get $B = K/J$ and

$$S_B = \sqrt{\frac{H - K^2/J}{(n - 2)J}} = \sqrt{\frac{HJ - K^2}{(n - 2)J^2}}$$

Then

$$\frac{B}{S_B} = \sqrt{\frac{K^2}{J^2} \div \frac{HJ - K^2}{(n - 2)J^2}} = \sqrt{\frac{K^2(n - 2)}{HJ - K^2}}$$

Dividing throughout the numerator and the denominator by HJ

$$\frac{B}{S_B} = \sqrt{\frac{\dfrac{K^2(n - 2)}{HJ}}{1 - \dfrac{K^2}{HJ}}} = \sqrt{\frac{r^2(n - 2)}{1 - r^2}} = r\sqrt{\frac{n - 2}{1 - r^2}}$$

which therefore follows a $t$ distribution with $(n - 2)$ degrees of freedom.

**10.3.** Suppose $r > 0$. Let us increment $r$ by some quantity $h$ such that $r + h \neq 1$. Then the expression in (10.6) becomes

$$(r + h)\sqrt{\frac{n - 2}{1 - (r + h)^2}}$$

Now both factors (as well as the numerator and the denominator) are positive and

$$\sqrt{\frac{n - 2}{1 - (r + h)^2}} > \sqrt{\frac{n - 2}{1 - r^2}}$$

since the left hand side has the smaller denominator; hence the increment in $r$ has increased the transform. The converse would apply if $h$ had been negative. Likewise if $r < 0$, a negative increment of $h$ makes (1) more negative and conversely.

**10.4.** On the scale of Fisher's transform (10.5), the sample correlation coefficient has an approximately Gaussian distribution with the approximate variance $1/(n - 3)$. In the present case, then, the standard deviation is 0.2, while the transform of the point estimate is 0.2027. The one-tailed rejection region at the 5% level would be $(1.645)(0.2)$ units above this point, that is, 0.5317. Let us take a few values of $\rho$ under the alternative hypothesis and evaluate what proportions of their respective distributions lie in the rejection region.

| $\rho$ | $H(\rho)$ | Number of standard above critical value | Power |
|---|---|---|---|
| 0.25 | 0.2554 | $-1.3816$ | 0.0835 |
| 0.3 | 0.3095 | $-1.1109$ | 0.1333 |
| 0.4 | 0.4236 | $-0.5404$ | 0.2945 |
| 0.5 | 0.5493 | 0.0879 | 0.5350 |
| 0.6 | 0.6931 | 0.8071 | 0.7902 |
| 0.7 | 0.8673 | 1.6779 | 0.9533 |
| 0.8 | 1.0986 | 2.8344 | 0.9977 |

The power function computed in this fashion may conveniently be plotted on the original scale (i.e., the values given in the first column). It is evident that at the upper limit of the sample space the power is 1.0.

## CHAPTER 11

**11.1.** The ratios are displayed in the third column of Table 8.2. The mean is 0.084,86. The three signs under the heading "Code" (the last column of that table) indicate whether the observed value is above $(+)$ or below $(-)$ the following quantities: (i) 0.084,86 times the height in centimeters, (ii) the value predicted by the linear regression equation from Chapter 8, and (iii) the value predicted by the cubic regression equation in Chapter 9. Thus we may compare the concordances. Note the following points. Despite the difference in their mathematical form, the two regression equations give concordant values in 37 of the 41 cases, compared with 30 concordances for each regression equation with the ratio. Note further, that according to Figure 11.3 (middle diagram) we would expect the greatest incongruities with the ratio at the extreme ends, and the least in the intermediate values of height. If we divide the data as nearly as possible into three equal groups, making the center group of size 13, and the others of size 14, we find the following concordances:

|                    | Group | | |
|--------------------|:-----:|:------:|:-------:|
|                    | Lowest | Middle | Highest |
|                    | (14)  | (13)   | (14)    |
| Ratio and linear   | 9     | 13     | 8       |
| Ratio and cubic    | 9     | 13     | 8       |
| Linear and cubic   | 12    | 13     | 12      |

Despite the fact that it is the smallest, the middle group shows the highest number of concordances. The concordances for the two regression models show an almost uniform distribution of concordances. As one can see from examining the last column of Table 8.2 in detail, the discordances are predominantly of the type predicted from Figure 11.3, i.e., high values of the height lead to $(-)$ values for the ratio, and conversely.

This example is open to the criticism that there is no good reason to suppose that the serum protein level increases with the height and therefore nobody would ever consider expressing it as such a ratio. Quite so. But if there had been a statistically significant, but biologically trivial, positive regression, would it have made much more sense to use a ratio in that case either?

**11.2.** As we noted previously (problem 4.6) the appropriate transformation to normalize the data and to stabilize the variance is the square root. We have the following results for the transformed values.

|                              | Control subjects | Heterozygotes |
|------------------------------|:----------------:|:-------------:|
| $n$                          | 27               | 13            |
| $\Sigma\sqrt{x}$             | 333.874,856,5    | 115.571,284,4 |
| $m_{\sqrt{x}}$               | 12.365,735,43    | 8.890,098,797 |
| $\Sigma(\sqrt{x})^2$         | 4,157            | 1049.4        |
| $\Sigma(\sqrt{x} - m_{\sqrt{x}})^2$ | 28.391,859,23 | 21.959,863,94 |
| $s_{\sqrt{x}}$               | 1.044,985        | 1.352,771     |

Since the samples are small, we shall use a pooled estimate of the variance

$$\hat{\sigma}^2 = \frac{28.391,859,23 + 21.959,863,94}{27 + 13 - 2}$$

$$= 1.325,045\ldots$$

**a.** The fact that the proband is the *oldest* sib effectively disposes of genetic bias due to ascertainment. Moreover, we are, for practical purposes, sure that the parents are both carriers, and the probabilities of the genotypes for any sib of an affected person are then: wild type 1/4; carrier 1/2; affected 1/4. An unaffected adult sib must belong to one of the first two categories in the ratio 1:2. The relative densities of the competing distributions are thus estimated to be for *carriers:* Gaussian with mean 8.890 and standard deviation 1.151; and *controls:* Gaussian with the same standard deviation and a mean of 12.366. Taking the intersection of the two curves as the natural point of discrimination, canceling terms in common but conserving the prior factor of 2, and taking natural logarithms, we find that the intersection is the solution to the equation

$$\frac{(x - 8.890)^2}{2(1.325,045)} = \frac{(x - 12.366)^2}{2(1.325,045)} + \log_e 2 \tag{1}$$

Whence $x = 10.898$. This quantity is 1.726 standard deviations above the mean for the carriers, so that there will be 4.2% of false negatives among them and 1.262 standard deviations below the mean for the controls giving 10.3% of false positives.

**b.** If the frequency of the syndrome is 1 in 100,000, the square root of this, or 0.003,162 should be the frequency of the gene and the probability of one such gene and one wild type in a person taken at random is

$$2(0.003,162)(1 - 0.003,162) = 0.006,304,5\ldots$$

while the probability of a homozygous wild type will be

$$(1 - 0.003,162\ldots)^2 = 0.993,685\ldots.$$

Hence we have the same equation as in (1) above, but with the logarithm of $(0.006,304,5\ldots)/(0.993,685) = \log_e(0.006,344,6\ldots)$ rather than of 2 on the right hand side of the equation. The point of intersection proves to be 8.657, which lies 0.200 standard deviations above the mean for carriers and 3.188 below that for homozygotes. The misclassification rate is 42.1% false negatives and 0.07% false positives.

**11.3.** We have the following preliminary calculations.

|  | Inbred | Outbred |
|---|---|---|
| $n$ | 27 | 27 |
| $\Sigma x$ | 91.35 | 92.39 |
| $m$ | 3.3833 | 3.4219 |
| $\Sigma x^2$ | 313.8565 | 318.5369 |
| CT | 309.0675 | 316.1449 |
| $\Sigma(x - m_X)^2$ | 4.7890 | 2.3920 |
| $s^2$ | 0.18419 | 0.09200 |
| $s$ | 0.4292 | 0.3033 |
| Gurland-Tripathi correction factor | 1.009,615 | 1.009,615 |
| $\hat{\sigma}$ | 0.4333 | 0.3062 |

The sample standard deviations are sufficiently different to make it worthwhile to treat them separately. Being given no prior probabilities, we may carry through the calculations assuming the groups equally probable. Setting the densities equal, canceling common terms and taking natural logarithms, we reduce our problem to solving

$$-\log_e 0.4333 - \frac{(x - 3.3833)^2}{2(0.4292)} = -\frac{(x - 3.4219)^2}{2(0.09200)} - \log_e 0.3062$$

whence $x = 3.099$ or 3.822. (There are two roots because the variances are unequal and the curves intersect at two points.) The misclassifications about these points are: for inbred group 74.4% and 50.0%, respectively, and for the outbred group 14.6% and 65.5%.

**11.4.** This measure is commonly used, particularly by the epidemiologist, when data are available for several subjects at risk observed for varying periods of time. It is an admirable criterion where: (1) we may suppose that the system "has no memory": if, for example, those at risk do not become less liable to be affected with longer exposure (because they become immune, or more skilled) or more likely to be affected (because there is cumulative damage such as occurs in a "multiple-hit" process which involves scars or because the condition depends on some process of sensitization as in allergies); (2) all subjects are at the same risk; and (3) risks are independent in all time periods.

On these conditions, the information obtained by watching 20 men for 10 years is exactly as useful as watching 200 men for 1 year each. How far these are plausible assumptions in any particular case, I would leave to the good sense of the investigator. It is not difficult to think of plenty of instances where they would be preposterous. What matters is to realize that the assumptions *must* be confronted explicitly.

## CHAPTER 12

**12.1.** The three parts of this problem may be conveniently done in parallel. We shall compensate for ignoring $K$ by multiplying the (relative) prior density by $10^{15}$.

| $p$ | Relative prior probability of parameter $10^{15}p^{20}(1-p)^{20}$ | Conditional probability (likehood) of sample $210p^4(1-p)^6$ | Joint probability density $21 \times 10^{16}p^{24}(1-p)^{26}$ |
|---|---|---|---|
| 0.2 | 0.120,89 | 0.088,080 | 0.010,648 |
| 0.3 | 27.821,84 | 0.200,121 | 5.567,734 |
| 0.4 | 401.998,87 | 0.250,823 | 100.830,425 |
| 0.5 | 909.494,70 | 0.250,078 | 186.517,468 |
| 0.6 | 401.998,87 | 0.111,477 | 44.813,522 |
| 0.47 | 846.202,98 | 0.227,126 | 192.194,296 |
| 0.48 | 880.829,03 | 0.220,396 | 194.131,462 |
| 0.49 | 902.246,33 | 0.213,022 | 192.198,412 |
| 0.479 | 887.939,79 | 0.221,110 | 194.112,017 |
| 0.481 | 883.585,71 | 0.219,687 | 194.112,021 |

The maximum appears to be 0.48. Indeed the alert reader will have seen that it must be. The joint density is a constant times $p^{24}(1-p)^{26}$ which is equivalent to the likelihood for a binomial with 24 successes in 50 trials. The maximizing value of $p$ for this function is simply $24/50 = 0.48$.

The prior probability corresponds to considerably more information than is contained in the sample. The larger the sample the more it dominates the answer, and if it is large enough, the prior probability used here would have a negligible impact (the principle of "the donkey in the beer" [3]). If the reader tests this conclusion against his own common sense, it will be seen to be eminently reasonable. If biochemical findings in only five cases of Yo-Yo disease have been previously published, my series of 50 cases need scarcely take them into account. On the other hand if 50 cases have been published and I have a further 5, it is not reasonable that my findings should be given more weight than the collective experience.

**12.2.** For simplicity, let us write $w$ for $(|x - \mu| - \frac{1}{2})^2$.[2] Then the right hand side of (12.6) becomes

$$\frac{w}{np} + \frac{w}{nq} = \frac{wq + wp}{npq} = \frac{w(q + p)}{npq} = \frac{w}{npq}$$

**12.3.** Suppose that there are $x$ successes. Then the ratio of the likelihoods under the null hypothesis $(p)$ to that under the alternate $(p + h)$ is

$$\text{LR}(x) = \frac{\binom{n}{x} p^x (1 - p)^{n-x}}{\binom{n}{x} (p + h)^x (1 - p - h)^{n-x}} \tag{1}$$

which is simplified slightly by cancelling the combinatorial coefficients. Now what happens if we increase $x$ by the (smallest) increment 1? The ratio of the likelihood ratios is

$$\frac{\text{LR}(x + 1)}{\text{LR}(x)} = \frac{p^{x+1}(1 - p)^{n-x-1}}{(p + h)^{x+1}(1 - p - h)^{n-x-1}} \div \frac{p^x(1 - p)^{n-x}}{(p + h)^x(1 - p - h)^{n-x}}$$

This reduces with a little canceling to

$$\frac{p(1 - p - h)}{(1 - p)(p + h)} = \frac{p - p^2 - ph}{(p - p^2 - ph) + h} < 1$$

Hence, if we increase $x$ by 2 the ratio becomes yet smaller, and so on. Thus the likelihood ratio decreases monotonically with increasing $x$. Furthermore, the larger $h$, the faster the likelihood ratio decreases. Hence the smallest ratios will be in the upper part of the sample space, which is where the rejection region should lie.


## CHAPTER 13

**13.1.** We first note that $\Sigma O_i = \Sigma E_i = n$. The conventional form of the chi-square statistic is

$$\sum \frac{(O_i - E_i)^2}{E_i} = \sum \frac{O_i^2}{E_i} - \sum \frac{2O_i E_i}{E_i} + \sum \frac{E_i^2}{E_i}$$

$$= \sum \frac{O_i^2}{E_i} - 2\Sigma O_i + \Sigma E_i$$

$$= \sum \frac{O_i^2}{E_i} - n$$

The data on the ABO blood group give the following results:

| Phenotype | Number observed | Expected number | | $\dfrac{O^2}{E}$ | $\dfrac{(O - E)^2}{E}$ |
|---|---|---|---|---|---|
| A | 119 | $239(p^2 + 2pr) =$ | 119.6603 | 118.3433 | 0.003,664 |
| B | 21 | $239(q^2 + 2pr) =$ | 21.7690 | 20.2582 | 0.027,165 |
| O | 87 | $239(r^2) =$ | 86.3845 | 87.6199 | 0.004,386 |
| AB | 12 | $239(2pq) =$ | 11.1861 | 12.8731 | 0.059,219 |
| Total | 239 | | 238.9999 | 239.0944 | 0.094,4 |
| | | | | $-239$ | |
| | | | $\chi^2 =$ | 0.0944 | |

Note that formula 13.2 involves less rounding and hence makes for more exact calculations.

**13.2.a.** There is only a single degree of freedom after we have estimated the two parameters and ensured that the totals agree. The significance level is about 0.76. The fit is excellent. We could also have extracted the square root of the chi-square statistic and interpreted it as a (two-tailed) standard Gaussian variate.

**b.** There are a number of points of logical interest.

While we are satisfied about the high level of significance, it is not alarmingly high, the fit is not "too good." (This criterion is often applied when investigators wish to check whether the goodness of fit may be an artifact, or due to overzealous editing of the results.)

Second, the chi-square statistic here (0.0944) is not quite so small as that for the minimum chi-square estimator (0.0935). This is logically to be expected since the latter estimator is designed to minimize the chi-square; but readers may be led to suspect that the minimum chi-square is, after all, the better estimator. But on further reflection, it is obvious that it is "best" only by the chi-square criterion: MLE is the "best" by the likelihood criterion! In fact both estimators are in the limit efficient.

Finally the logical implications of our test are somewhat marred by the lack of any clear alternative hypothesis (a common state, I am sorry to say, in genetic analysis). Hence, we can make no satisfactory claims about power.

**13.3.** This exercise is a simple matter of checking the calculations. Note that one of the advantages of grouping by estimated deciles is that the $E_i$ are all large enough for comfort and lend themselves to a very easy procedure for computing the chi-square statistic:

$$\chi^2 = (25^2 + 60^2 + 11^2 + 4^2 + 1^2 + 2^2 + 2^2)/10.5 - 105$$
$$= 4371/10.5 - 105 = 311.2857$$

However, the advantage of the "long method" of computing chi-square used in Table 4.2 is that one can immediately tell, from the size of the contributions to the chi-square statistic, where the main discrepancies lie.

**13.4.** In the conditions stated we would have two classes by phenotype:

| Phenotype | Probability | Observed |
|---|---|---|
| 1 | $p^2 + 2pq$ | 55 |
| 2 | $q^2$ | 74 |
| | | 129 |

The method-of-moments estimate (and the MLE) is found by setting

$$\hat{q}^2 = 74/129 = 0.573,643,4\ldots$$

$$\hat{q} = 0.757,393$$

Whence

$$\hat{p} = 1 - 0.757,393 = 0.242,607$$

For our test we have

| Phenotype | Expected |
|---|---|
| 1 | $129[(0.242,607)^2 + 2(0.242,607)(0.757,393)] = 54.9999$ |
| 2 | $129(0.757,393)^2 = 74.0001$ |

An almost perfect fit! But in fact we have consumed two degrees of freedom and the test statistic is a chi-square with no degrees of freedom. It will always give a perfect fit (any discrepancies being due to rounding error).

**13.5.** The contradiction stems from the fact that two meanings of the word *independent* are being confused. When we talk of the nonindependence of outcomes from multinomial trials we mean stochastic nonindependence: that the probability of one outcome for a trial bears on the probability of another outcome for the same trial (as they certainly do since they are mutually exclusive). But independence of estimations of parameters is not a stochastic, but a mathematical, relationship, exactly as is the case when we talk about linear independence of equations (q.v.). If in a binomial we have a well-characterized estimator of $p$ then we automatically have an estimator of $q = (1 - p)$ that has exactly similar properties. The parameter space *spans* only one dimension. But this would not be true of the ABO blood group data in Table 13.2. If we know both $p$ and $q$, then we know $r$ exactly; hence any two, but only two, of the parameters can be estimated independently. The three parameters *span* only two dimensions. This would be something like saying that the *rank* (in the matrix sense of the term, q.v.) of the parameter space is two. But that does not mean that in the statistical sense $\hat{p}$ and $\hat{q}$ will be uncorrelated.


# CHAPTER 14

**14.1.** The secret of dealing with such problems is to recognize that confidence limits and standard deviations can (like means) be linearly scaled. Hence one may do the calculations in convenient terms and then rescale quite simply. We know that the convolution (sum) of Poisson variates is a Poisson variate. Hence, if the attack rate for a single day is Poisson, so will that for the total of all the days, which are, *ex hypothesi,* days of constant risk (there is no seasonal variation) and independent (there is no clustering). The best estimate of the expected number affected in 1,642 days is the observed, i.e., 462, and we may begin by setting confidence limits on it. We may use the square root transform (which has a variance of 0.25 or a standard deviation of 0.5) and take 1.96 standard deviations on either side, the two values being converted to the origin scale by squaring:

$$[\sqrt{462} - 1.96(0.5)]^2 \quad \text{and} \quad [\sqrt{462} + 1.96(0.5)]^2$$

$$\text{or} \ (21.49 - 0.98)^2 \quad \text{and} \quad (21.49 + 0.98)^2$$

$$\text{or} \ 420.83 \quad \text{and} \quad 505.09$$

To convert this to a rate per day, we simply divide by the total number of days or 1,642:

$$0.256 \quad \text{and} \quad 0.308$$

**14.2. a.** The foregoing method may be used here; but since the numbers are small, we may prefer to use the more exact method. Reference to the Pearson-Hartley tables or direct, if somewhat tedious, summation of the terms of a truncated Poisson series, gives the 95% confidence limits as 1.37 and 9.15; and on 12 as 6.92 and 19.44. We then have the following estimates per thousand:

|         |                  | Confidence limits |       |
| Group   | Point estimates  | Lower | Upper |
|---------|------------------|-------|-------|
| Control | 5.99             | 2.05  | 13.70 |
| DES treated | 17.32        | 9.99  | 28.05 |

There is some overlap in the confidence intervals.

**b.** The comparison of the two proportions must, of course, take sample size into consideration. The observed numbers affected may be thought of as realizations of Poisson variates. So to an excellent approximation we have for the first

$$\operatorname{var}(\sqrt{X} + \sqrt{X+1}\,) = 1$$

$$\operatorname{var}\left(\sqrt{\frac{X}{668}} + \sqrt{\frac{X+1}{668}}\,\right) = \frac{1}{668} = 0.001,497$$

and for the second

$$\operatorname{var}\left(\sqrt{\frac{X}{693}} + \sqrt{\frac{X+1}{693}}\,\right) = \frac{1}{693} = 0.001,443$$

and the variance of the difference of the transforms will be the sum of these quantities, i.e., 0.002,940... giving a standard deviation of 0.0542.... The appropriate aproximate $z$ test would be

$$\frac{\sqrt{\dfrac{12}{693}} + \sqrt{\dfrac{13}{693}} - \sqrt{\dfrac{4}{668}} - \sqrt{\dfrac{5}{668}}}{0.0542} = 1.930$$

A one-tailed test may be thought appropriate here; accordingly the null hypothesis would be rejected.

**14.3.** First we will do a broad test of heterogeneity of risk, using a chi-square contingency table.

| Mother's age |          | Normal Pregnancy | Trisomy | Total |
|--------------|----------|------------------|---------|-------|
| 34–36        | Observed | 866              | 10      | 876   |
|              | Expected | 862.0571         | 13.9429 |       |
| 37–38        | Observed | 495              | 6       | 501   |
|              | Expected | 493.0258         | 7.9742  |       |
| 39–40        | Observed | 302              | 7       | 309   |
|              | Expected | 304.0818         | 4.9182  |       |
| 41–42        | Observed | 130              | 6       | 136   |
|              | Expected | 133.8353         | 2.1647  |       |
| Total        |          | 1793             | 29      | 1822  |

$$\chi^2 = \frac{866^2}{862.0571} + \cdots + \frac{6^2}{2.1647} - 1822 = 9.4305$$

The statistic is distributed with 3 degrees of freedom and yields a significance level of 0.0248. We may decide to analyze the "trisomies 21" and the "other trisomies," although with some disquiet at the small sizes of the expected numbers. The results are (expected numbers following each in parentheses): Trisomy 21: 6

(7.2119); 3 (4.1246); 4 (2.5439); 2 (1.1196). Other trisomies: 4 (6.7311); 3 (3.8496); 3 (2.3743); and 4 (1.0450). The figures for those without trisomies remain unchanged. The chi-square statistic, now distributed with $(4 - 1)(3 - 1) = 6$ degrees of freedom, is 12.0024. The significance level is now slightly greater than 0.05.

## CHAPTER 15

**15.1. a.** We have the following results on the paired difference, $d$, denoting (low fat−vegetable fat): 0.06, 0.46, 0.84, 0.90, 0.54, 2.16, 0.38.

$$\Sigma d = 5.34$$
$$m_D = 0.7629$$
$$\Sigma d^2 = 6.8324$$
$$m_D \Sigma d = 4.0737$$
$$\Sigma(d - m_D)^2 = 2.7587$$
$$s^2 = 0.4598$$
$$s_M^2 = 0.06568$$
$$s_M = 0.2563$$
$$t = 2.9765$$
$$p = 0.0248 \text{ (two-tailed)}$$

**b.** All seven readings have the same sign. Hence on a bionomial test we reject the null hypothesis at

$$p = (\tfrac{1}{2})^7 \times 2 = 0.0156 \text{ (two-tailed)}$$

**Comment.** Here is an instance in which the binomial test gives the lower significance level. But there is no clear warrant for using the binomial rather than the $t$ test. In fact, extensive experience suggests that the distribution (among subjects) of the mean survival of platelets is close to Gaussian. The experimentor may, by special pleading, escape the criticism that the binomial test is inadmissible (q.v.); but this would be to burn one's boats. Over a great many actual examples of the effects of manipulating platelet survival, this is the only one in which I have ever found a lower significance level with the binomial than with the $t$ test. Whatever course the statistician takes it must be self-consistent. The admissibility of a test is not to be judged by the outcome.

**15.2.** Taking the results as they appear in Table 9.9 we have the following rankings.

| Mean size of family | Variance of family size | Probability of survival |
|---|---|---|
| 9 | 10 | 8 |
| 5 | 4 | 6 |
| 1 | 1 | 1 |
| 10 | 7 | 11 |
| 15 | 15 | 15 |
| 4 | 5 | 4 |
| 2 | 3 | 3 |
| 6 | 12 | 5 |

| Mean size<br>of family | Variance of<br>family size | Probability<br>of survival |
|:---:|:---:|:---:|
| 11 | 8 | 13 |
| 7 | 2 | 10 |
| 13 | 11 | 14 |
| 8 | 9 | 7 |
| 14 | 14 | 12 |
| 12 | 13 | 9 |
| 3 | 6 | 2 |

Rearranging them so that the means are in order, we get

| Mean | 1 | 2 | 3 | 4 | 5 | 6 | 7 | 8 | 9 | 10 | 11 | 12 | 13 | 14 | 15 |
|---|---|---|---|---|---|---|---|---|---|---|---|---|---|---|---|
| Variance | 1 | 3 | 6 | 5 | 4 | 12 | 2 | 9 | 10 | 7 | 8 | 13 | 11 | 14 | 15 |
| Survival | 1 | 3 | 2 | 4 | 6 | 5 | 10 | 7 | 8 | 11 | 13 | 9 | 14 | 12 | 15 |

The numbers in the "variance" line that are to the right of, and greater than, each number in the "mean" line (i.e., the noninversions) are

$$14 + 12 + 9 + 9 + 9 + 3 + 8 + 5 + 4 + 5 + 4 + 2 + 2 + 1 = 87$$

The denominator is $n(n - 1) = 210$. Thus the $Q$ statistic is $105 - 87 = 18$ and

$$\tau = 1 - 4(18)/210 = 0.657$$

For the mean and the survivals, the result is

$$14 + 12 + 12 + 11 + 9 + 9 + 5 + 7 + 6 + 4 + 2 + 3 + 1 + 1 = 96$$

$Q$ is 9   and   $\tau$ is $1 - 4(9)/210 = 0.829$.

Finally, ordering on the variance line, we get

| Variance | 1 | 2 | 3 | 4 | 5 | 6 | 7 | 8 | 9 | 10 | 11 | 12 | 13 | 14 | 15 |
|---|---|---|---|---|---|---|---|---|---|---|---|---|---|---|---|
| Survival | 1 | 10 | 3 | 6 | 4 | 2 | 11 | 13 | 7 | 8 | 14 | 5 | 9 | 12 | 15 |

The numbers of noninversions are

$$14 + 5 + 11 + 8 + 9 + 9 + 4 + 2 + 5 + 4 + 1 + 3 + 2 + 1 = 78$$

$Q$ is 27   and   $\tau$ is $1 - 4(27)/210 = 0.486$.

**15.3.** Most readers, I am sure, will never have seen such a distribution, and may be tempted to give the problem up in despair. But this would be a mistake. When all else fails, we may devise our own theory. Here there are only eight possible combinations, so an exhaustive listing is no great undertaking.

**a.** Denoting the phenotype at the particular locus by its order we get the following results, arranged in ascending order by total phenotype:

| Combina-<br>tion of<br>phenotypes | Probability<br>$P(x)$ | Pheno-<br>typic<br>effect $(x)$ | $xP(x)$ | $x^2P(x)$ |
|:---:|:---:|:---:|:---:|:---:|
| LLL | $(0.9)(0.5)(0.98) = 0.4410$ | 2 | 0.8820 | 1.7640 |
| LHL | $(0.9)(0.5)(0.98) = 0.4410$ | 3 | 1.3230 | 3.9690 |

| Combination of phenotypes | Probability $P(x)$ | Phenotypic effect ($x$) | $xP(x)$ | $x^2P(x)$ |
|---|---|---|---|---|
| HLL | $(0.1)(0.5)(0.98) = 0.0490$ | 4 | 0.1960 | 0.7840 |
| HHL | $(0.1)(0.5)(0.98) = 0.0490$ | 5 | 0.2450 | 1.2250 |
| LLH | $(0.9)(0.5)(0.02) = 0.0090$ | 6 | 0.0540 | 0.3240 |
| LHH | $(0.9)(0.5)(0.02) = 0.0090$ | 7 | 0.0630 | 0.4410 |
| HLH | $(0.1)(0.5)(0.02) = 0.0010$ | 8 | 0.0080 | 0.0640 |
| HHH | $(0.1)(0.5)(0.02) = 0.0010$ | 9 | 0.0090 | 0.0810 |
| Total | 1.0000 | | 2.7800 | 8.6520 |

**b.** The mean is simply the sum of the terms in the fourth column, 2.78. The variance is

$$8.652 - (2.78)^2 = 0.9236.$$

**c.** There are problems over two issues in this part of the problem. First, the discrete, and indeed the lumpy, character of the distribution will not allow us to construct tests of the exact size we would wish, and the significance level will in fact be less than the size. Second, we have no specified alternative hypothesis so we cannot set up a most powerful test. Accordingly, we follow the custom of arbitrarily putting the rejection region in the tails. We can in fact find a nontrivial rejection region in one tail only, as will be evident from this tabulation:

| | Lower tail | | Upper tail | |
|---|---|---|---|---|
| Size of test | Rejection region | Significance level | Rejection region | Significance level |
| 0.02 | None | 0 | 6 or more | 0.02 |
| 0.10 | None | 0 | 5 or more | 0.0690 |

**15.4.** Invoking the method of moments, we have two estimating equations. One is based on an exact relationship

$$a = 31.773v \tag{1}$$

The other is based on the approximate relationship

$$as = 6.4 \tag{2}$$

Substituting our empirical estimate of $s$ in (2), we have

$$0.092a = 6.4$$

$$a = 69.56$$

Substituting in (1) we get

$$69.56 = 31.773v$$

$$v = 2.19..$$

Now $v$ must be an integer; so we take the nearest value, 2, and substituting back in (1)

$$a = (2)(31.773) = 63.546$$

The 95% confidence limits on $w$ are (since the sample size is large) approximately

$$31.773 - (1.96)(0.02) \quad \text{and} \quad 31.773 + (1.96)(0.02)$$

i.e., 31.734 and 31.812

Those on $a$ will simply be twice as large:

$$63.468 \quad \text{and} \quad 63.624 \tag{3}$$

Now this is a strange result. We are so accustomed to thinking that the strength of a chain is its weakest link that it seems impossible that we should get this degree of precision in our answer when we are employing an approximation so crude as that in (2), where at least the last digit is suspect. The paradox lies in that the parameter space for $v$ is discrete. A chemist would be morally certain that the true valency is 2. If we were to entertain the possibility that the true value were 1, then the Dulong-Petit product would be 2.923; and for $v = 3$, it would be 8.769. Such values, if true, would suggest that the LDP is a very poor approximation—certainly nothing like good enough to warrant being expressed even tentatively to one decimal place. Of course theoretically the measurements of all the characteristics used in this problem are subject to experimental error. But even supposing that the error of measurement is faithfully Gaussian, however far out in the tails we go, the probability that the true value of $v$ is 3 or 1 is preposterously small. The only alternative is to suppose that the precision of measurement is in accord with the confidence limits cited. This result is more extensively discussed in reference (1).

## CHAPTER 16

**16.1** The grouped data appears as follows. (In each case, the class includes the lower limit but not the upper. For instance, "3-4" includes all cases of 3 or greater, but less than 4.)

| Cell | Number | Cell | Number |
|------|--------|--------|--------|
| 0-1 | 44 | 10-11 | 1 |
| 1-2 | 12 | 12-13 | 2 |
| 2-3 | 8 | 13-14 | 2 |
| 3-4 | 9 | 14-15 | 1 |
| 4-5 | 4 | 19-20 | 1 |
| 5-6 | 7 | 27-28 | 1 |
| 6-7 | 1 | 33-34 | 1 |
| 7-8 | 1 | 34-35 | 1 |
| 8-9 | 1 | 55-56 | 1 |
| 9-10 | 6 | 248-249 | 1 |

$$\Sigma x = 723.5$$
$$m_X = 6.8905$$
$$\Sigma x^2 = 70,482.25$$
$$CT = 4,985.2595$$

$$\begin{array}{rr}
\text{Mean square} = & 629.7788 \\
\text{Sheppard's correction} = & \underline{\phantom{00}0.0833} \\
\text{Adjusted estimate} = & 629.6954 \\
\hat{s} = & 25.0937
\end{array}$$

We compare these results with those found by ungrouped calculation given in Chapter 4. The mean is overestimated by $6.89 - 6.62 = 0.27$ months. The standard deviation is overestimated by $25.09 - 25.06 = 0.03$, surprisingly little, considering how J-shaped the distribution is.

**16.2.** This subject is an inexhaustible exercise in imagination. The following list is illustrative only, and touches on factors that operate in some generality. Readers will no doubt be able to generate plenty of instances peculiar to their own fields. (I should be interested in hearing of such examples.) Several chapters of (3) deal with various aspects of the problem.

**a.** The character of the population at large may have changed, e.g., with respect to age, sex, race, general state of health, etc.

**b.** The differential response to attempts at recruitment of cases may change. For instance, the prospect of an available treatment will tend to stimulate the interest of those interested in cures. Awareness of the disease may increase, because of publicity or education. The professional stature of the investigator may change—for better or for worse—and types of cases referred to him change. He may for example have more difficult diagnostic problems sent to him: or patients from a wider (and hence more heterogeneous) catchment. A newly introduced health scheme may mean that a different social class of patient is eligible for care.

**c.** Diagnostic standards of the referring doctors may change for several reasons: heightened awareness; the impact of new diagnostic techniques, especially if they change diagnostic criteria; changes in critical decision points because of new effective treatment or prevention, or a changing threshold of litigation, etc.

**16.3.** The distribution of time until polyps appear is $N(15, 16)$ and the further time from polyposis to cancer $N(12, 9)$. The distribution of the total time until cancer, since the convolution of two Gaussian variates is itself a Gaussian variate, is $N(27, 25)$. We shall evaluate the various functions of interest at intervals of five years.

| Time | Probability of polyps | Probability of cancer | Probability density of cancer | Hazard of cancer $r(x)$ | Prevalence of polyposis $F(x) - R(x)$ |
|---|---|---|---|---|---|
| | $F(x)/2$ | $R(x)/2$ | $r(x)/2$ | $2 - R(x)$ | $2 - R(x)$ |
| 0 | 0.0000 | 0.0000 | 0.0000 | 0.0000 | 0.0000 |
| 5 | 0.0031 | 0.0000 | 0.0000 | 0.0000 | 0.0031 |
| 10 | 0.0528 | 0.0002 | 0.0001 | 0.0001 | 0.0526 |
| 15 | 0.2500 | 0.0041 | 0.0022 | 0.0022 | 0.2469 |
| 20 | 0.4472 | 0.0404 | 0.0150 | 0.0156 | 0.4239 |
| 25 | 0.4969 | 0.1723 | 0.0368 | 0.0445 | 0.3922 |
| 30 | 0.5000 | 0.3629 | 0.0333 | 0.0523 | 0.2152 |
| 35 | 0.5000 | 0.4726 | 0.0111 | 0.0210 | 0.0519 |
| 40 | 0.5000 | 0.4977 | 0.0014 | 0.0027 | 0.0046 |
| 45 | 0.5000 | 0.4999 | 0.0001 | 0.0001 | 0.0002 |
| 50 | 0.5000 | 0.5000 | 0.0000 | 0.0000 | 0.0000 |

**16.4.** The densities at intervals of one unit are as follows. (Because of symmetry, pairs of readings are given for the same density.)

| Baseline units | | $f(x)$ | | Baseline units | | $f(x)$ |
|---|---|---|---|---|---|---|
| 20 | 40 | 0.00 | | 26 | 34 | 0.06 |
| 21 | 39 | 0.01 | | 27 | 33 | 0.07 |
| 22 | 38 | 0.02 | | 28 | 32 | 0.08 |
| 23 | 37 | 0.03 | | 29 | 31 | 0.09 |
| 24 | 36 | 0.04 | | | 30 | 0.10 |
| 25 | 35 | 0.05 | | | | |

The area of a trapezoid is the sum of the lengths of the parallel sides multiplied by half the distance between them.

**a.** Grouping by 5 units and assigning the probability as a mass at the midpoint, we have

| Range | Mid-point | Assigned probability | | $xP(x)$ | $x^2 P(x)$ |
|---|---|---|---|---|---|
| 20–25 | 22.5 | $5 \times (0.00 + 0.05)/2 = 0.125$ | | 2.8125 | 63.28125 |
| 25–30 | 27.5 | $5 \times (0.05 + 0.10)/2 = 0.375$ | | 10.3125 | 283.59375 |
| 30–35 | 32.5 | $5 \times (0.10 + 0.05)/2 = 0.375$ | | 12.1875 | 396.09375 |
| 35–40 | 37.5 | $5 \times (0.05 + 0.00)/2 = \underline{0.125}$ | | $\underline{4.6875}$ | $\underline{175.78125}$ |
| | | | 1.000 | 30.0000 | 918.75 |

Subtract $(30)^2$       900

Subtract Sheppard's correction, $5^2/12$       $\underline{2.0833}$

Approximated variance       16.666...

**b.** Likewise grouping by 4 units, we get the same mean and the probabilities are 0.08, 0.24, 0.36, 0.24, and 0.08 at 22, 26, 30, 34, and 38 respectively. The middle class in this case does not have a trapezoidal distribution, and its area is best estimated as the sum of two trapezoids. The crude sum of squares from the mean is 17.92, and Sheppard's correction is now $(4)^2/12 = 1.333$. The approximated variance is thus 16.586....

**c.** Finally, the same argument applied to unit intervals gives a crude sum of squares, from the origin, of 916.75. Subtracting the square of the mean and Sheppard's correction $(1/12)$ gives an approximation of 16.66....

**Comment.** Exact calculation (by the calculus) shows that the true variance is indeed 16.666.... It is noteworthy that the middle grouping (in which the maximum, 30, is included as the midpoint of the central class) is less accurate than the other two groupings. In any case, considering the unnaturally angular shape of the distribution, the approximation is good in all cases.

**16.5.** This problem has the same structure as the last except that the densities are back-to-front. For the first grouping (**a**), for instance, the approximated second raw moment is

$$(22.5)^2 \times (0.375) + (27.5)^2 \times (0.125) + (32.5)^2 \times (0.125) + (37.5)^2 \times (0.375) = 943.75.$$

The approximate variance is found as before by subtracting 900 and Sheppard's correction, which, as in the first grouping for problem 16.4 is 25/12. The result is 41.666....

**b.** As before, the distribution in the middle class is not trapezoidal, and its area is found, by summing two trapezoidal areas, to be 0.04. The approximated variance is then $946.08 - 900 - 16/12 = 44.746\ldots$

**c.** The third grouping gives a crude raw second moment of 949.75 and the approximated variance is $49.666\ldots$

**Comment.** The exact variance is 50 and the grouped variance corrected for grouping is in all cases less than 50, although the gap gets progressively smaller as the fineness of the grouping increases. In each case, therefore, Sheppard's correction makes the approximation from grouping less exact. The bizarre shape of the distribution, with its abrupt terminals, should have warned us that this result would occur.

**16.6.** Table 16.3 is easily condensed. The number of deaths in a year is found simply by adding the deaths month by month, and the same is true for those withdrawn alive. Using the same system for numbering the headings, we get the following abbreviated table.

| Year | Number at the outset | Withdrawn | Mean at risk | Deaths | Fraction dying | Fraction surviving | Cumulative probability of survival | $h$ | $H$ |
|---|---|---|---|---|---|---|---|---|---|
| 1 | 149 | 72 | 113 | 14 | 0.12389 | 0.87611 | 0.87611 | 0.00125 | 0.00125 |
| 2 | 63 | 19 | 53.5 | 1 | 0.01869 | 0.98131 | 0.85973 | 0.00035 | 0.00160 |
| 3 | 43 | 18 | 34 | 1 | 0.02941 | 0.97059 | 0.83444 | 0.00089 | 0.00249 |
| 4 | 24 | 14 | 19 | 1 | 0.05263 | 0.94737 | 0.79052 | 0.00292 | 0.00542 |
| 5 | 9 | 6 | 6 | 0 | 0.00000 | 1.00000 | 0.79052 | 0.00000 | 0.00542 |
| 6 | 3 | | | | | | | | |

**Comment.** The overall pattern corresponds reasonably well to that for the more refined grouping by time.

**16.7.** This argument, though unlikely to be taken seriously, is nevertheless illuminating to analyze. For what it states, it is quite sound, and no doubt could be applied, say, in particle physics. But it is the omissions in the argument—the assumptions and conclusions that are implied but not stated (technically known as enthymemes)—that may do the damage. Of these I recognize two major ones that I shall discuss.

**a.** The probability structure is ambiguous. The probability of a particular southbound car colliding with *any* east- or westbound car whatsoever (we assume random, independent times of arrival at the intersection) depends not only on the length of time that any *particular* east- or westbound car occupies the intersection, but on the proportion of time that *some* car does. To take the argument to the extreme, if the east-west road were replaced by a railroad along which a freight train is being continuously drawn there will be no time at all at which the southbound car can cross. The speed of either the train or the car is quite immaterial to this fact. Of course if the number of cars *per hour* remains constant, the argument is sound; but if the number of cars *per mile* is fixed, the argument is false. By analogy, one does not become less sunburnt by running randomly in the sun than by standing still, or avoid cancer of the lung by breathing slowly and deeply.

**b.** Nobody has ever been killed by a probability. In the argument, the implication lurks that what does the damage is the collision. But again, the fallacy is obvious when one considers the extreme case. If all cars are traveling at one mile per hour, we may be quite certain that nobody will die from any collision. Indeed, it is doubtful that at such a speed there would be perceptible damage to the car. A collision with both cars traveling at 90 miles per hour would be almost certainly fatal to both.

If we confine our statement to mortality, the criterion should not be the probability of collision ($C$) but death ($D$). Since the latter event is a proper subset of the former, we should concern ourselves with the joint probability of the event ($CD$), which is simply the product of a marginal and a conditional probability:

$$P(CD) = P(D|C)P(C)$$

If by increasing their velocity so as to spend half as long in the intersection, drivers increase the mortality rate tenfold if hit, the net impact will be to increase the mortality rate by a factor of five, in spite of the fact that the probability of an accident occurring is halved.

## CHAPTER 17

**17.1.** This represents a fundamental confusion about the whole nature of a distribution and of testing a hypothesis. Unfortunately, it is not an uncommon type of mistake. I can see three fundamental defects.

**i.** It is obvious by definition that the highest values will always be greater than the lowest and there is no point in trying to prove it. This would be true even with replicate determinations on aliquots from the same sample. The one exception would be the statistically degenerate case where the variance is zero, i.e., there is no statistical issue.

**ii.** The notion of a distribution as a prediction for the outcome of one future realization falls apart if one could show that (in some intelligible sense) the quantity being investigated is ambiguous. We may reasonably suppose in most cases that a variable has a mean. The proposed test, if it is about anything at all, seems to deal with whether the mean is equal to itself and (as noted under *i*) the answer furnished by this test is certain to be negative, which shows the very idea of doing the test to be absurd. Of course the investigator undoubtedly had the idea that his sample values represent a mixture of distributions and he is trying to show that the two components have differing means. But to address this problem, he needs a definition of his groups that differs from the test criterion. For instance, if he knew the sample were a mixture of males and females, or Blacks and Whites then he should separate them on sex or race and not on the test measurement.

**iii.** Assuming the distribution is bell-shaped, the reader will see from a simple diagram that the histograms in the tails do not look in the least like Gaussian distributions but positively and negatively J-shaped respectively. As we have seen in Figure 17.2, the central limit theorem operates sluggishly in such cases, and I would assume Gaussian normality only with some hundreds of cases in both groups. The total sample would have to be prodigiously large even to warrant the formal test of the hypothesis.

**17.2.** Traditionally, the notion of Mendelian inheritance is that almost all of the phenotype—as gauged, for instance, by the partition of variance—is ascribable to the operation of one genetic locus. No matter what the environment, or the other content of the genome, we would be able to identify unambiguously what the impact of the locus is. An age-dependent disorder such as Huntington's chorea would not be Mendelian and nothing more for instance. In the present problem, then we shall suppose two features: identical twins will have exactly the same phenotype; and nonidentical twins will be no more alike than any comparable pair of sibs. (It is not difficult to think of many mechanisms in the external or intrauterine environment which would undermine both assumptions.) On these terms we have the following proposals in the three cases.

**a.** The phenotypes for twins will be identical and completely dependent. Count them as one trial; i.e., discard one result.

**b.** The phenotypes are independently distributed over the sample space of progeny from those matings. Treat them as two trials.

**c.** If possible, determine the zygosity by means of other phenotypic features. Failing this, I would do my analysis under both suppositions. In the likely case, which analysis is used will make no material difference. Both will lead to acceptance or to rejection. We may then rest assured as to what the appropriate statistical conclusion is, whatever reservations we may have about biological detail. (However, one may note that the phenotype of the one may explode the hypothesis that is being tested; e.g., a normal twin born to two parents affected at the same locus explodes the hypothesis of an autosomal recessive.)

**17.3.** From Chapter 4 we are given the following results to which we shall appeal. The commonplace mean and variance are 6.62 and 25.06 respectively. The mean of the logarithms to the base 10 is 0.0780 and the standard deviation is 0.864.

**a.** The method-of-moments estimates of the mean and standard deviation are just the quantities cited—6.62 and 25.06. We may set them equal to the formulae given for the moments of the lognormal distribution in Chapter 3

$$e^{\hat{\mu}+\hat{\sigma}^2/2} = 6.62$$

$$e^{2\mu+\sigma^2}(e^{\sigma^2} - 1) = (25.06)^2$$

Then

$$e^{\hat{\sigma}^2} - 1 = \left(\frac{25.06}{6.62}\right)^2 = 14.329{,}998{,}81$$

$$e^{\hat{\sigma}^2} = 15.329{,}998{,}81$$

$$\hat{\sigma}^2 = 2.7298$$

Now taking logarithms for the first equation

$$\hat{\mu} + \hat{\sigma}^2/2 = 1.8901$$

$$\hat{\mu} = 1.8901 - \frac{2.7298}{2} = 0.52519$$

Using the central limit theorem (with some misgiving on such a small sample from such a grossly skewed population) we obtain for the 95% confidence limits points (25.06)(1.96)/105 on either side of 6.62, that is,

$$6.62 - 4.79 = 1.83$$

$$6.62 + 4.79 = 11.41$$

**b.** The MLE of the parameters are simply the mean and standard deviation of the logarithms of the survivals. Before substituting them back into the formula for the mean, we must convert them into natural logarithms, by multiplying them by $\log_e 10 = 2.3025\ldots$ to get 0.1796 for the mean and 1.9894 for the standard deviation. Then the mean of the original distribution is estimated as the natural antilogarithm of

$$0.1796 + (1.9894)^2/2 = 8.658.$$

The agreement with the MLE is reasonably good, considering that the confidence limits on the latter are so wide. Confidence limits on the method-of-moments estimator on the mean (other than the appeal to the central limit theorem that we have used) are a good deal more difficult to determine.

**17.4. a.** Since the samples are of equal size, the pooled estimate of the variance is a simple mean

$$S_p^2 = \frac{(0.2548)^2 + (0.2734)^2}{2} = 0.069,84$$

The standard deviation of the difference in the means is the square root of 2/16ths of this quantity or 0.093,43. The $t$ test is thus

$$\frac{1.0281 - 0.3423}{0.093,43} = 7.340$$

It is read with 30 degrees of freedom.

**b.** The weighting for the first group, $w_{X_1}$, is $(0.3675)^2/16 = 0.008,441$.
The weighting for the second is $(0.7718)^2/16 = 0.037,230$.
The sum of these two quantities is 0.045,671. Thus the degrees of freedom associated with the Aspin-Welch statistic is

$$\left[ \frac{1}{15}\left( \frac{0.008,441}{0.045,671} \right)^2 + \frac{1}{15}\left( \frac{0.037,230}{0.045,671} \right)^2 \right]^{-1} \cong 21.5$$

The Behrens-Fisher statistic is

$$\frac{2.8926 - 1.4515}{\sqrt{0.045,671}} = 6.743$$

Note that the Mickey-Brown criterion would assign not less than 15, or more than 30, degrees of freedom.

**c.** A convenient way in which to pair randomly is to leave the order of the data in the first column as they are (they happen to be arranged in ascending order) and to randomize the order of the second with a book of random numbers. An efficient method of doing so (7) is to take two-digit numbers between 1 and 96, divide by 16, and take the remainder as the random digit between 1 and 16. Exact multiples count as 16. Duplicates are destroyed. I got the following results:

| Saline $(c_i)$ | Random number | Remainder $x/16$ | Sulfinpyrazone $(s_i)$ | Random difference $(s_i - c_i)$ |
|---|---|---|---|---|
| 0.988 | 47 | 15 | 3.763 | 2.775 |
| 0.989 | 59 | 11 | 2.974 | 1.985 |
| 1.075 | 25 | 9 | 2.885 | 1.810 |

| Saline $(c_i)$ | Random number | Remainder $x/16$ | Sulfinpyrazone $(s_i)$ | Random difference $(s_i - c_i)$ |
|---|---|---|---|---|
| 1.116 | 60 | 12 | 3.327 | 2.211 |
| 1.126 | 2 | 2 | 1.953 | 0.827 |
| 1.132 | 21 | 5 | 2.539 | 1.407 |
| 1.255 | 6 | 6 | 2.565 | 1.310 |
| 1.438 | 90 | 10 | 2.943 | 1.505 |
| 1.473 | 81 | 1 | 1.516 | 0.043 |
| 1.534 | 7 | 7 | 2.722 | 1.188 |
| 1.639 | 52 | 4 | 2.446 | 0.807 |
| 1.681 | 88 | 8 | 2.735 | 1.054 |
| 1.874 | 61 | 13 | 3.376 | 1.502 |
| 1.890 | 32 | 0 = 16 | 4.695 | 2.805 |
| 1.925 | 46 | 14 | 3.706 | 1.781 |
| 2.089 | Remaining | 3 | 2.136 | 0.047 |

$\Sigma x = 23.057 \qquad m_X = 1.441,062,5$

$\Sigma x^2 = 42.923,731 \qquad CT = 33.226,578$

$s_{M_X} = 0.201,009$

$t = 7.169,1$

All three results lead to emphatic rejection of the null hypothesis of no effect. The result in this instance is of academic importance, for there is in fact overlap of only two results between the highest of the control values and the lowest of the treated ones.

**17.5.** Supposing that we are concerned with cases where alternatives may occur, we may dismiss as of no interest events with probabilities of either zero or unity. Let the sample space consist of $n$ (equiprobable) events; $a$ of them correspond to the event A, $b$ of them to event B, and $c$ of them to the joint event AB where

$$0 < a,b < n$$

$$0 < c < a,b$$

Then A and B are independent if and only if

$$P(A)P(B) = P(AB)$$

With a little rearrangement this becomes

$$ab = cn \qquad (1)$$

This equation has no solution in whole numbers if $n$ is prime. For any number can be factorized in only one way and if (1) were true the same factors would have to be present on both sides. But $n$ is not present on the left, since both $a$ and $b$ are less than $n$ and $n$ (being prime) cannot be the product of any combination of the factors of $a$ and $b$.

Hence, no two events (other than those involving at least one with probability zero or unity) in the same space can be independent.

**17.6.** Since each step works at its maximum rate, the rate of the entire system will be determined by its slowest step. Hence there are two states for the system:

**a.** All steps are fast and the system is fast. The probability is $(0.999)^{100} \cong e^{-0.1} = 0.9048$.

**b.** At least one step is slow and the system is slow. The probability is the complement of the foregoing = 0.0952.

And this is the distribution at equilibrium.

**Comment.** Here is an admirable example of a system to which the central limit theorem does *not* apply. For while the outcomes at each step are identically distributed and independent their effects are not additive. The distribution of the outcome will always represent a single Bernoulli trial and will never get any nearer being Gaussian. The only point at which the number of steps enters the answer is in the power to which 0.999 is raised.

Bear in mind the golden principle: multiplicity does not invariably lead to Gaussian normality.

E quindi uscimmo a riveder le stelle*
Dante, *Inferno* 34: 139.

*"And thence we went forth to gaze afresh upon the stars."

# APPENDIX 1
# MATRIX ALGEBRA

In what follows we will first discuss the basic operations and then certain applications. Finally, because many people derive help from it, we will look at geometric representation of some ideas in matrix algebra.


## BASIC DEFINITIONS AND OPERATIONS

### Vectors

**Example A.1.1.** In modern American hotels, it is customary to give the rooms numbers which do not have the ordinary meaning. Room 709 does not mean the 709th room—there may not in fact be more than 40 rooms altogether. It means the ninth room on the seventh flour. It directs the occupant to do two things:

1. To take the elevator to the seventh floor.
2. To walk along the corridor to the ninth room.

Such a system of instructions, which defines uniquely where he ends with respect to the reception desk, constitutes a vector and might be written in the form of a column

$$\begin{bmatrix} 7 \\ 9 \end{bmatrix}$$

where the order of the instructions is specified.

**Example A.1.2.** More elaborate ways may be used to define a position in space with respect to some reference point. For example, if we agree on some reference point, say the base of the Washington Monument, we might denote a point three miles north of it, two miles west, and one mile above it by

$$\begin{bmatrix} 3 \\ 2 \\ 1 \end{bmatrix}$$

For purposes of general algebraic discussion we shall represent a column vector by a single lowercase boldface symbol, e.g., $\mathbf{b}$. Other writers use various other symbols, e.g., $\bar{b}$. It will be convenient for certain purposes to write a column vector on its side as a row vector, i.e., in this example as [3  2  1]. In general notation we write this as $\mathbf{b}$ which is read "$\mathbf{b}$ prime" or "$\mathbf{b}$ transpose." It is often written $\mathbf{b}^\mathsf{T}$.

The entries in a vector are called elements. They are written in italic symbols. Ordinary algebra deals with symbols that represent a single element each. A vector with a single element, i.e., a single number, is a scalar, and classical algebra is therefore commonly called scalar algebra. In our first example, the first element is a floor number, the second a room number.

Not all the elements of a vector need be commensurable.

**Example A.1.3.** We might describe the room with the vector

$$\begin{bmatrix} 3 \\ 9 \\ 75 \\ 2 \\ 18.50 \end{bmatrix}$$

to denote "The room is the ninth on the 3rd floor, the temperature is 75 degrees F, the furnishings are second-class, and the cost is eighteen dollars and fifty cents per night."

**Definition 1.** *A vector is an ordered set of numerical elements in a single row or column.*

### Addition and subtraction of vectors

**Definition 2.** *Two vectors are conformable for addition if they contain the same number of elements and if corresponding elements refer to the same quality.*

**Example A.1.4.** Suppose a mountaineer was climbing Mount Everest in the traditional fashion by doing a series of journeys and setting up a base camp after each. The first journey takes him three miles north, six miles west, and one upwards, to the first base camp. Then if we call the starting point the origin, i.e., we represent it by a vector of zeros, the position of the first base camp may be represented by

$$\begin{bmatrix} 3 \\ 6 \\ 1 \end{bmatrix}$$

If in the second journey he travels two miles north, three west, and one and one-half upwards, the change would be represented by

$$\begin{bmatrix} 2 \\ 3 \\ 1.5 \end{bmatrix}$$

The relationship of the second base camp to the starting point would be five miles north, nine west, and two and one-half up. Thus

$$\begin{bmatrix} 3 \\ 6 \\ 1 \end{bmatrix} + \begin{bmatrix} 2 \\ 3 \\ 1.5 \end{bmatrix} = \begin{bmatrix} 5 \\ 9 \\ 2.5 \end{bmatrix}$$

This gives

**Rule 1.** Two vectors are added by adding together *corresponding* elements. For addition to be possible, obviously two vectors must contain the same number of elements. If we were faced with the problem of adding the two vectors

$$\begin{bmatrix} a \\ b \\ c \end{bmatrix} + \begin{bmatrix} d \\ e \end{bmatrix}$$

we would not know which are corresponding elements. The answer could be

$$\begin{bmatrix} a + d \\ b + e \\ c \end{bmatrix} \quad \text{or} \quad \begin{bmatrix} a + d \\ b \\ c + e \end{bmatrix} \quad \text{or} \quad \begin{bmatrix} a \\ b + d \\ c + e \end{bmatrix}$$

To distinguish them we would have to write the second vector as

$$\begin{bmatrix} d \\ e \\ 0 \end{bmatrix} \quad \text{or} \quad \begin{bmatrix} d \\ 0 \\ e \end{bmatrix} \quad \text{or} \quad \begin{bmatrix} 0 \\ d \\ e \end{bmatrix}$$

respectively. Obviously, too, the entries must refer to the same entity, i.e., the order must be preassigned. If we wished in Example A.1.3 to modify the vector because the temperature has gone up by $5°F$, it would not be appropriate to add five to the cost of the room.

It is probably self-evident from similar arguments that

**Rule 2.** One vector is subtracted from another by subtracting corresponding elements.

**Rule 3.** Two vectors are equal if and only if corresponding elements are equal. This naturally implies that they have the same dimension. For instance

$$\begin{bmatrix} a \\ b \\ c \end{bmatrix} = \begin{bmatrix} d \\ e \\ f \end{bmatrix}$$

implies that $a = d$ and $b = e$ and $c = f$.

**Multiplication of vectors**

**Example A.1.5.** Suppose a man is computing his income tax. We will suppose for simplicity that the rates are 20% on his salary, 30% on professional fees, and 40% on interest from investments. Then the rates can be written as a vector, i.e.,

$$\mathbf{b}' = [0.2 \quad 0.3 \quad 0.4]$$

We will further suppose his incomes from the three sources are $5,000, $2,000 and $500, respectively. These can be written in a vector

$$\mathbf{c} = \begin{bmatrix} 5{,}000 \\ 2{,}000 \\ 500 \end{bmatrix}$$

Ordinary arithmetic gives for his total tax $(0.2 \times 5000 + 0.3 \times 2000 + 0.4 \times 500) = 1800$. In vector operations this means that

**Rule 4.** To multiply two vectors together we multiply corresponding elements together and add these products.

For various reasons which will gradually become apparent, it is best to lay the operation out in a special way. First, it may be noted that the dimensions of a vector are given in terms of *first* the number of rows and *second* the number of columns. Thus in Example A.1.5, $\mathbf{c}$ has dimensions $3 \times 1$ whereas $\mathbf{b}'$ has the dimensions $1 \times 3$. The operation of multiplication in the above example written out in full is

$$[0.2 \quad 0.3 \quad 0.4] \begin{bmatrix} 5000 \\ 2000 \\ 500 \end{bmatrix} = 0.2 \times 5000 + 0.3 \times 2000 + 0.4 \times 500 = 1800$$

Dimensions $\quad 1 \times 3 \quad 3 \times 1$

**Definition 3.** *Two vectors are conformable for multiplication if they have the same contiguous dimensions.*

The product of the two vectors is a scalar, i.e., a vector of dimensions $1 \times 1$.

### Scalar multiplication

Suppose in Example A.1.5 that the income were measured in pounds but the tax is to be paid in dollars, the rate of exchange being \$2.00 per pound. Obviously we could make the exchange by multiplying each component of the income by 2.00, or the total product $(\mathbf{b}'\mathbf{c})$ by 2.00; or (though it has no common-sense meaning) each element in the tax vector by 2.00. We can think of this process as simply a change in scale and the operation is known as scalar multiplication. Note that no particular problem of conformableness arises.

### The length of a vector

If the elements of a vector represent distances all at right angles to each other (i.e., they are orthogonal) and if we may suppose that Pythagoras's theorem holds (i.e., we are in "Euclidean space") then it is easy to calculate the distance of the point represented by the vector from the origin.

**Example A.1.6.** If a point is three miles north and four miles east of the Washington Monument, then the distance of that point from the monument (as the crow flies) is $\sqrt{(3^2 + 4^2)} = 5$ miles (Fig. A.1.1$a$). If we now move the point vertically upwards into the air for 12 miles, the distance from the base of the monument is $\sqrt{(5^2 + 12^2)} = 13$ miles (Fig. A.1.1$b$). In the latter case the distance is clearly $\sqrt{(3^2 + 4^2 + 12^2)}$ which we can see to be simply the square root of the sum of the squares of the elements of the vector. Thus the position of the point, three miles north, four west, and twelve up is represented by the vector $\mathbf{a}' = [3 \quad 4 \quad 12]$. Then $\mathbf{a}'\mathbf{a}$ is given by

$$[3 \quad 4 \quad 12] \begin{bmatrix} 3 \\ 4 \\ 12 \end{bmatrix} = 3^2 + 4^2 + 12^2 = 169$$

We may generalize this to four- or higher-dimensional space (which it is impossible to illustrate graphically) to give

**Rule 5.** The *length* of a vector $\mathbf{a}$ is $\sqrt{(\mathbf{a}'\mathbf{a})}$ .

In certain circumstances we may wish to use the reciprocal of this length as a scalar multiplier to give the unit vector.

**Example A.1.7.** The vector $\mathbf{c}' = [4 \quad 4 \quad 7]$ has length $\sqrt{(4^2 + 4^2 + 7^2)} = 9$. If the reciprocal of this is used as a scalar multiplier we get a unit vector $\mathbf{d}'$. Thus

$$\mathbf{d}' = [\tfrac{4}{9} \quad \tfrac{4}{9} \quad \tfrac{7}{9}]$$

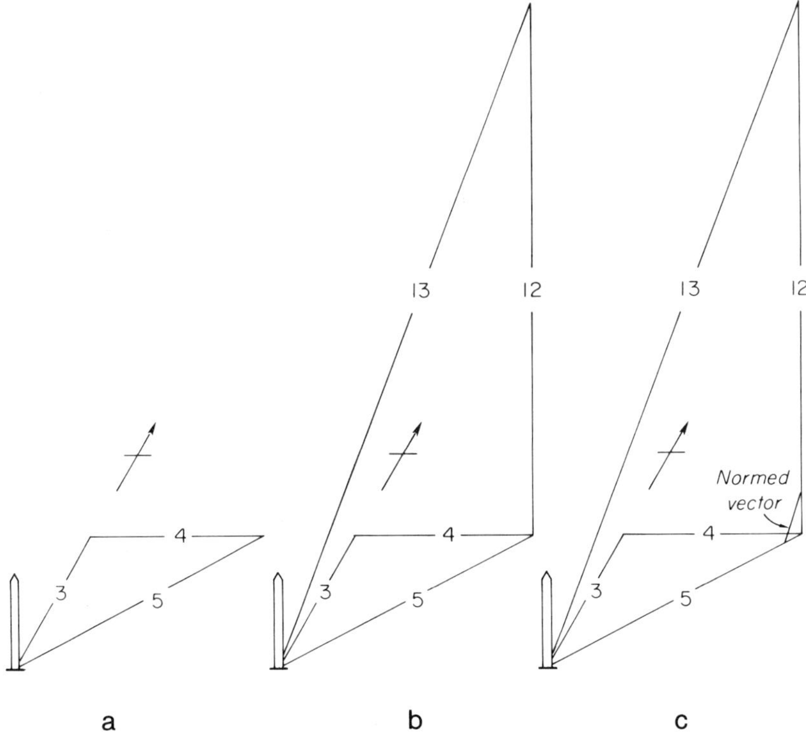


**Figure A.1.1.** Norming a vector. The coordinates of the point are three miles north, four east, and twelve above, the foot of the Washington Monument. Applying Pythagoras's theorem twice shows that the direct distance is thirteen miles. A line drawn parallel to this line and one unit in length is the normed vector, which preserves the proportions of the origin components of distance.

and

$$\mathbf{d'd} = (\tfrac{4}{9})^2 + (\tfrac{4}{9})^2 + (\tfrac{7}{9})^2 = 1$$

This provides us with a vector of unit length.

**Example A.1.8.** Thus in the two-dimensional case (Fig. A.1.1c) the vector

$$\begin{bmatrix} 0.5 \\ 1.2 \end{bmatrix}$$

has length 1.3. Let us draw a line from this point to the origin. A circle of unit radius with the origin as center cuts this line at approximately the point

$$\begin{bmatrix} 0.5 \div 1.3 \\ 1.2 \div 1.3 \end{bmatrix} = \begin{bmatrix} 0.3846 \\ 0.9231 \end{bmatrix}$$

The original vector is said to have been normalized.


## MATRICES

**Definition 4.** *A matrix (plural: matrices) is a rectangular array of quantities.* If it has $m$ rows and $n$ columns, it is said to be of dimensionality $m \times n$.

In Example A.1.5 we used the tax rates to solve a particular problem. But the method clearly has greater applicability.

**Example A.1.9.** We might wish to extend it to compute the income tax for husband and wife and yet to keep the amounts separate, as by law required. For instance

|  | Husband's income | Wife's income | Husband's tax | Wife's tax |
|---|---|---|---|---|

$$[0.2 \quad 0.3 \quad 0.4] \begin{bmatrix} 5000 & 3000 \\ 2000 & 1000 \\ 500 & 20 \end{bmatrix} = [1800 \quad 908]$$

| Dimensions | $1 \times 3$ | $3 \times 2$ | $1 \times 2$ |
|---|---|---|---|

As before the contiguous dimensions are equal so that the vector and the matrix are conformable for multiplication. The vector that results from multiplying them together has dimensions $1 \times 2$ which are the "outside" dimensions of the vector and the matrix. We can generalize this kind of operation still further to allow for state income tax which must also be kept separate.

**Example A.1.10.**

|  |  | Husband | Wife |  |
|---|---|---|---|---|

Federal tax $\begin{bmatrix} 0.2 & 0.3 & 0.4 \\ 0.01 & 0.02 & 0.05 \end{bmatrix}$ State tax  $\begin{bmatrix} 5000 & 3000 \\ 2000 & 1000 \\ 500 & 20 \end{bmatrix} = \begin{bmatrix} 1800 & 908 \\ 115 & 51 \end{bmatrix}$

| Dimensions | $2 \times 3$ | $3 \times 2$ | $2 \times 2$ |
|---|---|---|---|

Again the contiguous dimensions are the same, the product (this time a matrix) has the outside dimensions. We can now state

**Rule 6.** Two matrices which can be multiplied together must have the same contiguous dimensions, and their product will have dimensions equal to the noncontiguous dimensions.

Thus the product of two matrices of dimensions $(a \times b)$ and $(b \times c)$ will have dimensions $(a \times c)$.

Note that we can always treat a vector as a special case of a matrix in which one of the dimensions is unity. A scalar may be viewed as a special case of a matrix in which both dimensions are one.

*Notation:* When we wish to specify the elements of a matrix in general algebraic terms, we give the element two subscripts, the first to denote the row of the matrix to which it belongs, the second to denote the column. Thus $a_{ij}$ denotes the element in the $i^{\text{th}}$ row and the $j^{\text{th}}$ column.

A matrix is commonly written as a boldface uppercase letter, e.g., **A** or **B**. It is sometimes convenient to represent it by a typical element in square brackets $[a_{ij}]$.

**Definition 5.** *The null matrix is that for which* $a_{ij} = 0$ *for all values of i and j.*

Two matrices are *equal* if and only if corresponding elements are equal, i.e.,

$$\mathbf{A} = \mathbf{B} \quad \text{if and only if}$$

$$a_{ij} = b_{ij} \quad \text{for all } i \text{ and } j$$

**Addition and subtraction of matrices**

**Definition 6.** *Two matrices are conformable for addition if they have the same dimensions and corresponding elements refer to the same quantity.*

**Rule 7.** Two matrices may be added (or subtracted) only if they are conformable for addition. The operation is performed by adding (or subtracting) corresponding elements.

**Example A.1.11.**

$$\begin{bmatrix} 1 & 2 & 3 \\ -1 & 4 & -5 \\ 1 & 1 & -2 \end{bmatrix} + \begin{bmatrix} 7 & -3 & 4 \\ 1 & 9 & -7 \\ -1 & -1 & -4 \end{bmatrix} = \begin{bmatrix} 8 & -1 & 7 \\ 0 & 13 & -12 \\ 0 & 0 & -6 \end{bmatrix}$$

It will be noted that matrices are commutative for addition, i.e.,

**Rule 8.** $\mathbf{A} + \mathbf{B} = \mathbf{B} + \mathbf{A}$.

This can be shown in generality by considering the typical elements of each:

$$\mathbf{A} + \mathbf{B} = [a_{ij}] + [b_{ij}] = [a_{ij} + b_{ij}]$$

$$\mathbf{B} + \mathbf{A} = [b_{ij}] + [a_{ij}] = [b_{ij} + a_{ij}] = [a_{ij} + b_{ij}]$$

It will be evident from a similar argument that matrices obey the association law of addition, i.e.,

**Rule 9.** $\mathbf{A} + \mathbf{B} + \mathbf{C} = (\mathbf{A} + \mathbf{B}) + \mathbf{C} = \mathbf{A} + (\mathbf{B} + \mathbf{C})$.

Furthermore by considering each element separately we can see

**Rule 10.** $\mathbf{A}(\mathbf{B} + \mathbf{C}) = \mathbf{AB} + \mathbf{AC}$.

That is, matrices obey the distributive law of multiplication.

Rules 7 to 9 all appear trivial; but in general, for each system of algebra, these rules must be individually tested. As an illustration of how misleading the "obvious" may be, we may note that (although it may not be true for special cases)

**Rule 11. AB ≠ BA.**

That is, the commutative law of multiplication is *not* in general obeyed by matrices. The easiest way to see this is that the dimension of the product of two nonsquare matrices depends on the order in which the multiplication is performed.

**Example A.1.12.** Consider the multiplication of the two matrices.

$$\begin{bmatrix} 3 & 2 & 1 \\ 1 & -4 & 6 \end{bmatrix} \begin{bmatrix} 1 & 0 \\ 1 & 1 \\ 2 & -3 \end{bmatrix} = \begin{bmatrix} 7 & -1 \\ 9 & -22 \end{bmatrix}$$

Dimensions    $2 \times 3$    $3 \times 2$    $2 \times 2$

If we reverse the order of multiplication we get quite a different answer.

$$\begin{bmatrix} 1 & 0 \\ 1 & 1 \\ 2 & -3 \end{bmatrix} \begin{bmatrix} 3 & 2 & 1 \\ 1 & -4 & 6 \end{bmatrix} = \begin{bmatrix} 3 & 2 & 1 \\ 4 & -2 & 7 \\ 3 & 16 & -16 \end{bmatrix}$$

Dimensions    $3 \times 2$    $2 \times 3$    $3 \times 3$

However, it is easily verified that the associative law of multiplication is obeyed, i.e.,

**Rule 12. ABC = (AB)C = A(BC).**

This means that if we have 3 (or more) matrices to multiply together it does not matter whether we multiply the first two together and then multiply by the third, or whether we first multiply the latter two matrices and then multiply the product by the first, provided the first matrix is kept on the left.

**Example A.1.13.** We may write the same three matrices down twice.

$$\left\{ \begin{bmatrix} 2 & 1 \\ 1 & 0 \end{bmatrix} \begin{bmatrix} 1 & 3 \\ -4 & 6 \end{bmatrix} \right\} \begin{bmatrix} 1 & 9 \\ -4 & 3 \end{bmatrix} \text{ or } \begin{bmatrix} 2 & 1 \\ 1 & 0 \end{bmatrix} \left\{ \begin{bmatrix} 1 & 3 \\ -4 & 6 \end{bmatrix} \begin{bmatrix} 1 & 9 \\ -4 & 3 \end{bmatrix} \right\}$$

$$= \begin{bmatrix} -2 & 12 \\ 1 & 3 \end{bmatrix} \times \begin{bmatrix} 1 & 9 \\ -4 & 3 \end{bmatrix} \text{ or } \begin{bmatrix} 2 & 1 \\ 1 & 0 \end{bmatrix} \times \begin{bmatrix} -11 & 18 \\ -28 & -18 \end{bmatrix}$$

$$= \begin{bmatrix} -50 & 18 \\ -11 & 18 \end{bmatrix} = \begin{bmatrix} -50 & 18 \\ -11 & 18 \end{bmatrix}$$

**Identity operators**

In many systems of algebra, there are quantities, so-called identity operators which maintain the status quo when they are added or multiplied by any other quantity. Thus in scalar algebra

The additive identity operator is 0 since $n + 0 = n$ for all $n$.

The multiplicative identity operator is 1 since $n \times 1 = n$ for all $n$.

These are related in an interesting way. The logarithm of the multiplicative identity operator for scalar numbers is the additive operator for the logarithms of the numbers.

Analogous to these in matrix algebra we have

1. The additive identity operator is the null matrix, [0], every element of which is zero.

2. The multiplicative identity operator is the so-called identity matrix denoted by **I**. This has the property that for any matrix of dimensions $n \times m$ multiplying by the identity matrix gives the same matrix. Thus

$$\mathbf{I} \quad \times \quad \mathbf{A} \quad = \quad \mathbf{A}$$

Dimensions $\quad n \times n \quad\quad n \times m \quad\quad n \times m$

If **A** has dimensions $n \times m$, then the outside dimension of **I** must be $n$; but to be conformable with **A**, the contiguous dimension must also be $n$. Hence **I** must be square and only one dimension is required to characterize it; this is commonly written as a subscript, e.g., $\mathbf{I}_n$.

The structure of the identity matrix is that every element in it with two equal subscripts has value 1 and all other elements are zero. Hence

**Example A.1.14.**

$$\begin{bmatrix} 1 & 0 & 0 \\ 0 & 1 & 0 \\ 0 & 0 & 1 \end{bmatrix} \begin{bmatrix} 1 & 4 & -1 & 9 \\ 2 & 3 & -1 & 8 \\ -7 & 1 & 3 & 4 \end{bmatrix} = \begin{bmatrix} 1 & 4 & -1 & 9 \\ 2 & 3 & -1 & 8 \\ -7 & 1 & 3 & 4 \end{bmatrix}$$
$$\quad\quad \mathbf{I} \quad\quad\quad\quad\quad \mathbf{A} \quad\quad\quad\quad\quad\quad \mathbf{A}$$

**Definition 7.** *The principal diagonal of a square matrix is the set of elements in which the row and column numbers are equal, i.e., the elements on the diagonal from top left to bottom right.*

The other elements are called off-diagonal terms.

**Definition 8.** *A diagonal matrix is a square matrix in which the off-diagonal terms are all zero.* The terms on the principal diagonal of a diagonal matrix are called the eigenvalues. We shall find a more general meaning for this term later. The identity matrix as we have seen is a diagonal matrix, the elements on the principal diagonal all being 1.

**Definition 9.** *The transpose of a matrix* **A**, *commonly written* $\mathbf{A}'$ *or* $\mathbf{A}^{\mathsf{T}}$, *is that matrix obtained by interchanging the rows and columns,* i.e.,

$$[a_{ij}]' = [a_{ji}]$$

**Example A.1.15.**

$$\begin{bmatrix} 2 & 1 & 6 \\ 4 & 2 & 2 \end{bmatrix}' = \begin{bmatrix} 2 & 4 \\ 1 & 2 \\ 6 & 2 \end{bmatrix}$$

It will be evident that transposing twice restores the original matrix, i.e.,

**Rule 13.** $(\mathbf{A}')' = \mathbf{A}.$

Suppose the product of two matrices **A** and **B** is **C**, then the element $c_{ij}$ is the sum of the products of the elements of the $i^{\text{th}}$ row of **A** with the corresponding elements of the $j^{\text{th}}$ column of **B**, i.e.,

$$c_{ij} = \sum_k a_{ik} b_{kj}$$

Conversely, if we multiply $\mathbf{B}'$ by $\mathbf{A}'$, then we may write for the transposed matrices

$$c_{ji}^* = \sum_k b_{jk} a_{ki}$$

Now if we transpose the product matrix $\mathbf{AB}$ the element $c_{ij}$ in the product matrix becomes $c_{ji}^*$ in its transpose and we have shown that the element so obtained is the same as that of $\mathbf{B}'$ and $\mathbf{A}'$ multiplied together; hence, we may formulate

**Rule 14.** $(\mathbf{AB})' = \mathbf{B}'\mathbf{A}'$.

In words: the transpose of a product is the product of the transposes in the reverse order. It is readily shown that this is true for any number of matrices. Thus

$$(\mathbf{ABC})' = [(\mathbf{AB})\mathbf{C}]' \quad \text{by rule 12.}$$

$$= \mathbf{C}'(\mathbf{AB})' \quad \text{by rule 14.}$$

$$= \mathbf{C}'\mathbf{B}'\mathbf{A}' \quad \text{by rule 14.}$$

and so forth.

**Definition 10.** *A symmetrical matrix is one equal to its own transpose. It is necessarily square.*

Variance-covariance matrices (which we shall discuss later) are all symmetrical. *The power of a matrix* is an idea applicable to square matrices only. (Why?)

Thus   $(\mathbf{A})^2 = \mathbf{A}.\mathbf{A}.$

$$(\mathbf{A})^u = \mathbf{A}.\mathbf{A}.\mathbf{A}.\ldots.\mathbf{A}. \qquad u \text{ times}$$

In general, raising a matrix to a power involves tedious calculations. However, the special case of the diagonal matrix is much more tractable.

**Definition 11.** *If we write down a matrix $\mathbf{D}$ and to the right of it a matrix $\mathbf{A}$ and multiply them together, the matrix $\mathbf{A}$ is said to be premultiplied by $\mathbf{D}$ and $\mathbf{D}$ postmultiplied by $\mathbf{A}$.*

Premultiplying a matrix by a diagonal matrix is equivalent to multiplying each row of the right-hand matrix by the corresponding element of the diagonal matrix, e.g.,

**Example A.1.16.**

$$\begin{bmatrix} a & 0 & 0 \\ 0 & b & 0 \\ 0 & 0 & c \end{bmatrix} \begin{bmatrix} d_{11} & d_{12} & d_{13} & d_{14} \\ d_{21} & d_{22} & d_{23} & d_{24} \\ d_{31} & d_{32} & d_{33} & d_{34} \end{bmatrix} = \begin{bmatrix} ad_{11} & ad_{12} & ad_{13} & ad_{14} \\ bd_{21} & bd_{22} & bd_{23} & bd_{24} \\ cd_{31} & cd_{32} & cd_{33} & cd_{34} \end{bmatrix}$$

Now if the right-hand matrix were diagonal, this would be equivalent to multiplying the corresponding diagonal elements together, i.e.,

**Example A.1.17.**

$$\begin{bmatrix} a & 0 & 0 \\ 0 & b & 0 \\ 0 & 0 & c \end{bmatrix} \begin{bmatrix} d & 0 & 0 \\ 0 & e & 0 \\ 0 & 0 & f \end{bmatrix} = \begin{bmatrix} ad & 0 & 0 \\ 0 & be & 0 \\ 0 & 0 & cf \end{bmatrix}$$

From this we may easily generalize to two rules.

**Rule 15.** The product of any number of diagonal matrices is the same whatever the order of the matrices, i.e. (unlike matrices in general), they obey the commutative law of multiplication.

**Rule 16.** To raise a diagonal matrix to the power $u$, raise each element to the power $u$. Thus

**Example A.1.18.**

$$\begin{bmatrix} a & 0 & 0 \\ 0 & b & 0 \\ 0 & 0 & c \end{bmatrix}^u = \begin{bmatrix} a^u & 0 & 0 \\ 0 & b^u & 0 \\ 0 & 0 & c^u \end{bmatrix}$$

This relationship is readily proved by double induction on the value of $u$ and the dimension of the matrix.

**Definition 12.** *A matrix* **A** *is idempotent if and only if* $\mathbf{A}^n$ *is the same for all values of* $n \geq 1$.

A diagonal matrix is idempotent if and only if all the terms on the principal diagonal are zero or one.


## THE INVERSE OF A MATRIX

In scalar algebra we can define division as follows:

$$A/B = A \times 1/B$$

where by $1/B$ we mean the quantity such that

$$B \times 1/B = 1$$

$1/B$ is the inverse or reciprocal of $B$ and is commonly written $B^{-1}$. Of course $B^{-1}$ does not exist if $B = 0$.

In matrix algebra we define the inverse of a matrix **A**, analogously.

**Definition 13.** *If for any square matrix* **A**, *there exists a matrix* **B** *such that*

$$\mathbf{AB} = \mathbf{I}$$

*then* **B** *is the inverse of* **A**.

We may write

$$\mathbf{A} = \mathbf{B}^{-1}$$

We can then construct a matrix equivalent of division. Thus let

$$\mathbf{A} = \mathbf{BC}$$

and we wish to find **B**. In scalar algebra we would simply divide

$$\mathbf{B} = \mathbf{A}/\mathbf{C}$$

In matrix algebra we postmultiply by the inverse of **C**

$$\mathbf{AC}^{-1} = \mathbf{BCC}^{-1}$$

Now by definition $\mathbf{CC}^{-1} = \mathbf{I}$ and from rule 12 $\mathbf{BCC}^{-1} = \mathbf{BI} = \mathbf{B}$. Hence

$$\mathbf{AC}^{-1} = \mathbf{B}$$

Hence in transferring a matrix from one side of an equation to another we replace it by its inverse. For an inverse of a matrix to exist

1. The matrix must be square.

2. The matrix must be full of rank. Rank is an idea we will explore later. But trying to find the inverse of a matrix of less than full rank is like trying to find the reciprocal of zero in scalar algebra.

**Definition 14.** *A singular matrix is a square matrix that has no inverse.*

### Inversion of matrices

Certain matrices are easy to invert. For example, we can invert any diagonal matrix by replacing each term on the principal diagonal by its reciprocal.

**Example A.1.19.** It is easily verified, for example, that the inverse of

$$\begin{bmatrix} a & 0 & 0 \\ 0 & b & 0 \\ 0 & 0 & c \end{bmatrix} \text{ is } \begin{bmatrix} 1/a & 0 & 0 \\ 0 & 1/b & 0 \\ 0 & 0 & 1/c \end{bmatrix}$$

By multiplying them together we get $I_3$. Manifestly if any eigenvalue is zero, no inverse exists.

For the more general case, a wide variety of techniques for inverting matrices are available. One of the simplest is by performing linear operations on rows or columns (but not both). The idea is simple. We are given $\mathbf{A}$ and wish to find $\mathbf{B}$ such that

$$\mathbf{BA} = \mathbf{I}$$

and the method is to break $\mathbf{B}$ into separate parts.

If we can by premultiplying successively by a series of matrices $\mathbf{T}_1, \ldots, \mathbf{T}_n$ convert $\mathbf{A}$ into an identity matrix, i.e.,

$$\mathbf{T}_n, \ldots, \mathbf{T}_1\mathbf{A} = \mathbf{I}$$

then the product $(\mathbf{T}_n \ldots \mathbf{T}_1)$ is $\mathbf{B}$, the inverse of $\mathbf{A}$.

A trivial example from scalar arithmetic may help to fix ideas. What is the inverse of 60? We begin by writing down 60 (which can be considered a matrix of order 1) and in a separate column to the right, 1 (which is the identity matrix of order 1). Then we perform the same operations on both sides. Thus

**Example A.1.20.**

|  | 60 | 1 |
|---|---|---|
| $T_1$ Divide by 2 | 30 | $1/2$ |
| $T_2$ Divide by 2 | 15 | $1/4$ |
| $T_3$ Divide by 3 | 5 | $1/12$ |
| $T_4$ Divide by 5 | 1 | $1/60$ |

We have reduced the left hand term to 1 and the right hand term is the inverse we require. We could of course have performed this maneuver in one step (dividing both by 60). In principle we could invert a matrix in one step; but in practice it is usually necessary to do it in several steps (as here). In doing this, we can use only certain manipulations—multiplication, and division (which we can consider as a form of multiplication). We would not, for instance, arrive at the right answer if we were to add quantities to, or subtract them from, both of the columns above. The question arises, then, what kinds of change we can make in a matrix which can be represented as premultiplying by another matrix. On reflection we find that there are three:

1. We may interchange any two rows by multiplying by the identity matrix with the corresponding rows interchanged.

**Example A.1.21.**

$$\begin{bmatrix} 0 & 1 & 0 \\ 1 & 0 & 0 \\ 0 & 0 & 1 \end{bmatrix} \begin{bmatrix} 2 & 3 & 1 \\ 4 & 4 & 7 \\ 6 & 9 & -4 \end{bmatrix} = \begin{bmatrix} 4 & 4 & 7 \\ 2 & 3 & 1 \\ 6 & 9 & -4 \end{bmatrix}$$

2. We may multiply any row by a scalar by multiplying by a diagonal matrix with the scalar factor in the corresponding position.

**Example A.1.22.** Premultiplying by the matrix

$$\begin{bmatrix} \tfrac{1}{2} & 0 & 0 \\ 0 & -3 & 0 \\ 0 & 0 & 4 \end{bmatrix}$$

will multiply the first row by $\tfrac{1}{2}$, the second by $-3$, the third by 4.

3. We may add or subtract any row or any multiple of any row or rows to or from any other row.

**Example A.1.23.** Thus, if we wish to keep the first and third rows unchanged and add to the second row twice the first and subtract from it 5 times the third, we would multiply by

$$\begin{bmatrix} 1 & 0 & 0 \\ 2 & 1 & -5 \\ 0 & 0 & 1 \end{bmatrix}$$

We can do any or all of these maneuvers serially or simultaneously and we can keep track of what we have done by performing the same operations on the identity matrix. In the first example, we will write this out explicitly; with a little practice, the process can be carried through in a considerably abbreviated form.

**Example A.1.24.** We wish to invert the matrix

$$\begin{bmatrix} 2 & 9 \\ -2 & 1 \end{bmatrix}$$

The steps are laid out in detail in Table A.1.1.
The last matrix in the last column is now the inverse of the original matrix. This we verify by an identity check.

$$\begin{bmatrix} 2 & 9 \\ -2 & 1 \end{bmatrix} \begin{bmatrix} 0.05 & -0.45 \\ 0.1 & 0.1 \end{bmatrix} = \begin{bmatrix} 1 & 0 \\ 0 & 1 \end{bmatrix}$$

**Table A.1.1. Inversion of a 2 × 2 matrix**

| Step | Type of manipu-lation | Multiply by | Original matrix becomes | Identity matrix becomes |
|---|---|---|---|---|
| 1. Replace the second row by the sum of the rows | 3 | $\begin{bmatrix} 1 & 0 \\ 1 & 1 \end{bmatrix}$ | $\begin{bmatrix} 2 & 9 \\ 0 & 10 \end{bmatrix}$ | $\begin{bmatrix} 1 & 0 \\ 1 & 1 \end{bmatrix}$ |
| 2. Divide the first row by 2, the second by ten | 2 | $\begin{bmatrix} 0.5 & 1 \\ 0 & 0.1 \end{bmatrix}$ | $\begin{bmatrix} 1 & 4.5 \\ 0 & 1 \end{bmatrix}$ | $\begin{bmatrix} 0.5 & 0 \\ 0.1 & 0.1 \end{bmatrix}$ |
| 3. Take 4.5 times the second row from the first | 3 | $\begin{bmatrix} 1 & -4.5 \\ 0 & 1 \end{bmatrix}$ | $\begin{bmatrix} 1 & 0 \\ 0 & 1 \end{bmatrix}$ | $\begin{bmatrix} 0.05 & -0.45 \\ 0.1 & 0.1 \end{bmatrix}$ |

*Exercise:* Invert

$$\begin{bmatrix} 3 & 11 \\ 1 & 4 \end{bmatrix}$$

**Example A.1.25.** Invert the matrix

$$\begin{bmatrix} 1 & 2 & 2 \\ 3 & 1 & 3 \\ 1 & -1 & 1 \end{bmatrix}$$

*Step 1*—Take three times the first row from the second and once it from the third.

Matrix to be inverted          Identity matrix

$$\begin{bmatrix} 1 & 2 & 2 \\ 0 & -5 & -3 \\ 0 & -3 & -1 \end{bmatrix} \qquad \begin{bmatrix} 1 & 0 & 0 \\ -3 & 1 & 0 \\ -1 & 0 & 1 \end{bmatrix}$$

*Step 2*—Divide the second row by $-5$ and subtract twice the result from the first row and add three times it to the third.

$$\begin{bmatrix} 1 & 0 & 0.8 \\ 0 & 1 & 0.6 \\ 0 & 0 & 0.8 \end{bmatrix} \qquad \begin{bmatrix} -0.2 & 0.4 & 0 \\ 0.6 & -0.2 & 0 \\ 0.8 & -0.6 & 1 \end{bmatrix}$$

*Step 3*—Multiply the last row by 1.25 and subtract 0.6 of it from the second row and 0.8 of it from the first.

$$\begin{bmatrix} 1 & 0 & 0 \\ 0 & 1 & 0 \\ 0 & 0 & 1 \end{bmatrix} \quad \begin{bmatrix} -1 & 1 & -1 \\ 0 & 0.25 & -0.75 \\ 1 & -0.75 & 1.25 \end{bmatrix}$$

The right-hand matrix should now be multiplied by the original matrix to verify that it is its inverse.

The general procedure is as follows. By multiplying the first row by a suitable constant, we make its first element one (the first pivotal value) and by subtracting appropriate multiples of the first row from all the others, we ensure that all other elements in the first column are zeros. We then do the same for each column in turn. (This method does not work if any of the "pivotal" values is zero since there is no constant that will make 0 into unity.)

**The inverse of the product of two square nonsingular matrices**

If $\mathbf{A}$ and $\mathbf{B}$ are two nonsingular matrices

$$(\mathbf{AB})^{-1}(\mathbf{AB}) = \mathbf{I}$$

$$(\mathbf{AB})^{-1}\mathbf{AB} = \mathbf{I} \qquad \text{(by Rule 12)}$$

$$(\mathbf{AB})^{-1}\mathbf{A} = \mathbf{B}^{-1}$$

$$(\mathbf{AB})^{-1} = \mathbf{B}^{-1}\mathbf{A}^{-1}$$

Hence by induction

**Rule 17.** The inverse of a product of a set of matrices is the product of the inverses of the matrices in reverse order.

Note further, that if

$$\mathbf{AB} = \mathbf{I}$$

postmultiplying by $\mathbf{B}^{-1}$

$$\mathbf{A} = \mathbf{B}^{-1}$$

premultiplying by $\mathbf{B}$

$$\mathbf{BA} = \mathbf{BB}^{-1} = \mathbf{I}$$

hence

$$\mathbf{B} = \mathbf{A}^{-1}$$

which gives

**Rule 18.** If $\mathbf{A}$ is the inverse of $\mathbf{B}$, then $\mathbf{B}$ is the inverse of $\mathbf{A}$.

Furthermore if

$$\mathbf{AB} = \mathbf{I} = \mathbf{AC}$$

then premultiplying by $\mathbf{A}^{-1}$

$$\mathbf{A}^{-1}\mathbf{AB} = \mathbf{A}^{-1}\mathbf{AC}$$

$$\mathbf{B} = \mathbf{C}$$

so that

**Rule 19.** The inverse of a matrix is unique.

## ORTHOGONAL MATRICES

**Definition 15.** *If the inverse of a matrix is equal to its transpose, the matrix is said to be orthogonal.*

**Example A.1.26.** Consider the matrix

$$\mathbf{A} = \begin{bmatrix} 1 & 2 & -1 \\ 3 & -1 & 1 \\ -1 & 4 & 7 \end{bmatrix}$$

This particular matrix when postmultiplied by its transpose gives a diagonal matrix. In full

$$\underset{\mathbf{A}}{\begin{bmatrix} 1 & 2 & -1 \\ 3 & -1 & 1 \\ -1 & 4 & 7 \end{bmatrix}} \underset{\mathbf{A}'}{\begin{bmatrix} 1 & 3 & -1 \\ 2 & -1 & 4 \\ -1 & 1 & 7 \end{bmatrix}} = \underset{\mathbf{AA}'}{\begin{bmatrix} 6 & 0 & 0 \\ 0 & 11 & 0 \\ 0 & 0 & 66 \end{bmatrix}}$$

This we could have identified by inspection. Consider how we arrive at (for instance) the second element in the first row of $\mathbf{AA}'$. It is the sum of the products of the corresponding elements of the first row of $\mathbf{A}$ and the second column of $\mathbf{A}'$, i.e.,

$$1 \times 3 + 2 \times (-1) + (-1) \times 1 = 0$$

But the first column of $\mathbf{A}'$ is the same as the first row of $\mathbf{A}$ (from the definition of a transpose). Thus, without writing down the transpose we could simply have looked at the sum of the products of corresponding elements of the first and second rows of $\mathbf{A}$. Similarly we can work out the sums of products of the first and third and second and third and verify that they are both zero. Furthermore, it will be evident that the term on the diagonal is the sum of the squares of the elements in the corresponding rows. Thus for the third row

$$(-1)^2 + 4^2 + 7^2 = 66$$

If we now divide through each row by the square root of the sum of the squares of the elements in that row, the resulting matrix

$$\begin{bmatrix} 1/\sqrt{6} & 2/\sqrt{6} & -1/\sqrt{6} \\ 3/\sqrt{11} & -1/\sqrt{11} & 1/\sqrt{11} \\ -1/\sqrt{66} & 4/\sqrt{66} & 7/\sqrt{66} \end{bmatrix}$$

when multiplied by its transpose clearly gives the identity matrix and is therefore orthogonal. A way of looking at an orthogonal matrix is that each row (or column) is a unit vector (see Example 4.1.7).

**Properties of orthogonal matrices**

1. If **A** is orthogonal, then (by definition)

$$\mathbf{A}' = \mathbf{A}^{-1}$$

Hence

$$\mathbf{A}'\mathbf{A} = \mathbf{I}$$
$$\mathbf{A}\mathbf{A}' = \mathbf{I} \qquad \text{(from Rule 18)}$$

2. The sum of squares of the elements of any row or column is unity.

3. The sum of the products of corresponding elements in any two different rows (or any two different columns) is zero.

4. It can be shown that for any symmetric matrix **A** there exists an orthogonal matrix **U** such that

$$\mathbf{U}\mathbf{A}\mathbf{U}'$$

is diagonal.

**More on idempotency**

Arising out of this latter property, we can discern the character of a symmetric idempotent matrix **B**, i.e.,

$$\mathbf{B}\mathbf{B} = \mathbf{B}'\mathbf{B} = \mathbf{B}$$

Let us operate on **B** by the appropriate diagonalizing orthogonal matrix **U** and its transpose

$$(\mathbf{U}\mathbf{B}\mathbf{U}')'(\mathbf{U}\mathbf{B}\mathbf{U}') = \mathbf{U}\mathbf{B}'\mathbf{U}'\mathbf{U}\mathbf{B}\mathbf{U}' \qquad \text{(by Rule 17)}$$

Now $\mathbf{U}'\mathbf{U} = \mathbf{I}$ and $\mathbf{B}'\mathbf{B} = \mathbf{B}$.
Thus

$$\mathbf{U}\mathbf{B}'\mathbf{U}'\mathbf{U}\mathbf{B}\mathbf{U}' = \mathbf{U}\mathbf{B}'\mathbf{I}\mathbf{B}\mathbf{U}' = \mathbf{U}\mathbf{B}'\mathbf{B}\mathbf{U}' = \mathbf{U}\mathbf{B}\mathbf{U}'$$

Thus in diagonalized form ($\mathbf{U}\mathbf{B}'\mathbf{U}$), the matrix is still idempotent. Now we know a diagonal matrix is idempotent if and only if the values on the diagonal are zeros or ones. Thus we have

**Rule 20.** When a symmetrical idempotent matrix is reduced to diagonal form by orthogonal transformations, all the diagonal elements are zero or one.


**DETERMINANTS**

To develop theory further we shall have to consider what we mean by the determinant of a square matrix. The determinant of a matrix **A** is written $|\mathbf{A}|$. Its meaning can be illustrated in various ways, which will be used from time to time in the subsequent development. At this stage we will use a formal or manipulative approach.

Consider the matrix **A**

$$\mathbf{A} = \begin{bmatrix} a_{11} & a_{12} \\ a_{21} & a_{22} \end{bmatrix}$$

The determinant of this is found by taking all possible products of the elements, each product being composed of one element from each row so that the same column is not represented more than once in any one product. In the present instance there are only two such products, $a_{11}a_{22}$ and $a_{12}a_{21}$. These products take a sign which is determined by the subscripts of its elements. If we examine the term $a_{11}a_{22}$, the first subscript of the first term is less than that of the second; and the same is true of the second subscripts. The product is multiplied by $+1$. In the term $a_{12}a_{21}$, however, the first subscript of the first term (1) is less than that for the second (2). The second subscript of the first term (2) is greater than that for the second term (1), which constitutes an inversion. The product is multiplied by $(-1)$ once for each inversion.

**Example A.1.27.**

$$|\mathbf{A}| = \begin{vmatrix} 2 & 3 \\ 1 & 2 \end{vmatrix} = +(2 \times 2) - (3 \times 1) = 4 - 3 = 1.$$

It will be evident that

**Rule 21.** The determinant of a matrix is a scalar.

The determinant of a scalar is simply the scalar itself. For a matrix of dimensions $3 \times 3$ the determinant has 3! or 6 terms and in general for a $u \times u$ matrix there will be $u!$ terms. For each particular component the number of inversions when the component elements are paired in all possible ways is counted. If it is odd, the sign of the product is changed, but not if it is even. Thus we may consider the matrix

$$\begin{bmatrix} a_{11} & a_{12} & a_{13} \\ a_{21} & a_{22} & a_{23} \\ a_{31} & a_{32} & a_{33} \end{bmatrix}$$

The possible products may be grouped as follows

| *No inversions* | *One inversion* |
|---|---|
| $a_{11}a_{22}a_{33}$ | $a_{11}a_{23}a_{32}$ |
| | $a_{12}a_{21}a_{33}$ |

| *Two inversions* | *Three inversions* |
|---|---|
| $a_{12}a_{23}a_{31}$ | $a_{13}a_{22}a_{31}$ |
| $a_{13}a_{21}a_{32}$ | |

**Example A.1.28.**

$$\begin{vmatrix} 2 & 1 & 0 \\ 0 & 1 & 3 \\ 1 & 1 & 5 \end{vmatrix} = \begin{array}{l} +(2 \times 1 \times 5) + (1 \times 3 \times 1) + (0 \times 0 \times 1) \\ -(2 \times 3 \times 1) - (1 \times 0 \times 5) - (0 \times 1 \times 1) \end{array}$$

$$= 13 - 6 = 7$$

Note that if an element has a negative sign this is retained in the product. Thus

**Example A.1.29.**

$$\begin{vmatrix} 1 & -3 \\ 1 & 4 \end{vmatrix} = 1 \times 4 - [-3 \times 1] = 7$$

To expand a $3 \times 3$ matrix, it is convenient to copy the first two columns down twice, and compute the products of the elements in the lines shown, the product taking a positive sign if the line goes downwards and to the right, and vice versa.

**Example A.1.30.** To compute

$$\begin{vmatrix} 2 & 1 & 2 \\ 4 & 1 & -1 \\ 7 & 2 & -9 \end{vmatrix}$$

write down the first two columns, in order, to the right of the matrix

$$= [2 \times 1 \times (-9)] + [1 \times (-1) \times 7] + [2 \times 4 \times 2]$$

$$- [2 \times 1 \times 7] - [2 \times (-1) \times 2] - [1 \times 4 \times (-9)]$$

$$= -18 - 7 + 16 - 14 + 4 + 36$$

$$= 17$$

To find the determinant of a matrix of dimensions greater than $3 \times 3$ is tedious and rapidly becomes very cumbersome. It is considerably simplified by the use of cofactors.

**Definition 16.** *If we take an element $a_{ij}$ in an $n \times n$ matrix and delete the row and column in which it occurs, the determinant of the resulting $(n - 1) \times (n - 1)$ matrix is called the cofactor of $a_{ij}$ and is denoted by $\mathbf{A}_{ij}$.*

In computing the determinant we attach to the cofactor a sign which is positive if $(i + j)$ is even, negative if it is odd. It can be readily shown that

**Rule 22.** The determinant of a matrix is the sum of the products of the elements of any row (or column) each with its own cofactor.

**Example A.1.31.** Computing the result in Example A.1.30 by Rule 22, we may use the elements of the first row, remembering to change the sign of the middle term the subscripts of which are 12, their sum $1 + 2$ being odd.

$$2\begin{vmatrix} 1 & -1 \\ 2 & -9 \end{vmatrix} - 1\begin{vmatrix} 4 & -1 \\ 7 & -9 \end{vmatrix} + 2\begin{vmatrix} 4 & 1 \\ 7 & 2 \end{vmatrix}$$

$$= 2(-9 + 2) - (-36 + 7) + 2(8 - 7)$$

$$\qquad -14 \qquad\qquad + 29 \qquad\qquad + 2$$

$$= 17$$

as before.

Since every term in the expansion of a determinant contains one and only one element from each row and column and transposing a matrix does not alter an inversion it will be evident that

**Rule 23.** The determinant of a matrix is equal to the determinant of its transpose.

What effect does interchanging the first two rows of a matrix have on the determinant? Consider the matrix

$$\begin{bmatrix} 1 & 2 & 3 \\ 2 & 1 & 4 \\ 6 & 2 & 1 \end{bmatrix}$$

Interchanging the first two rows we get

$$\begin{bmatrix} 2 & 1 & 4 \\ 1 & 2 & 3 \\ 6 & 2 & 1 \end{bmatrix}$$

In expanding the first of these, we encounter a term made up of the elements on the principal diagonal $1 \times 1 \times 1$. Since the terms are of the form $a_{11}$, $a_{22}$, and $a_{33}$ and there is no inversion, we give it a positive sign. The corresponding term in the second determinant is of the form $a_{21}a_{12}a_{33}$ and contains one inversion; so it is given a negative sign. The same change in sign occurs with every term in the expansion, so that the entire determinant changes sign.

What if we introduce a second interchange? Now we have added another inversion for each term and the determinant is the same as that for the original matrix. It will be found that however we go about interchanging two and only two rows, we always end up with the equivalent of an odd number of interchanges of adjacent rows. For instance, to interchange the first and third rows, we may go through the arrangements 1  2  3 (at the start); interchange the first two, 2  1  3; then 2  3  1; and then 3  2  1, three changes in all. Hence

**Rule 24.** If we interchange any two rows (or columns) of a matrix, we change the sign of its determinant.

If any two rows of a matrix are identical, then interchanging them cannot change the value of the determinant and yet, by rule 24, the sign is changed. Thus

$$|\mathbf{A}| = -|\mathbf{A}|$$

which can only mean that $|\mathbf{A}| = 0$. Then

**Rule 25.** If any two rows (or any two columns) of a matrix are identical, the determinant is zero.

Since every term in the expansion of a determinant contains one element from each row (and column) it is evident that if we multiply (or divide) any row (or column) by a scalar factor, we increase the determinant by that factor.

**Example A.1.32.**

$$|\mathbf{A}| = \begin{vmatrix} 3 & 6 & 9 \\ 1 & 1 & 1 \\ 2 & 4 & 3 \end{vmatrix} = 9 + 12 + 36 - 12 - 18 - 18 = 9$$

We may factor out 3 from the elements of the first row

$$|\mathbf{A}| = 3 \times \begin{vmatrix} 1 & 2 & 3 \\ 1 & 1 & 1 \\ 2 & 4 & 3 \end{vmatrix} = 3[3 + 4 + 12 - 4 - 6 - 6]$$

$$= 9$$

From this and Rule 25 we have

**Rule 26.** If any row is a multiple of any other row or any column of another column, the determinant of the matrix is zero.

We are now in a position to throw some further light on the meaning of rank and singularity, which we have encountered in considering inverses. A singular matrix is one with a zero determinant.

**Definition 17.** A matrix of full rank is one with a nonzero determinant.

Suppose we add to one row a multiple of some row, and take its determinant. For example

$$\begin{vmatrix} a_{11} + ka_{21} & a_{12} + ka_{22} & a_{13} + ka_{23} \\ a_{21} & a_{22} & a_{23} \\ a_{31} & a_{23} & a_{33} \end{vmatrix}$$

can be expanded by the method of cofactors, the results grouped, and the multiplier factored out to give

$$\begin{vmatrix} a_{11} & a_{12} & a_{13} \\ a_{21} & a_{22} & a_{23} \\ a_{31} & a_{32} & a_{33} \end{vmatrix} + k \begin{vmatrix} a_{21} & a_{22} & a_{23} \\ a_{21} & a_{22} & a_{23} \\ a_{31} & a_{32} & a_{33} \end{vmatrix}$$

The first of these is the determinant of $\mathbf{A}$; the second (from Rule 25) has the value zero. Hence,

**Rule 27.** If to any row we add a multiple of any other row we do not alter the value of the determinant.

### Condensation

We may make use of rule 27 to simplify the evaluation of a determinant.

**Example A.1.33.**

$$\mathbf{A} = \begin{vmatrix} 1 & -3 & 7 \\ -2 & 4 & 5 \\ 3 & 1 & -1 \end{vmatrix}$$

If we multiply the first column by three and add it to the second, and multiply it by seven and subtract it from the third we get for $|\mathbf{A}|$

$$\begin{vmatrix} 1 & 0 & 0 \\ -2 & -2 & 19 \\ 3 & 10 & -22 \end{vmatrix}$$

and expanding this by cofactors we find that two of the terms are zero and the third is:

$$\begin{vmatrix} -2 & 19 \\ 10 & -22 \end{vmatrix} = 44 - 190 = -146$$

Since every row (and column) must be represented in every term, it is obvious that

**Rule 28.** If any row (or column) of a matrix consists of zeros the determinant is zero.

It can be shown that

**Rule 29.** The determinant of a product is the product of the determinants

$$|\mathbf{AB}| = |\mathbf{A}|\,|\mathbf{B}|$$

This throws light on the nature of singularity. Obviously $|\mathbf{I}| = 1$. Hence

$$|\mathbf{AA}^{-1}| = 1 \quad \text{or}$$

$$|\mathbf{A}|\,|\mathbf{A}^{-1}| = 1$$

But if $|\mathbf{A}| = 0$, we can find no matrix $\mathbf{A}^{-1}$ that will make this true. Thus a matrix with a zero determinant has no inverse.

**The use of determinants to invert a matrix**

Let us denote the cofactor of $a_{ij}$ by $\mathbf{A}_{ij}$. We have already seen (Rule 22) that

$$|\mathbf{A}| = \sum_j a_{ij}|\mathbf{A}_{ij}|.$$

Now suppose we replace each element of a matrix by its cofactor divided by the determinant, and transpose the resulting matrix. We get a new matrix represented by the typical term

$$A^* = \frac{|\mathbf{A}_{ij}|}{|\mathbf{A}|}$$

Let us now multiply by the original matrix to give a product matrix. The $i^{\text{th}}$ element in the $i^{\text{th}}$ row in the product matrix is

$$\frac{\sum_j a_{ij}|\mathbf{A}_{ij}|}{|\mathbf{A}|}$$

which from Rule 22 is clearly $|\mathbf{A}| \div |\mathbf{A}| = 1$. The same will be true for every term on the principal diagonal of the product matrix. Now let us take any off-diagonal term. Thus the second element in the first row will be

$$\frac{\sum_j a_{1j}|\mathbf{A}_{2j}|}{|\mathbf{A}|}$$

From Rule 22 the numerator is the same as

$$\begin{vmatrix} a_{11} & a_{12} & a_{13} & \cdots & a_{1n} \\ a_{11} & a_{12} & a_{13} & \cdots & a_{1n} \\ a_{31} & a_{32} & a_{33} & \cdots & a_{3n} \\ \cdot & & & & \\ \cdot & & & & \\ a_{n1} & a_{n2} & a_{n3} & \cdots & a_{nn} \end{vmatrix}$$

Now the determinant is zero from Rule 25. In general it will be found that all off-diagonal terms have as numerator the determinant of a matrix with two identical rows. Thus the product matrix has zeros for all off-diagonal terms and ones on the principal diagonal, i.e., it is the identity matrix. Thus

$$\mathbf{AA^*} = \mathbf{I}$$

whence by definition

$$\mathbf{A^*} = \mathbf{A}^{-1}$$

**Example A.1.34.** Let us invert the matrix in Example A.1.24 by this method. First we transpose the matrix and take the determinant $= 2 \times 1 - 9 \times (-2) = 20$. Since the cofactors are all scalars, their determinants are simply the numbers themselves and all we have to ensure is that the signs are carried. The inverse is thus clearly

$$\begin{bmatrix} 1/20 & -9/20 \\ 2/20 & 2/20 \end{bmatrix} = \begin{bmatrix} 0.05 & -0.45 \\ 0.1 & 0.1 \end{bmatrix}$$

**Example A.1.35.** Let us invert the matrix from Example A.1.25. The determinant is

$$1 + 6 - 6 + 3 - 6 - 2 = -4.$$

Transposing we get the cofactors

$$\begin{array}{ccc} [1 \times 1 - 3 \times (-1)] & -[2 \times 1 - 2 \times (-1)] & [2 \times 3 - 1 \times 2] \\ -[3 \times 1 - 3 \times 1] & [1 \times 1 - 2 \times 1] & -[1 \times 3 - 2 \times 3] \\ [3 \times (-1) - 1 \times 1] & -[1 \times (-1) - 1 \times 2] & [1 \times 1 - 2 \times 3] \end{array}$$

and dividing through by $-4$ (the determinant) we get the same answer as before.

**Definition 18.** *The rank of an* $n \times n$ *matrix is the dimension of the largest non-singular submatrix obtained from it by deleting rows and columns.*

**Example A.1.36.**

$$\begin{bmatrix} 3 & 2 & 1 \\ 0 & 2 & 2 \\ 1 & 1 & -1 \end{bmatrix} \quad \text{is of rank 3 since its determinant is } -10.$$

**Example A.1.37.**

$$\begin{bmatrix} 3 & 2 & 2 \\ 1 & 2 & 2 \\ 4 & 4 & 4 \end{bmatrix} \quad \text{is singular but contains the submatrix}$$

$$\begin{bmatrix} 3 & 2 \\ 1 & 2 \end{bmatrix} \quad \text{which is not, and the matrix is therefore of rank 2.}$$

## SOME APPLICATIONS OF MATRIX ALGEBRA

**Example A.1.38.** Consider the simultaneous equations

$$a + 2b = 5$$

$$2a + b = 7$$

The usual method of solving this is to multiply the top line by 2 and subtract the bottom line. We get $3b = 3$ whence $b = 1$; and substituting this in the bottom line we get

$$2a + 1 = 7$$

whence

$$a = 3$$

However, we can write this as a matrix equation

$$\begin{bmatrix} 1 & 2 \\ 2 & 1 \end{bmatrix} \begin{bmatrix} a \\ b \end{bmatrix} = \begin{bmatrix} 5 \\ 7 \end{bmatrix}$$

and as we have shown previously, if

$$\mathbf{Ab} = \mathbf{c}$$

$$\mathbf{A}^{-1}\mathbf{Ab} = \mathbf{A}^{-1}\mathbf{c}$$

$$\mathbf{b} = \mathbf{A}^{-1}\mathbf{c}$$

Thus

$$\begin{bmatrix} a \\ b \end{bmatrix} = \begin{bmatrix} 1 & 2 \\ 2 & 1 \end{bmatrix}^{-1} \begin{bmatrix} 5 \\ 7 \end{bmatrix}$$

The determinant of the matrix is $-3$ and we can easily write down the inverse

$$\begin{bmatrix} -1/3 & 2/3 \\ 2/3 & -1/3 \end{bmatrix}$$

and multiplying out

$$\begin{bmatrix} a \\ b \end{bmatrix} = \begin{bmatrix} (-1/3)5 + (2/3)7 \\ (2/3)5 + (-1/3)7 \end{bmatrix} = \begin{bmatrix} 3 \\ 1 \end{bmatrix}$$

as before.

To be sure, this method involves no less calculation than the first one used. And this is true if we generalize the problem to a system of $n$ simultaneous equations. However, from a practical viewpoint we can always set the problem up in matrix form and then do the arithmetic on a computer. This is probably worthwhile for four or more simultaneous equations.

But an added value in the matrix approach is the light it throws on the solution of systems of linear simultaneous equations. If we consider such a system

$$\mathbf{Ab} = \mathbf{c}$$

the solution

$$\mathbf{b} = \mathbf{A}^{-1}\mathbf{c}$$

is possible if and only if $\mathbf{A}$ has an inverse which implies—
  1. that $\mathbf{A}$ must be square and
  2. of full rank. The first condition implies that there must be as many equations as there are unknowns; the second that the equations must be linearly independent.

**Example A.1.39.** The equations

$$a + \phantom{2}b = 6$$

$$2a + 2b = 12$$

are not independent; indeed this is obvious since the second is merely twice the first. The matrix of coefficients

$$\begin{bmatrix} 1 & 1 \\ 2 & 2 \end{bmatrix}$$

has a zero determinant and therefore no inverse.

**Example A.1.40.** If we consider the equations

$$a + \phantom{1}2b + 3c = 12$$

$$3a + \phantom{1}4b + \phantom{3}c = \phantom{1}9$$

$$9a - 12b + 3c = \phantom{1}3$$

it is not immediately obvious that they are not independent. The determinant of the matrix however, is readily shown to be zero. The matrix is of rank 2.

If there are fewer equations than unknowns and they are linearly independent (i.e., the rank of the matrix is equal to the number of equations) an infinite number of solutions exist.

**Example A.1.41.**

$$a - b = 7$$

is satisfied by

$$a = 10 \qquad b = 3$$

$$a = 11 \qquad b = 4$$

$$a = 60 \qquad b = 53$$

and so on.

If there are more equations than unknowns then either

1. Some of the equations are not linearly independent and are therefore superfluous or

2. The solutions obtained will depend on which equations are used and the system of equations is inconsistent.

This leads us naturally to the least squares method which we have discussed in Chapters 3, 8, and 9 and which will occupy the next section.

## THE LINEAR MODEL

Relationships in real life can commonly be represented or at least approximated by linear equations.

**Example A.1.42.** The simplest such form is

$$y = ax$$

Thus the height in centimeters $(y)$ and height in inches $(x)$ are connected by the approximate relationship

$$y = 2.54x$$

**Example A.1.43.** Another simple form is

$$y = a + bx$$

Thus temperature in degrees Fahrenheit $(y)$ and centigrade $(x)$ are connected by the relationship

$$y = 32 + 1.8x$$

Consider the situation discussed in Chapter 3 where the values of $a$ and $b$ are to be found by experimental methods with error of measurement.

The error is usually as likely to be positive as negative and the experimenter is careful to ensure independence from reading to reading. We will suppose they have an equal variance. The least squares method is a way of finding $a$ and $b$ such that the sum of the squares of the errors

$$\sum_i h_i^2 = \sum_i (y_i - a - bx_i)^2$$

for each of the observations is a minimum.

The general linear model is of the form

$$y_i = \beta_0 + \beta_1 x_{11} + \beta_2 x_{12} + \cdots + \beta_k x_{1k} + h_1$$

.

.

.

$$y_n = \beta_0 + \beta_1 x_{n1} + \beta_2 x_{n2} + \cdots + \beta_k x_{nk} + h_n$$

where the $x$'s are fixed quantities, the $y$'s are observations. This model can be written

$$\begin{bmatrix} y_1 \\ \cdot \\ \cdot \\ \cdot \\ \cdot \\ y_n \end{bmatrix} = \begin{bmatrix} 1 & x_{11} & \cdots & x_{1k} \\ \cdot & \cdot & & \cdot \\ \cdot & \cdot & & \cdot \\ \cdot & \cdot & & \cdot \\ \cdot & \cdot & & \cdot \\ 1 & x_{n1} & & x_{nk} \end{bmatrix} \begin{bmatrix} \beta_0 \\ \cdot \\ \cdot \\ \cdot \\ \cdot \\ \beta_k \end{bmatrix} + \begin{bmatrix} h_1 \\ \cdot \\ \cdot \\ \cdot \\ \cdot \\ h_n \end{bmatrix}$$

or in matrix notation

$$\mathbf{y} = \mathbf{X}\beta + \mathbf{h}$$

where

    $\mathbf{y}$ is a vector of response variables and known
    $\mathbf{X}$ is a matrix of regressor variables and known with $n > k + 1$
    $\beta$ and $\mathbf{h}$ are unknown vectors
We wish to minimize

$$\Sigma h_i^2$$

or in matrix notation

$$\mathbf{h}'\mathbf{h}$$

or

$$(\mathbf{y} - \mathbf{X}\beta)'(\mathbf{y} - \mathbf{X}\beta)$$

and this can be achieved by choosing the values of $\beta$ given by $\mathbf{b}$ where

$$\mathbf{b} = (\mathbf{X}'\mathbf{X})^{-1}\mathbf{X}'\mathbf{y} \tag{1}$$

The derivation of this result, known as the Gauss-Markov theorem involves the use of matrix calculus and will here have to be taken on trust.

**Example A.1.44.** Suppose we have a set of readings of blood cholesterol on aliquots from the same sample of blood. We wish to find the least squares estimate of the mean, $\beta$ (in this degenerate case, a scalar). We suppose that each reading is estimating this quality but is subject to some experimental error. Our model is

$$y_i = \beta + h_i \quad i = 1, \ldots, n$$

In matrix form

$$
\begin{bmatrix} y_1 \\ y_2 \\ \cdot \\ \cdot \\ y_n \end{bmatrix}
=
\begin{bmatrix} 1 \\ 1 \\ \cdot \\ \cdot \\ 1 \end{bmatrix} [\beta]
+
\begin{bmatrix} h_1 \\ h_2 \\ \cdot \\ \cdot \\ h_n \end{bmatrix}
$$

which is of the form

$$\mathbf{y} = \mathbf{X}\beta + \mathbf{h}$$

Then according to formula (1) the least squares estimate of $\beta$ is

$$(\mathbf{X}'\mathbf{X})^{-1}\mathbf{X}'\mathbf{y}$$

Then

$$
\mathbf{X}'\mathbf{X} = [1\,1\cdots1]
\begin{bmatrix} 1 \\ 1 \\ \cdot \\ \cdot \\ 1 \end{bmatrix} = n
$$

and, being a simple scalar it has the inverse

$$(\mathbf{X}'\mathbf{X})^{-1} = 1/n$$

Also

$$
\mathbf{X}'\mathbf{y} = [1\cdots\cdots1]
\begin{bmatrix} y_1 \\ \cdot \\ \cdot \\ y_n \end{bmatrix} = \Sigma y
$$

so

$$\mathbf{b} = (1/n)\Sigma y$$

the usual sample average.

**Example A.1.45.** (The simple regression model.) Suppose serum cholesterol in part is constant and in part increases with age. With experimental error our model is

$$y_i = \beta_0 + \beta_1 x_i + h_i$$

where

$y = $ cholesterol level in appropriate units
$x = $ age in years
$\beta_0 = $ constant part of cholesterol level
$\beta_1 = $ the amount by which cholesterol goes up with each year of life
$h = $ random error about the predicted value

We can write this

$$\begin{bmatrix} y_1 \\ y_2 \\ \cdot \\ \cdot \\ y_n \end{bmatrix} = \begin{bmatrix} 1 & x_1 \\ 1 & x_2 \\ \cdot & \cdot \\ \cdot & \cdot \\ 1 & x_n \end{bmatrix} \begin{bmatrix} \beta_0 \\ \beta_1 \end{bmatrix} + \begin{bmatrix} h_1 \\ h_2 \\ \cdot \\ \cdot \\ h_n \end{bmatrix}$$

$$\mathbf{X'X} = \begin{bmatrix} 1 & 1 & \cdots & 1 \\ x_1 & x_2 & \cdots & x_n \end{bmatrix} \begin{bmatrix} 1 & x_1 \\ 1 & x_2 \\ \cdot & \\ \cdot & \\ 1 & x_n \end{bmatrix} = \begin{bmatrix} n & \Sigma x \\ \Sigma x & \Sigma x^2 \end{bmatrix}$$

$$\mathbf{X'y} = \begin{bmatrix} 1 & 1 & \cdots & 1 \\ x_1 & x_2 & \cdots & x_n \end{bmatrix} \begin{bmatrix} y \\ y_2 \\ \cdot \\ \cdot \\ y_n \end{bmatrix} = \begin{bmatrix} \Sigma y \\ \Sigma xy \end{bmatrix}$$

$$(\mathbf{X'X})^{-1}\mathbf{X'y} = \begin{bmatrix} n & \Sigma x \\ \Sigma x & \Sigma x^2 \end{bmatrix}^{-1} \begin{bmatrix} \Sigma y \\ \Sigma xy \end{bmatrix}$$

$$= \frac{1}{n\Sigma x^2 - (\Sigma x)^2} \begin{bmatrix} \Sigma x^2 & -\Sigma x \\ -\Sigma x & n \end{bmatrix} \begin{bmatrix} \Sigma y \\ \Sigma xy \end{bmatrix} = \begin{bmatrix} b_0 \\ b_1 \end{bmatrix}$$

where $b_0$ and $b_1$ are the least squares estimates of $\beta_0$ and $\beta_1$ respectively.
Multiplying

$$b_0 = \frac{\Sigma x^2 \, \Sigma y - \Sigma x \, \Sigma xy}{n\Sigma x^2 - (\Sigma x)^2} = m_Y - m_x b_1$$

$$b_1 = \frac{n\Sigma xy - \Sigma x \Sigma y}{n\Sigma x^2 - (\Sigma x)^2} = \frac{\Sigma(x - m_x)(y - m_Y)}{\Sigma(x - m_x)^2}$$

where by $m_x$ we mean $(\Sigma x)/n$ and by $m_Y$ we mean $(\Sigma y)/n$. These formulas will be familiar from Chapter 8.

The results so far may seem rather unrewarding, but we can now see how easily the method can be extended to any number of regressor variables. Thus the quadratic model

$$y_i = \beta_0 + \beta_1 x + \beta_2 x^2 + h_i$$

gives

$$\mathbf{X'X} = \begin{bmatrix} n & \Sigma x & \Sigma x^2 \\ \Sigma x & \Sigma x^2 & \Sigma x^3 \\ \Sigma x^2 & \Sigma x^3 & \Sigma x^4 \end{bmatrix}$$

and

$$\mathbf{X'y} = \begin{bmatrix} \Sigma y \\ \Sigma xy \\ \Sigma x^2 y \end{bmatrix}$$

and if we want to use cubic or higher powers of $x$ the corresponding expressions can be written down by inspection. Or we might wish to use models of the form

$$y_i = \beta_0 + \beta_1 x + \beta_2 z + \beta_3 xz + \beta_4 x^2 + \beta_5 z^2 + \cdots$$

all of which can be dealt with in the same way.


**Variance***

In scalar algebra we show (7) that if

$$\mathrm{Var}(X) = \sigma^2$$

then

$$\mathrm{Var}(aX) = a^2 \sigma^2$$

which we could write as $a\sigma^2 a$. In matrix algebra we have an analogous relationship. By the variance of a vector of random variables, we mean a matrix in which the terms on the principal diagonal are variances, and those off the diagonal are covariances. Thus

$$\mathrm{var}\begin{bmatrix} X \\ Y \end{bmatrix} = \begin{bmatrix} \sigma^2 & \xi \\ \xi & \varsigma^2 \end{bmatrix}$$

means

$$\mathrm{Var}(X) = \sigma^2$$

$$\mathrm{Var}(Y) = \varsigma^2$$

$$\mathrm{Cov}(X, Y) = \mathrm{Cov}(Y, X) = \xi$$

By the last property, a variance matrix is necessarily square and symmetrical.

If, as is often so, the observations are independent, the off-diagonal terms will be

---

*Unfortunately we encounter a clash in convention. In scalar algebra we denote random variables by uppercase letters, and nonrandom quantities (e.g., sample values after the fact) by lower case. I will adhere to the convention in scalar equations. But in matrix notation, (boldface) uppercase letters denote matrices, lowercase, vectors. I will let this convention override the former. It will be clear enough from the context when a quantity is a random variable.

zero; and if they all have the same variance, all the entries on the principal diagonal will be the same, i.e.,

$$\begin{bmatrix} \sigma^2 & 0 & 0 \\ 0 & \sigma^2 & 0 \\ 0 & 0 & \sigma^2 \end{bmatrix}$$

which may be written $\mathbf{I}\sigma^2$.

If we scalar-multiply a vector by some constant quantity we have a relationship analogous to the scalar case.

If $\mathrm{var}(\mathbf{X}) = \mathbf{V}$

$$\mathrm{var}(\mathbf{AX}) = \mathbf{AVA}'$$

If $\mathbf{A}$ is a matrix or a scalar multiplier, the result will be a matrix. If we replace it by a vector, the resulting product is a scalar.

### Variance of a set of least squares estimates

We can see that

$$\mathbf{b} = (\mathbf{X}'\mathbf{X})^{-1}\mathbf{X}'\mathbf{y}$$

is of the foregoing form if we set $\mathbf{A} = (\mathbf{X}'\mathbf{X})^{-1}\mathbf{X}'$.

Now consider the original model

$$\mathbf{y} = \mathbf{X}\beta + \mathbf{h}$$

$\mathbf{X}\beta$ contains no random element and therefore contributes nothing to the variance. Hence

$$\mathrm{var}(\mathbf{y}) = \mathrm{var}(\mathbf{h}) = \mathbf{I}\sigma^2$$

$$\mathrm{var}(\mathbf{b}) = \mathrm{var}[(\mathbf{X}'\mathbf{X})^{-1}\mathbf{X}'\mathbf{y}]$$

$$= (\mathbf{X}'\mathbf{X})^{-1}\mathbf{X}'\mathbf{I}\sigma^2\mathbf{X}(\mathbf{X}'\mathbf{X})^{-1}$$

(Note that $\mathbf{X}'\mathbf{X}$ is always symmetric and therefore equal to its own transpose.)

Now $\sigma^2$ is a scalar multiplier and may be factored out. Thus

$$\mathrm{var}(\mathbf{b}) = (\mathbf{X}'\mathbf{X})^{-1}\mathbf{X}'\mathbf{I}\mathbf{X}(\mathbf{X}'\mathbf{X})^{-1}\sigma^2$$

$$= (\mathbf{X}'\mathbf{X})^{-1}\mathbf{X}'\mathbf{X}(\mathbf{X}'\mathbf{X})^{-1}\sigma^2$$

$$\mathrm{var}(\mathbf{b}) = (\mathbf{X}'\mathbf{X})^{-1}\sigma^2$$

which is a simple but also very general result.

**Example A.1.46.** In Example A.1.43 we had the model

$$Y_i = \beta + H_i$$

and we formed

$$(\mathbf{X}'\mathbf{X})^{-1} = 1/n$$

$$\mathrm{var}(b) = \sigma^2/n$$

**Example A.1.47.** In Example A.1.45 we have the model

$$Y_i = \beta_0 + \beta_1 x_1 + H_i$$

The variance matrix for the sample of $n$ such values is

$$\frac{1}{n \Sigma x^2 - (\Sigma x)^2} \begin{bmatrix} \Sigma x^2 & -\Sigma x \\ -\Sigma x & n \end{bmatrix} \sigma^2$$

This process may be extended indefinitely. Note that in general it will not be necessary to invert $\mathbf{X}'\mathbf{X}$ algebraically. If the computer program is properly arranged it will print out the inverse of this matrix as an intermediate step.

### Analysis of variance

It is convenient to note that the error sum of squares (*ESS*) which we have minimized is a scalar. $ESS = (\mathbf{y} - \mathbf{X}\beta)'(\mathbf{y} - \mathbf{X}\beta) = \mathbf{y}'\mathbf{y} - \beta'\mathbf{X}'\mathbf{y} - \mathbf{y}'\mathbf{X}\beta + \beta'\mathbf{X}'\mathbf{X}\beta$

Two facts help us to simplify this. All the terms on the right-hand side are scalars and therefore equal to their own transposes. Thus

$$\beta'\mathbf{X}\mathbf{y} = \mathbf{y}'\mathbf{X}'\beta$$

Also when we substitute our least squares estimate $\mathbf{b}$ for $\beta$ the last term becomes

$$\mathbf{b}'\mathbf{X}'\mathbf{X}\mathbf{b} = \mathbf{b}'\mathbf{X}'\mathbf{X}(\mathbf{X}'\mathbf{X})^{-1}\mathbf{X}'\mathbf{y} = \mathbf{b}'\mathbf{X}'\mathbf{y}$$

Thus we have

$$ESS = \mathbf{y}'\mathbf{y} - \mathbf{b}'\mathbf{X}'\mathbf{y} - \mathbf{b}'\mathbf{X}'\mathbf{y} + \mathbf{b}'\mathbf{X}'\mathbf{y}$$

$$= \mathbf{y}'\mathbf{y} - \mathbf{b}'\mathbf{X}'\mathbf{y}$$

Now we have the general relationship from Chapter 7:

Total sum of squares = Error sum of squares + Regression sum of squares.

$\mathbf{y}'\mathbf{y}$ is the total sum of squares.
Thus $\mathbf{b}'\mathbf{X}'\mathbf{y}$ is clearly the regression sum of squares.

*Desirable output from the computer* comprises
1. $(\mathbf{X}'\mathbf{X})$
2. $\mathbf{X}'\mathbf{y}$
3. $(\mathbf{X}'\mathbf{X})^{-1}$
4. $(\mathbf{X}'\mathbf{X})^{-1}\mathbf{X}'\mathbf{y}$, the least squares estimate of $\beta$
5. $\mathbf{y}'\mathbf{y}$ the total sum of squares
6. $\mathbf{b}'\mathbf{X}'\mathbf{y}$ the regression sum of squares

## SOME GEOMETRICAL INTERPRETATIONS

Many people find algebraic manipulations easier to grasp if they can attach a geometrical meaning to them. In the present section I shall attempt to do this briefly; it is obvious that I must confine our efforts to two- and three-dimensional space.

We have seen that a vector corresponds to a point in space, the elements being the coordinates. Let us explore this further.

**Example A.1.48.** For definiteness let the vector be

$$\mathbf{b} = \begin{bmatrix} 6 \\ 10 \end{bmatrix}$$

If we premultiply **b** with the matrix

$$\begin{bmatrix} 1 & 0 \\ 0 & 1 \end{bmatrix}$$

the vector remains unchanged. However, the matrix

$$\begin{bmatrix} 1 & 0 \\ 0 & -1 \end{bmatrix}$$

(operation 1) gives the vector

$$\begin{bmatrix} 6 \\ -10 \end{bmatrix}$$

which (in the conventional notation) corresponds to rotating the plane about the $x$ axis. It is a mirror image of the original vector.

$$\begin{bmatrix} -1 & 0 \\ 0 & 1 \end{bmatrix}$$

(operation 2) gives the vector

$$\begin{bmatrix} -6 \\ 10 \end{bmatrix}$$

i.e., is equivalent to a rotation about the $y$ axis. Furthermore, the matrix

$$\begin{bmatrix} -1 & 0 \\ 0 & -1 \end{bmatrix}$$

(operation 3) gives

$$\begin{bmatrix} -6 \\ -10 \end{bmatrix}$$

which is the equivalent to successive rotations about both axes. It is noteworthy that

$$\begin{bmatrix} -1 & 0 \\ 0 & 1 \end{bmatrix} \begin{bmatrix} 1 & 0 \\ 0 & -1 \end{bmatrix} = \begin{bmatrix} -1 & 0 \\ 0 & -1 \end{bmatrix}$$

so that multiplication of two matrices is equivalent to condensing two steps into one.

If we premultiply by

$$\begin{bmatrix} 3 & 0 \\ 0 & 3 \end{bmatrix}$$

(operation 4) we get

$$\begin{bmatrix} 18 \\ 30 \end{bmatrix}$$

which is equivalent to "stretching" the system by a factor of three in both directions. We can think of this as simply a change in scale, and the operation is in fact simply a scalar multiplication by 3. However, if the matrix is

$$\begin{bmatrix} 2 & 0 \\ 0 & 3 \end{bmatrix}$$

(operation 5) the resulting vector

$$\begin{bmatrix} 12 \\ 30 \end{bmatrix}$$

represents unequal stretching in the two axes.

Two other matrices are of interest:

$$\begin{bmatrix} 2 & 2 \\ 0 & 0 \end{bmatrix}$$

(operation 6) gives

$$\begin{bmatrix} 32 \\ 0 \end{bmatrix}$$

i.e., it moves this point (and in fact any point) to the $x$ axis. If we go further,

$$\begin{bmatrix} 0 & 0 \\ 0 & 0 \end{bmatrix}$$

(operation 7) gives

$$\begin{bmatrix} 0 \\ 0 \end{bmatrix}$$

i.e., it moves any point into the origin.

What happens if we multiply a vector by a nondiagonal matrix?

**Example A.1.49.** To make the effects clear we will consider the changes on four vectors which represent the corners of a unit square. We shall be repeatedly using them from now on:

$$\begin{bmatrix} 0 \\ 0 \end{bmatrix}, \quad \begin{bmatrix} 1 \\ 0 \end{bmatrix}, \quad \begin{bmatrix} 0 \\ 1 \end{bmatrix} \quad \text{and} \quad \begin{bmatrix} 1 \\ 1 \end{bmatrix}$$

The first of those represents the origin, the second a point on one axis, the third a point on the other, the fourth a point on neither.

We now premultiply by the matrix

$$\begin{bmatrix} 1 & 0.5 \\ 0.5 & 2 \end{bmatrix}$$

This produces the vectors

$$\begin{bmatrix} 0 \\ 0 \end{bmatrix}, \quad \begin{bmatrix} 1 \\ 0.5 \end{bmatrix}, \quad \begin{bmatrix} 0.5 \\ 2 \end{bmatrix} \quad \text{and} \quad \begin{bmatrix} 1.5 \\ 2.5 \end{bmatrix}$$

The origin is unchanged; furthermore, if for example, we had doubled each element in each vector, we would merely "magnify" the resulting pattern without changing its shape. We now see that multiplying by this matrix alters scale and also changes the angle between the axes. It will be found that any point on the old axis when operated on by the matrix will lie on the new axis.

It is worthy of note that if we premultiply the same four vectors by the non-diagonal matrix

$$\begin{bmatrix} 0 & 1 \\ 1 & 0 \end{bmatrix}$$

we get

$$\begin{bmatrix} 0 \\ 0 \end{bmatrix}, \quad \begin{bmatrix} 0 \\ 1 \end{bmatrix}, \quad \begin{bmatrix} 1 \\ 0 \end{bmatrix} \quad \text{and} \quad \begin{bmatrix} 1 \\ 1 \end{bmatrix}$$

which means that the points are moved but the angles are preserved, i.e., the initial configuration is a square and so is the new one. This is due to the fact (as can be readily verified) that the matrix is orthogonal and it can now be seen whence the name derives; *orthogonal* means "at right angles"; and multiplying by an orthogonal matrix is equivalent to rotating the axis, but maintaining the right angles.

The inverse of a matrix can now be considered. The inverse of the matrix used in Example A.1.49 is

$$\begin{bmatrix} {}^{8}/_{7} & -{}^{2}/_{7} \\ -{}^{2}/_{7} & {}^{4}/_{7} \end{bmatrix}$$

and operating on the *results* of Example A.1.49 we get

$$\begin{bmatrix} 0 \\ 0 \end{bmatrix}, \quad \begin{bmatrix} 1 \\ 0 \end{bmatrix}, \quad \begin{bmatrix} 0 \\ 1 \end{bmatrix} \quad \text{and} \quad \begin{bmatrix} 1 \\ 1 \end{bmatrix}$$

that is, if a matrix moves a point $a$ to a point $b$, then the inverse of the matrix will move $b$ to $a$. If we now consider Example A.1.48 operation 7, we find that *any* point is moved to the origin, so that no "uniqueness" is preserved and every point loses its identity. It is clearly impossible to reverse this process, and this is reflected algebraically by the fact that the matrix is not of full rank and therefore has no inverse.

We can now get a geometric insight into rank.

**Example A.1.50.** Consider the matrix

$$\begin{bmatrix} 1 & 1 \\ 2 & 2 \end{bmatrix}$$

This is of rank 1. If we multiply the same four points as in Example A.1.49 by it we get

$$\begin{bmatrix} 0 \\ 0 \end{bmatrix}, \quad \begin{bmatrix} 1 \\ 2 \end{bmatrix}, \quad \begin{bmatrix} 1 \\ 2 \end{bmatrix} \quad \text{and} \quad \begin{bmatrix} 2 \\ 4 \end{bmatrix}$$

For each point (and for every point) the second element is twice the first. So every point comes to lie on a straight line passing through the origin with a gradient of 2 (Fig. A.1.2).

Thus from Examples A.1.48 operation 7, A.1.49, and A.1.50 we get the following results for a plane.

A matrix of rank 2 moves every unique point to a unique point in that plane.

A matrix of rank 1 moves every point on to one line.

A matrix of rank 0 moves every point to one point (the origin). (Fig. A.1.3). Clearly it takes a plane (which is of dimension 2) to accommodate the first, whereas the second can be accommodated in one dimension and the third in none. We can generalize this to say that the rank of a matrix is the minimum number of dimensions that will accommodate all points operated on by the matrix.

## Determinants

**Example A.1.51.** Let us begin by operating on the usual four points

$$\begin{bmatrix} 0 \\ 0 \end{bmatrix}, \quad \begin{bmatrix} 1 \\ 0 \end{bmatrix}, \quad \begin{bmatrix} 0 \\ 1 \end{bmatrix} \quad \text{and} \quad \begin{bmatrix} 1 \\ 1 \end{bmatrix}$$

with the matrix

$$\begin{bmatrix} 2 & 0 \\ 0 & 4 \end{bmatrix}$$

They become

$$\begin{bmatrix} 0 \\ 0 \end{bmatrix}, \quad \begin{bmatrix} 2 \\ 0 \end{bmatrix}, \quad \begin{bmatrix} 0 \\ 4 \end{bmatrix} \quad \text{and} \quad \begin{bmatrix} 2 \\ 4 \end{bmatrix}$$

The original four points enclosed a square of unit area, the new four points a rectangle of area 8. The ratio of the two is the determinant of the matrix.

**Example A.1.52.** Let us try a more general matrix (Figure A.1.4). The matrix

$$\begin{bmatrix} 1 & 0.5 \\ 0.5 & 2 \end{bmatrix}$$

has a determinant of $7/4$. It converts the usual unit square to the parallelogram bounded by the four points

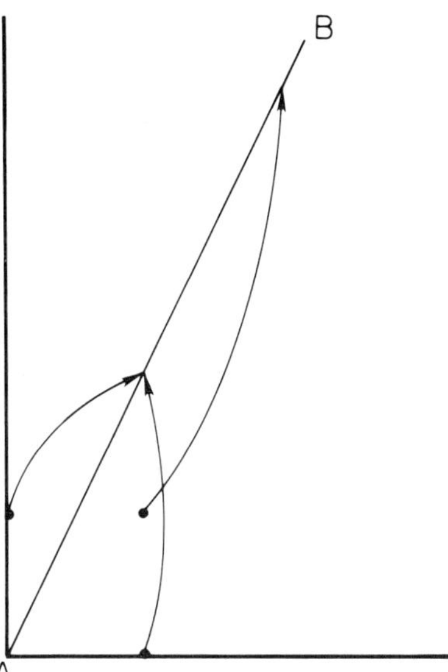



**Figure A.1.2.** The results of operating on a $2 \times 1$ vector by a $2 \times 2$ matrix of rank one. All the points fall on one straight line.

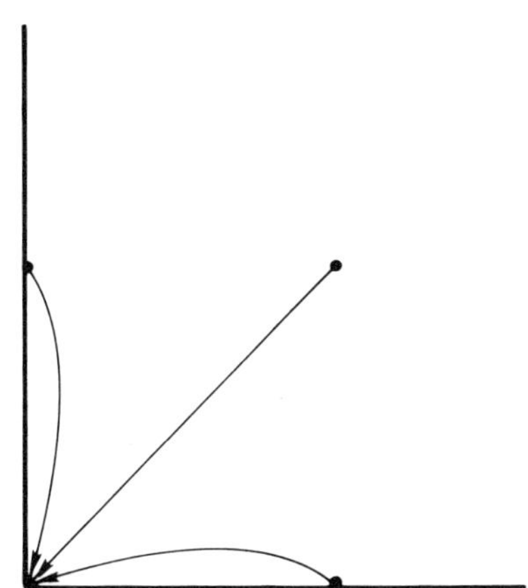

**Figure A.1.3.** Result of operating on a $2 \times 1$ vector by a $2 \times 2$ matrix of rank 0. All the points fall into the origin.

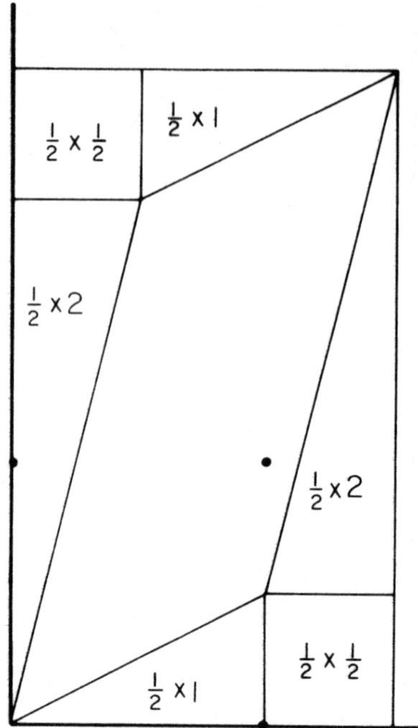


**Figure A.1.4.** The representation of the determinant of a matrix as the area of a rhomboid given by the matrix operating on the unit square.

$$\begin{bmatrix} 0 \\ 0 \end{bmatrix}, \quad \begin{bmatrix} 1 \\ 1/2 \end{bmatrix}, \quad \begin{bmatrix} 1/2 \\ 2 \end{bmatrix} \quad \text{and} \quad \begin{bmatrix} 3/2 \\ 5/2 \end{bmatrix}$$

and the area of this is best thought of as the area of a rectangle minus the area of two pairs of similar triangles and two small squares, i.e.,

$$(1\tfrac{1}{2}) \times (2\tfrac{1}{2}) - 2[\tfrac{1}{2} \times \tfrac{1}{2} \times 2 + \tfrac{1}{2} \times \tfrac{1}{2}] = {}^{7}/_{4}$$

It is not hard to prove the analogous result for $3 \times 3$ matrices (where the determinant represents the ratio of volumes) and it can be extended to—though not illustrated for—higher dimensions (where we are dealing with hyper-volumes). In the light of this we can see why the determinant of a matrix less than full rank must be zero. If any point in three space can be mapped on to a plane, the unit volume in three space must reduce to a figure in a plane which has no volume, or to a line or a point.

**Eigenvalues (Fig. A.1.5)**

Consider the equation

$$x/a + y/b = 1$$

This represents a straight line, cutting the $x$ axis at $(a, 0)$ and the $y$ axis at $(0, b)$. If we square each term

$$(x/a)^2 + (y/b)^2 = 1$$

we have the equation of an ellipse, cutting the axes at the same points. We can write the former as a pair of vectors,

$$[1/a \quad 1/b] \begin{bmatrix} x \\ y \end{bmatrix} = 1$$

and the latter as

$$[x \quad y] \begin{bmatrix} 1/a & 0 \\ 0 & 1/b \end{bmatrix} \begin{bmatrix} 1/a & 0 \\ 0 & 1/b \end{bmatrix} \begin{bmatrix} x \\ y \end{bmatrix} = 1$$

or

$$[x \quad y] \begin{bmatrix} 1/a^2 & 0 \\ 0 & 1/b^2 \end{bmatrix} \begin{bmatrix} x \\ y \end{bmatrix} = 1$$

The left hand slide of this equation is known as a quadratic form. Manifestly the eigenvalues are reciprocals of the squares of the half axes of the ellipse. We can show

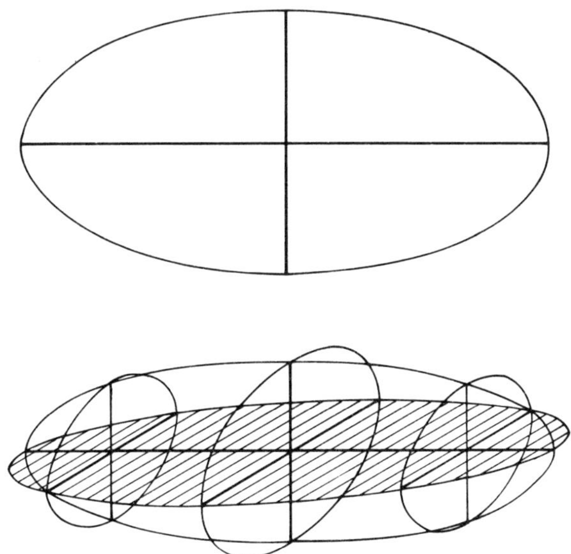

**Figure A.1.5.** The representation of the eigenvalues of a symmetrical matrix as the axes of an ellipsoid. *Upper* diagram in a 2 × 2 matrix. *Lower* diagram in a 3 × 3 matrix. In each case the eigenvalues are shown by a half axis. In the three-dimensional ellipsoid any section through the figure at right angles to an axis will appear as an ellipse. (Two such sections orthogonal to the main axis are shown.)

that any quadratic form with a symmetric $2 \times 2$ matrix corresponds to an ellipse. We know that for any symmetric matrix $\mathbf{Q}$, there exists an orthogonal matrix $\mathbf{U}$ such that $\mathbf{UQU}'$ is diagonal; furthermore, we know that operating with an orthogonal matrix is equivalent to a rotation of axes.

Diagonalization of a matrix consists of rotating the axes of the corresponding ellipse (the so-called eigenvectors) so that they are in line with the axes of the coordinate system. For this reason it is commonly called the principal axis transformation.

# APPENDIX 2
# TABLES OF STANDARD DISTRIBUTION FUNCTIONS

**Four-point interpolation**

Most readers will be familiar with ordinary linear interpolation. To make the most of the tables that follow, it is worthwhile to discuss briefly a more exact procedure. For theoretical details of the topic see standard books on numerical analysis. Suppose that for four equally-spaced arguments,

$$x, \quad (x + d), \quad (x + 2d), \quad \text{and} \quad (x + 3d)$$

the corresponding values of the function $f(x)$ are

$$y_1, \quad y_2, \quad y_3, \quad \text{and} \quad y_4$$

We wish to find an approximation to the value of the function at $(x + d + kd)$, where $k$ is some fraction between 0 and 1. Then we compute the following four weights:

$$w_1 = -k(2 - k)(1 - k)/6$$

$$w_2 = (2 - k)(1 - k)(1 + k)/2$$

$$w_3 = k(1 + k)(2 - k)/2$$

$$w_4 = -k(1 + k)(1 - k)/6$$

Then the desired approximation is

$$f(x + d + kd) = w_1 y_1 + w_2 y_2 + w_3 y_3 + w_4 y_4$$

**Example.** For the standard Gaussian distribution, what is $F(1.645)$? In Table A.2.1, $F(x)$ is tabled at intervals of $d = 0.1$, and here $k$ is 0.45 of such an interval. The weights are therefore

$$-0.0639375, \quad 0.6180625, \quad 0.5056875 \quad \text{and} \quad -0.0598125.$$

The resulting approximation is 0.95001498 compared with the directly calculated value 0.95001509.

# Table A.2.1. The standard Gaussian ("normal") distribution

These values show the probability that a random value of a Gaussian variate with mean zero and unit variance will not exceed the argument.

| $x$ | $F(x)$ | $x$ | $F(x)$ | $x$ | $F(x)$ |
|-----|--------|-----|--------|-----|--------|
| 0.1 | 0.539 827 84 | 1.6 | 0.945 200 71 | 3.1 | 0.999 032 40 |
| 0.2 | 0.579 259 71 | 1.7 | 0.955 434 54 | 3.2 | 0.999 312 86 |
| 0.3 | 0.617 911 42 | 1.8 | 0.964 069 68 | 3.3 | 0.999 516 58 |
| 0.4 | 0.655 421 74 | 1.9 | 0.971 283 44 | 3.4 | 0.999 663 07 |
| 0.5 | 0.691 462 46 | 2.0 | 0.977 249 87 | 3.5 | 0.999 767 37 |
| 0.6 | 0.725 746 88 | 2.1 | 0.982 135 58 | 3.6 | 0.999 840 89 |
| 0.7 | 0.758 036 35 | 2.2 | 0.986 096 55 | 3.7 | 0.999 892 20 |
| 0.8 | 0.788 144 60 | 2.3 | 0.989 275 89 | 3.8 | 0.999 927 65 |
| 0.9 | 0.815 939 87 | 2.4 | 0.991 802 46 | 3.9 | 0.999 951 90 |
| 1.0 | 0.841 344 75 | 2.5 | 0.993 790 33 | 4.0 | 0.999 968 33 |
| 1.1 | 0.864 333 94 | 2.6 | 0.995 338 81 | 4.1 | 0.999 979 34 |
| 1.2 | 0.884 930 33 | 2.7 | 0.996 533 03 | 4.2 | 0.999 986 65 |
| 1.3 | 0.903 199 52 | 2.8 | 0.997 444 87 | 4.3 | 0.999 991 46 |
| 1.4 | 0.919 243 34 | 2.9 | 0.998 134 19 | 4.4 | 0.999 994 59 |
| 1.5 | 0.933 192 80 | 3.0 | 0.998 650 10 | 4.5 | 0.999 996 60 |

*Note 1.* From symmetry,

$$F(-x) = 1 - F(x).$$ Thus

$$F(-0.1) = 1 - 0.53982784 = 0.46017216$$

In particular $F(0) = F(-0) = 0.5$.

*Note 2.* Four-point interpolation is accurate to at least five decimal places.

## Table A.2.2. Critical values of the chi-square statistic

Each entry shows the value of the chi-square variate that is exceeded in a random realization with the probability shown at the head of the column, according to the number of degrees of freedom. Four-point interpolation among degrees of freedom ensures an accuracy of at least three decimal places.

| | Size of test (Significance level) | | | | | | | | | | | |
|---|---|---|---|---|---|---|---|---|---|---|---|---|
| DF | 0.05 | 0.04 | 0.03 | 0.02 | 0.01 | 0.008 | 0.006 | 0.004 | 0.002 | 0.001 | 0.0005 | 0.0001 |
| 1 | 3.8415 | 4.2179 | 4.7093 | 5.4119 | 6.6349 | 7.0335 | 7.5503 | 8.2838 | 9.5495 | 10.8276 | 12.1157 | 15.1367 |
| 2 | 5.9915 | 6.4378 | 7.0131 | 7.8241 | 9.2103 | 9.6566 | 10.2320 | 11.0429 | 12.4292 | 13.8155 | 15.2018 | 18.4207 |
| 3 | 7.8147 | 8.3112 | 8.9473 | 9.8374 | 11.3449 | 11.8270 | 12.4466 | 13.3164 | 14.7955 | 16.2662 | 17.7300 | 21.1074 |
| 4 | 9.4877 | 10.0255 | 10.7119 | 11.6679 | 13.2767 | 13.7890 | 14.4458 | 15.3656 | 16.9238 | 18.4668 | 19.9974 | 23.5127 |
| 5 | 11.0705 | 11.6443 | 12.3746 | 13.3882 | 15.0863 | 15.6251 | 16.3150 | 17.2790 | 18.9074 | 20.5150 | 22.1053 | 25.7442 |
| 6 | 12.5916 | 13.1978 | 13.9676 | 15.0332 | 16.8119 | 17.3748 | 18.0946 | 19.0988 | 20.7912 | 22.4577 | 24.1028 | 27.8563 |
| 7 | 14.0671 | 14.7031 | 15.5091 | 16.6224 | 18.4753 | 19.0605 | 19.8079 | 20.8491 | 22.6007 | 24.3218 | 26.0176 | 29.8760 |
| 8 | 15.5073 | 16.1708 | 17.0105 | 18.1682 | 20.0902 | 20.6961 | 21.4693 | 22.5452 | 24.3521 | 26.1245 | 27.8680 | 31.8276 |
| 9 | 16.9190 | 17.6083 | 18.4796 | 19.6790 | 21.6660 | 22.2914 | 23.0888 | 24.1973 | 26.0564 | 27.8770 | 29.6645 | 33.7040 |
| 10 | 18.3070 | 19.0208 | 19.9219 | 21.1608 | 23.2093 | 23.8531 | 24.6735 | 25.8130 | 27.7216 | 29.5883 | 31.4198 | 35.5640 |
| 11 | 19.6751 | 20.4120 | 21.3416 | 22.6179 | 24.7250 | 25.3864 | 26.2286 | 27.3977 | 29.3536 | 31.2640 | 33.1324 | 37.3670 |
| 12 | 21.0261 | 21.7851 | 22.7418 | 24.0540 | 26.2170 | 26.8952 | 27.7585 | 28.9558 | 30.9570 | 32.9095 | 34.8213 | 39.1344 |
| 13 | 22.3620 | 23.1423 | 24.1250 | 25.4715 | 27.6882 | 28.3826 | 29.2661 | 30.4904 | 32.5360 | 34.5212 | 36.4778 | 40.8707 |
| 14 | 23.6848 | 24.4855 | 25.4931 | 26.8728 | 29.1412 | 29.8512 | 30.7540 | 32.0046 | 34.0913 | 36.1233 | 38.1094 | 42.5793 |
| 15 | 24.9958 | 25.8162 | 26.8479 | 28.2595 | 30.5779 | 31.3030 | 32.2243 | 33.4992 | 35.6280 | 37.6973 | 39.7188 | 44.2632 |
| 16 | 26.2962 | 27.1356 | 28.1907 | 29.6332 | 31.9999 | 32.7395 | 33.6792 | 34.9795 | 37.1461 | 39.2524 | 41.3081 | 45.9249 |
| 17 | 27.5871 | 28.4450 | 29.5227 | 30.9950 | 33.4084 | 34.1621 | 35.1201 | 36.4438 | 38.6485 | 40.7902 | 42.8792 | 47.5664 |
| 18 | 28.8693 | 29.7451 | 30.8447 | 32.3462 | 34.8053 | 35.5727 | 36.5471 | 37.8943 | 40.1361 | 42.3124 | 44.4338 | 49.1894 |
| 19 | 30.1435 | 31.0366 | 32.1576 | 33.6873 | 36.1909 | 36.9717 | 37.9626 | 39.3323 | 41.6103 | 43.8202 | 45.9731 | 50.7955 |
| 20 | 31.4104 | 32.3206 | 33.4624 | 35.0196 | 37.5662 | 38.3600 | 39.3672 | 40.7589 | 43.0720 | 45.3147 | 47.4985 | 52.3860 |

**Table A.2.2. Critical values of the chi-square statistic** (*cont'd.*)

Each entry shows the value of the chi-square variate that is exceeded in a random realization with the probability shown at the head of the column, according to the number of degrees of freedom. Four-point interpolation among degrees of freedom ensures an accuracy of at least three decimal places.

| | Size of test (Significance level) | | | | | | | | | | | |
| DF | 0.05 | 0.04 | 0.03 | 0.02 | 0.01 | 0.008 | 0.006 | 0.004 | 0.002 | 0.001 | 0.0005 | 0.0001 |
|---|---|---|---|---|---|---|---|---|---|---|---|---|
| 21 | 32.6705 | 33.5970 | 34.7595 | 36.3435 | 38.9322 | 39.7387 | 40.7617 | 42.1747 | 44.5223 | 46.7970 | 49.0108 | 53.9620 |
| 22 | 33.9244 | 34.8673 | 36.0492 | 37.6595 | 40.2894 | 41.1083 | 42.1468 | 43.5807 | 45.9618 | 48.2679 | 50.5111 | 55.5246 |
| 23 | 35.1724 | 36.1311 | 37.3323 | 38.9683 | 41.6384 | 42.4694 | 43.5231 | 44.9775 | 47.3915 | 49.7282 | 52.0002 | 57.0746 |
| 24 | 36.4150 | 37.3891 | 38.6093 | 40.2704 | 42.9798 | 43.8227 | 44.8912 | 46.3656 | 48.8118 | 51.1786 | 53.4788 | 58.6130 |
| 25 | 37.6525 | 38.6417 | 39.8804 | 41.5661 | 44.3141 | 45.1687 | 46.2516 | 47.7457 | 50.2234 | 52.6197 | 54.9475 | 60.1403 |
| 26 | 38.8851 | 39.8891 | 41.1461 | 42.8558 | 45.6417 | 46.5076 | 47.6049 | 49.1181 | 51.6269 | 50.0520 | 56.4069 | 61.6573 |
| 27 | 40.1133 | 41.1318 | 42.4066 | 44.1400 | 46.9629 | 47.8401 | 48.9513 | 50.4834 | 53.0226 | 55.4760 | 57.8576 | 63.1645 |
| 28 | 41.3371 | 42.3699 | 43.6622 | 45.4188 | 48.2782 | 49.1664 | 50.2913 | 51.8420 | 54.4110 | 56.8923 | 59.3000 | 64.6624 |
| 29 | 42.5570 | 43.6038 | 44.9133 | 46.6927 | 49.5879 | 50.4868 | 51.6252 | 53.1942 | 55.7925 | 58.3012 | 60.7347 | 66.1517 |
| 30 | 43.7730 | 44.8336 | 46.1599 | 47.9618 | 50.8922 | 51.8018 | 52.9534 | 54.5403 | 57.1674 | 59.7031 | 62.1619 | 67.6326 |
| 31 | 44.9853 | 46.0595 | 47.4025 | 49.2264 | 52.1914 | 53.1114 | 54.2761 | 55.8806 | 58.5362 | 61.0983 | 63.5820 | 69.1057 |
| 32 | 46.1943 | 47.2817 | 48.6411 | 50.4867 | 53.4858 | 54.4161 | 55.5936 | 57.2155 | 59.8990 | 62.4872 | 64.9955 | 70.5712 |
| 33 | 47.3999 | 48.5005 | 49.8760 | 51.7429 | 54.7755 | 55.7160 | 56.9062 | 58.5451 | 62.2562 | 63.8701 | 66.4025 | 72.0296 |
| 34 | 48.6024 | 49.7159 | 51.1072 | 52.9952 | 56.0609 | 57.0113 | 58.2140 | 59.8698 | 62.6080 | 65.2472 | 67.8035 | 73.4812 |
| 35 | 49.8019 | 50.9281 | 52.3351 | 54.2438 | 57.3421 | 58.3023 | 59.5173 | 61.1897 | 63.9546 | 66.6188 | 69.1986 | 74.9262 |
| 36 | 50.9985 | 52.1373 | 53.5597 | 55.4889 | 58.6192 | 59.5892 | 60.8162 | 62.5050 | 65.2963 | 67.9852 | 70.5881 | 76.3650 |
| 37 | 52.1923 | 53.3435 | 54.7811 | 56.7305 | 59.8925 | 60.8720 | 62.1110 | 63.8160 | 66.6333 | 69.3465 | 71.9722 | 77.7977 |
| 38 | 53.3835 | 54.5470 | 55.9995 | 57.9688 | 61.1621 | 62.1511 | 63.4018 | 65.1227 | 67.9657 | 70.7029 | 73.3512 | 79.2247 |
| 39 | 54.5722 | 55.7477 | 57.2151 | 59.2040 | 62.4281 | 63.4264 | 64.6888 | 66.4254 | 69.2938 | 72.0547 | 74.7253 | 80.6462 |
| 40 | 55.7585 | 56.9459 | 58.4279 | 60.4361 | 63.6907 | 64.6982 | 65.9721 | 67.7243 | 70.6177 | 73.4020 | 76.0946 | 82.0623 |

| 41 | 56.9424 | 58.1415 | 59.6379 | 61.6654 | 64.9501 | 65.9667 | 67.2519 | 69.0194 | 71.9375 | 74.7449 | 77.4593 | 83.4733 |
| 42 | 58.1240 | 59.3348 | 60.8455 | 62.8918 | 66.2062 | 67.2318 | 68.5283 | 70.3110 | 73.2535 | 76.0838 | 78.8197 | 84.8793 |
| 43 | 59.3035 | 60.5257 | 62.0505 | 64.1156 | 67.4594 | 68.4938 | 69.8013 | 71.5990 | 74.5658 | 77.4186 | 80.1757 | 86.2806 |
| 44 | 60.4809 | 61.7144 | 63.2531 | 65.3367 | 68.7095 | 69.7527 | 71.0712 | 72.8838 | 75.8744 | 78.7495 | 81.5277 | 87.6773 |
| 45 | 61.6562 | 62.9010 | 64.4535 | 66.5553 | 69.9568 | 71.0087 | 72.3381 | 74.1653 | 77.1795 | 80.0767 | 82.8757 | 89.0695 |
| 46 | 62.8296 | 64.0855 | 65.6516 | 67.7714 | 71.2014 | 72.2619 | 73.6019 | 75.4436 | 78.4813 | 81.4003 | 84.2198 | 90.4574 |
| 47 | 64.0011 | 65.2679 | 66.8475 | 68.9852 | 72.4433 | 73.5123 | 74.8629 | 76.7190 | 79.7797 | 82.7204 | 85.5603 | 91.8412 |
| 48 | 65.1708 | 66.4485 | 68.0413 | 70.1968 | 73.6826 | 74.7600 | 76.1212 | 77.9914 | 81.0750 | 84.0371 | 86.8972 | 93.2209 |
| 49 | 66.3387 | 67.6271 | 69.2331 | 71.4061 | 74.9195 | 76.0052 | 77.3767 | 79.2610 | 82.3673 | 85.3506 | 88.2305 | 94.5967 |
| 50 | 67.5048 | 68.8039 | 70.4230 | 72.6133 | 76.1539 | 77.2478 | 78.6296 | 80.5278 | 83.6566 | 86.6607 | 89.5605 | 95.9688 |
| 51 | 68.6693 | 69.9789 | 71.6109 | 73.8184 | 77.3860 | 78.4881 | 79.8800 | 81.7920 | 84.9430 | 87.9680 | 90.8872 | 97.3371 |
| 52 | 69.8322 | 71.1522 | 72.7970 | 75.0214 | 78.6158 | 79.7259 | 81.1280 | 83.0536 | 86.2266 | 89.2722 | 92.2108 | 98.7019 |
| 53 | 70.9935 | 72.3238 | 73.9813 | 76.2226 | 79.8433 | 80.9615 | 82.3736 | 84.3127 | 87.5074 | 90.5734 | 93.5312 | 100.0632 |
| 54 | 72.1532 | 73.4938 | 75.1639 | 77.4218 | 81.0688 | 82.1949 | 83.6168 | 85.5694 | 88.7857 | 91.8718 | 94.8487 | 101.4211 |
| 55 | 73.3115 | 74.6622 | 76.3447 | 78.6191 | 82.2921 | 83.4261 | 84.8578 | 86.8237 | 90.0614 | 93.1675 | 96.1632 | 102.7758 |
| 56 | 74.4683 | 75.8291 | 77.5239 | 79.8147 | 83.5134 | 84.6551 | 86.0966 | 88.0756 | 91.3345 | 94.4605 | 97.4749 | 104.1273 |
| 57 | 75.6238 | 76.9944 | 78.7015 | 81.0085 | 84.7328 | 85.8822 | 87.3333 | 89.3253 | 92.6053 | 95.7510 | 98.7838 | 105.4756 |
| 58 | 76.7778 | 78.1584 | 79.8776 | 82.2007 | 85.9502 | 87.1073 | 88.5679 | 90.5729 | 93.8736 | 97.0388 | 100.0901 | 106.8210 |
| 59 | 77.9305 | 79.3209 | 81.0521 | 83.3911 | 87.1657 | 88.3304 | 89.8005 | 91.8183 | 95.1397 | 98.3242 | 101.3937 | 108.1634 |
| 60 | 79.0820 | 80.4820 | 82.2251 | 84.5800 | 88.3794 | 89.5516 | 91.0311 | 93.0616 | 96.4035 | 99.6072 | 102.6948 | 109.5029 |
| 70 | 90.5312 | 92.0241 | 93.8813 | 96.3875 | 100.4252 | 101.6694 | 103.2389 | 105.3911 | 108.9295 | 112.3169 | 115.5776 | 122.7547 |
| 80 | 101.8795 | 103.4588 | 105.4221 | 110.3928 | 112.3288 | 113.6401 | 115.2934 | 117.5591 | 121.2805 | 124.8392 | 128.2613 | 135.7825 |
| 90 | 113.1453 | 114.8057 | 116.8688 | 119.6484 | 124.1163 | 125.4906 | 127.2227 | 129.5951 | 133.4885 | 137.2084 | 140.7823 | 148.6273 |
| 100 | 124.3421 | 126.0794 | 128.2367 | 131.1417 | 135.8067 | 137.2407 | 139.0473 | 141.5206 | 145.5769 | 149.4493 | 153.1670 | 161.3187 |
| 110 | 135.4802 | 137.2904 | 139.5375 | 142.5617 | 147.4143 | 148.9050 | 150.7826 | 153.3520 | 157.5633 | 161.5807 | 165.4353 | 173.8791 |
| 120 | 146.5674 | 148.4474 | 150.7802 | 153.9182 | 158.9502 | 160.4952 | 162.4405 | 165.1017 | 169.4612 | 173.6174 | 177.6029 | 186.3260 |

## Table A.2.3. Critical values of the Gosset $t$ statistic

| DF | Size of test (Significance level) | | | | | |
|---|---|---|---|---|---|---|
|  | 0.2 | 0.05 | 0.01 | 0.001 | 0.0001 | 0.000,001 |
| 1 | 3.061048 | 12.706205 | 63.656741 | 636.619246 | 1061.031643 | 1051050 |
| 2 | 1.885618 | 4.302653 | 9.924843 | 31.599055 | 99.992502 | 999.998 |
| 3 | 1.637744 | 3.182446 | 5.840909 | 12.923979 | 28.000131 | 130.154 |
| 4 | 1.533206 | 2.776445 | 4.604095 | 8.610302 | 15.544101 | 49.458 |
| 5 | 1.475884 | 2.570582 | 4.032143 | 6.868827 | 11.177710 | 28.479 |
| 6 | 1.439756 | 2.446912 | 3.707428 | 5.958816 | 9.082347 | 20.0479 |
| 7 | 1.414924 | 2.364624 | 3.499483 | 5.407883 | 7.884584 | 15.7670 |
| 8 | 1.396815 | 2.306004 | 3.355387 | 5.041305 | 7.120004 | 13.2572 |
| 9 | 1.383029 | 2.262157 | 3.249835 | 4.780913 | 6.593683 | 11.63680 |
| 10 | 1.372184 | 2.228139 | 3.169273 | 4.586894 | 6.211051 | 10.51651 |
| 11 | 1.363430 | 2.200985 | 3.105807 | 4.436979 | 5.921194 | 9.70143 |
| 12 | 1.356217 | 2.178813 | 3.054540 | 4.317791 | 5.694466 | 9.08472 |
| 13 | 1.350171 | 2.160369 | 3.012276 | 4.220832 | 5.512515 | 8.60332 |
| 14 | 1.345030 | 2.144787 | 2.976843 | 4.140454 | 5.363413 | 8.21806 |
| 15 | 1.340606 | 2.131450 | 2.946713 | 4.072765 | 5.239088 | 7.90323 |
| 16 | 1.366757 | 2.119905 | 2.920782 | 4.014996 | 5.133894 | 7.64153 |
| 17 | 1.333379 | 2.109816 | 2.898231 | 3.965126 | 5.043765 | 7.42070 |
| 18 | 1.330391 | 2.100922 | 2.878440 | 3.921646 | 4.965706 | 7.23208 |
| 19 | 1.327728 | 2.093024 | 2.860935 | 3.883406 | 4.897461 | 7.06912 |
| 20 | 1.325341 | 2.085963 | 2.845340 | 3.849516 | 4.837301 | 6.92705 |
| 25 | 1.316345 | 2.059539 | 2.787436 | 3.725144 | 4.619136 | 6.42414 |
| 30 | 1.310415 | 2.042272 | 2.749996 | 3.645959 | 4.482417 | 6.11910 |
| 35 | 1.306212 | 2.030108 | 2.723806 | 3.591147 | 4.388796 | 5.91477 |
| 40 | 1.303077 | 2.021075 | 2.704459 | 3.550966 | 4.320700 | 5.76848 |
| 45 | 1.300649 | 2.014103 | 2.689585 | 3.520251 | 4.268955 | 5.65868 |
| 50 | 1.298714 | 2.008559 | 2.677793 | 3.496013 | 4.228310 | 5.57324 |
| 55 | 1.297134 | 2.004045 | 2.668216 | 3.476398 | 4.195542 | 5.50489 |
| 60 | 1.295821 | 2.000298 | 2.660283 | 3.460200 | 4.168565 | 5.44897 |
| 65 | 1.294712 | 1.997138 | 2.653604 | 3.446598 | 4.145970 | 5.40239 |
| 70 | 1.293763 | 1.994437 | 2.647905 | 3.435015 | 4.126770 | 5.36297 |
| 80 | 1.292224 | 1.990063 | 2.638691 | 3.416337 | 4.095895 | 5.29994 |
| 90 | 1.291029 | 1.986675 | 2.631565 | 3.401935 | 4.072156 | 5.25177 |
| 100 | 1.290075 | 1.983972 | 2.625891 | 3.390491 | 4.053334 | 5.21376 |
| a | −1.30 | −2.33 | −3.70 | −5.40 | −7.00 | — |
| b | −0.057 | −0.594 | −1.389 | −1.50 | −4.809 | — |
| ∞ | 1.281551 | 1.959964 | 2.575829 | 3.290527 | 3.890592 | 4.89164 |

Note that for more than 20 degrees of freedom one may approximate the appropriate $t$ value by dividing the appropriate Gaussian value (i.e., the $t$ with infinitely many degrees of freedom) by an appropriate factor

$$\sqrt{(1 + a/n + b/n^2)}$$

No such formula is sufficiently accurate to use for the last column.

**Table A.2.4. The F Distribution. Number of degrees of freedom for numerator: 2**

| DF for denominator | Size (Significance level) | | | | |
|---|---|---|---|---|---|
| | 0.1 | 0.05 | 0.02 | 0.01 | 0.001 |
| 1 | 49.5000 | 199.500 | 1249.50 | 4999.50 | 500000 |
| 2 | 9.00000 | 19.0000 | 49.0000 | 99.0000 | 999.000 |
| 3 | 5.46238 | 9.55209 | 18.8581 | 30.8165 | 148.500 |
| 4 | 4.32455 | 6.94427 | 12.1421 | 18.0000 | 61.2456 |
| 5 | 3.77971 | 5.78613 | 9.45440 | 13.2739 | 37.1223 |
| 6 | 3.46330 | 5.14325 | 8.05209 | 10.9248 | 27.0000 |
| 7 | 3.25744 | 4.73741 | 7.20256 | 9.54657 | 21.6890 |
| 8 | 3.11311 | 4.45897 | 6.63659 | 8.64911 | 18.4937 |
| 9 | 3.00645 | 4.25649 | 6.23399 | 8.02151 | 16.3871 |
| 10 | 2.92446 | 4.10282 | 5.93362 | 7.55943 | 14.9054 |
| 11 | 2.85951 | 3.98229 | 5.70124 | 7.20571 | 13.8116 |
| 12 | 2.80679 | 3.88529 | 5.51629 | 6.92660 | 12.9737 |
| 13 | 2.76316 | 3.80556 | 5.36569 | 6.70096 | 12.3127 |
| 14 | 2.72646 | 3.73889 | 5.24075 | 6.51488 | 11.7789 |
| 15 | 2.69517 | 3.68232 | 5.13544 | 6.35889 | 11.3391 |
| 16 | 2.66817 | 3.63372 | 5.04551 | 6.22623 | 10.9710 |
| 17 | 2.64463 | 3.59153 | 4.96783 | 6.11211 | 10.6584 |
| 18 | 2.62394 | 3.55455 | 4.90006 | 6.01290 | 10.3899 |
| 19 | 2.60561 | 3.52189 | 4.84044 | 5.92587 | 10.1568 |
| 20 | 2.58925 | 3.49282 | 4.78757 | 5.84893 | 9.95262 |
| 25 | 2.52830 | 3.38518 | 4.59336 | 5.56799 | 9.92251 |
| 30 | 2.48871 | 3.31582 | 4.46955 | 5.39034 | 8.77339 |
| 35 | 2.46093 | 3.26742 | 4.38376 | 5.26794 | 8.46968 |
| 40 | 2.44036 | 3.23172 | 4.32083 | 5.17850 | 8.25075 |
| 45 | 2.42452 | 3.20431 | 4.27270 | 5.11031 | 8.08551 |
| 50 | 2.41195 | 3.18260 | 4.23471 | 5.05661 | 7.95641 |
| 60 | 2.39325 | 3.15041 | 4.17854 | 4.97743 | 7.76776 |
| 70 | 2.38001 | 3.12767 | 4.13902 | 4.92187 | 7.63657 |
| 80 | 2.37014 | 3.11076 | 4.10971 | 4.88073 | 7.54008 |
| 90 | 2.36251 | 3.09769 | 4.08710 | 4.84905 | 7.46614 |
| 100 | 2.35642 | 3.08729 | 4.06913 | 4.82390 | 7.40768 |
| 120 | 2.34733 | 3.07177 | 4.04237 | 4.78650 | 7.32110 |
| 140 | 2.34087 | 3.06075 | 4.02340 | 4.76003 | 7.26008 |
| 160 | 2.33604 | 3.05252 | 4.00925 | 4.74029 | 7.21475 |
| 180 | 2.33229 | 3.04614 | 3.99829 | 4.72570 | 7.17976 |
| 200 | 2.32929 | 3.04105 | 3.98955 | 4.71285 | 7.15193 |

| DF for denominator | Size (Significance level) | | | | |
|---|---|---|---|---|---|
| | 0.1 | 0.05 | 0.02 | 0.01 | 0.001 |
| 1 | 53.5932 | 215.707 | 1350.50 | 5403.35 | 540379 |
| 2 | 9.16179 | 19.1643 | 49.1657 | 99.1662 | 999.167 |
| 3 | 5.39077 | 9.27662 | 18.1097 | 29.4567 | 141.108 |
| 4 | 4.19086 | 6.59138 | 11.3435 | 16.6944 | 56.1772 |
| 5 | 3.61947 | 5.40945 | 8.67019 | 12.0600 | 33.2025 |
| 6 | 3.28876 | 4.75706 | 7.28698 | 9.77953 | 23.7033 |
| 7 | 3.07407 | 4.34683 | 6.45394 | 8.45128 | 18.7723 |
| 8 | 2.92379 | 4.06618 | 5.90138 | 7.59099 | 15.8295 |
| 9 | 2.81286 | 3.86254 | 5.50966 | 6.99191 | 13.9018 |
| 10 | 2.72767 | 3.70826 | 5.21819 | 6.55231 | 12.5527 |
| 11 | 2.66022 | 3.58743 | 4.99321 | 6.21672 | 11.5611 |
| 12 | 2.60552 | 3.49029 | 4.81448 | 5.95254 | 10.8042 |
| 13 | 2.56027 | 3.41053 | 4.66916 | 5.73938 | 10.2089 |
| 14 | 2.52222 | 3.34388 | 4.54876 | 5.56388 | 9.72936 |
| 15 | 2.48978 | 3.28738 | 4.44741 | 5.41696 | 9.33525 |
| 16 | 2.46181 | 3.23887 | 4.36094 | 5.29221 | 9.00593 |
| 17 | 2.43743 | 3.19677 | 4.28631 | 5.18499 | 8.72685 |
| 18 | 2.41600 | 3.15990 | 4.22127 | 5.09188 | 8.48745 |
| 19 | 2.39702 | 3.12735 | 4.16407 | 5.01028 | 8.27993 |
| 20 | 2.38008 | 3.09839 | 4.11340 | 4.93819 | 8.09837 |
| 25 | 2.31701 | 2.99124 | 3.92751 | 4.67546 | 7.45107 |
| 30 | 2.27607 | 2.92227 | 3.80923 | 4.50973 | 7.05445 |
| 35 | 2.24735 | 2.87418 | 3.72739 | 4.39574 | 6.78696 |
| 40 | 2.22609 | 2.83874 | 3.66742 | 4.31256 | 6.59453 |
| 45 | 2.20972 | 2.21154 | 3.62158 | 4.24920 | 6.44953 |
| 50 | 2.19672 | 2.79000 | 3.58542 | 4.19934 | 6.33637 |
| 60 | 2.17741 | 2.75807 | 3.53200 | 4.12589 | 6.17123 |
| 70 | 2.16373 | 2.73554 | 3.49444 | 4.07439 | 6.05655 |
| 80 | 2.15354 | 2.71878 | 3.46660 | 4.03629 | 5.97230 |
| 90 | 2.14566 | 2.70583 | 3.44513 | 4.00696 | 5.90778 |
| 100 | 2.13937 | 2.69553 | 3.42807 | 3.98369 | 5.85680 |
| 120 | 2.12999 | 2.68016 | 3.40267 | 3.94909 | 5.78136 |
| 140 | 2.12331 | 2.66925 | 3.38468 | 3.92461 | 5.72823 |
| 160 | 2.11832 | 2.66110 | 3.37126 | 3.90637 | 5.68878 |
| 180 | 2.11445 | 2.65479 | 3.36087 | 3.89226 | 5.65834 |
| 200 | 2.11136 | 2.64975 | 3.35258 | 3.88102 | 5.63413 |

**The F Distribution.** (*cont'd.*) **Number of degrees of freedom for numerator: 4**

| DF for denominator | Size (Significance level) | | | | |
|---|---|---|---|---|---|
| | 0.1 | 0.05 | 0.02 | 0.01 | 0.001 |
| 1 | 55.8330 | 224.583 | 1405.83 | 5624.58 | 562500 |
| 2 | 9.24341 | 19.2468 | 49.2487 | 99.2494 | 999.250 |
| 3 | 5.34264 | 9.11718 | 17.6938 | 28.7099 | 137.100 |
| 4 | 4.10724 | 6.38823 | 10.8994 | 15.9770 | 53.4358 |
| 5 | 3.52019 | 5.19216 | 8.23303 | 11.3919 | 31.0850 |
| 6 | 3.18076 | 4.53367 | 6.85943 | 9.14830 | 21.9235 |
| 7 | 2.96053 | 4.12031 | 6.03472 | 7.84664 | 17.1980 |
| 8 | 2.80642 | 3.83785 | 5.48891 | 7.00607 | 14.3916 |
| 9 | 2.69268 | 3.63308 | 5.10265 | 6.42208 | 12.5603 |
| 10 | 2.60533 | 3.47804 | 4.81564 | 5.99433 | 11.2828 |
| 11 | 2.53618 | 3.35669 | 4.59434 | 5.66830 | 10.3461 |
| 12 | 2.48010 | 3.25916 | 4.41869 | 5.41195 | 9.63272 |
| 13 | 2.43370 | 3.17911 | 4.27600 | 5.20533 | 9.07273 |
| 14 | 2.39469 | 3.11224 | 4.15785 | 5.03537 | 8.62231 |
| 15 | 2.36143 | 3.05556 | 4.05844 | 4.89320 | 8.25268 |
| 16 | 2.33274 | 3.00691 | 3.97368 | 4.77257 | 7.94420 |
| 17 | 2.30774 | 2.96470 | 3.90056 | 4.66896 | 7.68306 |
| 18 | 2.28577 | 2.92774 | 3.83685 | 4.57903 | 7.45927 |
| 19 | 2.26630 | 2.89510 | 3.78085 | 4.50025 | 7.26546 |
| 20 | 2.24893 | 2.86608 | 3.73125 | 4.43069 | 7.09603 |
| 25 | 2.18424 | 2.75871 | 3.54942 | 4.17742 | 6.49305 |
| 30 | 2.14223 | 2.68962 | 3.43383 | 4.01787 | 6.12452 |
| 35 | 2.11276 | 2.64146 | 3.35391 | 3.90824 | 5.87640 |
| 40 | 2.09094 | 2.60597 | 3.29537 | 3.82829 | 5.69813 |
| 45 | 2.07415 | 2.57873 | 3.25064 | 3.76742 | 5.56393 |
| 50 | 2.06081 | 2.55717 | 3.21537 | 3.71954 | 5.45928 |
| 60 | 2.04098 | 2.52521 | 3.16328 | 3.64904 | 5.30670 |
| 70 | 2.02694 | 2.50265 | 3.12666 | 3.59964 | 5.20084 |
| 80 | 2.01648 | 2.48588 | 3.09952 | 3.56310 | 5.12312 |
| 90 | 2.00839 | 2.47292 | 3.07860 | 3.53499 | 5.06363 |
| 100 | 2.00193 | 2.46261 | 3.06198 | 3.51268 | 5.01665 |
| 120 | 1.99230 | 2.44723 | 3.03724 | 3.47953 | 4.94715 |
| 140 | 1.98544 | 2.43631 | 3.01971 | 3.45607 | 4.89822 |
| 160 | 1.98032 | 2.42816 | 3.00664 | 3.43860 | 4.86191 |
| 180 | 1.97635 | 2.42184 | 2.99652 | 3.42508 | 4.83390 |
| 200 | 1.97317 | 2.41679 | 2.98845 | 3.41432 | 4.81163 |

| DF for denominator | Size (Significance level) | | | | |
|---|---|---|---|---|---|
| | 0.1 | 0.05 | 0.02 | 0.01 | 0.001 |
| 1 | 57.2401 | 231.162 | 1440.61 | 5763.65 | 576405 |
| 2 | 9.29262 | 19.2964 | 49.2986 | 99.2993 | 999.300 |
| 3 | 5.30915 | 9.01345 | 17.4288 | 28.2371 | 134.580 |
| 4 | 4.05057 | 6.25605 | 10.6157 | 15.5219 | 51.7116 |
| 5 | 3.45298 | 5.05032 | 7.95293 | 10.9670 | 29.7524 |
| 6 | 3.10751 | 4.38737 | 6.58474 | 8.74589 | 20.8027 |
| 7 | 2.88334 | 3.97152 | 5.76472 | 7.46043 | 16.2058 |
| 8 | 2.72644 | 3.68749 | 5.22271 | 6.63182 | 13.4847 |
| 9 | 2.61061 | 3.48165 | 4.83950 | 6.05694 | 11.7137 |
| 10 | 2.52164 | 3.32583 | 4.55496 | 5.63632 | 10.4807 |
| 11 | 2.45118 | 3.20387 | 4.33569 | 5.31600 | 9.57837 |
| 12 | 2.39402 | 3.10587 | 4.16174 | 5.06434 | 8.89210 |
| 13 | 2.34672 | 3.02543 | 4.02048 | 4.86162 | 8.35408 |
| 14 | 2.30694 | 2.95824 | 3.90355 | 4.69496 | 7.92180 |
| 15 | 2.27302 | 2.90129 | 3.80520 | 4.55561 | 7.56739 |
| 16 | 2.24375 | 2.85240 | 3.72136 | 4.43742 | 7.27185 |
| 17 | 2.21825 | 2.80999 | 3.64905 | 4.33593 | 7.02186 |
| 18 | 2.19582 | 2.77285 | 3.58605 | 4.24788 | 6.80777 |
| 19 | 2.17595 | 2.74005 | 3.53069 | 4.17076 | 6.62246 |
| 20 | 2.15822 | 2.71088 | 3.48166 | 4.10268 | 6.46056 |
| 25 | 2.09216 | 2.60298 | 3.30198 | 3.85495 | 5.88506 |
| 30 | 2.04924 | 2.53355 | 3.18780 | 3.69901 | 5.53391 |
| 35 | 2.01912 | 2.48514 | 3.10887 | 3.59191 | 5.29777 |
| 40 | 1.99681 | 2.44946 | 3.05107 | 3.51383 | 5.12826 |
| 45 | 1.97963 | 2.42208 | 3.00691 | 3.45441 | 5.00073 |
| 50 | 1.96599 | 2.40040 | 2.97209 | 3.40767 | 4.90134 |
| 60 | 1.94571 | 2.36827 | 2.92067 | 3.33888 | 4.75652 |
| 70 | 1.93134 | 2.34558 | 2.88454 | 3.29068 | 4.65610 |
| 80 | 1.92063 | 2.32872 | 2.85775 | 3.25504 | 4.58241 |
| 90 | 1.91234 | 2.31568 | 2.83711 | 3.22762 | 4.52603 |
| 100 | 1.90574 | 2.30531 | 2.82071 | 3.20587 | 4.49150 |
| 120 | 1.89587 | 2.28985 | 2.79630 | 3.17354 | 4.41567 |
| 140 | 1.88885 | 2.27886 | 2.77901 | 3.15067 | 4.36934 |
| 160 | 1.88361 | 2.27066 | 2.76611 | 3.13364 | 4.33496 |
| 180 | 1.87953 | 2.26430 | 2.75613 | 3.12047 | 4.30844 |
| 200 | 1.87628 | 2.25923 | 2.74817 | 3.10997 | 4.28736 |

**The F Distribution.** (*cont'd.*) **Number of degrees of freedom for numerator: 6**

| DF for denominator | Size (Significance level) | | | | |
| --- | --- | --- | --- | --- | --- |
| | 0.1 | 0.05 | 0.02 | 0.01 | 0.001 |
| 1 | 58.2044 | 233.986 | 1464.45 | 5858.99 | 585937 |
| 2 | 9.32553 | 19.3295 | 49.3318 | 99.3326 | 999.333 |
| 3 | 5.28473 | 8.94064 | 17.2451 | 27.9107 | 132.847 |
| 4 | 4.00974 | 6.16313 | 10.4186 | 15.2069 | 50.5250 |
| 5 | 3.40450 | 4.95028 | 7.75770 | 10.6723 | 28.8344 |
| 6 | 3.05455 | 4.28386 | 6.39277 | 8.46612 | 20.0296 |
| 7 | 2.82739 | 3.86596 | 5.57561 | 7.19140 | 15.5208 |
| 8 | 2.66833 | 3.58058 | 5.03588 | 6.37068 | 12.8580 |
| 9 | 2.55085 | 3.37375 | 4.65449 | 5.80177 | 11.1281 |
| 10 | 2.46058 | 3.21717 | 4.37141 | 5.38581 | 9.92561 |
| 11 | 2.38906 | 3.09461 | 4.15334 | 5.06921 | 9.04662 |
| 12 | 2.33102 | 2.99612 | 3.98037 | 4.82057 | 8.37881 |
| 13 | 2.28297 | 2.91526 | 3.83993 | 4.62036 | 7.85572 |
| 14 | 2.24255 | 2.84772 | 3.72369 | 4.45582 | 7.43576 |
| 15 | 2.20808 | 2.79046 | 3.62594 | 4.31827 | 7.09168 |
| 16 | 2.17832 | 2.74131 | 3.54261 | 4.20163 | 6.80493 |
| 17 | 2.15239 | 2.69865 | 3.47074 | 4.10150 | 6.56249 |
| 18 | 2.12958 | 2.66130 | 3.40814 | 4.01463 | 6.35497 |
| 19 | 2.10936 | 2.62831 | 3.35313 | 3.93857 | 6.17542 |
| 20 | 2.09132 | 2.59897 | 3.30441 | 3.87142 | 6.01860 |
| 25 | 2.02406 | 2.49041 | 3.12587 | 3.62717 | 5.46168 |
| 30 | 1.98033 | 2.42052 | 3.01243 | 3.47347 | 5.12225 |
| 35 | 1.94962 | 2.37178 | 2.93401 | 3.36793 | 4.89418 |
| 40 | 1.92687 | 2.33585 | 2.87658 | 3.29101 | 4.73056 |
| 45 | 1.90935 | 2.30827 | 2.83271 | 3.23247 | 4.60753 |
| 50 | 1.89543 | 2.28643 | 2.79812 | 3.18643 | 4.51167 |
| 60 | 1.87472 | 2.25405 | 2.74703 | 3.11867 | 4.37205 |
| 70 | 1.86004 | 2.23119 | 2.71112 | 3.07120 | 4.27529 |
| 80 | 1.84911 | 2.21419 | 2.68451 | 3.03611 | 4.20429 |
| 90 | 1.84064 | 2.20105 | 2.66400 | 3.00910 | 4.14999 |
| 100 | 1.83389 | 2.19060 | 2.64770 | 2.98768 | 4.10712 |
| 120 | 1.82381 | 2.17500 | 2.62345 | 2.95585 | 4.04374 |
| 140 | 1.81663 | 2.16393 | 2.60626 | 2.93333 | 3.99915 |
| 160 | 1.81127 | 2.15566 | 2.59345 | 2.91656 | 3.96606 |
| 180 | 1.80711 | 2.14924 | 2.58353 | 2.90359 | 3.94055 |
| 200 | 1.80378 | 2.14413 | 2.57562 | 2.89326 | 3.92027 |

| DF for denominator | Size (Significance level) | | | | |
|---|---|---|---|---|---|
| | 0.1 | 0.05 | 0.02 | 0.01 | 0.001 |
| 1 | 59.4390 | 238.883 | 1494.99 | 5981.07 | 598144 |
| 2 | 9.36677 | 19.3710 | 49.3734 | 99.3742 | 999.375 |
| 3 | 5.25167 | 8.84523 | 17.0071 | 27.4892 | 130.619 |
| 4 | 3.95493 | 6.04104 | 10.1622 | 14.7989 | 48.9962 |
| 5 | 3.33927 | 4.81831 | 7.50295 | 10.2893 | 27.6495 |
| 6 | 2.98303 | 4.14680 | 6.14148 | 8.10165 | 19.0303 |
| 7 | 2.75157 | 3.72572 | 5.32733 | 6.84004 | 14.6340 |
| 8 | 2.58934 | 3.43810 | 4.78999 | 6.02887 | 12.0455 |
| 9 | 2.46940 | 3.22958 | 4.41045 | 5.46712 | 10.3680 |
| 10 | 2.37715 | 3.07165 | 4.12882 | 5.05669 | 9.20414 |
| 11 | 2.30399 | 2.94799 | 3.91189 | 4.74446 | 8.35478 |
| 12 | 2.24457 | 2.83856 | 3.73984 | 4.49936 | 7.71035 |
| 13 | 2.19534 | 2.76691 | 3.60015 | 4.30206 | 7.20614 |
| 14 | 2.15390 | 2.69867 | 3.48453 | 4.13994 | 6.80173 |
| 15 | 2.11852 | 2.64079 | 3.38729 | 4.00445 | 6.47067 |
| 16 | 2.08798 | 2.59109 | 3.30440 | 3.88957 | 6.19498 |
| 17 | 2.06133 | 2.54795 | 3.23290 | 3.79096 | 5.96204 |
| 18 | 2.03788 | 2.51015 | 3.17061 | 3.70542 | 5.76276 |
| 19 | 2.01709 | 2.47677 | 3.11587 | 3.63052 | 5.59043 |
| 20 | 1.99853 | 2.44706 | 3.06738 | 3.56441 | 5.43999 |
| 25 | 1.92924 | 2.33705 | 2.88966 | 3.32393 | 4.90626 |
| 30 | 1.88412 | 2.26616 | 2.77669 | 3.17262 | 4.58142 |
| 35 | 1.85239 | 2.21667 | 2.69856 | 3.06871 | 4.36336 |
| 40 | 1.82886 | 2.18017 | 2.64133 | 2.99298 | 4.20703 |
| 45 | 1.81071 | 2.15213 | 2.59760 | 2.93534 | 4.08954 |
| 50 | 1.79629 | 2.12992 | 2.56311 | 2.89000 | 3.99804 |
| 60 | 1.77482 | 2.09696 | 2.51216 | 2.82328 | 3.86482 |
| 70 | 1.75960 | 2.07369 | 2.47634 | 2.77653 | 3.77254 |
| 80 | 1.74825 | 2.05637 | 2.44978 | 2.74196 | 3.70486 |
| 90 | 1.73945 | 2.04298 | 2.42931 | 2.71536 | 3.65311 |
| 100 | 1.73244 | 2.03232 | 2.41304 | 2.69426 | 3.61226 |
| 120 | 1.72195 | 2.01642 | 2.38883 | 2.66290 | 3.55188 |
| 140 | 1.71449 | 2.00512 | 2.37167 | 2.64072 | 3.50940 |
| 160 | 1.70891 | 1.99669 | 2.35887 | 2.62420 | 3.47790 |
| 180 | 1.70458 | 1.99014 | 2.34896 | 2.61141 | 3.45360 |
| 200 | 1.70112 | 1.98492 | 2.34106 | 2.60123 | 3.43429 |

**The F Distribution.** (*cont'd.*) **Number of degrees of freedom for numerator: 10**

| DF for denominator | Size (Significance level) | | | | |
|---|---|---|---|---|---|
| | 0.1 | 0.05 | 0.02 | 0.01 | 0.001 |
| 1 | 60.1950 | 241.882 | 1513.69 | 6055.85 | 605621 |
| 2 | 9.39157 | 19.3959 | 49.3984 | 99.3992 | 999.400 |
| 3 | 5.23041 | 8.78552 | 16.8596 | 27.2287 | 129.247 |
| 4 | 3.91987 | 5.96437 | 10.0027 | 14.5459 | 48.0526 |
| 5 | 3.29740 | 4.73506 | 7.34380 | 10.0510 | 26.9166 |
| 6 | 2.93693 | 4.05996 | 5.9839 | 7.87411 | 18.4109 |
| 7 | 2.70251 | 3.63652 | 5.17113 | 6.62006 | 14.0833 |
| 8 | 2.53803 | 3.34716 | 4.63483 | 5.81429 | 11.5401 |
| 9 | 2.41631 | 3.13728 | 4.25605 | 5.25654 | 9.89430 |
| 10 | 2.32260 | 2.97823 | 3.97497 | 4.84914 | 8.75386 |
| 11 | 2.24822 | 2.85362 | 3.75843 | 4.53928 | 7.92239 |
| 12 | 2.18776 | 2.75338 | 3.58667 | 4.29605 | 7.29202 |
| 13 | 2.13763 | 2.67102 | 3.44719 | 4.10026 | 6.79916 |
| 14 | 2.09539 | 2.60215 | 3.33172 | 3.93939 | 6.40406 |
| 15 | 2.05931 | 2.54371 | 3.23458 | 3.80493 | 6.08077 |
| 16 | 2.02814 | 2.49351 | 3.15175 | 3.69093 | 5.81166 |
| 17 | 2.00093 | 2.44991 | 3.08030 | 3.59306 | 5.58437 |
| 18 | 1.97698 | 2.44170 | 3.01803 | 3.50816 | 5.38997 |
| 19 | 1.95572 | 2.37793 | 2.96330 | 3.43381 | 5.22191 |
| 20 | 1.93673 | 2.34787 | 2.91482 | 3.36818 | 5.07524 |
| 25 | 1.86578 | 2.23647 | 2.73701 | 3.12940 | 4.55513 |
| 30 | 1.81948 | 2.16457 | 2.62388 | 2.97909 | 4.23879 |
| 35 | 1.78688 | 2.11433 | 2.54560 | 2.87583 | 4.02652 |
| 40 | 1.76268 | 2.07724 | 2.48823 | 2.80054 | 3.87438 |
| 45 | 1.74400 | 2.04873 | 2.44437 | 2.74322 | 3.76006 |
| 50 | 1.72914 | 2.02614 | 2.40975 | 2.69813 | 3.67105 |
| 60 | 1.70700 | 1.99259 | 2.35861 | 2.63175 | 3.54147 |
| 70 | 1.69129 | 1.96887 | 2.32263 | 2.58522 | 3.45173 |
| 80 | 1.67956 | 1.95122 | 2.29595 | 2.55081 | 3.38591 |
| 90 | 1.67047 | 1.93756 | 2.27537 | 2.52432 | 3.33558 |
| 100 | 1.66322 | 1.92669 | 2.25902 | 2.50331 | 3.29586 |
| 120 | 1.65237 | 1.91046 | 2.23467 | 2.47207 | 3.23716 |
| 140 | 1.64465 | 1.89892 | 2.21741 | 2.44997 | 3.19586 |
| 160 | 1.63887 | 1.89030 | 2.20453 | 2.43351 | 3.16524 |
| 180 | 1.63438 | 1.88361 | 2.19456 | 2.42077 | 3.14161 |
| 200 | 1.63080 | 1.87828 | 2.18661 | 2.41062 | 3.12284 |

**The F Distribution.** (*cont'd.*) **Number of degrees of freedom for numerator: 20**

| DF for denominator | Size (Significance level) | | | | |
|---|---|---|---|---|---|
| | 0.1 | 0.05 | 0.02 | 0.01 | 0.001 |
| 1 | 61.7403 | 248.013 | 1551.92 | 6208.73 | 620908 |
| 2 | 9.44130 | 19.4458 | 49.4483 | 99.4491 | 999.450 |
| 3 | 5.18448 | 8.66018 | 16.5533 | 26.6898 | 126.418 |
| 4 | 3.84433 | 5.80254 | 9.66958 | 14.0196 | 46.1003 |
| 5 | 3.20665 | 4.55813 | 7.00938 | 9.55264 | 25.3946 |
| 6 | 2.83633 | 3.87418 | 5.65089 | 7.39583 | 17.1201 |
| 7 | 2.59473 | 3.44452 | 4.83928 | 6.15543 | 12.9316 |
| 8 | 2.42463 | 3.15032 | 4.30357 | 5.35909 | 10.4797 |
| 9 | 2.29832 | 2.93645 | 3.92492 | 4.80799 | 8.89761 |
| 10 | 2.20074 | 2.77401 | 3.64365 | 4.40539 | 7.80374 |
| 11 | 2.12304 | 2.64644 | 3.42671 | 4.09904 | 7.00759 |
| 12 | 2.05967 | 2.54358 | 3.25440 | 3.85843 | 6.40480 |
| 13 | 2.00698 | 2.45888 | 3.11429 | 3.66460 | 5.93397 |
| 14 | 1.96245 | 2.38789 | 2.99813 | 3.50522 | 5.55683 |
| 15 | 1.92431 | 2.32753 | 2.90028 | 3.37189 | 5.24842 |
| 16 | 1.89127 | 2.27556 | 2.81673 | 3.25873 | 4.99180 |
| 17 | 1.86236 | 2.23035 | 2.74456 | 3.16151 | 4.77513 |
| 18 | 1.83684 | 2.19064 | 2.68159 | 3.07709 | 4.58986 |
| 19 | 1.81415 | 2.15549 | 2.62617 | 3.00310 | 4.42972 |
| 20 | 1.79384 | 2.12415 | 2.57701 | 2.93773 | 4.28996 |
| 25 | 1.71752 | 2.00747 | 2.39615 | 2.69932 | 3.79436 |
| 30 | 1.66730 | 1.93165 | 2.28050 | 2.54865 | 3.49278 |
| 35 | 1.63171 | 1.87837 | 2.20013 | 2.44480 | 3.29026 |
| 40 | 1.60515 | 1.83885 | 2.14101 | 2.36887 | 3.14498 |
| 45 | 1.58455 | 1.80837 | 2.09568 | 2.31092 | 3.03573 |
| 50 | 1.56810 | 1.78412 | 2.05982 | 2.26524 | 2.95060 |
| 60 | 1.54348 | 1.74798 | 2.00666 | 2.19780 | 2.82655 |
| 70 | 1.52592 | 1.72232 | 1.96913 | 2.15040 | 2.74052 |
| 80 | 1.51276 | 1.70316 | 1.94123 | 2.11527 | 2.67737 |
| 90 | 1.50253 | 1.68829 | 1.91966 | 2.08817 | 2.62905 |
| 100 | 1.49434 | 1.67643 | 1.90248 | 2.06664 | 2.59088 |
| 120 | 1.48207 | 1.65868 | 1.87686 | 2.03458 | 2.53441 |
| 140 | 1.47330 | 1.64602 | 1.85865 | 2.01186 | 2.49466 |
| 160 | 1.46672 | 1.63655 | 1.84505 | 1.99490 | 2.46515 |
| 180 | 1.46160 | 1.62919 | 1.83450 | 1.98177 | 2.44238 |
| 200 | 1.45751 | 1.62330 | 1.82608 | 1.97129 | 2.42428 |

**The F Distribution.** *(cont'd.)* **Number of degrees of freedom for numerator: 30**

| DF for denominator | Size (Significance level) | | | | |
|---|---|---|---|---|---|
| | 0.1 | 0.05 | 0.02 | 0.01 | 0.001 |
| 1 | 62.2650 | 250.0950 | 1564.90 | 6260.65 | 626099 |
| 2 | 9.45792 | 19.4624 | 49.4650 | 99.4758 | 999.467 |
| 3 | 5.16811 | 8.61657 | 16.4477 | 26.5045 | 125.449 |
| 4 | 3.81742 | 5.74587 | 9.55399 | 13.8377 | 45.4286 |
| 5 | 3.17408 | 4.49571 | 6.89258 | 9.37932 | 24.8688 |
| 6 | 2.79996 | 3.80816 | 5.53384 | 7.22853 | 16.6722 |
| 7 | 2.55545 | 3.37580 | 4.72192 | 5.99201 | 12.5304 |
| 8 | 2.38301 | 3.07940 | 4.18573 | 5.19812 | 10.1087 |
| 9 | 2.25472 | 2.86365 | 3.80649 | 4.64858 | 8.54755 |
| 10 | 2.15542 | 2.69955 | 3.52453 | 4.24693 | 7.46879 |
| 11 | 2.07621 | 2.57048 | 3.30687 | 3.94113 | 6.68394 |
| 12 | 2.01149 | 2.46627 | 3.13381 | 3.70078 | 6.08983 |
| 13 | 1.95757 | 2.38033 | 2.99294 | 3.50704 | 5.62583 |
| 14 | 1.91193 | 2.30820 | 2.87604 | 3.34759 | 5.25415 |
| 15 | 1.87277 | 2.24678 | 2.77746 | 3.21411 | 4.95018 |
| 16 | 1.83879 | 2.19384 | 2.69320 | 3.10073 | 4.69722 |
| 17 | 1.80901 | 2.14770 | 2.62034 | 3.00324 | 4.48359 |
| 18 | 1.78268 | 2.10714 | 2.55671 | 2.91851 | 4.30088 |
| 19 | 1.75924 | 2.07118 | 2.50064 | 2.84420 | 4.14290 |
| 20 | 1.73822 | 2.03908 | 2.45086 | 2.77848 | 4.00499 |
| 25 | 1.65894 | 1.91918 | 2.26725 | 2.53830 | 3.51549 |
| 30 | 1.60647 | 1.84087 | 2.14933 | 2.38596 | 3.21709 |
| 35 | 1.56909 | 1.78559 | 2.06708 | 2.28062 | 3.01634 |
| 40 | 1.54107 | 1.74443 | 2.00637 | 2.20338 | 2.87210 |
| 45 | 1.51926 | 1.71256 | 1.95970 | 2.14428 | 2.76346 |
| 50 | 1.50179 | 1.68715 | 1.92268 | 2.09759 | 2.67869 |
| 60 | 1.47553 | 1.64914 | 1.86763 | 2.02847 | 2.55494 |
| 70 | 1.45672 | 1.62203 | 1.82864 | 1.97974 | 2.46894 |
| 80 | 1.44257 | 1.60173 | 1.79956 | 1.94352 | 2.40570 |
| 90 | 1.43154 | 1.58593 | 1.77702 | 1.91553 | 2.35724 |
| 100 | 1.42269 | 1.57330 | 1.75905 | 1.89325 | 2.31890 |
| 120 | 1.40937 | 1.55434 | 1.73216 | 1.86000 | 2.26212 |
| 140 | 1.39983 | 1.54079 | 1.71301 | 1.83637 | 2.22207 |
| 160 | 1.39266 | 1.53061 | 1.69867 | 1.81871 | 2.19231 |
| 180 | 1.38707 | 1.52270 | 1.68753 | 1.80502 | 2.16932 |
| 200 | 1.38258 | 1.51636 | 1.67863 | 1.79408 | 2.15102 |

# APPENDIX 3

# UNBIASED ESTIMATORS OF MOMENTS FROM A SAMPLE OF SIZE $n$

**Raw moments of all orders of $X$**

$$\hat{\mu}_r' = \frac{\sum\limits_i x_i^r}{n}$$

**Central moments and cumulants**

First moment: always zero.

First mixed moment (covariance between $X$ and $Y$):

$$\hat{\sigma}_{XY} = \frac{\sum\limits_i (x_i - m_X)(y_i - m_Y)}{n - 1} = \frac{\sum x_i y_i - \sum x_i \sum y_i / n}{n - 1}$$

Second moment and cumulant (variance) of $X$:

$$\hat{\sigma}^2 = \frac{\sum\limits_i (x_i - m_X)^2}{n - 1} = \frac{\sum\limits_i x_i^2 - [\sum x_i]^2 / n}{n - 1}$$

Third moment and cumulant of $X$:

$$\hat{\mu}_3 = \frac{n \sum\limits_i (x_i - m)^3}{(n - 1)(n - 2)}$$

$$= \frac{n(\hat{\mu}_3' - 3m\hat{\mu}_2' + 2m^3)}{(n - 1)(n - 2)}$$

# GLOSSARY

**Acceptance region:** An arbitrary segment (or sometimes multiple segments) of the sample space of a random variable comprising a proportion of the distribution under the null hyposthesis (q.v.). It is the complement of the size of a test (q.v.). If the sample statistic falls within the acceptance region the null hypothesis is "accepted." In Neyman-Pearson hypothesis testing (q.v.) the region is chosen so as to maximize the power for some specified class of alternative hypotheses.

**Accuracy:** The smaller the absolute value of the bias of a (not necessarily efficient) estimator (q.v.) the greater its accuracy. (*CF.* precision.)

**Alleles:** Diverse types of genes that may occupy the same locus (q.v.).

**Asymptotic:** In the statistical context, used of a property (e.g., unbiasedness, efficiency, Gaussian normality) of an estimator (q.v.) that is more closely attained as the sample size increases and in the limit is exactly attained.

**Average:** *See* expectation.

**Bayes's theorem:** The posterior probability of a hypothesis, given a set of data, is proportional to the product of the probability of the data given the hypothesis, and the prior probability of the hypothesis being true. For details see (3).

**Bernoulli trial:** A random variable that may assume only two values, 0 and 1. (The outcome associated with the latter is arbitrarily termed a *success,* with the former, a *failure.*)

**Bias (of an estimator):** The expected value of the estimator of a parameter minus the true value of the parameter.

**Binomial variable of order** $n$: The convolution (q.v.) of $n$ independent Bernoulli trials (q.v.) all having the same probability of success.

**Bioassay:** The assessment of the action of a drug or other treatment by biological rather than chemical or physical means. The so-called *indirect* method consists of giving fixed doses of the drug and finding what proportion of the subjects show some qualitative result (usually death, or sometimes cure). *Hence* in the statistical context, bioassay connotes the study of quantal (q.v.) traits where the probability of an outcome is a continuous function of one or more regressor variables.

**Black box:** An object characterized by the empirical relationship between externally applied manipulations and externally observed responses, without any attempt to speculate about its inner structure or workings.

**Chi-square distribution with 1 degree of freedom:** The distribution of the square of a standard normal variate (q.v.).

**Chi-square distribution with $n$ degrees of freedom:** The convolution (q.v.) of $n$ independent chi-square variates each with one degree of freedom.

**Cofactor** of an element of a square matrix is the determinant (suitably signed) of the submatrix remaining after the row and column in which the element occurs have been deleted. It is multiplied by $(-1)$ if the sum of the subscripts of the element is odd.

**Conditional probability:** A probability (q.v.) that applies when some condition is imposed on some well-defined sample space. For example, "the probability of being over six feet tall *if a man*" is a conditional probability about height (conditioned on sex).

**Confidence interval:** An interval about a point estimate of a parameter within which (with some specified degree of assurance) the statistician asserts the parameter to lie. The statement, if it were made before the data were collected, would be a probability.

**Conformable:** (used of vectors and matrices) having the appropriate relationship between their dimensions such that the one can operate on the other. They are conformable for addition if they have identical dimensions; they are conformable for multiplication if the contiguous dimensions are equal.

**Confounding:** The relationship in an experiment between two factors, the several effects of which cannot be logically separated by analysis of the data. Confounding may be deliberate, avoidable, or neither (see reference 3).

**Conservative:** Used of a statistical criterion that understates some merit of a procedure. For instance a conservative interval-estimate is one that has a greater probability than it purports to have of including an estimated parameter. (*Contrast with* radical.)

**Consistent estimator:** If for an estimator of a parameter an integer, $n$, can be found such that for random samples of size $n$ or greater the probability is arbitrarily small that the estimate will not be arbitrarily close to the true value, the estimator is consistent. A sufficient, but not a necessary, set of conditions for consistency is that in the limit the bias and the variance are both zero.

**Continuous distribution:** A distribution (q.v.) for which the probabilities of all particular values are zero, but some values (which comprise the *domain*) have a probability density (q.v.) greater than zero.

**Continuous function:** A function that may be represented graphically without lifting the pen; one for which the limiting value as the argument approaches any particular point equals the function evaluated at that point.

**Convolution** of two or more random variables is the distribution of their sum.

**Correlation:** A relationship between two random variables, $X$ and $Y$, such that the expectation of the product of their deviations from their respective means is not zero. A correlation of zero is a necessary, but not a sufficient condition for independence (q.v.). (*Contrast with* regression.)

**Covariance:** The average, or expected, value of the product of the signed deviations of two random variables from their respective means. A covariance of zero is a necessary, but not a sufficient, condition for statistical independence (q.v.) of two variates.

**Cumulants:** A set of quantities related to the *central moments* of a variate (*see* moment of a random variable) and used as criteria for identifying the distribution to which it belongs.

**(Cumulative) distribution function**, written $F(x)$, is a formula stating the probability that the random variable $X$ assumes the value $x$ or less.

**Dependence:** The contrary of independence (q.v.).

**Determinant** of a (square) matrix, **A**, and written $|\mathbf{A}|$, is the sum of all possible products each comprising one and only one element from each row and column and signed negatively if the number of inversions of all possible pairs of subscripts of the elements is odd.

**Diachronic:** Applied to data observed over a period of time (*contrast with* synchronic). It may give rise to the problems associated with a time series (q.v.).

**Diagonal matrix:** A square matrix of which all the elements except for those on the principal diagonal (q.v.) are zero.

**Dichotomous:** Having two, and only two, types of outcome. (Less commonly, the word *trichotomous* is used if there are three possible outcomes, or *polytomous* if there are many.)

**Discrete distribution:** A distribution in which no value has a finite probability density greater than zero. (This implies that all "possible" values of the random variable have a probability greater than zero.)

**Discrimination:** The capacity to distinguish between members of two groups on the basis of their statistical characteristics (*cf.* power).

**Dishonest:** Used of a random variable that has a nonzero probability of assuming an infinite value. The term is usually applied to the outcome of a stochastic (q.v.) process.

**Distribution, Probability:** A predictive statement about the outcome of a single realization of a random process, comprising a random variable, the set of values that it may assume, and the probability or probability density associated with each.

**Domain:** The set of values that a random variable (q.v.) may assume (*contrast with* range).

**Dominant:** Used in genetics of the phenotype (q.v.) of a gene that obscures the phenotype of another allele (q.v.).

**Efficiency** in a consistent estimator is measured by its precision. If the variance of the estimator attains the lower bound (q.v.) for at least large samples, it is "absolutely efficient." If it attains the lower bound for all sizes of samples it is "always absolutely efficient." If it is not absolutely efficient but has a smaller variance than any other estimator, it is "relatively efficient."

**Eigenvalue:** An element on the main diagonal in the diagonal form of a matrix $A$. More generally, any root of $\lambda$ in the equation $|A - I\lambda| = 0$ is an eigenvalue of $A$.

**Element** (used of matrices and vectors): A scalar quantity occupying a single location in an array of quantities.

**Error of the first kind:** The probabilistic decision to reject the null hypothesis (q.v.) when it is true.

**Error of the second kind:** The probabilistic decision to accept the null hypothesis when it is false.

**Estimate:** A guess (after the event) at the value of an unknown parameter generated from a particular realization of a random process by means of an estimator. Estimates are not random variables. They are written as lowercase letters, commonly the Roman equivalent of the Greek letters used to denote parameters (e.g., $r$ for $\rho$, $m$ for $\mu$, etc.).

**Estimator:** A systematic procedure for obtaining an estimate (q.v.). An estimator, written symbolically as an uppercase letter, is a random variable. Also it usually has an expectation, variance, bias, and other statistical properties.

**Expectation** (also the average or mean value) of a random variable, written $E(.)$, is the sum of the products of each value it may assume and the probability associated with it. (For a continuous variable read *integral* for *sum* and *probability density* for *probability.*)

**$n$-Factorial, or $n!$:** Used of a positive integer, $n$, is the product of $n$ and each of the positive integers less than it. (Noninteger values of $n$ call for a more complicated definition.)

**Failure:** *See* success.

**False negative:** Used in connection with a dichotomous variable (one state being "positive" and one "negative") for a positive result incorrectly identified as negative.

**False positive:** The converse of a false negative (q.v.).

**Full rank:** A matrix with an inverse, i.e., one that is not singular (q.v.), is said to be "of full rank."

**Gaussian distribution:** A continuous distribution with infinite limits having the probability density function given on p. 10.

**Gauss-Markov theorem:** Shows that under specified assumptions (see p. 479) the least squares estimator is unbiased and has the lowest variance among all possible linear combinations of the data.

**Genotype:** The genetic constitution of a patient, commonly, but not necessarily, used of a single locus (q.v.). *Cf.* phenotype. McKusick has paraphrased the difference: "Phenotype is reputation; genotype is character."

**Hazard** of a continuous variate, $X$, is the ratio of its probability density to the complement of its distribution function, $f(x)/[1 - F(x)]$. Where diseases are concerned, over a short period of time the hazard is approximated by the expected incidence per unit time of new cases among those previously unaffected.

**Histogram:** A diagram in which the numbers of realizations of a random variate falling within a set of assigned intervals are proportionally represented by the areas of rectangles contained in the interval. (It is commonly, but not necessarily, used of empirical data.) If the histogram has unit area, the intervals are small, and the data are representative, the histogram may approximate the probability distribution.

**Homoscedastic:** Having equal variances. (Used in connection with comparisons among groups.)

**Hypothesis:** Any formal surmise designed for empirical testing. In the statistical context, hypotheses always deal with the values of parameters.

**Idempotent matrix:** A matrix that is unchanged by raising it to any finite positive power.

**Identity matrix:** A diagonal matrix of which all the elements on the main diagonal are unity. Any matrix conformable for multiplication is unchanged when pre- or post-multiplied by it.

**Incidence:** The rate (usually per unit time) at which eligible members of a population are converted from one state to another (usually from health to disease).

**Independence:** (1) In a conceptual sense, two events are independent if there is no correspondence (e.g., causal or associative) between them. (2) Statistically, two events are independent if their joint probability (or joint probability density) is equal to the product of their individual probabilities (or probability densities).

**Interaction:** (1) In a conceptual sense, an interaction between two processes exists if their joint effect is other than would have been predicted from an analysis of their separate effects. It is invariant under any nondegenerate transformation. (2) Statistically, interaction between two processes means that their joint effect differs from the sum of their individual effects. It is not necessarily invariant under nonlinear transformation.

**Inverse of a matrix** is that matrix, written $A^{-1}$, such that the products $A^{-1}A$ and $AA^{-1}$ are identity matrices.

**Inversion:** Any perturbation of a natural ordering. Thus in the sequence 12354, "54" represents an inversion of the natural order "45." Each possible combination of elements may represent an individual inversion. For instance 12543 exhibits three inversions: 5 before 4, 5 before 3, and 4 before 3. This pattern is of major interest in nonparametric statistics. Two elements in a matrix exhibit an inversion if the orderings of their corresponding subscripts are not the same. Thus the elements with subscripts 24 and 36 respectively show no inversion, whereas those with subscripts 24 and 31 do, the first element of the former being less than that of the latter, but conversely for the second elements. Clearly if one element lies below and to the left of the other, there is an inversion.

**Kurtosis** of a distribution is a conventional criterion of the conformity of it to the Gaussian (q.v.). The criterion is the ratio of the fourth central moment to the square of the variance. For the Gaussian variate, this ratio is 3. Anything more is called *positive kurtosis,* anything less, *negative kurtosis.*

**Lexis variate:** the convolution (q.v.) of independent but nonidentical Bernoulli trials (q.v.).

**Likelihood:** A statistical term denoting the plausibility of a surmised set of parameters, given a particular model and a set of data. Formally, the same expression is a *probability* when viewed as a function of the data, and a *likelihood* when viewed as a function of the parameters. Some writers use the terms interchangeably.

**Linkage:** The relationship between two loci (q.v.) such that regardless of the alleles involved, the genes transmitted from parent to child are not independently inherited.

**Locus** (*plural* **loci**): The pair of regions of two homologous chromosomes occupied by a single pair of alleles. (In the X chromosome of the male it is an unpaired structure.)

**Lower bound of the variance of an estimator:** Under weak (i.e., undemanding) assumptions there is, for any given size of sample, a lower limit to the variance of a consistent estimator of a parameter. It is the inverse of minus the expectation of the second derivative of the natural logarithm of the likelihood with respect to the parameter. The principle generalizes readily to several parameters where *inverse* is applied to the matrix of second, and mixed partial, derivatives.

**Lyonization:** In the mammalian female (who normally has two $X$ chromosomes) only one gene per X-linked locus may be phenotypically expressed, the other being inactivated or *Lyonized*. Which gene is inactivated is apparently randomly selected.

**Matrix:** A rectangular array of separate scalar elements.

**Mean:** *See* expectation.

**Median:** The fiftieth percentile of a continuous distribution. It may not be unique.

**Mode:** The value or values of a discrete random variable at which the probability is higher than for any adjacent value. For a continuous distribution an analogous definition is expressed in probability density. It may not be unique.

**Moment of a random variable:** The sum (or integral) of a function multiplied by the probability or probability density. Usually the function consists of the random variable minus some arbitrary constant, $a$, all raised to some particular power, $r$. If $a = 0$, we speak of *raw moments*. If $a$ is the expectation (q.v) of the random variable, we speak of *central moments*. The order of moment (second, third, etc.) is denoted by the value of $r$.

**Monotonic:** If for a function, $g(x)$, of a variable there exist *both* some values of $x$ such that the smallest possible nonzero increase in $x$ leads to an increase in $g(x)$ *and* other values such that a like increase in $x$ leads to a decrease in $g(x)$, then $g(x)$ is a nonmonotonic function of $x$. Otherwise it is monotonic. For example, the function $g(x) = sin(x)$ is nonmonotonic. The natural logarithm of $x$ is a monotonic *increasing*, the reciprocal is a monotonic *decreasing*, function.

**Neyman-Pearson hypothesis testing:** A method of proof of a hypothesis about the value of a parameter where rational prior probabilities are not available; the test is to be based on the data only and there is one hypothesis "in possession" (the null hypothesis), that will be rejected in favor of an alternative hypothesis, should it prove a sufficiently implausible explanation for the data against which it is to be tested. The system is used as proof of a model by goodness of fit, or of the truth of some contrary assertion by badness of fit *("reductio ad absurdum")*.

**Normalize:** Used in several senses. *(a)* To scale a probability space in such a way that the probability of the whole is unity (as in the last step of a Bayesian analysis). *(b)* To convert a continuous random variable into Gaussian form by transformation (see Chapter 4). *(c)* To rescale the elements of a vector, **b**, such that it is of unit length, ie., $\mathbf{b}'\mathbf{b} = 1$.

**Null hypothesis:** A conjecture about the value of some unknown parameter. It is given a privileged status at the start of a Neyman-Pearson test of hypothesis (q.v.) in that the onus of proof rests on those that hold any contrary hypothesis. If it is rejected as untenable, the alternative hypothesis will be accepted.

**Objective indeterminacy:** Uncertainty (about some state or event) that is in the nature of knowledge (principally because the event has not yet occurred).

**Orthogonal matrix:** A matrix of which the transpose (q.v.) and the inverse (q.v.) are equal.

**Outcome variable:** *See* response variable.

**Pathognomonic:** The finding $A$ is pathognomonic of the disease $B$ if $A$ occurs in no other condition, i.e., $A$ is a sufficient condition for the diagnosis of $B$.

**Percentile:** For any particular distribution, $x$ is the $a^{\text{th}}$ percentile if $F(x) = a/100$.

**Phenotype:** The appearance of one of a class of subjects of variable genotypes (q.v.).

The phenotype (unlike the genotype) varies with the sophistication of the method of observation.

**Point estimate:** A single value that, in some sense, is the most apt guess at a parameter.

**Population** (also Universe): The group of elements on which a sampling process operates.

**Power of a test:** The complement of the probability of an error of the second kind (q.v.).

**Precision:** The smaller the variance of a (not necessarily unbiased or consistent) estimator, the greater the precision.

**Prevalence:** The proportion of a population that exhibit a characteristic. Unlike incidence (q.v.) it is not a rate (with respect, for instance, to time).

**Principal diagonal:** Those elements of a square matrix having the two conventional subscripts (denoting row and column of the element, respectively) equal; those elements on the line from the extreme top left to the extreme bottom right.

**Prior probability:** The probability (of an outcome of a random process or, less strictly, of the value of an unknown parameter) that embodies all the information that is deemed pertinent before the process is realized.

**Probability:** A measure of the assurance about the occurrence of some particular event on a particular sample space. There are two criteria in  common use: the proportion of outcomes in a very large sample of realizations of the process in which the event occurs (the *objective* criterion); and the odds that an ideally rational, well-informed man would be prepared to give or take on the outcome occurring (the *subjective* criterion). Some writers distinguish a third type, *personal* probability, which comprises idiosyncratic components of conviction that cannot be explicated.

**Probability density:** Suppose that for a continuous distribution we draw a curve in such a way that the area bounded by vertical lines at any two arbitrarily chosen distinct values of the variate, and the segments of the curve and the baseline between them correspond to the probability that the random variable will assume some value between the two points. Then the height of the curve at any one value is the probability density of that value. (The actual probability of any particular value of a continuous variate is, by definition of a continuous variate, zero. Hence the need for this elaborate definition.)

**Proband:** A unit (usually a person), exhibiting some characteristic of interest, through whom a subset (e.g., family, epidemic) is first brought to the attention of an inquirer.

**Quantal:** Used of data that may assume two possible states only (*cf.* Bernoulli trial).

**Quartile:** The $25^{th}$ (lower quartile) or $75^{th}$ (upper quartile) percentiles of a continuous distribution.

**Radical:** The converse of conservative (q.v.).

**Random variable:** A quantity that varies in accordance with some probability distribution.

**Range:** The set of values that a function of a variable (e.g., probability density function) may assume. (*Contrast* domain. The domain of a binomial variate of order $n$ is the set of integers 0, 1, . . . , $n$. The range is the semiopen interval (0,P] where P is the modal probability.)

**Rank:** *(a)* Of a sample value, the position occupied by that value when all the values in the sample are arranged in ascending order of magnitude; *(b)* of a (square) matrix, the dimension of the largest nonsingular submatrix obtained by deleting selected rows and columns.

**Reference set:** Sample space or universe (q.v.). The term is used in probability algebra rather than in statistical theory.

**Regression:** The relationship of a random variable (the response variable) to another variable (the regressor variable, q.v.) on which it is conditioned.

**Regressor variable:** A mathematical variable that may also be a random variable, from particular values of which another (random) variable (the response variable) is to be predicted.

**Rejection region:** The complement of the acceptance region (q.v.). If the sample value falls within it, the null hypothesis will be rejected in favor of the alternative hypothesis.

**Replication:** One of a number of observations made under conditions that are, in some sense, considered identical. In principle, variation among replicates is ascribed to pure error of observation or measurement, and not to any systematic effects.

**Residual error:** That putatively pure random error in data after all systematic effects have been taken into account.

**Response variable:** A random variable which is being predicted from a regressor variable (q.v.).

**Robustness** of a statistical procedure is the degree to which it is reliable despite departures of an area of application from the strict assumptions on which the procedure is logically based. (For example, the Gaussian distribution is very *robust* when applied to sample means.)

**Sample space:** The set of all possible outcomes from a random experiment. In a statistical (as distinct from a probabilistic) context, it is usually referred to as a universe (q.v.).

**Scalar:** A number (of however many digits) that represents a single quantity; an element in a vector or matrix.

**Scalar multiplication** of a matrix or a vector by a scalar quantity $k$ consists of multiplying every element by $k$. It is equivalent to multiplying by a conformable diagonal matrix of which all diagonal elements are $k$.

**Sensitivity:** The probability that a test will identify as positive any randomly chosen member of a population of positive elements.

**Significance level:** The (random) size of the test where the rejection region is chosen after the event, such that the null hypothesis would just be rejected. (Some statisticians mean by significance level the same as the size of a test, q.v.)

**Singular:** Used of a square matrix with no inverse; of which at least one eigenvalue is zero; of which the determinant is zero. (These three criteria are all equivalent.)

**Size of a test:** The probability of an error of the first kind selected a priori, and independently of the data. It is not a random variable, as the significance level is.

**Skewness of a distribution** is the degree of asymmetry it exhibits. The usual criterion is the third central moment scaled by dividing by the cube of the standard deviation. If the result is positive, the skewness is termed positive, and conversely.

**Specificity** of a diagnostic test for a diseased state is the probability that any randomly chosen person in whom the test is deemed positive will in fact have the disease. (*Contrast with* sensitivity.)

**Spline:** A point, in a regression function that is everywhere continuous (q.v.), at which there is an angulation so that the gradients immediately to the left and to the right of the point are unequal.

**Standard deviation:** The square root of the variance (q.v.).

**Standard error:** A term commonly used for the standard deviation of an estimate of a parameter from random data. (Throughout this text, I have used the preferred term *standard deviation of the estimate.*)

**Standard normal variate:** A Gaussian variate (q.v.) with mean zero and a variance of unity. Any Gaussian variate may be transformed into such a variate. (*See also* chi-square distribution with one degree of freedom.)

**Stochastic:** Probabilistic; aleatory; chance. Usually used of processes represented as functions of time or distance.

**Subjective indeterminacy:** Uncertainty (about a state or about the outcome of a random process) that reflects no objective indeterminacy (q.v.) but merely the corporate ignorance of the observers.

**Success:** The opposing outcomes for a set of Bernoulli trials (q.v.) may be arbitrarily designated *successes* and *failures.* These terms imply no merit or value to a particular outcome.

**Superpopulation:** A more or less fictitious reference set from which a naturally limited set of elements is construed to be a sample. The superpopulation is a device used to rescue data from the charge of idiosyncrasy and to admit of conclusions that have some generality. (Some philosophical aspects of this notion are analyzed in detail in reference 70.)

**Symmetrical matrix:** A matrix equal to its own transpose (q.v.).

**Synchronic:** Used of a set of observations made simultaneously on all the members of a sample (*contrast* diachronic). It furnishes only indirect and specious evidence as to how the individual units would behave diachronically.

**Synteny:** The state of loci carried on the same pair of chromosomes. They may or may not exhibit linkage (q.v.)

**Thrombosis:** An aggregation on a macroscopic scale in vivo of the solid elements of the blood.

**Time series:** A diachronic (q.v.) study of a sample of repeated measurements on subjects (elements) in which there is serial correlation of residual errors in the combined sample space and from the common regression line.

**Transpose** of a matrix (or vector) is that matrix (or vector) of which each row is the corresponding column of the original, and conversely.

**Type 1 regression model:** A regression in which the regressor variable is a mathematical, but not a random, variable.

**Type 2 regression model:** A regression in which the regressor variable is a random variable.

**Universe:** *see* population; sample space.

**Validation:** A valid conclusion is one attained by formally correct means. It is not necessarily true if based on data that are erroneous for stochastic or other reasons. Validation is the process whereby the valid nature of a process is established. It should not be confused with *verification,* which is the means by which the truth of a conclusion is corroborated.

**Variance:** The mean square deviation of a random variate from its mean.

**Variate:** Random variable.

**Vector:** A column or row comprising two or more scalar (q.v.) quantities. It may be viewed as a matrix of which one of the dimensions is unity.

# REFERENCES


1. Murphy, E. A. "Classification and its alternatives." In *Clinical Judgment: A Critical Appraisal*, edited by, H. T. Engelhardt, S. F. Spicker, and B. Towers, pp. 59–85. Dordrecht: D. Reidel Publishing Co., 1979.
2. Finney, D. J. "The truncated binomial distribution." *Ann. Eugen.* (London) 14:319–28, 1949.
3. Murphy, E. A. *The Logic of Medicine*. Baltimore: Johns Hopkins University Press, 1976.
4. Feinstein, A. R. *Clinical Biostatistics*. St. Louis: Mosby, 1977.
5. Ingle, D. J. *Is It Really So? A Guide to Clear Thinking*. Philadelphia: Westminister Press, 1976.
6. Wulff, H. R. *Rational Diagnosis and Treatment*. Oxford: Blackwell, 1976.
7. Murphy, E. A. *Probability in Medicine*. Baltimore: Johns Hopkins University Press, 1979.
8. Hoel, P. *Elementary Statistics*. 2d ed. New York: Wiley, 1966.
9. Gardner, E. J. "A genetic and clinical study of intestinal polyposis, a predisposing factor for carcinoma of the colon and rectum." *Amer. J. Hum. Genet.* 3:167–76, 1951.
10. Reed, T. E., and Neel, J. V. "A genetic study of multiple polyposis of the colon (with an appendix deriving a method of estimating relative fitness)." *Amer. J. Hum. Genet.* 7:236–63, 1955.
11. Bussey, J. H. R. *Familial Polyposis Coli*. Baltimore: Johns Hopkins University Press, 1973.
12. Pavlides, G. P.; Milligan, F. D.; Clarke, D. N.; Cohen, S. B.; Wennstrom, C. J.; Burbige, E. J.; Krush, A. J.; and Murphy, E. A. "Hereditary polyposis coli: The diagnostic value of colonoscopy, barium enema, and fecal occult blood." *Cancer* 40:2632–39, 1977.
13. Fried, K., and Davies, A. M. "Some effects on the offspring of uncle-niece marriage in the Moroccan Jewish Community in Jerusalem." *Amer. J. Hum. Genet.* 26:65–72, 1974.
14. Murphy, E. A., and Chase, G. A. *Principles of Genetic Counseling*. Chicago: Year Book Publishers, 1975.
15. Stevens, S. S. "To honor Fechner and repeal his law." *Science* 133:80–86, 1961.
16. Clarke, C. A.; Donohoe, W. T. A.; McConnell, R. B.; Woodrow, J. C.; Finn, R.; Krevans, J. R.; Kulke, W.; Lehane, D.; and Sheppard, P. M. "Further experimental studies on the prevention of Rh haemolytic disease." *Brit. Med. J. 1*: 979–84, 1963.
17. Jorgenson, R. J.; Bolling, D. R.; Yoder, O. C.; and Murphy, E. A. "Blood pressure studies in the Amish." *Johns Hopkins Med. J.* 131:329–350, 1972.

18. Bøe, J.; Humerfelt, S. B.; and Wederwang, F. "The blood pressure in a population." *Acta Med. Scand.* 1957, Supp. 321.

19. Murphy, E. A.; Thomas, C. B.; and Bolling, D. R. "The precursors of hypertension and coronary disease: Statistical consideration of distributions in a population of medical students. II. Blood pressure." *Johns Hopkins Med. J.* 120:1–21, 1967.

20. Krush, A. J., and others. "Serum carcinoembryonic antigen in hereditary polyposis coli." Unpublished.

21. Kincaid-Smith, P.; McMichael, J.; and Murphy, E. A. "The clinical course and pathology of hypertension with papilloedema (malignant hypertension)." *Quart. J. Med.* 27:117–53, 1958.

22. Morton, N. E. "Sequential tests for the detection of linkage." *Amer. J. Hum. Genet.* 7:277–318, 1955.

23. Mustard, J. F.; Murphy, E. A.; Rowsell, H. C.; and Downie, H. G. "Factors influencing thrombus formation *in vivo*." *Amer. J. Med.* 33:621–47, 1962.

24. Begent, N., and Born, G. V. R. "Growth rate *in vivo* of platelet thrombi, produced by iontophoresis of ADP, as a function of mean blood flow velocity." *Nature* (London) 227:926–30, 1970.

25. Murphy, E. A. "Evaluation of clinical data. Improvement of efficiency by simple transformation." *J. Chronic Dis.* 15:795–809, 1962.

26. Shanoff, H. M.; Little, A.; Murphy, E. A.; and Rykert, H. E. "Studies of male survivors of myocardial infarction due to 'essential' atherosclerosis. 1. Characteristics of the patients." *Canad. Med. Ass. J.* 84:519–30, 1961.

27. Florey, C. du V. "The use and interpretation of ponderal index and other weight-height ratios in epidemiological studies." *J. Chronic Dis.* 23:93–103, 1970.

28. *Weight, Height and Selected Body Dimensions of Adults, United States 1960–1962.* U.S. National Center for Health Statistics, Series 11, no. 8. Washington, D.C.: U.S. Department of Health, Education and Welfare, 1965.

29. Rao, C. R. *Advanced Statistical Methods in Biometric Research.* New York: Wiley, 1952.

30. Freeman, M. F., and Tukey, J. W. "Transformations related to the angular and the square root." *Ann. Math. Stat.* 21:607–11, 1950.

31. Raimi, R. A. "The peculiar distribution of first digits." *Scientific American,* December 1969, pp. 109–20.

32. Raimi, R. A. "The first digit problem." *Amer. Math. Monthly* 83:521–38, 1976.

33. Downie, H. G.; Murphy, E. A.; Rowsell, H. C.; and Mustard, J. F. "Extracorporeal circulation: A device for the quantitative study of thrombus formation in flowing blood." *Circulation Res.* 12:441–48, 1963.

34. Murphy, E. A.; Robinson, G. A.; Rowsell, H. C.; and Mustard, J. F. "The pattern of platelet disappearance." *Blood* 30:26–38, 1967.

35. Gajl-Peczalska, K. J.; Park, B. J.; Bigger, W. D.; and Good, R. A. "*B* and *T* lymphocytes in primary immunodeficiency in man." *J. Clin. Invest.* 52:919–28, 1973.

36. Murphy, E. A. "The normal." *Amer. J. Epidemiol.* 98:403–11, 1973.

37. MacMahon, B., and Pugh, T. F. *Epidemiology: Principles and Methods.* Boston: Little, Brown, 1970.

38. Morris, J. N. *Uses of Epidemiology.* 2d ed. Baltimore: Williams & Wilkins, 1964.

39. Feinstein, A. R. "The derangements of 'the range of normal.'" *Clin. Pharm. Ther.* 15:528-40, 1974.

40. Lau, L. Personal communication.

41. The Lipid Research Clinics Program Epidemiology Committee. "Plasma lipid distribution in selected North American populations. The lipid research clinics program prevalance study." *Circulation* 60:427-39. 1979.

42. Healy, M. J. R. "Normal values from a statistical standpoint." *Bull. Acad. Roy. Med. Belg.* 9:703-18, 1969.

43. Murphy, E. A. and Abbey, H. "The normal range—a common misuse." *J. Chronic Dis.* 20:79-88, 1967.

44. Migeon, C. G. Personal communication.

45. Elveback, L. R. "How high is high? A proposed alternative to the normal range." *Mayo Clinic Proc.* 47:93-97, 1972.

46. Cochran, W. G. *Sampling Techniques.* New York: Wiley, 1963.

47. Stuart, A. *Basic Ideas of Scientific Sampling.* New York: Hafner, 1964.

48. Royall, R. M. *The Prediction Approach to Finite Population Sampling Theory: Application to the Hospital Discharge Services.* National Center for Health Sciences, Series 2, no. 55. Washington, D.C.: U.S. Department of Health, Education and Welfare, 1973.

49. Murphy, E. A.; Rowsell, H. C.; and Mustard, J. F. "The effects of sitosterol on serum cholesterol, platelet economy, thrombogenesis and atherosclerosis in the rabbit." *Atherosclerosis* 17:257-68, 1973.

50. Mishkel, M. A. "Neonatal plasma lipids as measured in cord blood." *Canad. Med. Ass. J.* 111:775-80, 1974.

51. Kwiterovich, P. O. "Pediatric aspects of hyperlipoproteinemia." In *Hyperlipidemia: Diagnosis and Therapy*, edited by B. M. Rifkind and R. I. Levy, p. 258. New York: Grune and Stratton, 1977.

52. Chern, C. J., and Beutler, E. "Biochemical and electrophoretic studies of erythrocyte pyridoxine kinase in white and black Americans." *Amer. J. Hum. Genet.* 28:9-17, 1976.

53. Falconer, D. S. "The inheritance of liability to certain diseases estimated from the incidence among relatives." *Ann. Hum. Genet.* (London) 29:51-76, 1965.

54. Gale, A. N., and Murphy, E. A. "The use of serum creatine phosphokinase in genetic counseling for Duchenne muscular dystrophy. I. Analysis of results from 29 studies." *J. Chronic Dis.* 31:101-9, 1978.

55. Gale, A. N., and Murphy, E. A. "The use of serum creatine phosphokinase in genetic counseling for Duchenne muscular dystrophy. II. Review of methods of assay and factors which may be relevant in the interpretation of serum creatine phosphokinase activity." *J. Chronic Dis.* 32: 639-51, 1979.

56. Gale, A. N., and Murphy, E. A. "Prospects for the extraordinary." (In preparation)

57. Fisher, R. A. "On the mathematical foundation of theoretical statistics." *Phil. Trans. Roy. Soc. of London* A, 222:309-68, 1922.

58. Murphy, E. A., and Denson, K. W. E. "Statistical analysis of experiments performed in connection with thromboplastin standardization." *Thromb. Diath. Haemorr.* Suppl. 39:377-95, 1970.

59. Gurland, J., and Tripathi, R. C. "A simple approximation for unbiased estimation of the standard deviation." *Amer. Statistician* 25:30-32, 1971.

60. Reed, S. C., and Anderson, V. E. "Effect of changing sexuality on the gene

pool." In *Human Sexuality and the Mentally Retarded*, edited by F. F. de la Cruz and G. D. LaVeck, pp. 111–25. New York: Brunner Mazel, 1973.

61. Tukey, J. W. Quoted in Snedecor, G. W. *Statistical Methods*, 5th ed., pp. 251–54. Ames, Ia.: Iowa State University Press, 1956.

62. Duncan, D. B. "Multiple range and multiple F tests." *Biometrics* 11:1–42, 1955.

63. Scheffé, H. *Regression and Analysis of Variance.* New York: Wiley, 1959.

64. Miller. R. G. "Developments in multiple comparisons." *J. Amer. Stat. Assoc.* 72:779–88, 1978.

65. Mustard, J. F.; Rowsell, H. C.; Murphy, E. A.; and Downie, H. G. "Diet and thrombus formation. Quantitative studies using an extracorporeal circulation in pigs." *J. Clin. Invest.* 42:1783–89, 1963.

66. Federer, W. T. *Experimental Design. Theory and Application.* New York: Macmillan, 1955.

67. Cochran, W. G., and Cox, G. M. *Experimental Designs.* New York: Wiley, 1950.

68. Eddington, A. S. *The Nature of the Physical World.* P. 21. London: Cambridge University Press, 1928.

69. Chesterton, G. K. *The Everlasting Man.* New York: Image Books, 1955.

70. Murphy E. A. *Skepsis, Dogma and Belief. Uses and Abuses in Medicine.* Baltimore: Johns Hopkins University Press, 1981.

71. Murphy E. A. "The quantitative genetics of disease: Ambiguities." *Amer. J. Med. Genet.* 7:103–13, 1980.

72. Mustard, J. F., and Murphy, E. A. Unpublished data.

73. Bolling, D. R., and Murphy, E. A. "The estimation of blood platelet survival. VI. Evaluation of the graphical method." *Johns Hopkins Med. J.* 143:25–31, 1978.

74. Graybill, F. A. *An Introduction to Linear Statistical Models.* New York: McGraw-Hill, 1961.

75. Dronamraju, K. R.; Murphy, E. A.; and Schulze, J. "Blood group frequencies in Andhra Pradesh (1955–1960)." *Acta Genet.* (Basel) 17:446–53, 1967.

76. Murphy, E. A.; Francis, M. E.; and Bolling, D. R. "Estimation of blood platelet survival. V. A method for the analysis of population data." *J. Chronic Dis.* 134:797–815, 1973.

77. Baertl, J. Personal communication.

78. Little, J. A., and Shanoff, H. M. "Studies of male survivors of myocardial infarction due to 'essential' atherosclerosis. II. Lipids and lipoproteins." *Canad. Med. Assn. J.* 89:961–74, 1963.

79. Nelson, W. N. E.; Vaughan, V. C.; and McKay, R. J., eds. *Textbook of Pediatrics.* 9th ed., pp. 42–47. Philadelphia: Saunders, 1964.

80. Konigsmark, B. W., and Murphy, E. A. "Volume of the ventral cochlear nucleus in man: Its relationship to neuronal population and age." *J. Neuropathol. Exp. Neurol.* 31:304–16, 1972.

81. Mann, G. V. "Diet-heart: End of an era." *New Eng. J. Med.* 297:644–50, 1977.

82. Murphy, E. A. "Genetic and evolutionary fitness." *Amer. J. Med. Genetics.* 2:51–79, 1978.

83. David, F. N. *Tables of the Correlation Coefficient.* Cambridge: Cambridge University Press, 1938.

84. Olkin, I., and Pratt, J. W. "Unbiased estimation of certain correlation coefficients." *Ann. Math. Stat.* 29:201-11, 1958.

85. Mustard, J. F., and Murphy, E. A. "Effects of different dietary fats on blood coagulation, platelet economy and blood lipids." *Brit. Med. J.* 1: 1651-55, 1962.

86. Ratnoff, O. D., and Jones, P. K. "The laboratory diagnosis of the carrier state of classic hemophilia." *Ann. Int. Med.* 86:521-28, 1977.

87. McKusick, V. A. *Mendelian Inheritance in Man. Catalogs of Autosomal Dominant, Autosomal Recessive, and X-Linked Phenotypes.* 5th ed. Baltimore: Johns Hopkins University Press, 1978.

88. Bolling, D. R.; Borgaonkar, D. S.; Herr, H. M.; and Davis, M. "Evaluation of dermal patterns in Down's syndrome by predictive discrimination. II. Composite score based on the combination of left and right pattern areas." *Clin. Genet.* 2:163-69, 1971.

89. Roberts, J. A. F.; Norman, R. M.; and Griffiths, R. "Studies on a child population. 1. Definition of the sample, method of ascertainment, and analysis of the results of a group of intelligence tests." *Ann. Eugen.* (London) 6:319-38, 1936.

90. Hamilton, M.; Pickering, G. W.; Roberts, J. A. F.; and Sowry, G. S. C. "The aetiology of essential hypertension. 2. Scores for arterial blood pressures adjusted for differences in age and sex." *Clin. Sci.* 13:37-49, 1954.

91. Murphy, E. A. "Segregation of noisy Mendelian traits and the effect of age-dependence: a prolegomenon." *Amer J. Med. Genet.* 4:173-90, 1979.

92. Platt, R. "Heredity in hypertension." *Quart. J. Med.* 16:111-33, 1947.

93. Gold, R. J. M.; Maag, U. R.; Neal, J. L.; and Scriver, C. R. "The use of biochemical data in screening for mutant alleles and in genetic counseling." *Ann. Hum. Genet.* (London) 37:315-26, 1974.

94. Penrose, L. S. *An Introduction to Human Biochemical Genetics. Eug. Lab. Memoirs 37.* London: Cambridge University Press, 1955.

95. Murphy, E. A. "The normal, eugenics, and racial survival." *Johns Hopkins Med. J.* 136:98-106, 1975.

96. Renwick, J. H. "Progress in mapping human autosomes." *Brit. Med. Bull.* 25:65-73, 1969.

97. Elston, R. C., and Lange, K. "The prior probability of autosomal linkage." *Ann. Hum. Genet.* (London) 38:341-50, 1975.

98. Omoto, K; Misawa, S.; Harada, S.; Sumpaico, J. S.; Medado, P. M.; and Ogonuki, H. "Population genetic studies of the Philipino Negritos. I. A pilot survey of red cell enzyme and serum protein groups." *Amer. J. Hum. Genet.* 30:190-201, 1978.

99. Zanardi, P.; Dell'Acqua, G.; Menini, C.; and Barrai, I. "Population genetics in the Province of Ferrara. I. Genetic distances and geographical distances." *Amer. J. Hum. Genet.* 29:169-77, 1977.

100. Keys, A.; Aravanis, C.; Blackburn, H.; van Buchem, F. S. P.; Buzina, R.; Djordjevic, B. S.; Fidanza, F.; Karvonen, M. J.; Menotti, A.; Puddu, V.; and Taylor, H. L. "Coronary heart disease: Overweight and obesity as risk factors." *Ann. Int. Med.* 77:15-27, 1972.

101. Cochran, W. G. "Some methods for strengthening the common $\chi^2$ tests." *Biometrics* 10:417-51, 1954.

102. Pearson, E. S., and Hartley, H. O. *Biometrika Tables for Biostatisticians. Volume 1.* Cambridge: Cambridge University Press, 1962.

103. Lewitus, Z., and Neumann, J. "On the distribution of coronary thrombosis attacks." *Amer. Heart J.* 53:339–42, 1957.

104. Brain, D. D.; Tilley, B. C.; Labarthe, D. R.; O'Fallon, W. M.; Noller, K. L.; and Kurland, L. T. "Breast cancer in DES-exposed mothers. Absence of association." *Mayo Clin. Proc.* 55:89–93, 1980.

105. Crandall, B. F.; Lebhertz, T. B.; Rubinstein, L.; Robertson, R. D.; Sample, W. F.; Sarti, D.; and Howard, J. "Chromosomal findings in 2,500 second trimester amniocenteses." *Amer J. Med. Genet.* 5:345–56, 1980.

106. Siegel, S. *Nonparametric Statistics for the Behavioral Sciences.* New York: McGraw-Hill, 1956.

107. Tukey, J. W. *Exploratory Data Analysis.* Reading, Mass.: Addison-Wesley, 1977.

108. *Tables of the Binomial Probability Distribution. National Bureau of Standards Applied Mathematics Series 6,* Washington, D.C.: U.S. Government Printing Office, 1949.

109. Venter, J. H. "On estimation of the mode." *Ann. Math. Stat.* 38:1446–55, 1967.

110. Owen, D. B. *Handbook of Statistical Tables.* Reading, Mass.: Addison-Wesley, 1962.

111. Milton, R. C. "An extended table of critical values for the Mann-Whitney (Wilcoxon) two-sample statistic." *J. Amer. Stat. Assoc.* 59:925–34, 1964.

112. Rowsell, H. C.; Murphy, E. A.; and Mustard, J. F. "Heparin dosage and atherogenesis in the rabbit." *Arch. Path.* 80:63–39, 1965.

113. Murphy, E. A. "Hereditary polyposis coli. II. Genetic counseling." *Johns Hopkins Med. J.* 145:177–86, 1979.

114. Ashley, D. J. B. "Colonic cancer arising in polyposis coli." *J. Med. Genet.* 6:376–78, 1969.

115. Knudson, A. G. "Genetic and environmental interaction in the origin of human cancer." In *Genetics of Human Cancer. Progress in Cancer Research and Therapy Vol. 3,* edited by J. J. Mulvihill, R. W. Miller, and J. F. Fraumeni, pp. 391–97. New York: Raven, 1977.

116. Feinstein, A. R. "The epidemiologic trohoc, the ablative risk ratio, and 'retrospective' research." *Clin. Pharm. Therap.* 14:291–307, 1973.

117. Hawkins, M. R.; Murphy, E. A.; and Abbey, H. "The familial component in longevity. A study of offspring of nonagenarians. I. Methods and preliminary report." *Bull. Johns Hopkins Hosp.* 117:24–26, 1965.

118. Abbott, M. H.; Murphy, E. A.; Bolling. D. R.; and Abbey, H. "The familial component in longevity. A study of offspring of nonagenarians. II. Preliminary analysis of the completed study." *Johns Hopkins Med. J.* 134:1–16, 1974.

119. Abbott, M. H.; Abbey, H.; Bolling, D. R.; and Murphy, E. A. "The familial component in longevity. A study of offspring of nonagenarians. III. Intrafamilial studies." *Amer. J. Med. Genet.* 2:105–20, 1978.

120. Pearl, R., and Pearl, R. D. *The Ancestry of the Long-Lived.* Baltimore: Johns Hopkins Press, 1934.

121. Chiang, C. L. *Introduction to Stochastic Processes in Biostatistics.* New York: Wiley, 1968.

122. Gross, A. J., and Clark, V. A. *Survival Distributions: Reliability Applications in the Biomedical Sciences.* New York: Wiley, 1975.

123. Murphy, E. A., and Francis, M. E. "The estimation of blood platelet survival. II. The multiple hit model." *Thromb. Diath. Haemorr.* 25:53–80, 1971.

124. Murphy, E. A.; Francis, M. E.; and Mustard, J. F. "The estimation of blood platelet survival. IV. Characteristics of the residual errors from regression." *Thromb. Diath. Haemorr.* 28:447–56, 1972.

125. Murphy, E. A., and Kwiterovich, P. O. "Genetics of the hyperlipoproteinemias." In *Hyperlipidemia. Diagnosis and Therapy*, edited by B. F. Rifkind and R. I. Levy, pp. 218–47. New York: Grune and Stratton, 1977.

126. Rowsell, H. C.; Glynn, M. F.; Mustard, J. F.; and Murphy E. A. "Effects of heparin on platelet economy in dogs." *Amer. J. Physiol.* 213:915–22, 1967.

127. Murphy, E. A., and Salzman, E. W. "Platelet adhesiveness in von Willebrand's disease: a cooperative study." *Thromb. Diath. Haemorr.* Suppl. 51, 35: 341–76, 1972.

128. Murphy, E. A., and Mustard, J. F. "Coagulation tests and platelet economy in atherosclerotic and control subjects." *Circulation* 25:114–25, 1962.

129. Carter, T. C., and Falconer, D. S. "Stocks for detecting linkage in the mouse, and the theory of their design." *J. Genet.* 50:307–10, 1951.

130. Bolling, D. R., and Murphy, E. A. "Finite sample properties of maximum likelihood estimates of the recombination fraction in double backcross matings of man." *Amer. J. Med. Genet.* 3:81–95, 1979.

131. Aspin, A. A. "Tables for use in comparisons whose accuracy involves two variances separately estimated. (With an appended note by B. L. Welch.)" *Biometrika* 36:290–96, 1949.

132. Mickey, M. R., and Brown, M. B. "Bounds on the distribution functions of the Behrens-Fisher statistic." *Ann. Math. Stat.* 37:639–42, 1966.

# INDEX